Trends in Mathematics

Trends in Mathematics is a series devoted to the publication of volumes arising from conferences and lecture series focusing on a particular topic from any area of mathematics. Its aim is to make current developments available to the community as rapidly as possible without compromise to quality and to archive these for reference.

Proposals for volumes can be submitted using the Online Book Project Submission Form at our website www.birkhauser-science.com.

Material submitted for publication must be screened and prepared as follows:

All contributions should undergo a reviewing process similar to that carried out by journals and be checked for correct use of language which, as a rule, is English. Articles without proofs, or which do not contain any significantly new results, should be rejected. High quality survey papers, however, are welcome.

We expect the organizers to deliver manuscripts in a form that is essentially ready for direct reproduction. Any version of TeX is acceptable, but the entire collection of files must be in one particular dialect of TeX and unified according to simple instructions available from Birkhäuser.

Furthermore, in order to guarantee the timely appearance of the proceedings it is essential that the final version of the entire material be submitted no later than one year after the conference.

More information about this series at http://www.springer.com/series/4961

Mark Agranovsky • Anatoly Golberg •
Fiana Jacobzon • David Shoikhet •
Lawrence Zalcman
Editors

Complex Analysis and Dynamical Systems

New Trends and Open Problems

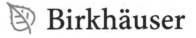 Birkhäuser

Editors

Mark Agranovsky
Department of Mathematics
Bar-Ilan University
Ramat-Gan, Israel

Anatoly Golberg
Department of Mathematics
Holon Institute of Technology
Holon, Israel

Fiana Jacobzon
Department of Mathematics
ORT Braude College
Karmiel, Israel

David Shoikhet
Department of Mathematics
Holon Institute of Technology
Holon, Israel

Lawrence Zalcman
Department of Mathematics
Bar-Ilan University
Ramat-Gan, Israel

ISSN 2297-0215   ISSN 2297-024X   (electronic)
Trends in Mathematics
ISBN 978-3-319-88893-4   ISBN 978-3-319-70154-7   (eBook)
https://doi.org/10.1007/978-3-319-70154-7

Mathematics Subject Classification (2010): 30-XX, 31-XX, 32-XX, 37-XX

Printed on acid-free paper

This book is published under the trade name Birkhäuser, www.birkhauser-science.com
The registered company is Springer International Publishing AG
The registered company address is: Gewerbestrasse 11, 6330 Cham, Switzerland

Preface

This volume was originally conceived as a collection of articles surveying selected open problems in complex analysis and adjacent areas, aimed at a broad audience of researchers ranging from PhD students to seasoned experts. Among the initiators of this project, Alexander Vasil'ev (known affectionately as "Sasha" to his many friends) played a dominant role. Sasha passed away, quite unexpectedly, on October 19, 2016, while the volume was still in the initial stages of preparation. As the remaining members of the editorial committee, we have seen the work to completion.

During the editorial process, the scope of the collection has been modified and enlarged somewhat. In addition to presenting open questions and problems, many of the papers here assembled emphasize possible new trends in the development of the areas involved. Actually, this idea, suggested by Sasha himself, provides a unifying theme connecting the various contributions. We have also included Sasha's final papers, prepared in collaboration with his students and colleagues.

We dedicate this volume to the memory of our friend, Sasha Vasil'ev.

Ramat-Gan, Israel   Mark Agranovsky
Holon, Israel   Anatoly Golberg
Karmiel, Israel   Fiana Jacobzon
Holon, Israel   David Shoikhet
Ramat-Gan, Israel   Lawrence Zalcman

Professor Alexander Vasil'ev (1962-2016)

Contents

On Polynomially Integrable Domains in Euclidean Spaces

Mark Agranovsky

To the memory of Alexander Vasiliev

Abstract Let D be a bounded domain in $\mathbb{R}^n$, with smooth boundary. Denote $V_D(\omega, t)$, $\omega \in S^{n-1}, t \in \mathbb{R}$, the Radon transform of the characteristic function χ_D of the domain D, i.e., the $(n-1)$-dimensional volume of the intersection D with the hyperplane $\{x \in \mathbb{R}^n :< \omega, x >= t\}$. If the domain D is an ellipsoid, then the function V_D is algebraic and if, in addition, the dimension n is odd, then $V(\omega, t)$ is a polynomial with respect to t. Whether odd-dimensional ellipsoids are the only bounded smooth domains with such a property? The article is devoted to partial verification and discussion of this question.

Keywords Radon transform • Fourier transform • Cross-section • Polynomial • Ellipsoid

2010 MSC 44A12, 51M99

1 Introduction

Vassiliev proved [1, 2, 10] that if D is a bounded domain in even-dimensional space $\mathbb{R}^n$, with C^∞ boundary, then the two-valued function, evaluated the n-dimensional volumes $\widehat{V}^\pm(\omega, t)$ of the two complementary portions of D which are cut-off by the section of D by the hyperplane $\{< \omega, x >= t\}$, is not an algebraic function of the parameters of the hyperplane. That means that no functional equation $Q(\omega_1, \cdots, t, \omega_n, \widehat{V}^\pm(\omega, t)) = 0$, where Q is a polynomial, is fulfilled. The result of

---

M. Agranovsky (✉)
Department of Mathematics, Bar-Ilan University, Ramat-Gan, Israel
e-mail: agranovs@math.biu.ac.il

M. Agranovsky et al. (eds.), *Complex Analysis and Dynamical Systems*,
Trends in Mathematics, https://doi.org/10.1007/978-3-319-70154-7_1

[10] is a multidimensional generalization of a celebrated Newton's Lemma XXVIII in Principia [9, 11] for convex ovals in $\mathbb{R}^2$. The smoothness condition is essential. The main object of interest in this article is the *section-volume function*

$$V_D(\omega, t) = \int\limits_{<x, \omega> = t} f(x) dV_{n-1}(x),$$

which is the t-derivative of $\widehat{V}^{\pm}$.

Contrary to the multi-valued cut-off-volume function $\widehat{V}_D$, its derivative $V_D = \frac{d}{dt} \widehat{V}_D$ is single-valued and can be algebraic in any dimension of the ambient space (the differentiation can "improve" the algebraicity, just keep in mind the algebraic function $\frac{1}{1+x^2}$ which is the derivative of the non-algebraic function $\tan x$).

For example, if D is the unit ball in $\mathbb{R}^n$ then

$$V_D(\omega, t) = c(1 - t^2)^{\frac{n-1}{2}}$$

is algebraic in any dimension. If n is odd then, even better, V_D is a polynomial in t. Here and further, c will denote a nonzero constant, which exact value is irrelevant.

Ellipsoids in odd-dimensional spaces also have the same property. For instance, if

$$E = \{\sum_{j=1}^{n} \frac{x_j^2}{b_j^2} = 1\}$$

then

$$V(\omega, t) = c \frac{b_1 \cdots b_n}{h^n(\omega)} (h^2(\omega) - t^2)^{\frac{n-1}{2}},$$

where

$$h(\omega) = \sqrt{\sum_{j=1}^{n} b_j^2 \omega_j^2}.$$

Definition 1 We will call a domain D *polynomially integrable* if its section-volume function coincides with a polynomial in t :

$$V_D(\omega, t) = \sum_{j-1}^{N} a_j(\omega) t^j$$

in the domain $V_D(\omega, t) > 0$ of all (ω, t) such that the hyperplane $< \omega, x > = t$ hits the domain D.

We assume that the leading coefficient $a_N(\omega)$ is not identical zero and in this case we call N *the degree* of the polynomially integrable domain D.

Note that in [10] the term "algebraically integrable" refers to the cut-off volume function $\widehat{V}_D$, rather than to V_D but polynomial integrability with respect to either function is the same.

It was conjectured in [10] that ellipsoids are the only smoothly bounded algebraically integrable domains in odd-dimensional Euclidean spaces. We are concerned with the similar question about the section-volume function $V_D(\omega, t)$ and the condition is that this function is a polynomial in t. Although polynomials are definitely algebraic functions, we do not impose conditions on the dependence on ω. Still, our conjecture is that the odd-dimensional ellipsoids are the only bounded polynomially integrable domains in odd-dimensional Euclidean spaces.

The present article contains some partial results and observations on the conjecture.

Namely, in Sects. 1–4 we prove that there are no bounded polynomially integrable domains with smooth boundaries in Euclidean spaces of even dimensions (Theorem 2), while in odd dimensions the following is true: (1) polynomially integrable bounded domains with smooth boundaries are convex (Theorem 5), (2) there are no polynomially integrable bounded domains in $\mathbb{R}^n$, with smooth boundaries, of degree strictly less than $n - 1$ (Theorem 7), (3) polynomially integrable bounded domains in $\mathbb{R}^n$, with smooth boundaries, of degree $\leq n - 1$ are only ellipsoids (Theorem 8), (4) polynomially integrable bounded domains in $\mathbb{R}^3$, with smooth boundaries, having axial and central symmetry are only ellipsoids (Theorem 11). In Sect. 7 a relation of polynomial integrability and stationary phase expansions is discussed. Some open questions are formulated in Sect. 8.

2 Preliminaries

2.1 Support Functions

For any bounded domain D define the support functions

$$
\begin{aligned}
h_D^+(\omega) &= \sup_{x \in D} <\omega, x>, \\
h_D^-(\omega) &= \inf_{x \in D} <\omega, x>,
\end{aligned}
\tag{1}
$$

where ω is a vector in the unit sphere S^{n-1} and $\langle, \rangle$ is the standard inner product in $\mathbb{R}^n$

The two support functions are related by

$$
h_D^+(-\omega) = -h_D^-(\omega).
$$

The relation with translations is given by

$$h_{D+a}^{\pm}(\omega) = h_D^{\pm}(\omega) + <a, \omega> . \tag{2}$$

If $t \notin [h_D^-(\omega), h_D^+(\omega)]$ then the $< \omega, x >= t$ is disjoint from the closed domain $\overline{D}$ (and therefore $V_D(\omega, t) = 0$).

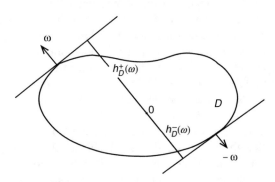

2.2 Radon Transform

We will be using several standard facts about the Radon transform. They can be found in many books or articles on Radon transform, see e.g., [6, 7].

The Radon transform of a continuous compactly supported function f in $\mathbb{R}^n$ is defined as the integral

$$Rf(\omega, t) = \int_{<\omega, x>=t} f(x) dV_{n-1}(x)$$

of f over the $(n-1)$-dimensional plane $< \omega, x >= t$, with the unit normal vector ω, on the distance t from the origin, against the $(n - 1)$-dimensional volume element dV_{n-1}.

There is a nice relation, called **Projection-Slice Formula**, between Fourier and Radon transforms:

$$\widehat{f}(r\omega) = F_{t \to r} Rf(\omega, t), \tag{3}$$

where $\widehat{f}$ is the n-dimensional Fourier transform and $F_{t \to r}$ is the one-dimensional Fourier transform with respect to the variable t. The formula follows immediately by writing the integral as a double integral against dV_{n-1} and dt.

The Radon transform can be inverted in different ways. The following formula is called back-projection inversion formula [7, Thm 2.13] :

$$f(x) = c\Delta^{\frac{n-1}{2}} \int_{\omega \in S^{n-1}} Rf(\omega, <\omega, x>)d\omega,$$

where f is sufficiently smooth compactly supported function.

There is a crucial difference between the inversion formulas in even and odd dimensions. If n is even then the exponent $\frac{n-1}{2}$ is fractional and $\Delta^{\frac{n-1}{2}}$ becomes an integral operator, so that the inversion formula is not local. However, when n is odd then $\frac{n-1}{2}$ is integer and the operator in front of the integral is differential so that the formula is local.

Due to the relation

$$\widehat{\frac{\partial f}{\partial x_i}}(\omega, t) = \omega_i \frac{\partial \widehat{f}}{\partial t}(\omega, t), \tag{4}$$

the inversion formula can be rewritten in odd dimensions as:

$$f(x) = c \int_{\omega \in S^{n-1}} \left(\frac{d^{n-1}}{dt^{n-1}} Rf\right)(\omega, <\omega, x>)d\omega, \tag{5}$$

where $d\omega$ is the Lebesgue measure on the unit sphere S^{n-1}. The Plancherel formula for the Radon transform in odd dimensions is:

$$c \int_{\mathbb{R}^n} f(x)g(x)dV(x) = \int_{S^{n-1} \times \mathbb{R}} \left(\frac{d^{n-1}}{dt^{n-1}} Rf\right)(\omega, t)Rg(\omega, t)d\omega dt. \tag{6}$$

The following conditions characterize the range of Radon transform in class of Schwartz functions:

1. $g(-\omega, -t) = g(\omega, t)$,
2. The k-moment $\int_{\mathbb{R}^n} g(\omega, t)t^k dt, k = 0, 1, \cdots$ extends from the unit sphere S^{n-1} to $\mathbb{R}^n$ as a polynomial of degree at most k.

The immediate corollary of property 1 is that if $V_D(\omega, t) = \sum_{k=0}^{N} a_k(\omega)t^k$ then

$$a_k(-\omega) = (-1)^k a_k(\omega),$$

i.e. a_k is an even function on S^{n-1} when k is even and is an odd function when k is odd.

In the sequel, the above facts will be used for the section-volume function V_D which is just the Radon transform of the characteristic function of the domain D :

$$V_D(\omega, t) = (R\chi_D)(\omega, t), \ t \in [h_D^-(\omega), h_D^+(\omega)].$$

All domains under consideration will be assumed bounded, with C^∞ boundary, although some statements are true even under weaker smooth assumptions. As it is shown in [1, 10], smoothness plays an important role in that circle of questions.

3 There Are No Polynomially Integrable Domains in Even Dimensions

The condition of algebraic integrability in [10] involves both variables ω and t, while the polynomial integrability imposes a condition with respect to t only, so that the statement formulated in the title on this section is not a straightforward corollary of the result in [10]. However, this statement easily follows from the asymptotic behavior of the section-volume function near the boundary points of D :

Theorem 2 *There are no polynomially integrable domain with C^2-smooth boundary in $\mathbb{R}^n$ with even n.*

Proof Let D be such a domain. We will show that $V_D(\omega, t)$ can not behave polynomially when the section of D by the hyperplanes $< x, \omega > = t$ shrink to an elliptical point on ∂D. The elliptical point of the surface ∂D is a point at which the principal curvatures are all nonzero and of the same sign.

To find such a point, consider the maximally distant, from the origin, boundary point $a \in \partial D$:

$$|a| = max_{x \in \partial D}|x|.$$

Using rotation and translation, we can move a to 0 and make the tangent plane at 0 the coordinate plane

$$T_{\partial D}(0) = \{x_n = 0\}$$

so that the domain D is in the upper half-plane $x_n \geq 0$. Let b the image of the point a under those transformations. Then b is located on the x_n-axis and

$$D \subset B(b, |b|),$$

where $B(b, r)$ is the ball of radius r with center b. The boundary surface ∂D can be represented, in a neighborhood of $0 \in \partial D$, as the graph

$$\partial D \cap U = \{x_n = f(x'), x' = (x_1, \ldots, x_{n-1})\}.$$

Since the tangent plane at 0 is $x_n = 0$ the first differential vanishes at the origin, $df_0 = 0$. Using rotations around the x_n axis we can diagonalize the second differential and then the equation of ∂D near 0 becomes

$$\partial D \cap U = \{x_n = \frac{1}{2} \sum_{j=1}^{n-1} \lambda_j x_j^2 + S(x')\},$$

where $S(x') = o(|x'|^2), x' \to 0$.

The coefficients λ_j are the principal curvatures of the surface ∂D and since $D \subset B(b, |b|)$ all of them are not less than the curvature $\frac{1}{|b|}$ of the sphere $S(b, |b|)$:

$$\lambda_j \geq \frac{1}{|b|} > 0, j = 1, \ldots, n - 1.$$

Indeed, the equation of the sphere $S(b, |b|)$ is

$$x_n = |b| - \sqrt{|b|^2 - |x'|^2}$$

and since $D \subset B(b, |b|)$ we have

$$\frac{1}{2} \sum_{j=1}^{n-1} \lambda_j x_j^2 + o(|x'|^2) \geq |b| - \sqrt{|b|^2 - |x'|^2}$$

$$= \frac{|x'|^2}{|b| + \sqrt{|b|^2 - |x'|^2}}.$$

and then the inequalities follow by the passage to the limit.

Let us turn further to the function V_D. The value $V_D(e_n, t), e_n = (0, \ldots, 0, 1)$, $t > 0$, is the $(n-1)$-dimensional volume of the section $D \cap \{x_n = t\}$ and the leading term of the asymptotic when $t \to 0$ is defined by the leading, quadratic, term in the decomposition of f. This leading term of $V_D(e_n, t)$ is the volume of the ellipsoid

$$\frac{1}{2} \sum_{j=1}^{n-1} \lambda_j x_j^2 = t,$$

which is $c \frac{t^{\frac{n-1}{2}}}{\lambda_1 \cdots \lambda_{n-1}}, c = \frac{(2\pi)^{\frac{n-1}{2}}}{\Gamma(\frac{n+1}{2})}$. Therefore,

$$V_D(e_n, t) = c \frac{t^{\frac{n-1}{2}}}{\lambda_1 \cdots \lambda_{n-1}} + o(t^{\frac{n-1}{2}}), t \to 0 + .$$

Thus, the function $V_D(e_n, t)$ has zero at $t = 0$ of order $\frac{n-1}{2}$. If this function is a polynomial then the order must be integer, which is not the case when n is even. Thus, $V_D(e_n, t)$ is not a polynomial in t. In fact, all we have used is that a is an elliptical point of the convex hypersurface ∂D. Since the points of ∂D, sufficiently close to a, are elliptical as well, we conclude that, moreover, for an open set of directions ω, the function $V_D(\omega, t)$ is not a polynomial in t. This proves the theorem.

4   Polynomially Integrable Domains in $\mathbb{R}^{2m+1}$ Are Convex

Lemma 3 *Suppose that the section-volume function V_D coincides with a polynomial in t :*

$$V_D(\omega, t) = a_0(\omega) + a_1(\omega)t + \ldots + a_N(\omega)t^N,$$

when $t \in [h_D^-(\omega), h_D^+(\omega)]$. Then

$$\int\limits_{|\omega|=1} a_k(\omega)p(\omega)dA(\omega) = 0$$

for any polynomial p of $\deg p \leq k - n + 1$ and any $k > n - 1$.

Proof Regarding the function V_D as the Radon transform of the characteristic function χ_D of the domain D, we can write by the inversion formula (5):

$$\chi_D(x) = c \int\limits_{|\omega|=1} \frac{d^{n-1}}{dr^{n-1}} V_D(\omega, <x, \omega>)dA(\omega)$$

$$= c \sum_{k=n-1}^{N} \frac{k!}{(k-n+1)!} \int\limits_{|\omega|=1} a_k(\omega) <x, \omega>^{k-(n-1)} dA(\omega).$$

Since $\chi_D(x) = 1$ for $x \in D$, the power series in x in the left hand side equals 1 identically and hence each term with $k - (n - 1) > 0$ vanishes:

$$\int\limits_{|\omega|=1} a_k(\omega) <x, \omega>^{k-(n-1)} dA(\omega) = 0, x \in D.$$

Since x is taken from an open set, the functions $y \rightarrow <x, y>^{k-n+1}$ span the space of all homogeneous polynomials of degree at most $k - n + 1$. Those polynomials, being restricted on the unit sphere, generate restrictions of all polynomials p, $\deg p \leq k - n + 1$. Lemma is proved.

Remark 4 The assertion of Lemma remains true also under the assumption that V_D expands as an infinite power series in t:

$$V_D(\omega, t) = \sum_{k=0}^{\infty} a_j(\omega) t^k,$$

in a neighborhood of $[h^-(\omega), h^+(\omega)]$.

Theorem 5 *If a smoothly bounded domain D in $\mathbb{R}^n$, with n odd, is polynomially integrable then it is convex.*

Proof Let D be such a domain and denote $\widehat{D}$ the convex hull of D. If $D \neq \widehat{D}$ then there is an open portion Γ of the boundary ∂D which is disjoint from a neighborhood of $\partial \widehat{D}$. Then one can construct a function which vanishes near $\partial \widehat{D}$ and behaves in a prescribed way on Γ. In particular, one can construct a compactly supported C^∞ function ψ in $\mathbb{R}^n$ with the following properties:

1. *supp* $\psi \subset \widehat{D}$,
2. $\int\limits_{\partial D} \psi(x) v_1(x) dS(x) \neq 0$, where $v(x), x \in \partial D$, is the external unit normal field on the boundary of D.

Denote $\Psi(\omega, t)$ the Radon transform of ψ :

$$\Psi(\omega, t) = (R\psi)(\omega, t) = \int\limits_{<\omega, x>=t} \psi(x) dV_{n-1}(x).$$

Formula (4) implies

$$\Psi_1(\omega, t) = \omega_1 \Psi_t'(\omega, t). \tag{7}$$

Using (7), write the Plancherel formula (6) for the pair of functions: χ_D and $\frac{\partial \psi}{\partial x_1}$. Although formula (6) works for Schwartz functions, we can extend the equality to our case approximating χ_D by smooth functions, if the differentiation in t in the left hand side applies to the second, smooth, factor $\Psi = R\psi$. We obtain

$$\int_D \frac{\partial \psi}{\partial x_1}(x) dV(x) = \int_{\mathbb{R}^n} \chi_D(x) \frac{\partial \psi}{\partial x_1}(x) dV(x)$$

$$= \int\limits_{|\omega|=1} \int_{\mathbb{R}} R\chi_D(\omega, t) (\Psi_1(\omega, t))_t^{(n-1)} dA\omega) dt. \tag{8}$$

By the condition, the Radon transform $R\chi_D(\omega, t)$ of the characteristic function of D coincides with the polynomial $V_D(\omega, t)$ as long as the hyperplane $<\omega, x> = t$ hits D. The latter occurs, if and only if it hits $\widehat{D}$. In turn, that is equivalent to the inequality $h_-(\omega) \le t \le h_+(\omega)$. Thus, the integration in t in the right hand side of (8) is performed on the segment $[h_D^-(\omega), h_D^+(\omega)]$.

The left hand side of (8) can be rewritten, using (7) and Green's formula, as:

$$\int_D \frac{\partial \psi}{\partial x_1}(x)dV(x) = \int_{\partial D} \psi(x)\nu_1(x)dS(x),$$

where $\nu_1(x)$ is the first coordinate of $\nu(x)$.

Finally, (8) takes the form:

$$I := \int_{\partial D} \psi(x)\nu_1(x)dS(x)$$

$$= \int_{\omega \in S^{n-1}} \int_{h_{(\omega)}}^{h_+(\omega)} V_D(\omega, t)\omega_1 \Psi^{(n)}(\omega, t)dtdA(\omega).$$

Furthermore, the function $(\psi)'_{x_1}$ vanishes in a neighborhood of the boundary $\partial \widehat{D}$. Therefore, its integrals over hyperplanes $<\omega, x> = t$ with

$$t \notin I_\varepsilon := [h_-(\omega) + \varepsilon, h_+(\omega) - \varepsilon],$$

vanish as well for sufficiently small $\varepsilon > 0$ and

$$\Psi(\omega, t) = 0, \text{ for } t \notin I_\varepsilon. \tag{9}$$

Hence, all the derivatives of $\Psi(\omega, t)$ vanish at the end points $t = h_\pm(\omega)$ of the integration interval and then integration by parts in t yields:

$$I = (-1)^n \int_{\omega \in S^{n-1}} \int_{h_{(\omega)}}^{h_+(\omega)} V_D^{(n)}(\omega, t)\omega_1 \Psi(\omega, t)dtdA(\omega).$$

More explicitly:

$$I = (-1)^n \sum_{k=n}^{N} \frac{k!}{(k-n)!} t^{k-n} \int_{\omega \in S^{n-1}} \int_{-\infty}^{+\infty} a_k(\omega)\omega_1 \Psi(\omega, t)dtd\omega.$$

We have replaced here the integral on $[h_D^-(\omega), h_D^+(\omega)]$ by integration on $(-\infty, \infty)$, due to (9). Consider an arbitrary term in the right hand side:

$$I_k := \int_{\omega \in S^{n-1}} \int_{\mathbb{R}} a_k(\omega)t^{k-n}\omega_1 \Psi(\omega, t)dtd\omega,$$

$k = n, n + 1, \ldots N$. The function Ψ is the Radon transform of a smooth compactly supported function ψ. Therefore, according to the description of the range of Radon transform in Sect. 2, the moment condition (2) is fulfilled. This means that the function

$$Q_k(\omega) := \int_{\mathbb{R}} \Psi(\omega, t) t^{k-n} dt$$

extends from the unit sphere $|\omega| = 1$ as a polynomial of degree at most $k - n$. Therefore, the function $\omega_1 Q_k(\omega)$ extends as a polynomial of degree at most $k - n + 1$. But then

$$I_k = \int_{\omega \in S^{n-1}} a_k(\omega) \omega_1 Q_k(\omega) d\omega = 0$$

by Lemma 3.

Thus, we have proven that all $I_k = 0$ and therefore I which is a sum of I_k is zero as well, $I = 0$. But $I = \int_{\partial D} \psi(x) \nu_1(x) dS(x) \neq 0$ by the choice of the function ψ. This contradiction is obtained under assumption that D is strictly smaller than its convex hull $\widehat{D}$. Therefore, the opposite, they coincide and thus D is convex. Theorem is proved.

Remark 6 It can be seen from the proof that Theorem 5 is valid also in the case when $V_D(\omega, t)$ decomposes into infinite power series in t. More precisely, if for every fixed $\omega \in S^{n-1}$ the function $V_D(\omega, \cdot)$ extends, in a neighborhood of the segment $[h_D^{-1}(\omega), h_D^+(\omega)]$, as a power series.

5 Polynomially Integrable Domains in $\mathbb{R}^n$, $n = 2m + 1$, of Degree $\leq n - 1$ Are Ellipsoids

In this section, we characterize odd-dimensional ellipsoids by the property of the functions $V_D(\omega, t)$ being polynomials in t of degree at most $n - 1$. In fact, $n - 1$ is also the lower bound for the degrees of those polynomials. Everywhere in this section the dimension n is assumed being odd.

Theorem 7 *If D is a polynomially integrable domain in $\mathbb{R}^n$ with smooth boundary then for almost all $\omega \in S^{n-1}$ the degree of the polynomial $V_D(\omega, \cdot)$ is at least $n - 1$. In particular, there are no polynomially integrable domains of degree less than $n - 1$.*

Proof For every point $x \in \partial D$ denote by $\kappa(x)$ the Gaussian curvature, i.e., the product if the principal curvatures of ∂D at the point x. The Gaussian curvature coincides with the Jacobian of the Gaussian (spherical) map

$$\gamma : \partial D \to S^{n-1},$$

which maps any $x \in \partial D$ to the external normal vector $\gamma(x)$ to ∂D at x. Respectively, the points of zero Gaussian curvature is exactly the set $Crit(\gamma)$ of critical points of the Gaussian map.

Pick any $x \in \partial D \setminus Crit(\gamma)$. Denote $\omega = \gamma(x)$. Since by Theorem 5 the boundary ∂D is convex, all the principal curvatures $\kappa_1(x), \ldots, \kappa_{n-1}(x)$ at x are nonnegative. Since $x \notin Crit(\gamma)$, the product $\kappa(x) = \kappa_1(x) \cdots \kappa_{n-1}(x) \neq 0$ and therefore $\kappa_1(x) > 0, \ldots, \kappa_{n-1}(x) > 0$.

This implies that $V_D(\omega, t)$ has zero at $t = h_D^+(\omega)$ of order $\frac{n-1}{2}$ (see ([6, 1.7]) or the arguments of the proof of Theorem 2). Therefore

$$V_D(\omega, t) = [h_D^+(\omega) - t]^{\frac{n-1}{2}} Q(t),$$

where Q is a polynomial.

Repeating the argument for the point $x' \in \partial D$ with the normal vector $\gamma(x') = -\omega$ we conclude that the polynomial V_D has also zero of order $\frac{n-1}{2}$ at the point $t = h_D^-(\omega) = -h_D^+(-\omega)$. Therefore

$$V_D(\omega, t) = [h_D^+(\omega) - t]^{\frac{n-1}{2}} [t - h_D^-(\omega)]^{\frac{n-1}{2}} Q_1(t) \tag{10}$$

and hence

$$\deg V_D(\omega, \cdot) \geq n - 1.$$

By the choice of x, the estimate holds for all regular values $\omega = \gamma(x), x \in \partial D \setminus Crit(\gamma)$. By Sard's theorem the Lebesgue measure $mes\{\gamma(Crit(\gamma)\} = 0$ and hence the estimate for the degree is fulfilled for almost all $\omega \in S^{n-1}$. This proves the theorem.

Theorem 8 *Let n be odd. Let D be a bounded domain in $\mathbb{R}^n$ with smooth boundary. Suppose that for almost all $\omega \in S^{n-1}$ the function $V_D(\omega, t)$ is a polynomial in t, of degree at most $n - 1$. Then D is an ellipsoid.*

Proof By our assumption and formula (10), for almost all $\omega \in S^{n-1}$ the following representation holds:

$$V(\omega, t) = A(\omega)[(h_D^+(\omega) - t)(h_D^-(\omega) - t)]^{\frac{n-1}{2}}, h_D^-(\omega) < t < h_D^+(\omega),$$

where $A(\omega) > 0$.

The function V_D is the Radon transform of the characteristic function χ_D of the domain D :

$$V_D(\omega, t) = (R\chi_D)(\omega, t) = \int_{<\omega, x>=t} \chi_D(x) dV_{n-1}(x)$$

Applying Projection-Slice Formula (3) we obtain:

$$\widehat{\chi}_D(r\omega) = \int_{\mathbb{R}} e^{irt} V_D(\omega, t) dt$$

$$= A(\omega) \int_{h_D^-(\omega)}^{h_D^+(\omega)} e^{irt} [(h_D^+(\omega) - t)(h_D^-(\omega) - t)]^{\frac{n-1}{2}} dt.$$

(11)

Consider the following functions on S^{n-1} :

$$B(\omega) = \frac{h_D^+(\omega) + h_D^-(\omega)}{2},$$

$$C(\omega) = \frac{h_D^+(\omega) - h_D^-(\omega)}{2}.$$

The function $C(\omega)$ expresses the half-width of the domain D in the direction ω.

Performing the following change of variable in the integral in the left hand side in (12):

$$t = s + B(\omega),$$

we obtain

$$\widehat{\chi}_D(r\omega) = (-1)^{\frac{n+1}{2}} A(\omega) e^{iB(\omega)r} \int_{-C(\omega)}^{C(\omega)} e^{irs} (C^2(\omega) - t^2)^{\frac{n-1}{2}} dt.$$

Performing the next change of variable $s = C(\omega)u$ we arrive at:

$$\widehat{\chi}_D(r\omega) = M(\omega) e^{iB(\omega)r} \int_{-1}^{1} e^{iC(\omega)ru} (1 - u^2)^{\frac{n-1}{2}} du,$$

(12)

where

$$M(\omega) = (-1)^{\frac{n+1}{2}} A(\omega)(C(\omega))^{n-1}.$$

Lemma 9 *The function $M(\omega)$ is a (positive) constant. The function $B(\omega)$ is either zero or a homogeneous polynomial of degree one. The function $C^2(\omega)$ is a homogeneous quadratic polynomial.*

Proof of Lemma The straightforward differentiation of both sides of (12) with respect to r at the point $r = 0$ gives:

$$\widehat{\chi}_D(0) = M(\omega),$$

$$\frac{d}{dr} \widehat{\chi}_D(r\omega)|_{r=0} = M(\omega) iB(\omega)\alpha,$$

$$\frac{d^2}{dr^2} \widehat{\chi}_D(r\omega)|_{r=0} = -M(\omega)\alpha[B^2(\omega) + C^2(\omega)],$$

(13)

where we have denoted

$$\alpha := \int_{-1}^{1} (1 - u^2)^{\frac{n-1}{2}} \, du.$$

On the other hand,

$$\widehat{\chi}_D(0) = vol(D),$$

$$\frac{d}{dr} \widehat{\chi}_D|_{r=0} = i \int_D < \omega, y > dy, \tag{14}$$

$$\frac{d^2}{dr^2} \widehat{\chi}_D|_{r=0} = - \int_D < \omega, y >^2 dy.$$

Therefore, from (13) and the first equality in (14) we obtain

$$M(\omega) = vol(D) > 0.$$

By (13) and the second equality in (14) we have

$$B(\omega) = \frac{1}{\alpha vol(D)} \int_D < \omega, y > dy,$$

and hence B is a linear form. In particular, it can be identically zero, if, for instance, the domain D has a central symmetry.

Finally, (13), the third equality in (14) imply

$$B^2(\omega) + C^2(\omega) = \frac{1}{\alpha vol(D)} \int_D < \omega, y >^2 dy$$

and then

$$C^2(\omega) = \frac{1}{(volD)^2} \int_D < \omega, y >^2 dV(y) - B^2(\omega)$$

is a quadratic form, because B is a linear form. The lemma is proved.

Let us proceed with the proof of Theorem 8. Applying a suitable orthogonal transformation we can assume that the quadratic form $C^2(\omega)$ has a diagonal form:

$$C^2(\omega) = c_1 \omega_1^2 + \ldots + c_n \omega_n^2.$$

Also, according to Lemma 9 one has

$$M(\omega) = M = const \text{ and } B(\omega) = < b, \omega >,$$

for some vector b.

Remember that

$$B(\omega) = \frac{1}{2}(h_+(\omega) + h_-(\omega)),$$

$$C(\omega) = \frac{1}{2}(h_+(\omega) - h_-(\omega)).$$

Applying translation by vector b and a dilatation of the domain D, and using property (2), we may assume that $M = 1, B(\omega) = 0$. This means that $h_D^+ = h_D^-$ and then $C = h_D^+$. Thus,

$$h_+(\omega) = C(\omega) = \sqrt{\sum_{j=1}^{n} c_j \omega_j^2}.$$

But the function in the right hand side is exactly the support function h_E^+ of the ellipsoid

$$E = \{\sum_{j=1}^{n} \frac{x_j^2}{c_j} = 1\}$$

and hence

$$D = E.$$

This completes the proof of Theorem 8.

The following result is a combination of Theorem 8 and Lemma 3:

Theorem 10 *Let D be a domain in $\mathbb{R}^n$, with a smooth boundary, where n is odd. Suppose that the section-volume function V_D is a polynomial, $V_D(\omega, t) = a_0(\omega) + a_1(\omega)t + \cdots a_N(\omega)t^N$, $h_D^-(\omega) \leq t \leq h_D^+(\omega)$ and for each $k > n - 1$ the coefficient $a_k(\omega)$ coincides with the restriction to S^{n-1} of a polynomial of degree at most $k - n + 1$. Then D is an ellipsoid.*

Proof Lemma 3 and the condition imply that $a_k = 0$ for $k > n - 1$. Theorem 8 implies that D is an ellipsoid.

6 Axially Symmetric Polynomially Integrable Domains in 3D Are Ellipsoids

In this section we confirm the conjecture about ellipsoids as the only polynomially integrable domains, however under the pretty strong condition of axial symmetry. To avoid additional technical difficulties, we consider the three-dimensional case.

The following theorem asserts that, in this case, to conclude that the domain is an ellipsoid, it suffices essentially to demand the polynomial integrability only in two directions:

Theorem 11 *Let the D be an axially and centrally symmetric domain in $\mathbb{R}^3$. Suppose that the section-volume function $V_D(\omega, t)$ is a polynomial in t, when ω is one of the two orthogonal directions $\omega = \xi$ and $\omega = \eta$, where ξ is the symmetry axis and η is any orthogonal direction. Assume also that $V_D(\xi, t) = b_0 + b_1 t + \cdots + b_{2N} t^{2N}$ and $b_{2N} < 0$. Then D is an ellipsoid.*

Proof We assume that the center of symmetry of D is 0 and the symmetry axis ξ the z-axis, $\xi = (0, 0, 1)$.

Due to the axial symmetry, the cross-section of D by the plane $z = t$ is a two-dimensional disc $D(0, r(t))$ of radius $r(t)$ and therefore the volume of the corresponding cross-section is

$$V_D(\xi, t) = \pi [r(t)]^2.$$

By our assumption, $V_D(\xi, t)$ is a polynomial

$$P(t) := V_D(\xi, t).$$

The central symmetry implies that it is an even polynomial, i.e.

$$V_D(\xi, t) = P(t) = b_0 + b_2 t^2 + \cdots + b_{2N} t^{2N}$$

and hence

$$P(t) = \pi [r(t)]^2.$$

Since $t = z$ and $r(t)^2 = x^2 + y^2$, we conclude that the equation of the boundary ∂D of the domain D is given by:

$$x^2 + y^2 = P(z) = b_0 + b_2 z^2 + \cdots + b_{2N} z^{2N}, \tag{15}$$

where we have incorporated the constant $\frac{1}{\pi}$ into P.

Consider now the cross-section of the domain D in the orthogonal direction η. This vector lies in x, y-plane and due to rotational symmetry can be taken $\eta = (0, 1, 0)$. Intersect D by the hyperplane $y = t$. The equation of that section can be written as:

$$x^2 - (b_2 z^2 + \ldots + b_{2N} z^{2N}) = b_0 - t^2. \tag{16}$$

Now introduce the family $\Omega(\alpha), \alpha > 0$ of domains in $\mathbb{R}^2_{x,z}$:

$$\Omega(\alpha) = \{(x, z) \in \mathbb{R}^2 : x^2 - (b_2 z^2 + \ldots + b_{2N} z^{2N}) \leq \alpha^2\}.$$

Every domain $\Omega(\alpha)$ is non-empty, as $0 \in \Omega(\alpha)$. Also, the domains $\Omega(\alpha)$ are bounded because the condition $b_{2N} < 0$ implies that the left hand side of the inequality tends to ∞ when $(x, z) \to \infty$ and hence $\mathbb{R}^2 \setminus B_R$ is free of points from $\Omega(\alpha)$, for any fixed α and R large enough.

Note that $b_0 = vol(D \cap \{z = 0\}) > 0$. The parameter $t = y$ varies in a neighborhood of 0 and therefore $b_0 - t^2 > 0$ for small t and, correspondingly,

$$D \cap \{y = t\} = \Omega(\sqrt{b_0 - t^2}),$$

for $b_0 - t^2 > 0$. The function $A(\alpha) := vol\Omega(\alpha)$ is real analytic in α. On the other hand, due to the central symmetry the function

$$vol(D \cap \{y = t\}) = V_D(\eta, t)$$

is a polynomial, $B(t^2)$, in t^2,

Since

$$A(\alpha) = B(b - \alpha^2)$$

for $\alpha > 0$ from an open set, we conclude that the volume function $A(\alpha) = vol\Omega(\alpha)$ is a polynomial on the whole half-line $\alpha > 0$.

On the other hand, by the change of variables

$$u = \frac{x}{\alpha}, v = \frac{z}{\alpha^{\frac{1}{N}}}$$

we obtain

$$vol\Omega(\alpha) = \int_{x^2 - (b_2 z^2 + \ldots + b_{2N} z^{2N}) \leq \alpha^2} dx dz$$

$$= \alpha^{1 + \frac{1}{N}} \int_{u^2 - (b_2 v^2 \alpha^{\frac{2}{N} - 2} + \ldots + b_{2N} v^{2N}) \leq 1} du dv.$$

Since

$$\lim_{\alpha \to \infty} \int_{u^2 - (b_2 v^2 \alpha^{\frac{2}{N} - 2} + \ldots + b_{2N} v^{2N}) \leq 1} du dv = \int_{u^2 - b_{2N} v^{2N} \leq 1} du dv < \infty,$$

we conclude that

$$A(\alpha) = vol\Omega(\alpha) = O(\alpha^{1 + \frac{1}{N}}), \alpha \to \infty.$$

Since $A(\alpha)$ is a polynomial, the exponent $1 + \frac{1}{N}$ must be integer, which implies $N = 1$.

Thus, Eq. (15) of the domain D is given by the second order equation

$$x^2 + y^2 = b_0 + b_2 z^2, b_2 < 0,$$

and D is an ellipsoid.

7 Finite Stationary Phase Expansion Point of View

Let D be a convex polynomially integrable domain with the section-volume function

$$V_D(\omega, t) = \sum_{j=0}^{N} a_j(\omega) t^j.$$

The relation (3) between Fourier and Radon transforms yields

$$\widehat{\chi}_D(r\omega) = \sum_{j=0}^{N} a_j(\omega) \int_{h_D^-(\omega)}^{h_+(\omega)} t^j e^{irt} dt$$

$$= \sum_{j=0}^{N} a_j(\omega) \frac{d^j}{dr^j} \frac{e^{h_D^+(\omega)r} - e^{h_D^+(\omega)r}}{r}. \tag{17}$$

On the other hand, by the Green's formula

$$\widehat{\chi}_D(r\omega) = \int_D e^{ir<\omega,x>} dV(x)$$

$$= -\frac{1}{r^2} \int_D \Delta e^{ir<\omega,x>} dV(x) \tag{18}$$

$$= -\frac{1}{r^2} \int_{\partial D} e^{ir<\omega,x>} <\omega, v(x)> dS(x),$$

where dS is the surface measure on ∂D and $v(x)$ is the external unit normal vector to ∂D.

The right hand side in (17) can be written, after performing differentiation in r, in the form

$$e^{irh_D^+(\omega)} Q^+(\omega, \frac{1}{r}) + e^{irh_D^-(\omega)} Q^-(\omega, \frac{1}{r}),$$

where $Q^{\pm}(\omega, t)$ are polynomials with respect to t. Thus, we have

$$\int_{\partial D} e^{ir<\omega,x>} < \omega, \nu(x) > dS(x) = e^{irh_D^+(\omega)} r^2 Q^+(\omega, \frac{1}{r}) - e^{irh_D^-(\omega)} r^2 Q^-(\omega, \frac{1}{r}).$$

(19)

The critical points of the function

$$\varphi_\omega(x) := < \omega, x >$$

are the points x^+ and x^- such that

$$< \omega, x^+ > = \max_{x \in \bar{D}} < \omega, x > = h_D^+(\omega),$$

$$< \omega, x^- > = \min_{x \in \bar{D}} < \omega, x > = h_D^-(\omega).$$

(20)

Equation (19) is the stationary phase expansion (cf. [12]) in the powers of $\frac{1}{r}$ of the oscillatory integral in the left hand side, over the manifold ∂D with the large parameter r, at the critical points $x^{\pm}$ of the phase function $\varphi_\omega(x)$.

In a general setting, the stationary phase expansion is an asymptotic series. In our particular case, however, there is only finite number of terms in the expansion, which is rather an exact equality.

The phenomenon of stationary phase expansions with finite number of terms was studied in [3–5]. The simplest example of such expansion delivers the unit ball B^3 in $\mathbb{R}^3$. The Fourier transform of the characteristic function of the ball is

$$\frac{1}{\pi}\hat{\chi}_{B^n}(r\omega) = \int_{-1}^{1} e^{irz}(1 - z^2)dz$$

(21)

$$= e^{ir}(\frac{1}{ir} - \frac{2}{(ir)^2} + \frac{2}{(ir)^3}) + e^{-ir}(-\frac{1}{ir} - \frac{2}{(ir)^2} - \frac{2}{(ir)^3}).$$

The right hand side is the stationary phase expansion, at the critical points $z = \pm 1$, of the oscillatory integral on S^2, resulted from the Green's formula for the integral over the ball in the left hand side. The phase function of this oscillatory integral is the height-function $\varphi(x, y, z) = z$.

It was shown in [5] that finite (single-term) stationary phase expansion appears for the phase functions—moment maps of Hamiltonian actions on symplectic manifolds. This result was generalized in [4] for the even-dimensional, not necessarily symplectic, manifolds acted on by S^1 with isolated fixed points. The situation studied in [4] leads to multi-term expansions.

It is an open problem to describe all the situations when the stationary phase expansions are finite (e.g., see the discussion in [4]). Balls and ellipsoids do deliver such examples, but apparently do not exhaust them. We saw that polynomially integrable domains and the phase functions $\varphi_\omega(x) = <\omega, x>$ (the height-function in the direction ω) also provide finite stationary phase expansions (19).

However, in the case of polynomially integrable domains the situation is much more restrictive. Indeed, we have finite expansions, on the manifold ∂D, but for the whole family ω-height functions φ_ω, parametrized by the unit vector $\omega \in S^{n-1}$.

In fact, this property fully characterizes polynomially integrable domains:

Proposition 12 *Let D be a convex domain in $\mathbb{R}^n$ with smooth boundary. Then D is polynomially integrable if and only if the Fourier transform $\widehat{\chi}_D(r\omega)$ has, for any direction $\omega \in S^{n-1}$, a finite stationary phase expansion with respect to the large parameter r.*

Proof The part "if" is already proved by formula (19). We have to show that if (19) is valid then D is a polynomially integrable domain. By Projection-Slice Formula (3) the section-volume function $V_D(\omega, t)$ is the inverse one-dimensional Fourier transform with respect to r of $\widehat{\chi}(r\omega)$:

$$V_D(\omega, t) = F_{r\to t}^{-1}\{e^{ih_D^+(\omega)r}Q^+(\omega, \frac{1}{r}) + e^{ih_D^-(\omega)r}Q^-(\omega, \frac{1}{r})\}.$$

The general terms in the right hand side are

$$v_k^\pm(\omega, r) := q_k^\pm(\omega)e^{ih_D^\pm(\omega)r}\frac{1}{r^k}.$$

Consider one of them, say, v_k^+.

If

$$f_k(\omega, t) = F_{r\to t}^{-1}v_k^+(\omega, r)$$

then its kth derivative in t is

$$\frac{d^k}{dt^k}f_k(\omega, t) = F_{r\to t}^{-1}(ir)^k v_k(\omega, r)$$

$$= i^k q_k^+(\omega)F_{r\to t}^{-1}e^{irh_D^\pm(\omega)} = i^k\delta(t - h^+(\omega)).$$

$$(22)$$

Consequent integration in t yields that $f_k(\omega, t)$ is a piecewise polynomial, namely, it is a polynomial in t on $t < h_D^+(\omega)$, and is a polynomial in t on $t > h_D^+(\omega)$.

Likewise, the functions $F_{r\to t}^{-1}v_k^-(\omega, t)$ are polynomials on each half-line $t < h_D^-(\omega)$ and $t > h_D^-(\omega)$. Then the section-volume function V_D is a polynomial on the segment $h_D^-(\omega) \leq t \leq h_D^+(\omega)$, as a sum of $F^{-1}v_k^\pm$. Thus, D is polynomially integrable.

8 Open Questions

We conclude with formulating open questions.

- Are ellipsoids the only domains with infinitely smooth boundaries in the Euclidean spaces $\mathbb{R}^n$ of odd dimensions n for which the section-volume function $V_D(\omega, t)$ is a polynomial in t? The same question refers to the cut-off function $\widehat{V}_D(\omega, t)$ [10].
- What geometric properties of a domain can be derived from the finiteness of stationary phase expansion for the Fourier transform of its characteristic function?
- What is the role of the smoothness condition for the boundary? Are there polynomially integrable domains with a relaxed condition of smoothness, different from ellipsoids?

Acknowledgements I am grateful to Mikhail Zaidenberg for drawing my attention to the subject of this article and stimulating initial discussions.

Remark After the Submission After this article was submitted, Koldobsky, Merkurjev and Yaskin proved [8] that polynomially integrable infinitely smooth domains in odd-dimensional spaces are ellipsoids.

References

1. V.I. Arnold, V.A. Vassiliev, Newton's Principia read 300 years later. Not. Am. Math. Soc. **36**(9), 1148–1154 (1989)
2. V.I. Arnold, V.A. Vassiliev, Addendum to [3]. Not. Am. Math. Soc. **37**(1), 144
3. M.Atiyah, R. Bott, The moment map and equivariant cohomologies. Topology **23**, 1–28 (1964)
4. J. Bernard, Finite stationary phase expansions. Asian J. Math. **9**(2), 187–198 (2005)
5. J.J. Duistermatt, G.J. Heckman, On the variation in the cohomology of the symplectic form of the reduced phase space. Invent. Math. **69**, 259–268 (1982)
6. I.M. Gelfand, M.I. Graev, N.Ya. Vilenkin, *Generalized Functions, Volume 5: Integral Geometry and Representation Theory*, vol. 381 (AMS Chelsea Publishing, Providence, RI, 1966), p. 449
7. S. Helgason, *Groups and Geometric Analysis* (Academic, Cambridge, MA, 1984)
8. A. Koldobsky, A. Merkurjev, V. Yaskin, *On polynomially integrable convex bodies*. Preprint, https://arxiv.org/pdf/1702.00429.pdf
9. I. Newton, *Philosophiae Naturalis Principia Mathematica* (Benjamin Motte, London, 1687)
10. V.A. Vassiliev, Newton's lemma XXVIII on integrable ovals in higher dimensions and reflection groups. Bull. Lond. Math. Soc. **47**(2), 290–300 (2015)
11. Wikipedia, Newton's theorem about ovals, https://en.wikipedia.org/wiki/Newton's_theorem_about_ovals
12. R. Wong, *Asymptotic Approximations of Integrals* (Academic, Cambridge, MA, 1989)

A Survey on the Maximal Number of Solutions of Equations Related to Gravitational Lensing

Catherine Bénéteau and Nicole Hudson

Abstract This paper is a survey of what is known about the maximal number of solutions of the equation $f(z) = \bar{z}$, in particular when f is the Cauchy transform of a compactly supported positive measure. When f is a rational function, the number of solutions of this equation is equal to the number of images seen by an observer of a single light source deflected by a gravitational lens (such as a galaxy). We will discuss what is known in the context of harmonic polynomials, rational functions, polynomials in z and $\bar{z}$ (but not harmonic!) and even transcendental functions that arise in situations involving continuous mass distributions for different shapes. In particular, we discuss an example related to the lens equation for a limaçon-shaped gravitational lens.

Keywords Cauchy integrals • Gravitational lensing • Harmonic polynomials • Schwarz function

2010 Mathematics Subject Classification 30E10, 30E20, 85-02

1 Introduction

Studying and counting roots of systems of equations has been an important question in mathematics for centuries. In the complex plane, the Fundamental Theorem of Algebra, whose first proofs (with some gaps) are usually attributed to d'Alembert in 1746 and Gauss in 1799, tells us that every polynomial of degree n of a single complex variable z has exactly n complex roots (counting multiplicity). However, much less is known about the number of roots of polynomials containing both z and $\bar{z}$. A special case of interest is that of *harmonic* polynomials, that is, functions of the type $p(z) + \overline{q(z)}$, where p and q are complex polynomials. As an example of the many open problems in this area, it is still not known what the maximum number of roots of the equation $p(z) = \bar{z}^2$ is, if p is an arbitrary complex polynomial of degree

C. Bénéteau (✉) • N. Hudson
Department of Mathematics and Statistics, University of South Florida, Tampa, FL 33620, USA
e-mail: cbenetea@usf.edu; nicolehudson@mail.usf.edu

© Springer International Publishing AG 2018   23
M. Agranovsky et al. (eds.), *Complex Analysis and Dynamical Systems*,
Trends in Mathematics, https://doi.org/10.1007/978-3-319-70154-7_2

Fig. 1 Four separate images
of a single quasar located
behind a galaxy, with an
additional image too dim to
see. Credit: NASA, ESA, and
STScI. Source: http://
hubblesite.org/image/22/
news_release/1990-20

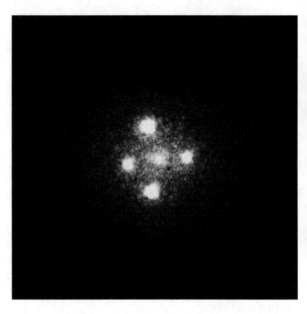

$n > 3$. If we extend this realm of questioning to include rational functions, some of
these mathematical problems become questions in gravitational microlensing.

Indeed, astrophysicists have long known the equation that relates the position of
a light source and the mass distribution of a gravitational lens (e.g., a galaxy) to
the images of the source formed by the lens (see Fig. 1). This equation depends on
the mass distribution of the lens and can be written in complex form in such a way
that positions and numbers of *roots* of rational functions occurring in the equation
correspond precisely to those of the *images* formed by the gravitational lens.

Over the last 15 years, there have been many articles published investigating
these questions from a mathematical point of view. In spite of this, there are still
many remaining open questions. This paper is a survey of some of the known results
and conjectures related to counting numbers of solutions of equations of this type.
We will be particularly interested in harmonic polynomials, rational functions arising from gravitational lenses consisting of point masses, non-harmonic polynomials
in z and $\bar{z}$, and certain transcendental functions that arise in situations involving
continuous mass distributions for different shapes. In particular, we discuss an
example related to the lens equation for a limaçon-shaped gravitational lens.

In Sect. 2, we introduce polynomials in z and $\bar{z}$ and discuss the particular case
of harmonic polynomials together with known results and open problems. We then
turn to the case of rational functions and explore the connection with gravitational
lensing. In Sect. 3, we address the case of continuous distributions: we first
examine what is known for gravitational lenses that are ellipses with uniform mass
distributions; we then examine a similar situation for a limaçon-shaped gravitational
lens. Our main contribution there is the lens equation in that case, together with
upper estimates on number of images produced. Although the estimates are not at
all sharp in that context, it may be that this approach will provide a model that is
useful for numerics and may offer insight into appropriate conjectures for future

research. In closing, we survey some results related to isothermal densities, which give rise to lens equations involving transcendental functions.

Thanks and Dedication The authors would like to thank Dmitry Khavinson for many helpful discussions during the writing of this paper. The first author would also like to dedicate this paper to her dear friend Sasha Vasiliev, a wonderful mathematician and person, who is dreadfully missed.

2 Harmonic Polynomials, Rational Functions, and Lensing

2.1 Polynomials in z and $\bar{z}$

The Fundamental Theorem of Algebra counts the exact number of (complex) roots of a complex polynomial of one variable in terms of the degree of the polynomial; however, if the polynomial involves the complex variable z and its conjugate $\bar{z}$, the situation immediately becomes much more complicated. Consider, for complex numbers $a_{k,j}$, the polynomial equation

$$f(z,\bar{z}) = \sum_{k=0}^{n}\sum_{j=0}^{m} a_{k,j}\, z^k \bar{z}^j = 0.$$

One (perhaps naive) approach is to rewrite f in terms of the real variables x and y, where $z = x + iy$. This results in

$$f(z,\bar{z}) = f(x+iy, x-iy)$$

$$= \sum_{i=0}^{n+m}\sum_{l=0}^{n+m} c_{i,l}\, x^i y^l = g(x,y).$$

Separating the equation $g(x,y) = 0$ into its real and imaginary parts yields a system of two real equations in x and y, and counting the common intersection points (in the real plane) of this system becomes a problem in algebraic geometry.

Notice that it is not always the case that there are any solutions at all or that number of roots is finite! Take the simple example $z\bar{z} = 1$, which of course gives the whole unit circle as a solution, while the equation $z\bar{z} = -1$ has no solutions. On the other hand, consider the following trivial example: the polynomial $f(z,\bar{z}) = z - 4i\bar{z} + 5 + 3i$ can be rewritten as $g(x,y) = x - 4y + 5 + i(y - 4x + 3)$, which gives rise to a system of two linear equations:

$$0 = x - 4y + 5 \qquad 0 = y - 4x + 3.$$

This system can be immediately solved and leads to a single complex root $z = \frac{1}{15}(17 + 23i)$.

One way in which these two examples differ is that the second is a harmonic function (that is, it satisfies the Laplace equation), and the roots of harmonic polynomials have been studied intensively in the last 15 years. Their rational counterparts also arise naturally in the study of gravitational lenses. Let us first turn to a discussion of what is known for harmonic polynomials.

2.2 Harmonic Polynomials

We begin with a definition.

Definition 2.1 A harmonic polynomial $f(z)$ is a function that can be written as $f(z) = p(z) + \overline{q(z)}$, where p and q are complex (analytic) polynomials in z.

One can ask whether there is a version of the Fundamental Theorem of Algebra that holds for these harmonic polynomials, that is, how many solutions can the following equation have, where p and q are both (analytic) polynomials, of degree n and m, respectively?

$$p(z) = \overline{q(z)} \tag{1}$$

Notice that if $n = m$, then the number of solutions might be infinite, as previously discussed, but one can show (see [24, 25]) that if $n > m$, the number of solutions is finite. Sheil-Small became interested in this question from the point of view of harmonic mappings and questions related to the Bieberbach Conjecture, and conjectured in 1992 [22] that if $n > m$, Eq. (1) has at most n^2 solutions. In his 1998 PhD thesis [24, 25], Wilmshurst used Bézout's Theorem to prove this conjecture. Intuitively, one can reduce (1) to two systems of real polynomial equations of degree n, and counting intersections of that system of curves gives n^2. Wilmshurst showed that n^2 is sharp by considering examples where $m = n - 1$. He then conjectured that if $1 \leq m < n - 1$, an even lower bound might occur, namely, that the largest number of zeros of (1) in that case is $3n - 2 + m(m - 1)$. In particular, for $m = 1$, this intriguing conjecture stated that the maximum number of zeros of an equation of the type $p(z) = \bar{z}$, where p is an (analytic) polynomial of degree $n > 1$, is equal to $3n - 2$. Several people (Bshouty, Crofoot, Lizzaik, Sarason, and others) worked on this and related questions in harmonic mappings, and made partial progress (see, e.g., [6]). Significant progress occurred on this problem in 2002, when Khavinson and Świątek [15] proved Wilmshurst's Conjecture for $m = 1$. They used the fact that fixed points of $\overline{p(z)}$ are also fixed points of the analytic polynomial $Q(z) = \overline{p(\overline{p(z)})}$ (of degree n^2), and then used complex dynamics and the argument principle for harmonic functions to count the different types of fixed points that can occur. They noticed that the bound is sharp by considering quadratic polynomials ($n = 2$). Geyer [9] later showed this bound to be sharp for all n. In 2016, Khavinson et al. [16] employed a clever method of comparing Minkowski areas of Newton polygons generated by harmonic polynomials to refine known results about maximal numbers of zeros of (1) when the degree m of q satisfies $1 < m < n - 1$, where n is the degree of p. For example, they show that for each $n > m$, there exists a harmonic polynomial $h(z) = p(z) + \overline{q(z)}$ with at least $m^2 + m + n$ roots.

Instead of considering *maximal* numbers of roots of harmonic polynomials, one can also think of probabilistic interpretations, that is: how many roots does such an equation have on average? In 2009, Li and Wei analyzed the probabilistic distribution of random harmonic polynomials [19]. This resulted in an exact equation for the expected number of roots of (1), a rather complicated Lebesgue integral. Li and Wei used this formula to show that, when the degree n of p is equal to the degree m of q, the expected number behaves like $\frac{\pi}{4} n^{3/2}$ as $n \to \infty$, while if $m = \alpha n + o(n)$ with $0 \le \alpha < 1$, the expected number is asymptotically n as $n \to \infty$. Others began trying to come up with examples, either to prove or disprove Wilmshurst's Conjecture for $m \ge 2$. In 2013, Lee et al. [17] used methods of algebraic geometry to conclude that for $m = n - 3$ and $n \ge 4$, there exists harmonic polynomials p_n and q_m for which the number of roots exceeds

$$ n^2 - 4n + 4 \left\lfloor \frac{n-2}{\pi} \arctan \frac{\sqrt{n^2 - 2n}}{n} \right\rfloor + 2, $$

which yields an infinite number of counterexamples to Wilmshurst's conjecture in general. Hauenstein, Lerario, Lundberg, and Mehta generalized these counterexamples in 2014 [10] by using an experimental, certified-counting approach. In particular, they conjectured that for all even n, there exist polynomials p with degree n and q with degree $n/2$ such that Eq. (1) has exactly $n^2/2 - n + 12$ roots. In 2016 [18], Lerario and Lundberg confirmed and sharpened the results of Li and Wei that had previously been obtained by computer experiments. Thomack [23] recently showed, using a slightly different probabilistic model, that if m is fixed, the expectation of the number of zeros is asymptotically n as $n \to \infty$. It is worth noting in addition that by the Argument Principle, if p has degree $n > 1$, then the number of zeros of $p(z) = \bar{z}$ is *at least* n.

In spite of all of this progress, the following basic open question remains.

Open Question 2.1 If p is a polynomial of degree $n > 3$ and q is a polynomial of degree m with $1 < m < n-1$, what is the maximal number of roots that the equation $p(z) + \overline{q(z)} = 0$ can have?

Any progress on this question would be of interest, in particular, showing whether or not the maximal number of roots is linear in n for m fixed would be a large step forward.

2.3 Rational Functions and Gravitational Lensing

In 2006, Khavinson and Neumann (see [13]) had the idea to replace the polynomial p in the first simplest harmonic polynomial equation $p(z) = \bar{z}$ by a rational function $r(z) := \frac{p(z)}{q(z)}$ of degree n, where p and q are polynomials, and by definition $n = \max\{\deg p, \deg q\}$. That is, they investigated the maximal number of solutions of the equation

$$ r(z) = \bar{z}. \tag{2} $$

They proved that the maximal number of solutions to (2) is $5n - 5$. In the process of working on this problem, they discovered an amazing connection with a problem in mathematical physics related to gravitational lensing. In fact, the astrophysicist Rhie [20] had conjectured this upper bound and had a beautiful simple geometric construction that showed that this bound can actually occur for all $n > 1$, thus confirming that the result is sharp. In [5], the authors studied intermediate numbers of solutions of Eq. (2), and showed that any number of simple zeros allowed by the Argument Principle occurs.

Gravitational lensing is the bending of light by a gravitational field. This phenomenon was predicted by general relativity. In fact, Einstein's theory was first proven by Sir Arthur Eddington via observation of a lensing event during a total solar eclipse in 1919. One way of formulating the gravitational lensing problem involves using multiple complex planes as the stage for the physical configuration of objects involved—a light-emitting source, a massive object (the lens), and the observer. More specifically, to mathematically model the lens, we consider three parallel planes (Fig. 2): the *observation plane* where the observer is located, the *lens plane* containing the galaxy (or other massive object) acting as the lens, and, located on the opposite side of the lens plane from the observer, the *source plane* where the light source is positioned.

The lens equation takes the following form:

$$w = z - \int_\Omega \frac{d\mu(\zeta)}{\bar{z} - \bar{\zeta}} - \gamma \bar{z}, \tag{3}$$

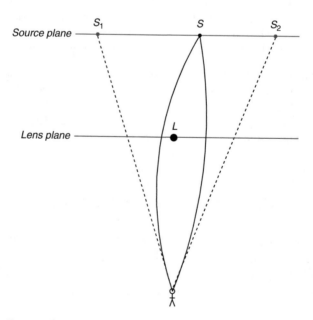

Fig. 2 Simple diagram of the source and lens planes

where w is the position of the light source, z is the position of a lensed image, γ is a gravitational shearing term, $d\mu$ is a measure related to the gravitational masses, and Ω is the domain containing the support of μ. The shearing term accounts for the effects of an external gravitational field on the lens system, such as the pull of a nearby galaxy. Many astrophysicists (Witt, Mao, Peters, Rhie, Burke, among others) have worked on estimates of numbers of solutions of the lens equation and related problems. For a nice discussion of the history of the discovery of the lensing effect and on early estimates from astrophysicists on number of images that can occur, see for example [8, 14] and the references therein.

The measure $d\mu$ can be discrete (representing point masses), or a continuous distribution of mass on a compact set. Neither model is completely physically accurate: galaxies are not actually continuous objects, and stars are not point masses, but this model is accurate enough to predict known physical phenomena, as seen in some of the pictures below. The discrete form of the lens equation produced by a discrete measure at n point masses becomes the following:

$$
w = z - \sum_{j=1}^{n} \frac{\sigma_j}{\bar{z} - \bar{z}_j} - \gamma \bar{z} \tag{4}
$$

where w, z, and γ play the same roles as before, and σ_j is a coefficient related to the mass and the so-called "optical depth" of each point mass on the lens plane.

When a gravitational lens, a light source, and an observer are colinear, the lensed image does not appear as a point, but a circle or ellipse! The halo of light that is formed is called an 'Einstein Ring' (see Fig. 3). Interestingly, this result can be obtained from either the discrete configuration, or the continuous distribution of an

(a)   (b)

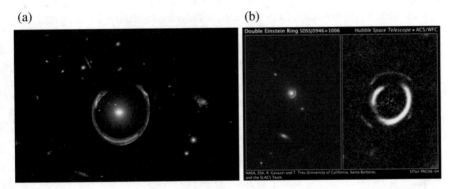

Fig. 3 Images of Einstein rings captured by the Hubble space telescope. (**a**) Light from a blue galaxy distorted by a red galaxy. Credit: NASA, ESA. Source: https://www.nasa.gov/sites/ default/files/thumbnails/image/15861603283_3579db3fc6_o.jpg. (**b**) An elliptical galaxy which is lensing light from two different background galaxies. Credit: NASA, ESA, R. Gavazzi and T. Treu (University of California, Santa Barbara). Source: https://www.nasa.gov/images/content/ 207624main_double_einstein_full.jpg

elliptical (or circular) galaxy. In the discrete case, simply set w, γ, and $\bar{z}_1$ equal to zero, and what follows immediately is the complex equation of a circle with radius $\sqrt{\sigma_1}$. The proof that in the continuous case, you obtain only a circle or an ellipse is more difficult and can be found in [8]. Again, although maximal numbers of solutions of Eq. (2) are known, as soon as the right hand side has degree larger than 1, the question regarding the exact number of images remains open.

3   Gravitational Lensing: Continuous Mass Distributions

Several authors have studied the question of number of roots of the lens equation for continuous densities in different contexts (see [2–4, 8]). Let us first examine the case of ellipses with uniform mass densities.

3.1   Ellipse with Uniform Mass Density

Fassnacht et al. [8] explored the case in which the gravitational lens was an ellipse with uniform mass density. In what follows, we describe their method and results. Conjugating lens equation (3) and using the fact that $d\mu(\zeta) = \frac{1}{\pi}dA(\zeta)$, a uniform mass density, we rewrite the lens equation in the form:

$$\bar{w} = \bar{z} - \frac{1}{\pi}\int_{\Omega} \frac{dA(\zeta)}{z - \zeta} - \gamma z$$

where $\Omega := \{z = x + iy : \frac{x^2}{a^2} + \frac{y^2}{b^2} < 1\}$, and $dA := dA(\zeta) = \frac{d\zeta\,d\bar{\zeta}}{2i}$ is area measure.

One can use the complex form of Green's Theorem to rewrite the integral in the lens equation for the ellipse as a line integral. The result will vary depending on whether $z \in \Omega$ or $z \notin \Omega$. In this context, z being inside or outside of Ω corresponds to the observed image appearing inside the galaxy (a "dim" image) or outside (a "bright" image).

For $z \notin \Omega$:

$$-\frac{1}{\pi}\int_{\Omega} \frac{dA}{z - \zeta} = -\frac{1}{2\pi i}\iint_{\Omega} \frac{d\zeta\,d\bar{\zeta}}{z - \zeta} = \frac{1}{2\pi i}\int_{\partial\Omega} \frac{\bar{\zeta}\,d\zeta}{\zeta - z}.$$

For $z \in \Omega$:

$$-\frac{1}{\pi}\int_{\Omega} \frac{dA}{z - \zeta} = -\frac{1}{2\pi i}\iint_{\Omega} \frac{d\zeta\,d\bar{\zeta}}{z - \zeta} = -\bar{z} + \frac{1}{2\pi i}\int_{\partial\Omega} \frac{\bar{\zeta}\,d\zeta}{\zeta - z}.$$

Both cases involve the same Cauchy integral:

$$\frac{1}{2\pi i} \int_{\partial\Omega} \frac{\bar{\zeta}\, d\zeta}{\zeta - z}.$$

In order to rewrite this integral, Fassnacht et al. [8] used the Schwarz function for the ellipse to express $\bar{\zeta}$ as an analytic function along the boundary. Indeed, suppose C is an analytic arc given by $f(x, y) = 0$. Rewriting this equation in complex notation gives

$$f\left(\frac{z + \bar{z}}{2}, \frac{z - \bar{z}}{2i}\right) \equiv g(z, \bar{z}) = 0.$$

If $\nabla f = \frac{\partial g}{\partial \bar{z}} \neq 0$ along C, then there exists a function $S(z)$, analytic in a neighborhood of C such that $S(z) = \bar{z}$ on C. Such a function $S(z)$ is called the *Schwarz function* of C. For more detail on the Schwarz function, see [7, 21].

After substituting the Schwarz function of the ellipse into the integral, applying Cauchy's formula, and performing some algebraic manipulations, Fassnacht et al. [8] transformed the lens equation into a polynomial in z and $\bar{z}$. Rewriting this lens equation as a system of two real equations, they derived that the maximum number of images produced by an elliptical lens is *four* outside and *one* inside.

3.2 Investigating the Limaçon

Let us now use this same method to investigate the number of images formed by a limaçon lens.

3.2.1 The Lens Equation

The limaçon (Fig. 4) can be thought of as the shape whose equation in polar coordinates is given by: $r = a + b \cos \theta$, where a and b are real constants.

The most familiar example is the cardioid (the dotted shape in Fig. 4), which occurs when $|a| = |b|$. (In some cases, the term limaçon is used interchangeably with cardioid, but in this paper, we will call the more general shape a limaçon.) It is easy to see that the area of a limaçon with no internal "loops" (i.e., $|a| \geq |b|$) is $\pi(a^2 + \frac{b^2}{2})$.

Let $\Omega := \{ z = re^{i\theta} \in \mathbb{C} \mid r < a + b \cos \theta \}$, where a, b are fixed. For simplicity, let us consider a and b positive. Let $\Gamma = \partial\Omega$ be the limaçon. The lens equation for a limaçon-shaped lens with uniform mass density $d\mu(\zeta) = \frac{2}{3\pi} dA(\zeta)$, is:

$$\bar{w} = \bar{z} - \frac{2}{3\pi} \int_{\Omega} \frac{dA(\zeta)}{z - \zeta} - \gamma z$$

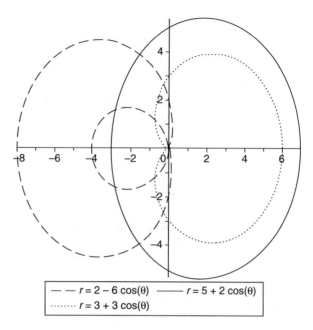

Fig. 4 Graph of three different limaçons

where $dA := dA(\zeta) = \frac{d\zeta\,d\bar{\zeta}}{2i}$, and the integral includes a normalizing factor so that the area of a limaçon with $a = b = 1$ is 1. Using the complex form of Green's Theorem to rewrite the limaçon lens equation as a line integral, as in [8], we notice that for $z \notin \Omega$:

$$-\frac{2}{3\pi} \int_{\Omega} \frac{dA}{z-\zeta} = -\frac{1}{3\pi i} \iint_{\Omega} \frac{d\zeta\,d\bar{\zeta}}{z-\zeta} = \frac{2}{3}\frac{1}{2\pi i} \int_{\partial\Omega} \frac{\bar{\zeta}\,d\zeta}{\zeta-z},$$

while for $z \in \Omega$:

$$-\frac{2}{3\pi} \int_{\Omega} \frac{dA}{z-\zeta} = -\frac{1}{3\pi i} \iint_{\Omega} \frac{d\zeta\,d\bar{\zeta}}{z-\zeta} = \frac{2}{3}\left[-\bar{z} + \frac{1}{2\pi i} \int_{\partial\Omega} \frac{\bar{\zeta}\,d\zeta}{\zeta-z}\right].$$

Note that this differs from the elliptical case only by the normalizing factor of $\frac{2}{3}$.

3.2.2 The Schwarz Function

Let us now revisit the Schwarz function of the limaçon (see also [7]). Squaring the equation $r = a + b\cos\theta$ so that $r^2 = ar + br\cos\theta$ and using complex coordinates

$(\zeta, \bar{\zeta})$ gives $\zeta\bar{\zeta} = a(\zeta\bar{\zeta})^{\frac{1}{2}} + \frac{b}{2}(\zeta + \bar{\zeta})$. Performing some algebraic manipulation leads to

$$\zeta\bar{\zeta} - \frac{b}{2}(\zeta + \bar{\zeta}) = a(\zeta\bar{\zeta})^{\frac{1}{2}}$$

$$(\zeta\bar{\zeta})^2 + \frac{b^2}{4}(\zeta^2 + \bar{\zeta}^2 + 2\zeta\bar{\zeta}) - b(\zeta^2\bar{\zeta} + \zeta\bar{\zeta}^2) = a^2\zeta\bar{\zeta}$$

$$\bar{\zeta}^2\left(\zeta - \frac{b}{2}\right)^2 - \zeta\bar{\zeta}\left(b\left(\zeta - \frac{b}{2}\right) + a^2\right) + \frac{b^2}{4}\zeta^2 = 0.$$

Since this equation is quadratic in $\bar{\zeta}$, we can apply the quadratic formula to obtain

$$\bar{\zeta} = \frac{1}{2(\zeta - \frac{b}{2})^2}\zeta\left(b\left(\zeta - \frac{b}{2}\right) + a^2\right)$$

$$\pm \frac{1}{2(\zeta - \frac{b}{2})^2}\sqrt{\zeta^2\left(b\left(\zeta - \frac{b}{2}\right) + a^2\right)^2 - 4\left(\zeta - \frac{b}{2}\right)^2\frac{b^2}{4}\zeta^2}$$

$$= \frac{\zeta}{2(\zeta - \frac{b}{2})^2}\left[b\left(\zeta - \frac{b}{2}\right) + a^2 \pm a\sqrt{2b\left(\zeta + \frac{a^2 - b^2}{2b}\right)}\right].$$

To simplify notation, let us define $A := \frac{a^2 - b^2}{2b}$, and notice that for $\zeta \in \Gamma$, the positive choice of sign in the above expression is necessary: that is, for $\zeta \in \Gamma$, we have

$$\bar{\zeta} = \varphi_1(\zeta) + \varphi_2(\zeta),$$

where $\varphi_1(\zeta) = \frac{\zeta}{2(\zeta - \frac{b}{2})^2}\left[b\left(\zeta - \frac{b}{2}\right) + a^2\right]$ and $\varphi_2(\zeta) = \frac{a\zeta\sqrt{2b(\zeta + A)}}{2(\zeta - \frac{b}{2})^2}$. Let us now further assume that the constants a and b satisfy $a > b > 0$, and make a choice of branch cut along the ray $(-\infty, -A]$. With that choice made, both φ_1 and φ_2 are analytic in a neighborhood of Γ, and thus the Schwarz function $S(\zeta) = \varphi_1(\zeta) + \varphi_2(\zeta)$. Note that φ_1 is analytic in $\mathbb{C} - \{b/2\}$, and φ_2 is analytic in $\Omega - \{b/2\}$, and that $b/2$ lies inside Ω.

3.2.3 Solving the Lens Equation and Counting Solutions

Recall that in order to find the lens equation for either z inside or outside the domain Ω, we need to calculate the integral

$$\int_{\partial\Omega} \frac{\bar{\zeta}\, d\zeta}{\zeta - z},$$

which can be rewritten using the Schwarz function from the previous section as

$$\int_{\partial\Omega} \frac{\varphi_1(\zeta) + \varphi_2(\zeta)}{\zeta - z} d\zeta.$$

Now for $z \notin \Omega$, then $\frac{\varphi_1(\zeta)+\varphi_2(\zeta)}{\zeta-z}$ is analytic inside Ω except at the pole $b/2$, and therefore by the Residue Theorem applied to the integrand as a function of ζ (with z fixed),

$$\frac{1}{2\pi i} \int_{\partial\Omega} \frac{\varphi_1(\zeta) + \varphi_2(\zeta)}{\zeta - z} d\zeta = \operatorname{Res}\left(\frac{\varphi_1(\zeta) + \varphi_2(\zeta)}{\zeta - z}; b/2\right).$$

A somewhat lengthy but straightforward calculation gives that

$$\operatorname{Res}\left(\frac{\varphi_1(\zeta) + \varphi_2(\zeta)}{\zeta - z}; b/2\right) = \frac{-2a^2 z + b^2(b/2 - z)}{2(b/2 - z)^2}.$$

On the other hand, if $z \in \Omega$, then the integrand has residues both at $b/2$ as well as at z, and thus, we need to add the residue at z to the previous expression, and we get that:

$$\frac{1}{2\pi i} \int_{\partial\Omega} \frac{\varphi_1(\zeta) + \varphi_2(\zeta)}{\zeta - z} d\zeta = \frac{b(b/2 - z)(b - z) - a^2 z + az\sqrt{2b(z + A)}}{2(z - b/2)^2}.$$

That is, we have proved the following.

Theorem 3.1 *Suppose* $\Omega := \{z = re^{i\theta} \in \mathbb{C} \mid r < a + b\cos\theta\}$ *is the interior of a limaçon with constants chosen so that* $a > b > 0$, *with constant mass density* $d\mu(\zeta) = \frac{2}{3\pi}dA(\zeta)$. *Then for* $z \notin \Omega$, *the lens equation is*

$$\bar{w} = \bar{z} + \frac{2}{3}\left[\frac{-2a^2 z + b^2(b/2 - z)}{2(b/2 - z)^2}\right] - \gamma z, \tag{5}$$

while for $z \in \Omega$, *the lens equation is:*

$$\bar{w} = \frac{1}{3}\bar{z} + \frac{2}{3}\left[\frac{b(b/2 - z)(b - z) - a^2 z + az\sqrt{2b(z + A)}}{2(z - b/2)^2}\right] - \gamma z. \tag{6}$$

Now we can represent z and $\bar{z}$ in terms of real variables x and y, and letting $\bar{w} = s - it$, where s and t are real constants, we can use these lens equations to count solutions. In particular, notice that Lens Equation (5) becomes a system of two real equations of degree 3, and therefore, by Bézout's Theorem, there are at most 9 images outside the limaçon. Notice that theorem of Khavinson and Neumann for rational functions gives $5n - 5$ for $n = 3$ in this case, thus 10 solutions, so this approach gives a slight improvement in that case. On the other hand, if there is no shearing term ($\gamma = 0$), then $n = 2$, and therefore the Khavinson-Neumann theorem gives 5 solutions, while Bézout's theorem still gives 9.

In the case of the Lens Equation (6), the Khavinson-Neumann theorem does not apply at all, since the equation does not involve a rational function. Rewriting the equation using real variables, squaring both sides, and canceling common factors of $(z - b/2)$ gives rise to a system of two real equations. The equation corresponding to the real part of the lens equation is of degree 4, regardless of the choice of γ. However, the imaginary part is of degree 3 if $\gamma = 0$ and of degree 4 otherwise. Hence, by Bézout's theorem, there are at most 16 solutions inside the limaçon if the shearing term doesn't vanish and at most 12 if it does. Thus, we have the following.

Theorem 3.2 *Suppose* $\Omega := \{ z = re^{i\theta} \in \mathbb{C} \mid r < a + b\cos\theta \}$ *is the interior of a limaçon with constants chosen so that $a > b > 0$, with constant mass density $d\mu(\zeta) = \frac{2}{3\pi}dA(\zeta)$. Then there are at most 9 solutions to the lens equation (5) if the shearing term $\gamma \neq 0$, and at most 5 solutions if $\gamma = 0$, while there are at most 16 solutions to the lens equation (6) if the shearing term $\gamma \neq 0$, and at most 12 solutions if $\gamma = 0$.*

It is clear that these are significant overestimates, especially for the images inside the limaçon, since information seems to be lost when squaring and moving to a generic estimate of real variable solutions. However, having a concrete expression for the lens equation depending on the parameters of the limaçon may lead to numerical experiments that will give rise to the appropriate conjecture for the number of images. Indeed, so far, numerical experiments show only 4 images inside with a shearing term present and 1 outside (see Figs. 5 and 6). Based on these numerical experiments, we state the following conjecture.

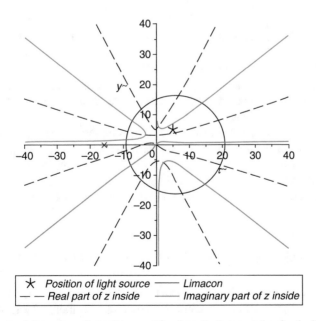

| ✴ Position of light source | —— Limacon |
| — — Real part of z inside | —— Imaginary part of z inside |

Fig. 5 Zero set of the system of equations resulting from the lens equation for the limaçon given by $r = 15 + 6\cos\theta$, a source position $w = 5(1 + i)$, and a shearing term of $\gamma = 2.5$. There are 4 images, located at the crossings of the solid and dashed curves that lie *within* the limaçon

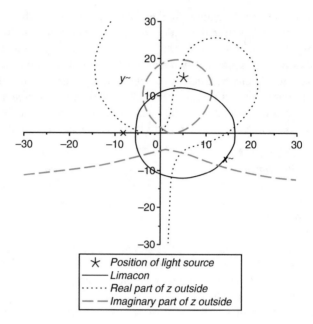

Fig. 6 Zero set of the system of equations resulting from the lens equation for the limaçon given by $r = 11 + 5.5 \cos\theta$, a source position $w = 5(1 + 3i)$, and a shearing term of $\gamma = 0$. There is 1 image, located at the crossing of the dotted and dashed curves that lies *outside* of the limaçon

Conjecture 3.3 Suppose $\Omega := \{ z = re^{i\theta} \in \mathbb{C} \mid r < a + b\cos\theta \}$ is the interior of a limaçon with constants chosen so that $a > b > 0$, with constant mass density $d\mu(\zeta) = \frac{2}{3\pi} dA(\zeta)$. Then there is at most 1 solution to the lens equation (5) and there are at most 4 solutions to the lens equation (6).

One might find this conjecture plausible due to the fact that we can think of the limaçon as a perturbation of the ellipse, and thus, the same estimates on the maximal number of zeros (see [8]) is reasonable.

Finally, the limaçon is an example of a quadrature domain, and it is known that for these domains, the lens equation for solutions inside the shape (the "dim" images) is of the form $a(z) = \bar{z}$ for an algebraic function a (see Remark 2 on p. 14 of [8]). Thus, finding maximal numbers of solutions for such shapes, or equivalently, for lens equations involving algebraic functions seems like a challenging problem.

3.3 Lens Equations Involving Transcendental Functions

Certain choices of mass distributions lead to lens equations involving transcendental functions. In [8], the authors discuss the situation of a more physically relevant density than the constant mass distribution, that is "isothermal" densities, which are densities that are constant on homothetic ellipses (rather than confocal). These densities are obtained by projecting the more realistic 3-dimensional density $1/r^2$ that is proportional to the square of the distance r from the origin. In this 3-dimensional context, the gas in the galaxy has constant temperature, hence the term "isothermal" (see, e.g., [11]). For the ellipse, such a choice results in a lens equation involving a branch of arcsin, a transcendental function! (Note that in the case of a circle, the situation is much simpler and gives rise to what is called a Chang-Refsdal lens, see [1].) In 2010, Khavinson and Lundberg [12] investigated the transcendental equation

$$\arcsin\left(\frac{k}{\bar{z}+\bar{w}}\right) = z, \tag{7}$$

which is the lens equation for the ellipse with isothermal density. They showed that an upper bound on the number of images outside the ellipse (the so-called "bright" images) is at most 8. In that same year, Bergweiler and Eremenko [3] showed that the sharp bound is actually 6, and indeed any number of solutions from 1 to 6 can occur.

In [2], the authors considered isothermal densities with an added twist factor, so that the mass density has a spiral structure controlled by a real parameter s. More specifically, they considered the density $d\mu(\zeta) = \rho(|\zeta|)dA(\zeta)$ where $\rho(r) = M/r$, $r = e^{\theta/s}$, with some restrictions on s, and here $r = |\zeta|$ and θ are polar coordinates. They calculated the lens equation in this case and showed that it involves the Gauss hypergeometric function (see Theorem 2.1 of [2]).

In general, very little is known about maximal numbers of solutions of equations involving transcendental functions. A recent result of Bergweiler and Eremenko [4] gives upper and lower bounds on the number of solutions N of the equation $p(z)\log|z| + q(z) = 0$, where p and q are (co-prime) polynomials of degree n and m respectively: they show that $\max\{n, m\} \leq N \leq 3\max\{n, m\} + 2n$, and that the estimate is sharp for many values of n and m. It is clear that investigating more situations stemming from a rich variety of transcendental functions will prove useful and important for the further development of the subject.

References

1. J. An, N. Evans, The Chang-Refsdal lens revisited. Mon. Not. R. Astron. Soc. **369**, 317–324 (2006)
2. S. Bell, B. Ernst, S. Fancher, C. Keeton, A. Komanduru, E. Lundberg, Spiral galaxy lensing: a model with twist. Math. Phys. Anal. Geom. **17**(3–4), 305–322 (2014)
3. W. Bergweiler, A. Eremenko, On the number of solutions of a transcendental equation arising in the theory of gravitational lensing. Comput. Methods Funct. Theory **10**(1), 303–324 (2010)
4. W. Bergweiler, A. Eremenko, On the number of solutions of some transcendental equations (2017). arXiv:1702.06453
5. P. Bleher, Y. Homma, L. Ji, R. Roeder, Counting zeros of harmonic rational functions and its application to gravitational lensing. Int. Math. Res. Not. **8**, 2245–2264 (2014)

6. D. Bshouty, W. Hengartner, T. Suez, The exact bound of the number of zeros of harmonic polynomials. J. dAnalyse Math. **67**, 207–218 (1995)
7. P. Davis, *The Schwarz Function and Its Applications*. Carus Mathematical Monographs, vol. 17 (Mathematical Association of America, Washington, DC, 1960)
8. C. Fassnacht, C. Keeton, D. Khavinson, Gravitational lensing by elliptical galaxies and the Schwarz function, in *Analysis and Mathematical Physics: Proceedings of the Conference on New Trends in Complex and Harmonic Analysis* ed. by B. Gustafsson, A. Vasil'ev (Birkhäuser, Basel, 2009), pp. 115–129
9. L. Geyer, Sharp bounds for the valence of certain harmonic polynomials. Proc. Am. Math. Soc. **136**(2), 549–555 (2008)
10. J. Hauenstein, A. Lerario, E. Lundberg, Experiments on the zeros of harmonic polynomials using certified counting. Exp. Math. **24**(2), 133–141 (2015)
11. C. Keeton, S. Mao, H. Witt, Gravitational lenses with more than 4 images, I. Classification of caustics. Astrophys. J. **537**, 697–707 (2000)
12. D. Khavinson, E. Lundberg, Transcendental harmonic mappings and gravitational lensing by isothermal galaxies. Compl. Anal. Oper. Theory **4**(3), 515–524 (2010)
13. D. Khavinson, G. Neumann, On the number of zeros of certain rational harmonic functions. Proc. Am. Math. Soc. **134**(4), 1077–1085 (2006)
14. D. Khavinson, G. Neumann, From the fundamental theorem of algebra to astrophysics: a 'Harmonious' path. Not. Am. Math. Soc. **55**(6), 666–675 (2008)
15. D. Khavinson, G. Świątek, On the number of zeros of certain harmonic polynomials. Proc. Am. Math. Soc. **131**(2), 409–414 (2002)
16. D. Khavinson, S.-Y. Lee, A. Saez, Zeros of harmonic polynomials, critical lemniscates and caustics (2016). arXiv 1508.04439
17. S.-Y. Lee, A. Lerario, E. Lundberg, Remarks on Wilmshurst's theorem. Indiana Univ. Math. J. **64**(4), 1153–1167 (2015)
18. A. Lerario, E. Lundberg, On the zeros of random harmonic polynomials: the truncated model. J. Math. Anal. Appl. **438**, 1041–1054 (2016)
19. W. Li, A. Wei, On the expected number of zeros of a random harmonic polynomial. Proc. Am. Math. Soc. **137**(1), 195–204 (2009)
20. S. Rhie, n-point gravitational lenses with $5n - 5$ images (2003). arXiv:astro-ph/0305166
21. H.S. Shapiro, *The Schwarz Function and Its Generalization to Higher Dimensions*. University of Arkansas Lecture Notes in the Mathematical Sciences, vol. 9 (Wiley, Hoboken, NJ, 1992)
22. T. Sheil-Small, *Complex Polynomials*. Cambridge Studies in Advanced Mathematics, vol. 73 (Cambridge University Press, Cambridge, 2002)
23. A. Thomack, On the zeros of random harmonic polynomials: the naive model (2016), https://arxiv.org/pdf/1610.02611.pdf
24. A. Wilmshurst, Complex harmonic polynomials and the valence of harmonic polynomials, D. Phil. thesis, University of York, York (1994)
25. A. Wilmshurst, The valence of harmonic polynomials. Proc. Am. Math. Soc. **126**, 2077–2081 (1998)

Boundary Interpolation by Finite Blaschke Products

Vladimir Bolotnikov

Abstract Given n distinct points $t_1, \ldots, t_n$ on the unit circle $\mathbb{T}$ and equally many target values $w_1, \ldots, w_n \in \mathbb{T}$, we describe all Blaschke products f of degree at most $n - 1$ such that $f(t_i) = w_i$ for $i = 1, \ldots, n$. We also describe the cases where degree $n - 1$ is the minimal possible.

Keywords Blaschke product • Interpolation

2010 Mathematics Subject Classification 30E05

1 Introduction

Let $\mathcal{S}$ denote the Schur class of all analytic functions mapping the open unit disk $\mathbb{D} = \{z \in \mathbb{C} : |z| < 1\}$ into the closed unit disk $\overline{\mathbb{D}}$. We denote by $\mathcal{RS}_k$ the set of all rational $\mathcal{S}$-class functions of degree at most k. The functions $f \in \mathcal{RS}_k$ that are unimodular on the unit circle $\mathbb{T}$ are necessarily of the form

$$f(z) = c \cdot \prod_{i=1}^{k} \frac{z - a_i}{1 - z\overline{a}_i}, \qquad |c| = 1, \; |a_i| < 1$$

and are called *finite Blaschke products*. They can be characterized as Schur-class functions that extend to a mapping from $\overline{\mathbb{D}}$ onto itself and then the degree $k = \deg f$ can be interpreted geometrically as the winding number of the image $f(\mathbb{T})$ of the unit circle $\mathbb{T}$ about the origin (i.e., the map $f : \overline{\mathbb{D}} \to \overline{\mathbb{D}}$ is k-to-1). We will write $\mathcal{B}_k$ for the set of all Blaschke products of degree at most k, and we will use notation $\mathcal{B}_k^\circ := \mathcal{B}_k \backslash \mathcal{B}_{k-1}$ for the set of Blaschke products of degree k.

V. Bolotnikov (✉)
Department of Mathematics, The College of William and Mary, Williamsburg,
VA 23187-8795, USA
e-mail: vladi@math.wm.edu

© Springer International Publishing AG 2018   39
M. Agranovsky et al. (eds.), *Complex Analysis and Dynamical Systems*,
Trends in Mathematics, https://doi.org/10.1007/978-3-319-70154-7_3

Given points $z_1, \ldots, z_n \in \mathbb{D}$ and target values $w_1, \ldots, w_n \in \overline{\mathbb{D}}$, the classical
Nevanlinna-Pick problem consists of finding a function $f \in \mathcal{S}$ such that

$$f(z_i) = w_i \quad \text{for} \quad i = 1, \ldots, n \quad (z_i \in \mathbb{D}, \ w_i \in \overline{\mathbb{D}}). \tag{1.1}$$

The problem has a solution if and only if the matrix $P = \left[\frac{1-w_i \overline{w}_j}{1-z_i \overline{z}_j} \right]_{i,j=1}^{n}$ is positive
semidefinite ($P \geq 0$) [17, 18]. If $\det P = 0$, the problem has a unique solution which
is a finite Blaschke product of degree $k = \operatorname{rank} P$. If P is positive definite ($P > 0$),
the problem is *indeterminate* (has infinitely many solutions), and its solution set
admits a linear-fractional parametrization with free Schur-class parameter. When
the parameter runs through the class $\mathcal{B}_\kappa^\circ$, the parametrization formula produces all
Blaschke-product solutions to the problem (1.1) of degree $n+\kappa$ for each fixed $\kappa \geq 0$.

If the problem (1.1) is indeterminate, it has no solutions in $\mathcal{B}_{n-1}$. However, it
still has solutions in $\mathcal{RS}_{n-1}$ (*low-degree solutions*). In case not all target values w_i's
are the same, the problem has infinitely many low-degree solutions which can be
parametrized by polynomials σ with $\deg \sigma < n$ and with all the roots outside $\mathbb{D}$; see
[7, 11, 12]. More precisely, for every such σ, there exists a unique (up to a common
unimodular constant factor) pair of polynomials $a(z)$ and $b(z)$, each of degree at
most $n - 1$ and such that

(1) $|a(z)|^2 - |b(z)|^2 = |\sigma(z)|^2$ for $|z| = 1$ and
(2) the function $f = b/a$ (which belongs to $\mathcal{RS}_{n-1}$ by part (1)) satisfies (1.1).

The question of finding a rational solution of the minimal possible degree $k_{\min}$ (and
even finding the value of $k_{\min}$) is still open.

The boundary version of the Nevanlinna-Pick problem interpolates preassigned
values $w_1, \ldots, w_n$ (interpreted as nontangential boundary limits in the non-rational
case) at finitely many points $t_1, \ldots, t_n$ on the unit circle $\mathbb{T}$. Obvious necessary
conditions $|w_i| \leq 1$ ($1 \leq i \leq n$) turn out to be sufficient, and a solvable problem
is always indeterminate. If at least one of the preassigned boundary values w_i is not
unimodular, the problem cannot be solved by a finite Blaschke product. However,
as in the classical "interior" case, the problem admits infinitely many low-degree
rational solutions $f \in \mathcal{RS}_{n-1}$ (unless all target values w_i's are equal to each other);
see e.g., [4]. As for now, the description of all low-degree solutions is not known.

In this paper we will focus on the "boundary-to-boundary" Nevanlinna-Pick
problem where all preassigned boundary values are unimodular. This problem can
be solved by a finite Blaschke product.

Theorem 1.1 *Given any points* $t_1, \ldots, t_n \in \mathbb{T}$ *and* $w_1, \ldots, w_n \in \mathbb{T}$, *there exists a
finite Blaschke product* $f \in \mathcal{B}_{n-1}$ *such that*

$$f(t_i) = w_i \quad \text{for} \quad i = 1, \ldots, n \quad (t_i, w_i \in \mathbb{T}). \tag{1.2}$$

Any rational function $f \in \mathcal{RS}_{n-1}$ *satisfying conditions* (1.2) *is necessarily a
Blaschke product (i.e.,* $f \in \mathcal{B}_{n-1}$).

The existence of a finite Blaschke product satisfying conditions (1.2) was first confirmed in [8] with no estimates for the minimal possible degree $k_{\min}$ of f. The estimate $k_{\min} \leq n^2 - n$ was obtained in [22]. Ruscheweyh and Jones [19] improved this estimate to $k_{\min} \leq n - 1$, which is sharp for some problems. Several different approaches to constructing interpolants in $\mathcal{B}_{n-1}$ have been presented in [15, 16, 21]. The paper [21] also contains interesting results concerning minimal degree solutions. The fact that all low-degree solutions to the problem (1.2) are necessarily Blaschke products was established in [4].

If $w_i = w \in \mathbb{T}$ for $i = 1, \ldots, n$, then the constant function $f(z) \equiv w$ is the only $\mathcal{B}_{n-1}$-solution to the problem; it is not hard to show that such a problem has no other *rational* solutions of degree less than n. Otherwise (i.e., when at least two target values are distinct), the set of all $\mathcal{B}_{n-1}$-solutions (or, which is the same, the set of all $\mathcal{RS}_{n-1}$-solutions) to the problem (1.2) is infinite. The parametrization of this set is presented in Theorem 3.5, the main result of the paper. In Theorem 3.6 we give a slightly different parametrization formula which is then used in Sect. 4 to characterize the problems (1.2) having no Blaschke product solutions of degree less than $n - 1$.

2 The Modified Interpolation Problem

Given a finite Blaschke product f and a collection $\mathbf{t} = \{t_1, \ldots, t_n\}$ of distinct points in $\overline{\mathbb{D}}$, let us define the associated *Schwarz-Pick matrix*

$$P^f(\mathbf{t}) = \left[p^f(t_i, t_j) \right]_{i,j=1}^n$$

by entry-wise formulas

$$p^f(t_i, t_j) = \begin{cases} t_i f'(t_i)\overline{f(t_i)} & \text{if } i = j \text{ and } t_i \in \mathbb{T}, \\ \dfrac{1 - f(t_i)\overline{f(t_j)}}{1 - t_i \overline{t_j}} & \text{otherwise.} \end{cases} \qquad (2.1)$$

In case $\mathbf{t} \subset \mathbb{T}$, we will refer to $P^f(\mathbf{t})$ as to the *boundary Schwarz-Pick matrix*. Observe that for a finite Blaschke product f, equalities

$$\lim_{z \to t} \frac{1 - |f(z)|^2}{1 - |z|^2} = tf'(t)\overline{f(t)} = |f'(t)| \qquad (2.2)$$

hold at every point $t \in \mathbb{T}$ and thus, the diagonal entries in $P^f(\mathbf{t})$ are all nonnegative. The next result is well-known (see e.g. [3, Lemma 2.1] for the proof).

Lemma 2.1 *For $f \in \mathcal{B}_k^\circ$ and a tuple $\mathbf{t} = \{t_1, \ldots, t_n\} \in \overline{\mathbb{D}}^n$, the matrix $P^f(\mathbf{t})$ (2.1) is positive semidefinite and* $\operatorname{rank} P^f(\mathbf{t}) = \min\{n, k\}$.

Boundary interpolation by Schur-class functions and, in particular, by finite Blaschke products becomes more transparent if, in addition to conditions (1.2), one prescribes the values of f' at each interpolation node t_i. We denote this modified problem by **MP**.

MP: *Given $t_i, f_i \in \mathbb{T}$ and $\gamma_i \geq 0$, find a finite Blaschke product f such that*

$$f(t_i) = w_i, \quad |f'(t_i)| = \gamma_i \quad \text{for} \quad i = 1, \ldots, n \quad (t_i, w_i \in \mathbb{T}, \; \gamma_i \geq 0). \tag{2.3}$$

The problem (2.3) is well-known in a more general context of rational functions $f : \mathbb{D} \to \mathbb{D}$ [2] and even in the more general context Schur-class functions [5, 20]. The results on finite Blaschke product interpolation presented in Theorem 2.2 below are easily derived from the general ones.

Theorem 2.2 *If the problem **MP** has a solution, then the* Pick *matrix P_n of the problem defined as*

$$P_n = \left[p_{ij}\right]_{i,j=1}^n, \quad \text{where} \quad p_{ij} = \begin{cases} \dfrac{1 - w_i \overline{w}_j}{1 - t_i \overline{t}_j} & \text{if } i \neq j, \\ \gamma_i & \text{if } i = j, \end{cases} \tag{2.4}$$

is positive semidefinite. Moreover,

(1) *If $P_n > 0$, then the problem **MP** has infinitely many solutions.*
(2) *If $P_n \geq 0$ and $\mathrm{rank}\,(P_n) = k < n$, then there is an $f \in \mathcal{B}_k$ such that*

$$f(t_i) = w_i, \quad |f'(t_i)| \leq \gamma_i \quad \text{for} \quad i = 1, \ldots, n \tag{2.5}$$

and no other rational function $f : \mathbb{D} \to \mathbb{D}$ meets conditions (2.5).

Remark 2.3 The second statement in Theorem 2.2 suggests a short proof of the first part of Theorem 1.1. Indeed, given $f_1, \ldots, f_n$, choose positive numbers $\gamma_1, \ldots, \gamma_{n-1}$ large enough so that the Pick matrix P_{n-1} defined via formula (2.4) is positive definite and extend P_{n-1} to $P_n = \begin{bmatrix} P_{n-1} & F \\ F^* & \gamma_n \end{bmatrix}$ by letting

$$\gamma_n := F P_{n-1}^{-1} F^*, \quad \text{where} \quad F = \left[p_{n,1} \; \cdots \; p_{n,n-1}\right], \quad p_{n,j} = \frac{1 - w_n \overline{w}_j}{1 - t_n \overline{t}_j}.$$

By the Schur complement argument, the matrix P_n constructed as above is singular. By Theorem 2.2, there is a finite Blaschke product $f \in \mathcal{B}_{n-1}^\circ$ satisfying conditions (2.5). Clearly, this f solves the problem (1.2).

It is seen from (2.1)–(2.4), that for every solution f to the problem **MP**, the Schwarz-Pick matrix $P^f(\mathbf{t})$ based on the interpolation nodes $t_1, \ldots, t_n$ is equal to the matrix P_n constructed in (2.4) from the interpolation data set. Hence, the first statement in Theorem 2.2 follows from Lemma 2.1. The last statement in the theorem can be strengthened as follows: *if P_n is singular, there is an $f \in \mathcal{B}_k$ satisfying conditions*

(2.5), *but there is no another Schur-class function g (even non-rational) such that*

$$\lim_{r \to 1} g(rt_i) = w_i, \quad |\lim_{r \to 1} g'(rt_i)| \le \gamma_i \quad \text{for} \quad i = 1, \ldots, n.$$

In any event, if P_n is singular, there is only one candidate which may or may not be a solution to the problem **MP**. The determinacy criterion for the problem **MP** in terms of the Pick matrix P_n is recalled in Theorem 2.6 below.

Definition 2.4 A positive semidefinite matrix of rank r is called *saturated* if every its $r \times r$ principal submatrix is positive definite. A positive semidefinite matrix is called *minimally positive* if none of its diagonal entries can be decreased so that the modified matrix will be still positive semidefinite

Remark 2.5 The rank equality in Lemma 2.1 implies that for $f \in \mathcal{B}_k^\circ$ and $k < n$, the Schwarz-Pick matrix (2.1) is saturated.

Theorem 2.6 ([20]) *The problem MP has a unique solution if and only if P_n is minimally positive or equivalently, if and only if $P_n \ge 0$ is singular and saturated. The unique solution is a finite Blaschke product of degree equal the rank of P_n.*

3 Parametrization of the Set of Low-Degree Solutions

With the first $n - 1$ interpolation conditions in (1.2) we associate the matrices

$$T = \begin{bmatrix} t_1 & & 0 \\ & \ddots & \\ 0 & & t_{n-1} \end{bmatrix}, \quad M = \begin{bmatrix} w_1 \\ \vdots \\ w_{n-1} \end{bmatrix}, \quad E = \begin{bmatrix} 1 \\ \vdots \\ 1 \end{bmatrix}. \tag{3.1}$$

Definition 3.1 A tuple $\gamma = \{\gamma_1, \ldots, \gamma_{n-1}\}$ of positive numbers will be called *admissible* if the Pick matrix P_{n-1}^γ defined as in (2.4) is positive definite:

$$P_{n-1}^\gamma = [p_{ij}]_{i,j=1}^{n-1} > 0, \quad \text{where} \quad p_{ij} = \begin{cases} \dfrac{1 - w_i \overline{w}_j}{1 - t_i \overline{t}_j} & \text{if } i \ne j, \\ \gamma_i & \text{if } i = j. \end{cases} \tag{3.2}$$

From now on, the tuple $\gamma = \{\gamma_1, \ldots, \gamma_{n-1}\}$ will serve as a parameter. Observe the Stein identity

$$P_{n-1}^\gamma - TP_{n-1}^\gamma T^* = EE^* - MM^*, \tag{3.3}$$

which holds true for any choice of $\gamma = \{\gamma_1, \ldots, \gamma_{n-1}\} \in \mathbb{R}^{n-1}$. In what follows, $\mathbf{e}_i$ will denote the i-th column in the identity matrix I_{n-1}.

Remark 3.2 If P_{n-1}^{γ} is invertible, it satisfies the Stein identity

$$(P_{n-1}^{\gamma})^{-1} - T^*(P_{n-1}^{\gamma})^{-1}T = XX^* - YY^*, \tag{3.4}$$

where the columns $X = \begin{bmatrix} x_1 \\ \vdots \\ x_{n-1} \end{bmatrix}$ and $Y = \begin{bmatrix} y_1 \\ \vdots \\ y_{n-1} \end{bmatrix}$ are given by

$$\begin{aligned} X &= (I - t_n T^*)(P_{n-1}^{\gamma})^{-1}(t_n I - T)^{-1}E, \\ Y &= (I - t_n T^*)(P_{n-1}^{\gamma})^{-1}(t_n I - T)^{-1}M. \end{aligned} \tag{3.5}$$

Furthermore, the entries

$$\begin{aligned} x_i &= (1 - t_n \bar{t}_i)\,\mathbf{e}_i^*(P_{n-1}^{\gamma})^{-1}(t_n I - T)^{-1}E, \\ y_i &= (1 - t_n \bar{t}_i)\,\mathbf{e}_i^*(P_{n-1}^{\gamma})^{-1}(t_n I - T)^{-1}M, \end{aligned} \qquad (i = 1, \ldots, n-1) \tag{3.6}$$

in the columns (3.5) are subject to equalities

$$|x_i| = |y_i| \neq 0 \quad \text{for} \quad i = 1, \ldots, n-1. \tag{3.7}$$

Proof Making use of (3.5) and (3.3), we verify (3.4) as follows:

$$\begin{aligned} XX^* - YY^* &= (I - t_n T^*)(P_{n-1}^{\gamma})^{-1}(t_n I - T)^{-1}\left[P_{n-1}^{\gamma} - TP_{n-1}^{\gamma}T^*\right] \\ &\quad \times (\bar{t}_n I - T^*)^{-1}(P_{n-1}^{\gamma})^{-1}(I - \bar{t}_n T) \\ &= (I - t_n T^*)(P_{n-1}^{\gamma})^{-1}\left[P_{n-1}^{\gamma}(I - t_n T^*)^{-1} + \right. \\ &\quad \left. + (I - \bar{t}_n T)^{-1}\bar{t}_n TP_{n-1}^{\gamma}\right] \times (P_{n-1}^{\gamma})^{-1}(I - \bar{t}_n T) \\ &= (P_{n-1}^{\gamma})^{-1}(I - \bar{t}_n T) + (I - t_n T^*)(P_{n-1}^{\gamma})^{-1}\bar{t}_n T \\ &= (P_{n-1}^{\gamma})^{-1} - T^*(P_{n-1}^{\gamma})^{-1}T. \end{aligned}$$

Comparing the corresponding diagonal entries on both sides of (3.4) gives $0 = |x_i|^2 - |y_i|^2$ so that $|x_i| = |y_i|$ for $i = 1, \ldots, n-1$. To complete the proof of (3.7), it suffices to show that x_i and y_i cannot be both equal zero. To this end, we compare the i-th rows of both sides in (3.4) to get

$$x_i X^* - y_i Y^* = \mathbf{e}_i^*(P_{n-1}^{\gamma})^{-1} - \mathbf{e}_i^* T^*(P_{n-1}^{\gamma})^{-1}T = \mathbf{e}_i^*(P_{n-1}^{\gamma})^{-1}(I - \bar{t}_i T). \tag{3.8}$$

Let us assume that $x_i = y_i = 0$. Then the expression on the right side of (3.8) is the zero row-vector. Since $(I - \bar{t}_i T)$ is the diagonal matrix with the i-th diagonal entry equal zero and all other diagonal entries being non-zero, it follows that all entries in the row-vector $\mathbf{e}_i^* (P_{n-1}^\gamma)^{-1}$, except the i-th entry, are zeroes, so that $\mathbf{e}_i^* (P_{n-1}^\gamma)^{-1} = \alpha \mathbf{e}_i^*$ for some $\alpha \in \mathbb{C}$. Then it follows from (3.6) and (3.1) that

$$x_i = \alpha (1 - t_n \bar{t}_i) \mathbf{e}_i^* (t_n I - T)^{-1} E = -\alpha \bar{t}_i.$$

Since $x_i = 0$ and $t_i \neq 0$, it follows that $\alpha = 0$ and hence, $\mathbf{e}_i^* (P_{n-1}^\gamma)^{-1} = \alpha \mathbf{e}_i^* = 0$. The latter cannot happen since the matrix P_{n-1}^γ is invertible. The obtained contradiction completes the proof of (3.7). □

For each admissible tuple $\gamma = \{\gamma_1, \ldots, \gamma_{n-1}\}$, we define the 2×2 matrix function

$$\Theta^\gamma(z) = \begin{bmatrix} \theta_{11}^\gamma(z) & \theta_{12}^\gamma(z) \\ \theta_{21}^\gamma(z) & \theta_{22}^\gamma(z) \end{bmatrix} \tag{3.9}$$

$$= I + (z - t_n) \begin{bmatrix} E^* \\ M^* \end{bmatrix} (I - zT^*)^{-1} (P_{n-1}^\gamma)^{-1} (t_n I - T)^{-1} \begin{bmatrix} E & -M \end{bmatrix},$$

where T, M, E are defined as in (3.1). Upon making use of the columns (3.5) and of the diagonal structure of T, we may write the formula (3.9) for Θ^γ as

$$\Theta^\gamma(z) = I + (z - t_n) \begin{bmatrix} E^* \\ M^* \end{bmatrix} (I - zT^*)^{-1} (I - t_n T^*)^{-1} \begin{bmatrix} X & -Y \end{bmatrix} \tag{3.10}$$

$$= I + \sum_{i=1}^{n-1} \frac{z - t_n}{(1 - z\bar{t}_i)(1 - t_n \bar{t}_i)} \cdot \begin{bmatrix} 1 \\ \overline{w_i} \end{bmatrix} \begin{bmatrix} x_i & -y_i \end{bmatrix}$$

$$= I + \sum_{i=1}^{n-1} \left(\frac{t_i}{1 - z\bar{t}_i} - \frac{t_i}{1 - t_n \bar{t}_i} \right) \cdot \begin{bmatrix} 1 \\ \overline{w_i} \end{bmatrix} \begin{bmatrix} x_i & -y_i \end{bmatrix}.$$

It is seen from (3.10) that Θ^γ is rational with simple poles at $t_1, \ldots, t_{n-1}$. We next summarize some other properties of Θ^γ needed for our subsequent analysis.

Theorem 3.3 *Let $P_{n-1}^\gamma > 0$, let X, Y, Θ^γ be defined as in (3.5), (3.9), and let*

$$\Upsilon(z) = \prod_{i=1}^{n-1} (1 - z\bar{t}_i) \quad and \quad J = \begin{bmatrix} 1 & 0 \\ 0 & -1 \end{bmatrix}. \tag{3.11}$$

(1) *For every unimodular constant w_n, the functions*

$$p(z) = \Upsilon(z) \left(\theta_{11}^\gamma(z) w_n + \theta_{12}^\gamma(z) \right), \quad q(z) = \Upsilon(z) \left(\theta_{21}^\gamma(z) w_n + \theta_{22}^\gamma(z) \right) \tag{3.12}$$

are polynomials of degree $n - 1$ with all zeros in $\overline{\mathbb{D}}$ and in $\mathbb{C} \backslash \mathbb{D}$, respectively.

(2) *The following identities hold for any $z, \zeta \in \mathbb{C}$ $(z\bar{\zeta} \neq 1)$:*

$$\frac{J - \Theta^{\gamma}(z) J \Theta^{\gamma}(\zeta)^*}{1 - z\bar{\zeta}} = \begin{bmatrix} E^* \\ M^* \end{bmatrix} (I - zT^*)^{-1} (P_{n-1}^{\gamma})^{-1} (I - \bar{\zeta}T)^{-1} \begin{bmatrix} E & M \end{bmatrix},$$

(3.13)

$$\frac{J - \Theta^{\gamma}(\zeta)^* J \Theta^{\gamma}(z)}{1 - z\bar{\zeta}} = \begin{bmatrix} X^* \\ -Y^* \end{bmatrix} (I - \bar{\zeta}T)^{-1} P_{n-1}^{\gamma} (I - zT^*)^{-1} \begin{bmatrix} X & -Y \end{bmatrix}.$$

(3.14)

(3) $\det \Theta^{\gamma}(z) = 1$ *for all* $z \in \mathbb{C} \backslash \{t_1, \ldots, t_n\}$.

The proofs of (1)–(3) can be found in [2] for a more general framework where P_{n-1}^{γ} is an invertible Hermitian matrix satisfying the Stein identity (3.3) for some T, E and M (though, of the same dimensions as in (3.1)) and t_n is an arbitrary point in $\mathbb{T} \backslash \sigma(T)$. In this more general setting, p has $\pi(P_{n-1}^{\gamma})$ zeroes in $\overline{\mathbb{D}}$ and q has $\nu(P_{n-1}^{\gamma})$ zeroes in $\mathbb{C} \backslash \mathbb{D}$ where $\pi(P_{n-1}^{\gamma})$ and $\nu(P_{n-1}^{\gamma})$ are respectively the number of positive and the number of negative eigenvalues of P_{n-1}^{γ} counted with multiplicities. Straightforward verifications of (3.13) and (3.14) rely solely on the Stein identities (3.3) and (3.4), respectively. Another calculation based on (3.3) shows that

$$\det \Theta^{\gamma}(z) = \det \left[(zI - T)(\bar{t}_n I - T^*)(I - zT^*)^{-1} (1 - \bar{t}_n T)^{-1} \right]$$

which is equal to one due to a special form (3.1) of T.

Remark 3.4 Let $P \in \mathbb{C}^{n \times n}$ be a positive semidefinite saturated matrix with rank $P = k < n$. Let $G \in \mathbb{C}^{n \times n}$ be a diagonal positive semidefinite matrix with rank $G = m < n - k$. Then rank$(P + G) = k + m$.
Indeed, since the matrix $\Phi P \Phi^{-1}$ is saturated for any permutation matrix Φ, we may take P and G conformally decomposed as follows:

$$P = \begin{bmatrix} P_{11} & P_{12} \\ P_{21} & P_{22} \end{bmatrix}, \quad D = \begin{bmatrix} 0 & 0 \\ 0 & \widetilde{G} \end{bmatrix}, \quad P_{11} \in \mathbb{C}^{k \times k}.$$

Since rank $P = \text{rank} P_{11} = k$ (i.e., P_{11} is invertible), it follows that the Schur complement of P_{11} in P is equal to the zero matrix: $P_{22} - P_{21} P_{11}^{-1} P_{12} = 0$. By the formula for the rank of a block matrix, we then have

$$\text{rank}(P + G) = \text{rank} \begin{bmatrix} P_{11} & P_{12} \\ P_{21} & P_{22} + \widetilde{G} \end{bmatrix} = \text{rank} P_{11} + \text{rank}(P_{22} + \widetilde{G} - P_{21} P_{11}^{-1} P_{12})$$

$$= k + \text{rank} \widetilde{G} = k + \text{rank} G = k + m.$$

The next theorem is the main result of this section.

Theorem 3.5 *Given data* $(t_i, w_i) \in \mathbb{T}^2$ $(i = 1, \ldots, n)$*, let* $\boldsymbol{\gamma} = \{\gamma_1, \ldots, \gamma_{n-1}\}$ *be an admissible tuple and let*

$$f_{\boldsymbol{\gamma}}(z) = \frac{\theta_{11}^{\gamma}(z)w_n + \theta_{12}^{\gamma}(z)}{\theta_{21}^{\gamma}(z)w_n + \theta_{22}^{\gamma}(z)} \tag{3.15}$$

where the coefficients θ_{ij}^{γ} *in (3.15) are constructed from* $\boldsymbol{\gamma}$ *by formula (3.9). Let* x_i *and* y_i *be the numbers defined as in (3.6). Then*

(1) $f_{\boldsymbol{\gamma}}$ *is a finite Blaschke product and satisfies conditions (1.2).*
(2) $f_{\boldsymbol{\gamma}}$ *satisfies conditions* $|f_{\boldsymbol{\gamma}}'(t_i)| \leq \gamma_i$ *for* $i = 1, \ldots, n-1$*. Moreover,* $|f_{\boldsymbol{\gamma}}'(t_i)| = \gamma_i$ *if and only if* $x_i w_n \neq y_i$*.*
(3) $\deg f_{\boldsymbol{\gamma}} = n - 1 - \ell$*, where* $\ell = \#\{i \in \{1, \ldots, n-1\} : x_i w_n = y_i\}$*.*

Conversely, every Blaschke product $f \in \mathcal{B}_{n-1}$ *subject to interpolation conditions (1.2) admits a representation (3.15) for some admissible tuple* $\boldsymbol{\gamma} = \{\gamma_1, \ldots, \gamma_{n-1}\}$*. This representation is unique if and only if* $\deg f = n - 1$*.*

Proof Let $\boldsymbol{\gamma} = \{\gamma_1, \ldots, \gamma_{n-1}\}$ be an admissible tuple and let Θ^{γ} be defined as in (3.9). Since Θ^{γ} is rational, the function $f_{\boldsymbol{\gamma}}$ is rational as well. Let

$$N(z) = \theta_{11}^{\gamma}(z)w_n + \theta_{12}^{\gamma}(z) \quad \text{and} \quad D(z) = \theta_{21}^{\gamma}(z)w_n + \theta_{22}^{\gamma}(z) \tag{3.16}$$

denote the numerator and the denominator in (3.15) and let

$$\Psi(z) = (I - zT^*)^{-1}(Xw_n - Y) = \sum_{i=1}^{n-1} \mathbf{e}_i \frac{x_i w_n - y_i}{1 - z\bar{t}_i}. \tag{3.17}$$

Combining (3.16) and (3.10) gives

$$\begin{bmatrix} N(z) \\ D(z) \end{bmatrix} = \Theta^{\gamma}(z) \begin{bmatrix} w_n \\ 1 \end{bmatrix} = \begin{bmatrix} w_n \\ 1 \end{bmatrix} + (z - t_n) \begin{bmatrix} E^* \\ M^* \end{bmatrix} (I - t_n T^*)^{-1} \Psi(z). \tag{3.18}$$

Taking the advantage of the matrix J in (3.11) and of relation (3.18) we get

$$|D(z)|^2 - |N(z)|^2 = -\begin{bmatrix} N(z)^* & D(z)^* \end{bmatrix} J \begin{bmatrix} N(z) \\ D(z) \end{bmatrix}$$

$$= -\begin{bmatrix} \bar{w}_n & 1 \end{bmatrix} \Theta^{\gamma}(z)^* J \Theta^{\gamma}(z) \begin{bmatrix} w_n \\ 1 \end{bmatrix}$$

$$= \begin{bmatrix} \bar{w}_n & 1 \end{bmatrix} \{J - \Theta^{\gamma}(z)^* J \Theta^{\gamma}(z)\} \begin{bmatrix} w_n \\ 1 \end{bmatrix}, \tag{3.19}$$

where for the third equality we used

$$\begin{bmatrix} \overline{w}_n & 1 \end{bmatrix} J \begin{bmatrix} w_n \\ 1 \end{bmatrix} = 1 - |w_n|^2 = 0.$$

Substituting (3.14) into (3.19) and making use of notation (3.17), we conclude

$$|D(z)|^2 - |N(z)|^2 = (1 - |z|^2)\Psi(z)^* P_{n-1}^\gamma \Psi(z).$$

Combining the latter equality with (3.15) and (3.16) gives

$$\frac{1 - |f_\gamma(z)|^2}{1 - |z|^2} = \frac{|D(z)|^2 - |N(z)|^2}{(1 - |z|^2)|D(z)|^2} = \frac{\Psi(z)^* P_{n-1}^\gamma \Psi(z)}{|D(z)|^2}, \qquad (3.20)$$

which implies, in particular, that f_γ is inner. Since f_γ is rational, it extends by continuity to a finite Blaschke product. One can see from (3.17) that

$$\lim_{z \to t_i}(z - t_i) \cdot \Psi(z) = -\mathbf{e}_i t_i(x_i w_n - y_i) \quad \text{for} \quad i = 1, \ldots, n-1, \qquad (3.21)$$

which together with (3.18) and (3.1) implies

$$\lim_{z \to t_i}(z - t_i) \cdot \begin{bmatrix} N(z) \\ D(z) \end{bmatrix} = (t_n - t_i) \begin{bmatrix} E^* \\ M^* \end{bmatrix}(I - t_n T^*)^{-1} \mathbf{e}_i t_i(x_i w_n - y_i)$$

$$= -\begin{bmatrix} 1 \\ \overline{w}_i \end{bmatrix} t_i^2(x_i w_n - y_i) \quad \text{for} \quad i = 1, \ldots, n-1. \qquad (3.22)$$

To show that f_γ satisfies conditions (1.2), we first observe that $\Theta^\gamma(t_n) = I$ (by definition (3.9)); now the equality $f_\gamma(t_n) = w_n$ is immediate from (3.15). Verification of other equalities in (1.2) depends on whether or not $x_i w_n = y_i$.

Case 1: Let us assume that $x_i w_n \neq y_i$. Then we have from (3.15) and (3.22),

$$f_\gamma(t_i) = \frac{(z - t_i)N(z)}{(z - t_i)D(z)} = \frac{t_i^2(x_i w_n - y_i)}{\overline{w}_i t_i^2(x_i w_n - y_i)} = \frac{1}{\overline{w}_i} = w_i. \qquad (3.23)$$

Under the same assumption, we conclude from (2.2), (3.2) and (3.20)–(3.22),

$$|f_\gamma'(t_i)| = \lim_{z \to t_i} \frac{1 - |f_\gamma(z)|^2}{1 - |z|^2} = \lim_{z \to t_i} \frac{|z - t_i|^2 \Psi(z)^* P_{n-1}^\gamma \Psi(z)}{|z - t_i|^2 |D(z)|^2}$$

$$= \frac{|x_i w_n - y_i|^2 \mathbf{e}_i^* P_{n-1}^\gamma \mathbf{e}_i}{|x_i w_n - y_i|^2} = \gamma_i.$$

Case 2: Let us assume that $x_i w_n = y_i$. Then the functions N, D and Ψ are analytic at t_i. To compute the values of these functions at $z = t_i$, we first take the adjoints of both sides in (3.8):

$$X\bar{x}_i - Y\bar{y}_i = (I - t_i T^*)(P_{n-1}^y)^{-1}\mathbf{e}_i.$$

We next divide both sides by $\bar{y}_i$ and make use of equalities $w_n = y_i/x_i = \bar{x}_i/\bar{y}_i$ (by the assumption of Case 2) to get

$$Xw_n - Y = \frac{1}{\bar{y}_i}(I - t_i T^*)(P_{n-1}^y)^{-1}\mathbf{e}_i.$$

Substituting the latter equality into (3.17) results in

$$\Psi(z) = \frac{1}{\bar{y}_i}(I - zT^*)^{-1}(I - t_i T^*)(P_{n-1}^y)^{-1}\mathbf{e}_i,$$

which being evaluated at t_i, gives

$$\Psi(t_i) = \frac{1}{\bar{y}_i}(I - \mathbf{e}_i\mathbf{e}_i^*)(P_{n-1}^y)^{-1}\mathbf{e}_i = \frac{1}{\bar{y}_i}\left((P_{n-1}^y)^{-1}\mathbf{e}_i - \mathbf{e}_i\widetilde{p}_{ii}\right), \qquad (3.24)$$

where $\widetilde{p}_{ii}$ denotes the i-th diagonal entry of $(P_{n-1}^y)^{-1}$. Evaluating the formula (3.18) at $z = t_i$ gives, in view of (3.24),

$$\begin{bmatrix} N(t_i) \\ D(t_i) \end{bmatrix} = \begin{bmatrix} w_n \\ 1 \end{bmatrix} + \frac{t_i - t_n}{\bar{y}_i}\begin{bmatrix} E^* \\ M^* \end{bmatrix}(I - t_n T^*)^{-1}\left((P_{n-1}^y)^{-1}\mathbf{e}_i - \mathbf{e}_i\widetilde{p}_{ii}\right).$$

Making use of formulas (3.5), we have

$$\frac{t_i - t_n}{\bar{y}_i}\begin{bmatrix} E^* \\ M^* \end{bmatrix}(I - t_n T^*)^{-1}(P_{n-1}^y)^{-1}\mathbf{e}_i = \frac{t_i - t_n}{\bar{y}_i}\begin{bmatrix} X^* \\ Y^* \end{bmatrix}(t_n I - T)^{-1}\mathbf{e}_i$$

$$= -\frac{1}{\bar{y}_i}\begin{bmatrix} X^*\mathbf{e}_i \\ Y^*\mathbf{e}_i \end{bmatrix} = -\frac{1}{\bar{y}_i}\begin{bmatrix} \bar{x}_i \\ \bar{y}_i \end{bmatrix} = -\begin{bmatrix} w_n \\ 1 \end{bmatrix}.$$

Combining the two latter formulas and again making use of (3.1) leads us to

$$\begin{bmatrix} N(t_i) \\ D(t_i) \end{bmatrix} = \frac{t_n - t_i}{\bar{y}_i}\begin{bmatrix} E^* \\ M^* \end{bmatrix}(I - t_n T^*)^{-1}\mathbf{e}_i\widetilde{p}_{ii}$$

$$= \frac{t_n - t_i}{\bar{y}_i(1 - t_n\bar{t}_i)}\begin{bmatrix} E^*\mathbf{e}_i \\ M^*\mathbf{e}_i \end{bmatrix}\widetilde{p}_{ii} = -\begin{bmatrix} 1 \\ w_i \end{bmatrix}\frac{t_i\widetilde{p}_{ii}}{\bar{y}_i}.$$

Thus,

$$N(t_i) = -\frac{t_i\widetilde{p}_{ii}}{\bar{y}_i} \quad \text{and} \quad D(t_i) = -\frac{t_i\widetilde{p}_{ii}\bar{w}_i}{\bar{y}_i}, \qquad (3.25)$$

and subsequently, $f_\gamma(t_i) = \frac{N(t_i)}{D(t_i)} = \frac{1}{w_i} = w_i$. Furthermore, we have from (2.2) and (3.20),

$$|f'_\gamma(t_i)| = \lim_{z \to t_i} \frac{1 - |f_\gamma(z)|^2}{1 - |z|^2} = \lim_{z \to t_i} \frac{\Psi(z)^* P^\gamma_{n-1} \Psi(z)}{|D(z)|^2} = \frac{\Psi(t_i)^* P^\gamma_{n-1} \Psi(t_i)}{|D(t_i)|^2}. \qquad (3.26)$$

In view of (3.24) and (3.25),

$$\Psi(t_i)^* P^\gamma_{n-1} \Psi(t_i) = \frac{1}{|y_i|^2} \left(\mathbf{e}_i^* (P^\gamma_{n-1})^{-1} - \widetilde{p}_{ii} \mathbf{e}_i^* \right) P^\gamma_{n-1} \left((P^\gamma_{n-1})^{-1} \mathbf{e}_i - \mathbf{e}_i \widetilde{p}_{ii} \right)$$

$$= \frac{1}{|y_i|^2} \left(\mathbf{e}_i^* (P^\gamma_{n-1})^{-1} \mathbf{e}_i - 2\widetilde{p}_{ii} + \widetilde{p}_{ii}^2 \mathbf{e}_i^* P^\gamma_{n-1} \mathbf{e}_i \right) = \frac{\gamma_i \widetilde{p}_{ii}^2 - \widetilde{p}_{ii}}{|y_i|^2},$$

$$|D(t_i)|^2| = \frac{\widetilde{p}_{ii}^2}{|y_i|^2}.$$

Substituting the two latter equalities into the right hand side of (3.26) we get

$$|f'_\gamma(t_i)| = \frac{\gamma_i \widetilde{p}_{ii}^2 - \widetilde{p}_{ii}}{\widetilde{p}_{ii}^2} = \gamma_i - \frac{1}{\widetilde{p}_{ii}} < \gamma_i.$$

We have verified equalities (1.2) and we showed that $|f'_\gamma(t_i)| \leq \gamma_i$ with strict inequality if and only if $x_i w_n = y_i$ (i.e., in Case 2). This completes the proof of statements (1) and (2) of the theorem.

To prove part (3), we multiply the numerator and the denominator on the right hand side of (3.15) by Υ (see formula (3.11)) to get a linear fractional representation for f with polynomial coefficients $\widetilde{\theta}^\gamma_{ij} = \Upsilon \theta^\gamma_{ij}$:

$$f_\gamma(z) = \frac{\widetilde{\theta}^\gamma_{11}(z) w_n + \widetilde{\theta}^\gamma_{12}(z)}{\widetilde{\theta}^\gamma_{21}(z) w_n + \widetilde{\theta}^\gamma_{22}(z)} = \frac{\Upsilon(z) N(z)}{\Upsilon(z) D(z)} = \frac{p(z)}{q(z)}, \qquad (3.27)$$

where p and q are the polynomials given in (3.12). Since the resulting function f_γ extends to a finite Blaschke product (with no poles or zeroes on $\mathbb{T}$), it follows that p and q have the same (if any) zeroes on $\mathbb{T}$ counted with multiplicities. The common zeroes may occur only at the zeroes of the determinant of the coefficient matrix and since $\det(\widetilde{\Theta}^\gamma(z)) = [\Upsilon(z)]^2$ (by statement (3) in Theorem 3.3 and by analyticity of $\det (\widetilde{\Theta}^\gamma(z)))$, it follows that p and q may have common zeros only at $t_1, \ldots, t_{n-1}$. By (3.22) and (3.11), $p(t_i) = 0$ if and only if $x_i w_n = y_i$. On the other hand, if this is the case, $D(t_i) \neq 0$ (by formula (3.25)) and therefore, t_i is a *simple* zero of $p = \Upsilon D$. Thus, D may have only simple zeros at $t_1, \ldots, t_{n-1}$. We summarize: by statement (4) in Theorem 3.3, the numerator p in (3.27) has $n - 1$ zeroes in $\overline{\mathbb{D}}$. All zeroes of p and q on $\mathbb{T}$ are simple and common; they occur precisely at those t_i's for which $x_i w_n = y_i$. After zero cancellations, the function f_γ turns out to be a finite Blaschke product of degree $n - 1 - \#\{i \in \{1, \ldots, n - 1\} : x_i w_n = y_i\}$. This completes the proof of part (3).

To prove the converse statement, let us assume that f is a Blaschke product of degree $k \leq n-1$ that satisfies conditions (1.2). For a fixed permutation $\{i_1, \ldots, i_{n-1}\}$ of the index set $\{1, \ldots, n - 1\}$, we choose the integers $\gamma_1, \ldots, \gamma_{n-1}$ so that

$$\gamma_{i_j} = |f'(t_{i_j})| \quad (1 \leq j \leq k) \quad \text{and} \quad \gamma_{i_j} > |f'(t_{i_j})| \quad (k < j \leq n - 1). \tag{3.28}$$

Due to this choice, the diagonal matrix

$$G = \begin{bmatrix} \gamma_1 - |f'(t_1)| & & 0 \\ & \ddots & \\ 0 & & \gamma_{n-1} - |f'(t_{n-1})| \end{bmatrix} \tag{3.29}$$

is positive semidefinite and rank $G = n - k - 1$. We are going to show that the tuple $\gamma = \{\gamma_1, \ldots, \gamma_{n-1}\}$ is admissible and that $f = f_\gamma$ as in (3.15).

Let $P^f = P^f(t_1, \ldots, t_{n-1}, z)$ be the Schwarz-Pick matrix of f based on the interpolation nodes $t_1, \ldots, t_{n-1}$ and one additional point $z \in \mathbb{D}$. According to (2.1) and due to interpolation conditions (1.2), this matrix has the form

$$P^f = \begin{bmatrix} |f'(t_1)| & \frac{1-w_1\overline{w}_2}{1-t_1\overline{t}_2} & \cdots & \frac{1-w_1\overline{w}_{n-1}}{1-t_1\overline{t}_{n-1}} & \frac{1-w_1\overline{f(z)}}{1-t_1\overline{z}} \\ \frac{1-w_2\overline{w}_1}{1-t_2\overline{t}_1} & |f'(t_2)| & \cdots & \frac{1-w_2\overline{w}_{n-1}}{1-t_2\overline{t}_{n-1}} & \frac{1-w_2\overline{f(z)}}{1-t_2\overline{z}} \\ \vdots & \vdots & \ddots & \vdots & \vdots \\ \frac{1-w_{n-1}\overline{w}_1}{1-t_{n-1}\overline{t}_1} & \frac{1-w_{n-1}\overline{w}_2}{1-t_{n-1}\overline{t}_2} & \cdots & |f'(t_{n-1})| & \frac{1-w_{n-1}\overline{f(z)}}{1-t_{n-1}\overline{t}_n} \\ \frac{1-f(z)\overline{w}_1}{1-z\overline{t}_1} & \frac{1-f(z)\overline{w}_2}{1-z\overline{t}_2} & \cdots & \frac{1-f(z)\overline{w}_{n-1}}{1-z\overline{t}_{n-1}} & \frac{1-|f(z)|^2}{1-|z|^2} \end{bmatrix}. \tag{3.30}$$

Observe that the leading $(n - 1) \times (n - 1)$ submatrix of P^f is the boundary Schwarz-Pick matrix $P^f(t_1, \ldots, t_{n-1})$, while the bottom row in P^f (without the rightmost entry) can be written in terms of the matrices (3.1) as $(E^* - f(z)M^*)(I - zT^*)^{-1}$. Thus, P^f can be written in a more compact form

$$P^f = \begin{bmatrix} P^f(t_1, \ldots, t_{n-1}) & (I - \overline{z}T)^{-1}(E - M\overline{f(z)}) \\ (E^* - f(z)M^*)(I - zT^*)^{-1} & \dfrac{1 - |f(z)|^2}{1 - |z|^2} \end{bmatrix}. \tag{3.31}$$

Let P^γ_{n-1} be the matrix defined via formulas (3.2) and let

$$\mathbb{P}^\gamma(z) := \begin{bmatrix} P^\gamma_{n-1} & (I - \overline{z}T)^{-1}(E - M\overline{f(z)}) \\ (E^* - f(z)M^*)(I - zT^*)^{-1} & \dfrac{1 - |f(z)|^2}{1 - |z|^2} \end{bmatrix}. \tag{3.32}$$

Taking into account the formula (3.29) for G and comparing (3.2) and (3.31) with (2.1) and (3.32), respectively, leads us to equalities

$$P_{n-1}^{\gamma} = P^f(t_1, \ldots, t_{n-1}) + G \quad \text{and} \quad \mathbb{P}^{\gamma}(z) = P^f + \begin{bmatrix} G & 0 \\ 0 & 0 \end{bmatrix}. \tag{3.33}$$

By Remark 2.5, the Schwarz-Pick matrices P^f and $P^f(t_1, \ldots, t_{n-1})$ are positive semidefinite and saturated. Moreover, since $f \in \mathcal{B}_k^{\circ}$, we have

$$\text{rank}\, P^f = \text{rank}\, P^f(t_1, \ldots, t_{n-1}) = k,$$

by Lemma 2.1. Then it follows from (3.33) by Remark 3.4 that

$$\text{rank}\, P_{n-1}^{\gamma} = \text{rank}\, P^f(t_1, \ldots, t_{n-1}) + \text{rank}\, G = n - 1 \tag{3.34}$$

(i.e., P_{n-1}^{γ} is positive definite and hence $\gamma = \{\gamma_1, \ldots, \gamma_{n-1}\}$ is admissible) and

$$\text{rank}\, \mathbb{P}^{\gamma}(z) = \text{rank}\, P^f + \text{rank}\, \begin{bmatrix} G & 0 \\ 0 & 0 \end{bmatrix} = n - 1 \quad \text{for all} \quad z \in \mathbb{D}. \tag{3.35}$$

By (3.34) and (3.35), the Schur complement of the block P_{n-1}^{γ} in (3.31) is equal to zero for every $z \in \mathbb{D}$:

$$\frac{1 - |f(z)|^2}{1 - |z|^2} - (E^* - f(z)M^*)(I - zT^*)^{-1}(P_{n-1}^{\gamma})^{-1}(I - \bar{z}T)^{-1}(E - M\overline{f(z)}) = 0. \tag{3.36}$$

The rational matrix-function Θ^{γ} constructed from $\gamma = \{\gamma_1, \ldots, \gamma_{n-1}\}$ via formula (3.9) satisfies the identity (3.13). Multiplying the latter identity (with $\zeta = z$) by the row-vector $\begin{bmatrix} 1 & -f(z) \end{bmatrix}$ on the left and by its adjoint on the right gives

$$(E^* - f(z)M^*)(I - zT^*)^{-1}(P_{n-1}^{\gamma})^{-1}(I - \bar{z}T)^{-1}(E - M\overline{f(z)})$$

$$= \begin{bmatrix} 1 & -f(z) \end{bmatrix} \frac{J - \Theta^{\gamma}(z)J\Theta^{\gamma}(z)^*}{1 - |z|^2} \begin{bmatrix} 1 \\ -\overline{f(z)} \end{bmatrix}$$

$$= \frac{1 - |f(z)|^2}{1 - |z|^2} - \begin{bmatrix} 1 & -f(z) \end{bmatrix} \frac{\Theta^{\gamma}(z)J\Theta^{\gamma}(z)^*}{1 - |z|^2} \begin{bmatrix} 1 \\ -\overline{f(z)} \end{bmatrix}$$

which, being combined with (3.36), implies

$$\begin{bmatrix} 1 & -f(z) \end{bmatrix} \frac{\Theta^{\gamma}(z)J\Theta^{\gamma}(z)^*}{1 - |z|^2} \begin{bmatrix} 1 \\ -\overline{f(z)} \end{bmatrix} = 0 \quad \text{for all} \quad z \in \mathbb{D}. \tag{3.37}$$

Let us consider the functions

$$g = \theta_{11}^{\gamma} - f\theta_{21}^{\gamma} \quad \text{and} \quad \mathcal{E} = \frac{f\theta_{22}^{\gamma} - \theta_{12}^{\gamma}}{\theta_{11}^{\gamma} - f\theta_{21}^{\gamma}}. \tag{3.38}$$

The function g is rational and due to (3.9), $g(t_n) = \theta_{11}^{\gamma}(t_n) - f(t_n)\theta_{21}^{\gamma}(t_n) = 1$. Hence, $g \not\equiv 0$ and the rational function $\mathcal{E}$ in (3.38) is well defined. Again, due to (3.9) and the n-th interpolation condition in (1.1),

$$\mathcal{E}(t_n) = f(t_n)\theta_{22}^{\gamma}(t_n) - \theta_{12}^{\gamma}(t_n) = f(t_n) = w_n. \tag{3.39}$$

Using the functions (3.38) we now rewrite equality (3.37) as

$$0 = |g(z)|^2 \cdot \left[1 - \mathcal{E}(z)\right] \frac{J}{1 - |z|^2} \left[\begin{array}{c} 1 \\ -\mathcal{E}(z) \end{array}\right] = \frac{|g(z)|^2(1 - |\mathcal{E}(z)|^2)}{1 - |z|^2}.$$

Since the latter equality holds for all $z \in \mathbb{D}$ and $g \not\equiv 0$, it follows that $|\mathcal{E}(z)| = 1$ for all $z \in \mathbb{D}$ so that $\mathcal{E}$ is a unimodular constant. By (3.39), $\mathcal{E} \equiv w_n$. Now representation (3.15) follows from the second formula in (3.38).

Finally, if $k = \deg f < n - 1$, then $n - k - 1$ parameters in (3.28) can be increased to produce various admissible tuples $\boldsymbol{\gamma}$ such that $f = f_{\boldsymbol{\gamma}}$. On the other hand, if $f \in \mathcal{B}_{n-1}$ admits two different representations (3.15), then for one of them (say, based on an admissible tuple $\boldsymbol{\gamma} = \{\gamma_1, \ldots, \gamma_{n-1}\}$), we must have $\gamma_i \neq |f'(t_i)|$ for some $i \in \{1, \ldots, n - 1\}$. Then $x_i w_n = y_i$, by part (2) of the theorem, and hence, $\deg f < n - 1$, by part (3). Thus, the representation $f = f_{\boldsymbol{\gamma}}$ is unique if and only if $\deg f = n - 1$, which completes the proof of the theorem. $\qquad\square$

We now reformulate Theorem 3.5 in the form that is more convenient for numerical computations. To this end, we let

$$\mathbf{p}_n = \left[\begin{array}{c} p_{1,n} \\ \vdots \\ p_{n-1,n} \end{array}\right], \quad \text{where} \quad p_{i,n} = \frac{1 - w_i\overline{w}_n}{1 - t_i\overline{t}_n}, \tag{3.40}$$

and, for an admissible tuple $\boldsymbol{\gamma} = \{\gamma_1, \ldots, \gamma_{n-1}\}$ and the corresponding $P_{n-1}^{\gamma} > 0$, we let $\boldsymbol{\Delta}^{\gamma} = (P_{n-1}^{\gamma})^{-1}\mathbf{p}_n$. If we denote by $P_{n-1,i}^{\gamma}(\mathbf{p}_n)$ the matrix obtained from P_{n-1}^{γ} by replacing its i-th column by $\mathbf{p}_n$, then by Cramer's rule, we have

$$\boldsymbol{\Delta}^{\gamma} = \left[\begin{array}{c} \Delta_1^{\gamma} \\ \vdots \\ \Delta_{n-1}^{\gamma} \end{array}\right], \quad \Delta_i^{\gamma} = \frac{\det P_{n-1,i}^{\gamma}(\mathbf{p}_n)}{\det P_{n-1}^{\gamma}} \quad (i = 1, \ldots, n - 1). \tag{3.41}$$

Theorem 3.6 *Given data* $(t_i, w_i) \in \mathbb{T}^2$ *(i = 1, ..., n), let* $\mathbf{p}_n$ *be defined as in* (3.40). *For any admissible tuple* $\boldsymbol{\gamma} = \{\gamma_1, \ldots, \gamma_{n-1}\}$, *the function*

$$f_{\boldsymbol{\gamma}}(z) = w_n \cdot \frac{1 - (1 - z\bar{t}_n) \cdot \sum_{i=1}^{n-1} \frac{\Delta_i^{\gamma}}{1 - z\bar{t}_i}}{1 - (1 - z\bar{t}_n) \cdot \sum_{i=1}^{n-1} \frac{\overline{w}_i w_n \Delta_i^{\gamma}}{1 - z\bar{t}_i}} \tag{3.42}$$

with the numbers Δ_i^{γ} *defined as in* (3.41), *is the Blaschke product of degree* $\deg f_{\boldsymbol{\gamma}} = n - 1 - \#\{i \in \{1, \ldots, n-1\} : \Delta_i^{\gamma} = 0\}$ *and satisfies conditions* (1.2). *Moreover,* $|f_{\boldsymbol{\gamma}}'(t_i)| = \gamma_i$ *if and only if* $\Delta_i^{\gamma} \neq 0$ *and* $|f_{\boldsymbol{\gamma}}'(t_i)| < \gamma_i$ *otherwise.*

Proof Observe that the column (3.40) can be written in terms of the matrices (3.1) as $\mathbf{p}_n = (I - \bar{t}_n T)^{-1}(E - M\overline{w}_n)$. Let x_i and y_i be the numbers defined in (3.6). From formulas (3.6) and (3.41), we have

$$\begin{aligned}
x_i w_n - y_i &= (1 - t_n \bar{t}_i)\, \mathbf{e}_i^* (P_{n-1}^{\gamma})^{-1}(t_n I - T)^{-1}(E w_n - M) \\
&= (1 - t_n \bar{t}_i)\, \mathbf{e}_i^* (P_{n-1}^{\gamma})^{-1}(I - \bar{t}_n T)^{-1}(E - M\overline{w}_n) w_n \bar{t}_n \\
&= (1 - t_n \bar{t}_i)\, \mathbf{e}_i^* (P_{n-1}^{\gamma})^{-1} \mathbf{p}_n w_n \bar{t}_n \\
&= (\bar{t}_n - \bar{t}_i)\, \mathbf{e}_i^* \boldsymbol{\Delta}^{\gamma} w_n = (\bar{t}_n - \bar{t}_i) \Delta_i^{\gamma} w_n.
\end{aligned} \tag{3.43}$$

Since $t_n \neq t_i$ and $w_n \neq 0$, it now follows that

$$x_i w_n = y_i \iff \Delta_i^{\gamma} = 0 \iff \det P_{n-1,i}^{\gamma}(\mathbf{p}_n) = 0. \tag{3.44}$$

We next observe that by the second representation for Θ^{γ} in (3.10) and (3.43),

$$\begin{aligned}
\Theta^{\gamma}(z) \begin{bmatrix} w_n \\ 1 \end{bmatrix} &= \begin{bmatrix} w_n \\ 1 \end{bmatrix} + \sum_{i=1}^{n-1} \begin{bmatrix} 1 \\ \overline{w}_i \end{bmatrix} \cdot \frac{(z - t_n)(x_i w_n - y_i)}{(1 - z\bar{t}_i)(1 - t_n \bar{t}_i)} \\
&= \begin{bmatrix} w_n \\ 1 \end{bmatrix} + \sum_{i=1}^{n-1} \begin{bmatrix} 1 \\ \overline{w}_i \end{bmatrix} \cdot \frac{(z - t_n)(\bar{t}_n - \bar{t}_i) \Delta_i^{\gamma} w_n}{(1 - z\bar{t}_i)(1 - t_n \bar{t}_i)} \\
&= \begin{bmatrix} w_n \\ 1 \end{bmatrix} - \sum_{i=1}^{n-1} \begin{bmatrix} 1 \\ \overline{w}_i \end{bmatrix} \cdot \frac{(1 - z\bar{t}_n) \Delta_i^{\gamma} w_n}{1 - z\bar{t}_i},
\end{aligned}$$

from which it follows that formulas (3.42) and (3.15) represent the same function $f_{\boldsymbol{\gamma}}$. Now all statements in Theorem 3.6 follow from their counter-parts in Theorem 3.5, by (3.44). $\square$

4 Existence of $\mathcal{B}_{n-2}$-Solutions

We will write $(\zeta_1, \ldots, \zeta_k) \in \mathcal{O}$ if given k points $\zeta_1, \ldots, \zeta_k \in \mathbb{T}$ are counter clockwise oriented on $\mathbb{T}$. For example, if $\zeta_1 = 1$, then $(1, \zeta_2, \ldots, \zeta_k) \in \mathcal{O}$ means that $\arg \zeta_{i+1} > \arg \zeta_i$ for all $i = 1, \ldots, k-1$. From now on, we will assume that the interpolation nodes $t_1, \ldots, t_n$ in problem (1.2) are counter clockwise oriented.

Theorem 4.1 *The problem (1.2) has a non-constant solution $f \in \mathcal{B}_{n-2}$ if and only if there exist three target values w_i, w_j, w_k having the same orientation as t_i, t_j, t_k on $\mathbb{T}$.*

As was pointed out in [21], the "only if" part follows by the winding number argument: the absence of the requested triple means that (up to rotation of $\mathbb{T}$) $\arg w_n \leq \arg w_{n-1} \leq \ldots \leq w_1$ with at least one strict inequality, and then the degree of any Blaschke product interpolation this data is at least $n-1$. In this section we prove the "if" part in Theorem 4.1.

Lemma 4.2 *Given three points $\zeta_i = e^{\vartheta_i}$ $(i = 1, 2, 3)$, the quantity*

$$G(\zeta_1, \zeta_2, \zeta_3) := -i(1 - \zeta_1 \overline{\zeta_2})(1 - \zeta_2 \overline{\zeta_3})(1 - \zeta_3 \overline{\zeta_1})$$

is real. Moreover,

$$G(\zeta_1, \zeta_2, \zeta_3) > 0 \iff (\zeta_1, \zeta_2, \zeta_3) \in \mathcal{O}, \quad G(\zeta_1, \zeta_2, \zeta_3) < 0 \iff (\zeta_1, \zeta_3, \zeta_2) \in \mathcal{O}.$$

Proof Since G is rotation-invariant, we may assume without loss of generality that $\zeta_1 = 1$ (i.e., $\vartheta_1 = 0$). Then a straightforward computation shows that

$$G(1, \zeta_2, \zeta_3) = 8 \sin \frac{\vartheta_3 - \vartheta_2}{2} \sin \frac{\vartheta_2}{2} \sin \frac{\vartheta_3}{2},$$

which implies all the desired statements.   □

Corollary 4.3 *The product of any three off-diagonal entries p_{ij}, p_{jk}, p_{ki} in the matrix (2.4),*

$$p_{ij} p_{jk} p_{ki} = \frac{1 - w_i \overline{w_j}}{1 - t_i \overline{t_j}} \cdot \frac{1 - w_j \overline{w_k}}{1 - t_j \overline{t_k}} \cdot \frac{1 - w_k \overline{w_i}}{1 - t_k \overline{t_i}} = \frac{G(w_i, w_j, w_k)}{G(t_i, t_j, t_k)}$$

is positive if and only if w_i, w_j, w_k are all distinct and have the same orientation on $\mathbb{T}$ as t_i, t_j, t_k.

Proof of Theorem 4.1 Let us assume that there are three target values having the same orientation on $\mathbb{T}$ as their respective interpolation nodes. By re-enumerating, we may assume without loss of generality that these values are w_{n-2}, w_{n-1}, w_n so that

$$q := \frac{p_{n-1,n-2} p_{n-2,n}}{p_{n-1,n}} = \frac{p_{n-1,n-2} p_{n-2,n} p_{n,n-1}}{|p_{n-1,n}|^2} > 0, \tag{4.1}$$

by Corollary 4.3 and since $p_{n-1,n} = \bar{p}_{n,n-1}$. We will show that in this case, there is an admissible tuple $\boldsymbol{\gamma}$ so that the number Δ^{γ}_{n-1} defined in (3.41) equals zero. To this end, let

$$\mathbf{b} = \begin{bmatrix} p_{1,n-2} \\ \vdots \\ p_{n-3,n-2} \end{bmatrix}, \quad \mathbf{c} = \begin{bmatrix} p_{1,n-1} \\ \vdots \\ p_{n-3,n-1} \end{bmatrix}, \quad \mathbf{d} = \begin{bmatrix} p_{1,n} \\ \vdots \\ p_{n-3,n} \end{bmatrix}.$$

Let $\boldsymbol{\gamma} = \{\gamma_1, \ldots, \gamma_{n-1}\}$ be any admissible tuple. Then the matrix P^{γ}_{n-1} (3.2) and the column $\mathbf{p}_n$ (3.40) can be written as

$$P^{\gamma}_{n-1} = \begin{bmatrix} P^{\gamma}_{n-3} & \mathbf{b} & \mathbf{c} \\ \mathbf{b}^* & \gamma_{n-2} & p_{n-2,n-1} \\ \mathbf{c}^* & p_{n-1,n-2} & \gamma_{n-1} \end{bmatrix} \quad \text{and} \quad \mathbf{p}_n = \begin{bmatrix} \mathbf{d} \\ p_{n-2,n} \\ p_{n-1,n} \end{bmatrix}. \tag{4.2}$$

Replacing the rightmost column in P^{γ}_{n-1} by $\mathbf{p}_n$ produces

$$P^{\gamma}_{n-1,n-1}(\mathbf{p}_n) = \begin{bmatrix} P^{\gamma}_{n-3} & \mathbf{b} & \mathbf{d} \\ \mathbf{b}^* & \gamma_{n-2} & p_{n-2,n} \\ \mathbf{c}^* & p_{n-1,n-2} & p_{n-1,n} \end{bmatrix}. \tag{4.3}$$

For any matrix A, we can make the entries of $(P^{\gamma}_{n-3} - A)^{-1}$ as small in modulus as we wish by choosing the diagonal entries $\gamma_1, \ldots, \gamma_{n-3}$ in P^{γ}_{n-3} big enough. Thus, we choose $\gamma_1, \ldots, \gamma_{n-3}$ so huge that $P^{\gamma}_{n-3} > 0$,

$$\det(P^{\gamma}_{n-3} - \mathbf{d}\mathbf{c}^* p^{-1}_{n-1,n}) \neq 0, \quad \mathbf{b}^*(P^{\gamma}_{n-3})^{-1}\mathbf{b} < \frac{q}{3}, \quad |X| < \frac{q}{3}, \tag{4.4}$$

where $q > 0$ is specified in (4.1) and where

$$X = \left(\mathbf{b}^* - \frac{p_{n-2,n}}{p_{n-1,n}}\mathbf{c}^*\right)\left(P^{\gamma}_{n-3} - \mathbf{d}p^{-1}_{n-1,n}\mathbf{c}^*\right)^{-1}\left(\mathbf{b} - \mathbf{d}\frac{p_{n-1,n-2}}{p_{n-1,n}}\right). \tag{4.5}$$

Then we choose

$$\gamma_{n-2} = q + X \quad \text{and} \quad \gamma_{n-1} \geq \mathbf{c}^*(P^{\gamma}_{n-3})^{-1}\mathbf{c} + \frac{3}{q} \cdot |p_{n-1,n-2} - \mathbf{c}^*(P^{\gamma}_{n-3})^{-1}\mathbf{b}|^2. \tag{4.6}$$

Then the tuple $\boldsymbol{\gamma} = \{\gamma_1, \ldots, \gamma_{n-1}\}$ is admissible. Indeed, by (4.4) and (4.6), the Schur complement of P^{γ}_{n-3} in the matrix $P^{\gamma}_{n-2} = \begin{bmatrix} P^{\gamma}_{n-3} & \mathbf{b} \\ \mathbf{b}^* & \gamma_{n-2} \end{bmatrix}$ is positive:

$$\gamma_{n-2} - \mathbf{b}^*(P^{\gamma}_{n-3})^{-1}\mathbf{b} = q + X - \mathbf{b}^*(P^{\gamma}_{n-3})^{-1}\mathbf{b} > q - \frac{q}{3} - \frac{q}{3} = \frac{q}{3} > 0, \tag{4.7}$$

and therefore, P^γ_{n-2} is positive definite. We next use (4.7) and the second relation in (4.6) to show that the Schur complement of P^γ_{n-2} in the matrix P^γ_{n-1} is also positive:

$$\gamma_{n-1} - \begin{bmatrix} \mathbf{c}^* & p_{n-1,n-2} \end{bmatrix} (P^\gamma_{n-2})^{-1} \begin{bmatrix} \mathbf{c} \\ p_{n-2,n-1} \end{bmatrix}$$

$$= \gamma_{n-1} - \mathbf{c}^*(P^\gamma_{n-3})^{-1}\mathbf{c} - (\gamma_{n-2} - \mathbf{b}^*(P^\gamma_{n-3})^{-1}\mathbf{b})^{-1} \cdot |p_{n-1,n-2} - \mathbf{c}^*(P^\gamma_{n-3})^{-1}\mathbf{b}|^2$$

$$> \gamma_{n-1} - \mathbf{c}^*(P^\gamma_{n-3})^{-1}\mathbf{c} - \frac{3}{q} \cdot |p_{n-1,n-2} - \mathbf{c}^*(P^\gamma_{n-3})^{-1}\mathbf{b}|^2 > 0.$$

Therefore, P^γ_{n-1} is positive definite. Finally, we have from (4.3), (4.1) and (4.5),

$$\det P^\gamma_{n-1,n-1}(\mathbf{p}_n)$$

$$= p_{n-1,n} \cdot \det \left(\begin{bmatrix} P^\gamma_{n-3} & \mathbf{b} \\ \mathbf{b}^* & \gamma_{n-2} \end{bmatrix} - \begin{bmatrix} \mathbf{d} \\ p_{n-2,n} \end{bmatrix} \begin{bmatrix} \mathbf{c}^* & p_{n-1,n-2} \end{bmatrix} p^{-1}_{n-1,n} \right)$$

$$= p_{n-1,n} \cdot (\gamma_{n-2} - q - X) \cdot \det(P^\gamma_{n-3} - \mathbf{d}\mathbf{c}^* p^{-1}_{n-1,n}) = 0.$$

where the last equality holds by the choice (4.6) of γ_{n-2}. By formula (3.42), $\Delta^\gamma_{n-1} = 0$. Then formula (3.42) will produce $f_\gamma \in \mathcal{B}_{n-2}$ solving the problem (1.2). This solution is not a constant function since w_{n-2}, w_{n-1}, w_n are all distinct. $\qquad\square$

5 Examples

In this section we illustrate Theorem 3.5 by several particular examples where the parametrization formula (3.15) (or (3.42)) is particularly explicit in terms of the interpolation data set.

5.1 Three-Points Problem

(cf. Example 3 in [21].) We want to find all $f \in \mathcal{B}_2$ satisfying conditions

$$f(t_i) = w_i \quad (t_i, f_i \in \mathbb{T}, \ i = 1, 2, 3). \tag{5.1}$$

We exclude the trivial case where $f_1 = f_2 = f_3$. By Theorem 3.6, all solutions $f \in \mathcal{B}_2$ to the problem (5.1) are given by the formula

$$f^\gamma(z) = w_3 \cdot \frac{1 - (1 - z\bar{t}_3) \cdot \left(\dfrac{\Delta^\gamma_1}{1 - z\bar{t}_1} + \dfrac{\Delta^\gamma_2}{1 - z\bar{t}_2} \right)}{1 - (1 - z\bar{t}_3) \cdot \left(\dfrac{w_3\overline{w}_1\Delta^\gamma_1}{1 - z\bar{t}_1} + \dfrac{w_3\overline{w}_2\Delta^\gamma_2}{1 - z\bar{t}_2} \right)}, \tag{5.2}$$

where Δ_1^γ and Δ_2^γ are given, according to (3.41), by

$$\Delta_1^\gamma = \frac{\gamma_2 P_{13} - P_{12}P_{23}}{\det P_2^\gamma} \quad \text{and} \quad \Delta_2^\gamma = \frac{\gamma_1 P_{23} - P_{21}P_{13}}{\det P_2^\gamma}, \tag{5.3}$$

where $p_{ij} = \frac{1-w_i\bar{w}_j}{1-t_i\bar{t}_j}$ and where $\gamma = \{\gamma_1, \gamma_2\}$ is any admissible pair, i.e.,

$$\gamma_1 > 0, \quad \gamma_2 > 0, \quad \gamma_1\gamma_2 > |p_{12}|^2.$$

Thus, any point (γ_1, γ_2) in the first quadrant $\mathbb{R}_+^2$ above the graph of $y = |p_{12}|^2 x^{-1}$ gives rise via formula (5.2) to a $\mathcal{B}_2$-solution to the problem (5.1). It follows immediately by the winding number argument that there are no solutions of degree one if w_1, w_2, w_3 do not have the same orientation on $\mathbb{T}$ as t_1, t_2, t_3. We can come to the same conclusion showing that in this case,

$$\Delta_1^\gamma \neq 0 \quad \text{and} \quad \Delta_2^\gamma \neq 0 \tag{5.4}$$

for any admissible $\{\gamma_1, \gamma_2\}$. Indeed, if $p_{13} = 0$ (i.e., $w_1 = w_3$), then (5.4) holds since p_{12}, p_{23} and γ_1 are all non-zero. Similarly, (5.4) holds if $p_{23} = 0$. If $p_{13} \neq 0$ and $p_{23} \neq 0$, then the numbers

$$\widetilde{\gamma}_2 := \frac{P_{12}P_{23}}{P_{13}} = \frac{P_{12}P_{23}P_{31}}{|P_{13}|^2} \quad \text{and} \quad \widetilde{\gamma}_1 := \frac{P_{21}P_{13}}{P_{23}} = \frac{P_{21}P_{13}P_{32}}{|P_{23}|^2} \tag{5.5}$$

are both non-positive, by Corollary 4.3. Hence, inequalities (5.4) hold for any positive γ_1, γ_2 and therefore, there are no zero cancellations in (5.2). On the other hand, if w_1, w_2, w_3 have the same orientation on $\mathbb{T}$ as t_1, t_2, t_3, then the numbers $\widetilde{\gamma}_2$ and $\widetilde{\gamma}_1$ in (5.5) are positive (again, by Corollary 4.3). Observe that $\widetilde{\gamma}_1\widetilde{\gamma}_2 = |p_{12}|^2$. Therefore, any pair $(\gamma_1, \widetilde{\gamma}_2)$ with $\gamma_1 > \widetilde{\gamma}_1$ is admissible, and since

$$\Delta_1^{\{\gamma_1, \widetilde{\gamma}_2\}} = 0, \quad \Delta_2^{\{\gamma_1, \widetilde{\gamma}_2\}} = \frac{\gamma_1 P_{23} - P_{21}P_{13}}{\gamma_1 \frac{P_{12}P_{23}}{P_{13}} - |P_{12}|^2} = \frac{P_{13}}{P_{12}},$$

the formula (5.2) amounts to

$$f^{\{\gamma_1, \widetilde{\gamma}_2\}}(z) = \frac{(1 - z\bar{t}_2)p_{12} - (1 - z\bar{t}_3)p_{13}}{(1 - z\bar{t}_2)\overline{w}_3 p_{12} - (1 - z\bar{t}_3)\overline{w}_2 p_{13}} \tag{5.6}$$

for any $\gamma_1 > \widetilde{\gamma}_1$. On the other hand, any pair $(\widetilde{\gamma}_1, \gamma_2)$ with $\gamma_2 > \widetilde{\gamma}_2$ is admissible, and since now

$$\Delta_1^{\{\widetilde{\gamma}_1, \gamma_2\}} = \frac{\gamma_2 P_{13} - P_{12}P_{23}}{\gamma_2 \frac{P_{21}P_{13}}{P_{23}} - |P_{12}|^2} = \frac{P_{23}}{P_{21}} \quad \text{and} \quad \Delta_2^{\{\widetilde{\gamma}_1, \gamma_2\}} = 0,$$

the formula (5.2) amounts to

$$f^{\{\widetilde{\gamma}_1,\gamma_2\}}(z) = \frac{(1 - z\bar{t}_1)p_{21} - (1 - z\bar{t}_3)p_{23}}{(1 - z\bar{t}_1)\overline{w}_3 p_{21} - (1 - z\bar{t}_3)\overline{w}_1 p_{23}} \tag{5.7}$$

for any $\gamma_2 > \widetilde{\gamma}_2$. A straightforward verification shows that formulas (5.6) and (5.7) define the same function (as expected, since the problem (5.1) has at most one rational solution of degree one).

5.2 Another Example

We next consider the n-point problem (1.2) where all target values but one are equal to each other (we assume without loss of generality that this common value is 1):

$$f(t_i) = 1 \quad (i = 1, \ldots, n-1) \quad \text{and} \quad f(t_n) = w_n \in \mathbb{T} \setminus \{1\}. \tag{5.8}$$

Proposition 5.1 *All functions* $f \in \mathcal{B}_{n-1}^{\circ}$ *subject to interpolation conditions* (5.8) *are parametrized by the formula*

$$f(z) = w_n \cdot \frac{1 - \displaystyle\sum_{i=1}^{n-1} \frac{(1 - z\bar{t}_n)(1 - \overline{w}_n)}{(1 - z\bar{t}_i)(1 - t_i\bar{t}_n)\gamma_i}}{1 - \displaystyle\sum_{i=1}^{n-1} \frac{(1 - z\bar{t}_n)(w_n - 1)\overline{w}_i}{(1 - z\bar{t}_i)(1 - t_i\bar{t}_n)\gamma_i}}, \tag{5.9}$$

where positive numbers $\gamma_1, \ldots, \gamma_{n-1}$ *are free parameters.*

Proof For the data set as in (5.8), $p_{i,j} = 0$ for all $1 \le i \ne j \le n-1$. Therefore, the matrix P_{n-1}^{γ} is diagonal and $\gamma = \{\gamma_1, \ldots, \gamma_{n-1}\}$ is admissible if and only if $\gamma_i > 0$ ($1 \le i \le n-1$). Furthermore, the numbers (3.41) are equal to

$$\Delta_i^{\gamma} = \frac{p_{i,n}}{\gamma_i} = \frac{1 - w_i\overline{w}_n}{\gamma_i(1 - t_i\bar{t}_n)} = \frac{1 - \overline{w}_n}{\gamma_i(1 - t_i\bar{t}_n)} \quad \text{for} \quad i = 1, \ldots, n-1.$$

Substituting the latter formulas in (3.42) gives (5.9). By Theorem 3.5, formula (5.9) parametrizes all $\mathcal{B}_{n-1}$-solutions to the problem (5.8). However, since $\Delta_i^{\gamma} \ne 0$ for all $i = 1, \ldots, n-1$, it follows that $\deg f = n-1$ for any f of the form (5.9). $\square$

Remark 5.2 Letting $w_n = 1$ in (5.9) we see that the only $f \in \mathcal{B}_{n-1}$ subject to equalities $f(t_i) = 1$ for $i = 1, \ldots, n$, is the constant function $f(z) \equiv 1$.

We now show how to get formula (5.9) using the approach from [6, 10] as follows. Let $f \in \mathcal{B}_{n-1}^{\circ}$ satisfy conditions (5.8) and let $\gamma_i = |f'(t_i)|$ ($1 \le i \le n-1$). Then $f^{-1}(\{1\}) = \{t_1, \ldots, t_{n-1}\}$ and the Aleksandrov-Clark measure $\mu_{f,1}$ of f at 1 is the sum of n point masses γ_i^{-1} at t_i. Therefore,

$$\frac{1 + f(z)}{1 - f(z)} = \int_{\mathbb{T}} \frac{\zeta + z}{\zeta - z} d\mu_{f,1}(\zeta) + ic = \sum_{i=1}^{n-1} \frac{1}{\gamma_i} \cdot \frac{t_i + z}{t_i - z} + ic \qquad (5.10)$$

for some $c \in \mathbb{R}$. Solving (5.10) for f and letting $\mathcal{E} = \frac{ic-1}{ic+1}$ (note that $\mathcal{E} \in \mathbb{T}\backslash\{1\}$) we get

$$f(z) = \frac{(1 - \Phi(z))\mathcal{E} + \Phi(z)}{-\Phi(z)\mathcal{E} + 1 + \Phi(z)}, \quad \text{where} \quad \Phi(z) = \frac{1}{2} \cdot \sum_{i=1}^{n-1} \frac{1}{\gamma_i} \cdot \frac{t_i + z}{t_i - z}. \qquad (5.11)$$

Evaluating (5.11) at $z = t_n$ and making use of the last condition in (5.8) we get

$$w_n = \frac{(1 - \Phi(t_n))\mathcal{E} + \Phi(t_n)}{-\Phi(t_n)\mathcal{E} + 1 + \Phi(t_n)} \iff \mathcal{E} = \frac{(1 + \Phi(t_n))w_n - \Phi(t_n)}{\Phi(t_n)w_n + 1 - \Phi(t_n)}.$$

Substituting the latter expression for $\mathcal{E}$ into (5.11) leads us to the representation

$$f(z) = \frac{w_n + (\Phi(z) - \Phi(t_n))(1 - w_n)}{1 + (\Phi(z) - \Phi(t_n))(1 - w_n)}$$

which is the same as (5.9), since $|w_n| = 1$ and since according to (5.11),

$$\Phi(z) - \Phi(t_n) = \frac{1}{2} \cdot \sum_{i=1}^{n-1} \frac{1}{\gamma_i} \cdot \left(\frac{t_i + z}{t_i - z} - \frac{t_i + t_n}{t_i - t_n} \right) = \sum_{i=1}^{n-1} \frac{1 - z\bar{t}_n}{\gamma_i (1 - z\bar{t}_i)(1 - t_i\bar{t}_n)}.$$

5.3 Boundary Fixed Points

By Theorem 1.1, there are infinitely many finite Blaschke products $f \in \mathcal{B}_{n-1}$ with given fixed boundary points $t_1, \ldots, t_n \in \mathbb{T}$, i.e., such that

$$f(t_i) = t_i \quad \text{for} \quad i = 1, \ldots, n. \qquad (5.12)$$

We will use Theorem 3.5 to parametrize all such Blaschke products. Since $w_i = t_i$ for $i = 1, \ldots, n$, we have $p_{i,j} = 1$ for all $1 \le i \ne j \le n$. By definition (3.2),

$$P_{n-1}^{\gamma} = \Gamma + EE^*, \quad \text{where} \quad \Gamma = \begin{bmatrix} \gamma_1 - 1 & & 0 \\ & \ddots & \\ 0 & & \gamma_{n-1} - 1 \end{bmatrix}, \quad E = \begin{bmatrix} 1 \\ \vdots \\ 1 \end{bmatrix}. \qquad (5.13)$$

If the tuple $\boldsymbol{\gamma} = \{\gamma_1, \ldots, \gamma_{n-1}\}$ contains two elements $\gamma_i \leq 1$ and $\gamma_j \leq 1$, then it is not admissible since the principal submatrix $\begin{bmatrix} \gamma_1 & 1 \\ 1 & \gamma_2 \end{bmatrix}$ of P_{n-1}^{γ} is not positive definite. Otherwise, that is, in one of the three following cases, the tuple $\boldsymbol{\gamma}$ is admissible:

(1) $\gamma_i > 1$ for all $i \in \{1, \ldots, n-1\}$.
(2) $\gamma_\ell = 1$ and $\gamma_i > 1$ for all $i \neq \ell$.
(3) $\gamma_\ell < 1$, $\gamma_i > 1$ for all $i \neq \ell$, and $\det P_{n-1}^{\gamma} > 0$.

Cases 1&3: Since $\gamma_i \neq 0$ for $i = 1, \ldots, n$, the matrix Γ is invertible. By basic properties of determinants,

$$\det P_{n-1}^{\gamma} = \det \Gamma \cdot \det(I + \Gamma^{-1}EE^*) = \det \Gamma \cdot (1 + E^*\Gamma^{-1}E),$$

which, on account (5.13), implies

$$\det P_{n-1}^{\gamma} = \left(1 + \sum_{i=1}^{n-1} \frac{1}{\gamma_i - 1}\right) \cdot \prod_{i=1}^{n-1}(\gamma_i - 1). \tag{5.14}$$

Using the latter formula, we can characterize Case 3 as follows:

$$\gamma_\ell < 1, \quad \gamma_i > 1 \quad \text{for all} \quad i \neq \ell, \quad \text{and} \quad \sum_{i=1}^{n-1} \frac{1}{\gamma_i - 1} < -1. \tag{5.15}$$

Since $p_{i,n} = 1$ for all $1 \leq i \leq n-1$, we have $\mathbf{p}_n = E$ in (3.40), and hence,

$$\det P_{n-1,j}^{\gamma}(E) = \lim_{\gamma_j \to 1} \det P_{n-1}^{\gamma} = \prod_{i \neq j}(\gamma_i - 1) \quad \text{for} \quad j = 1, \ldots, n-1.$$

Substituting the two latter formulas in (3.41) we get

$$\Delta_j^{\gamma} = \frac{\det P_{n-1,j}^{\gamma}(E)}{\det P_{n-1}^{\gamma}} = \frac{1}{\left(1 + \sum_{i=1}^{n-1} \frac{1}{\gamma_i - 1}\right) \cdot (\gamma_j - 1)} \qquad (j = 1, \ldots, n-1). \tag{5.16}$$

Substituting (5.16) into (3.42) (with $w_i = t_i$ for $i = 1, \ldots, n-1$) leads us (after straightforward algebraic manipulations) to the formula

$$f(z) = \frac{t_n + \displaystyle\sum_{i=1}^{n-1} \frac{z(1 - t_n\bar{t}_i)}{(1 - z\bar{t}_i)(\gamma_i - 1)}}{1 + \displaystyle\sum_{i=1}^{n-1} \frac{1 - t_n\bar{t}_i}{(1 - z\bar{t}_i)(\gamma_i - 1)}}. \tag{5.17}$$

Since $\Delta_j^{\gamma} \neq 0$ for all $j = 1, \ldots, n-1$, it follows by Theorem 3.5 that f is a Blaschke product of degree $n-1$ and $|f'(t_i)| = f'(t_i) = \gamma_i$ for $i = 1, \ldots, n-1$. Differentiating (5.17) and evaluating the obtained formula for $f'(z)$ at $z = t_n$ gives

$$\gamma_n := f'(t_n) = \left(\sum_{i=1}^{n-1} \frac{1}{\gamma_i - 1} \right) \left(1 + \sum_{i=1}^{n-1} \frac{1}{\gamma_i - 1} \right)^{-1},$$

which can be equivalently written as

$$\frac{\gamma_n}{1 - \gamma_n} = \sum_{i=1}^{n-1} \frac{1}{\gamma_i - 1} \quad \text{or} \quad \sum_{i=1}^{n} \frac{1}{\gamma_i - 1} = -1. \tag{5.18}$$

It follows from (5.18) that in Case 1, $0 < \gamma_n < 1$, so that t_n is the (hyperbolic) Denjoy-Wolff point of f. Alternatively, this conclusion follows from the Cowen-Pommerenke result [9]: *the inequality*

$$\sum_{i=1}^{n-1} \frac{1}{f'(t_i) - 1} \leq \frac{f'(t_n)}{1 - f'(t_n)} \tag{5.19}$$

for any analytic $f : \mathbb{D} \to \mathbb{D}$ *with boundary fixed points* $t_1, \ldots, t_{n-1}$ *and the (hyperbolic) boundary fixed point* t_n, *and equality prevails in* (5.19) *if and only if* $f \in \mathcal{B}_{n-1}^{\circ}$. Observe, that in Case 3, the point t_ℓ is the Denjoy-Wolff point of f, while t_n is a regular boundary fixed point with $\gamma_n = f'(t_n) > 1$.

Case 2: Direct computations show that in this case,

$$\det P_{n-1}^{\gamma} = \prod_{j \neq \ell} (\gamma_j - 1) = \det P_{n-1,\ell}^{\gamma}(E) \quad \text{and} \quad \det P_{n-1,i}^{\gamma}(E) = 0 \quad (i \neq \ell).$$

Therefore, according to (3.41), $\Delta_\ell^{\gamma} = 1$ and $\Delta_i^{\gamma} = 0$ for all $i \neq \ell$, which being substituted into (3.42), gives

$$f(z) = t_n \cdot \frac{1 - \frac{1-z\bar{t}_n}{1-z\bar{t}_\ell}}{1 - \frac{1-z\bar{t}_n}{1-z\bar{t}_\ell} \bar{t}_\ell t_n} = z.$$

We summarize the preceding analysis in the next proposition.

Proposition 5.3 *All functions* $f \in \mathcal{B}_{n-1}^{\circ}$ *satisfying conditions* (5.8) *are given by the formula* (5.17), *where the parameter* $\boldsymbol{\gamma} = \{\gamma_1, \ldots, \gamma_{n-1}\}$ *is either subject to relations* (5.15) *(in which case* t_ℓ *is the Denjoy-Wolff point of* f*) or* $\gamma_i > 1$ *for all* i *(in which case* t_n *is the Denjoy-Wolff point of* f*).*

In particular, it follows that the identity mapping $f(z) = z$ is the only function in $\mathcal{B}_{n-2}$ satisfying conditions (5.12). In fact, a more general result in [1] asserts that if the n-point Nevanlinna-Pick problem (1.2) (with $t_i, w_i \in \mathbb{C}$) has a rational solution of degree $k < n/2$ (in the setting of (5.12), $k = 1$), then it does not have other rational solutions of degree less than $n - k$.

6 Concluding Remarks and Open Questions

Some partial results on the problem (1.2) can be derived from general results on rational interpolation. Let $[x]$ denote the integer part of $x \in \mathbb{R}$ (the greatest integer not exceeding a given x). If we let

$$q = \text{rank} \left[\frac{w_{r+j} - w_i}{t_{r+i} - t_j} \right]_{i,j=1}^r = \text{rank} \left[\frac{1 - w_i \overline{w}_{r+j}}{1 - t_i \overline{t}_{r+j}} \right]_{i,j=1}^r, \qquad r = \left[\frac{n}{2} \right], \qquad (6.1)$$

then, by a result from [1], there are no rational functions f of degree less than q satisfying conditions (1.2). If $q \leq \frac{n-1}{2}$, then there is at most one rational f of degree equal q satisfying conditions (1.2), and the only candidate can be found in the form

$$f(z) = \frac{a_0 + a_1 z + \ldots + a_q z^q}{b_0 + b_1 z + \ldots + b_q z^q} \qquad (6.2)$$

by solving the linear system

$$a_0 + a_1 t_i + \ldots + a_q t_i^q = w_i (b_0 + b_1 t_i + \ldots + b_q t_i^q), \qquad i = 1, \ldots, n.$$

If there are no zero cancellations in the representation (6.2), f satisfies all conditions in (1.2). This f is a finite Blaschke product if and only if it has no poles in $\overline{D}$ and $b_i = \overline{a}_{q-i}$ for $i = 1, \ldots, n$. Alternatively, $f \in \mathcal{B}_q^\circ$ if and only if the boundary Schwarz-Pick matrix $P^f(t_1, \ldots, t_n)$ (see (2.1)) is positive semidefinite. Another result from [1] states that the next possible degree of a rational solution to the problem (1.2) is $n - q$ and moreover, there are infinitely many rational solutions of degree k for each $k \geq n - q$. Simple examples show that finite Blaschke product solutions of degree $n - q$ may not exist, so that the result from [1] provides a *lower bound* for the minimally possible degree of a Blaschke product solution. On the other hand (see e.g., [21]), if the problem (1.2) has a solution in $\mathcal{B}_{\kappa_0}^\circ$ for $\kappa_0 > \frac{n-1}{2}$, then it has infinitely solutions in $\mathcal{B}_k^\circ$ for each $k \geq \kappa_0$. Hence, the procedure verifying whether or not the problem (1.2) has a unique minimal degree Blaschke product solution (necessarily, $\deg f \leq \frac{n-1}{2}$) is simple. The question of some interest is to characterize the latter determinate case in terms of the original interpolation data set. A much more interesting question is:

Question 1 Find the minimally possible $\kappa_0 > \frac{n-1}{2}$ so that the problem (1.2) has a solution in $\mathcal{B}_{\kappa_0}^\circ$. For each $k \geq \kappa$, parametrize all $\mathcal{B}_k^\circ$-solutions to the problem.

In [13, 14], the boundary problem (1.2) was considered in the set $\mathcal{QB}_{n-1}$ of rational functions of degree at most $n-1$ that are unimodular on $\mathbb{T}$. Observe that any element of $\mathcal{QB}_{n-1}$ is equal to the ratio of two coprime Blaschke products g, h such that $\deg g + \deg h \leq n - 1$. In general, it not true that a rational function f of degree less than n and taking unimodular values at n points on $\mathbb{T}$ is necessarily unimodular on $\mathbb{T}$ (by Theorem 1.1, this is true, if a'priori, f is subject to $|f(t)| \leq 1$ for all $t \in \mathbb{T}$). Hence, the results concerning low-degree unimodular interpolation do not follow directly from the general results on the unconstrained rational interpolation. However, we were not able to find an example providing the negative answer for the next question:

Question 2 Let q be defined as in (6.1), so that there are infinitely many rational functions f, $\deg f = n - q$, satisfying conditions (1.2). Is it true that some (and therefore, infinitely many) of them are unimodular on $\mathbb{T}$?

We finally reformulate the boundary Nevanlinna-Pick problem (1.2) in terms of a positive semidefinite matrix completion problem. With $t_1, \ldots, t_n$ and $w_1, \ldots, w_n$ in hands, we specify all off-diagonal entries in the matrix P_n:

$$P_n = \begin{bmatrix} * & \frac{1-w_1\bar{w}_2}{1-t_1\bar{t}_2} & \cdots & \frac{1-w_1\bar{w}_{n-1}}{1-t_1\bar{t}_{n-1}} & \frac{1-w_1\bar{w}_n}{1-t_1\bar{t}_n} \\ \frac{1-w_2\bar{w}_1}{1-t_2\bar{t}_1} & * & \cdots & \frac{1-w_2\bar{w}_{n-1}}{1-t_2\bar{t}_{n-1}} & \frac{1-w_2\bar{w}_n}{1-t_2\bar{t}_n} \\ \vdots & \vdots & \ddots & \vdots & \vdots \\ \frac{1-w_{n-1}\bar{w}_1}{1-t_{n-1}\bar{t}_1} & \frac{1-w_{n-1}\bar{w}_2}{1-t_{n-1}\bar{t}_2} & \cdots & * & \frac{1-w_{n-1}\bar{w}_n}{1-t_{n-1}\bar{t}_n} \\ \frac{1-w_n\bar{w}_1}{1-t_n\bar{t}_1} & \frac{1-w_n\bar{w}_2}{1-t_n\bar{t}_2} & \cdots & \frac{1-w_n\bar{w}_{n-1}}{1-t_n\bar{t}_{n-1}} & * \end{bmatrix}. \tag{6.3}$$

Every choice of the (ordered) set $\gamma = \{\gamma_1, \ldots, \gamma_n\}$ of real diagonal entries produces a Hermitian completion of P_n which we have denoted by P_n^γ in (3.2). Completion question related to the problem (1.2) and to a similar problem in the class $\mathcal{QB}_{n-1}$ are the following:

Question 3 Given a partially specified matrix P_n (6.3), find a positive semidefinite completion P_n^γ with minimal possible rank.

The same question concerning finding the minimal rank *Hermitian* completion P_n^γ is related to minimal degree boundary interpolation by unimodular functions. Once the minimal rank completion is found, the finite Blaschke product f with the boundary Schwarz-Pick matrix $P^f(\mathbf{t}) = P_n^\gamma$ will be the minimal degree solution to the problem (1.2). Although Question 3 looks like an exercise in linear algebra, it turns out to be as difficult as the original interpolation problem.

References

1. A.C. Antoulas, B.D.O. Anderson, On the scalar rational interpolation problem. IMA J. Math. Control Inf. **3**(2–3), 61–88 (1986)
2. J.A. Ball, I. Gohberg, L. Rodman, *Interpolation of Rational Matrix Functions*. Operation Theory: Advances and Applications, vol. 45 (Birkhäuser Verlag, Basel, 1990)
3. V. Bolotnikov, A uniqueness result on boundary interpolation. Proc. Am. Math. Soc. **136**(5), 1705–1715 (2008)

4. V. Bolotnikov, S.P. Cameron, The Nevanlinna-Pick problem on the closed unit disk: minimal norm rational solutions of low degree. J. Comput. Appl. Math. **236**(13), 3123–3136 (2012)
5. V. Bolotnikov, A. Kheifets, The higher order Carathéodory–Julia theorem and related boundary interpolation problems, in *Operator Theory: Advances and Applications*, vol. 179 (Birkhauser, Basel, 2007), pp. 63–102
6. V. Bolotnikov, M. Elin, D. Shoikhet, Inequalities for angular derivatives and boundary interpolation. Anal. Math. Phys. **3**(1), 63–96 (2013)
7. C.I. Byrnes, T.T. Georgiou, A. Lindquist, A generalized entropy criterion for Nevanlinna-Pick interpolation with degree constraint. IEEE Trans. Autom. Control **46**(6), 822–839 (2001)
8. D.G. Cantor, R.R. Phelps, An elementary interpolation theorem. Proc. Am. Math. Soc. **16**, 523–525 (1965)
9. C.C. Cowen, C. Pommerenke, Inequalities for the angular derivative of an analytic function in the unit disk. J. Lond. Math. Soc. **26**(2), 271–289 (1982)
10. M. Elin, D. Shoikhet, N. Tarkhanov, Separation of boundary singularities for holomorphic generators. Annali di Matematica Pura ed Applicata **190**, 595–618 (2011)
11. T.T. Georgiou, A topological approach to Nevanlinna–Pick interpolation. SIAM J. Math. Anal. **18**, 1248–1260 (1987)
12. T.T. Georgiou, The interpolation problem with a degree constraint. IEEE Trans. Autom. Control **44**(3), 631–635 (1999)
13. C. Glader, Rational unimodular interpolation on the unit circle. Comput. Methods Funct. Theory **6**(2), 481–492 (2006)
14. C. Glader, Minimal degree rational unimodular interpolation on the unit circle. Electron. Trans. Numer. Anal. **30**, 88–106 (2008)
15. C. Glader, M. Lindström, Finite Blaschke product interpolation on the closed unit disc. J. Math. Anal. Appl. **273**, 417–427 (2002)
16. P. Gorkin, R.C. Rhoades, Boundary interpolation by finite Blaschke products. Constr. Approx. **27**(1), 75–98 (2008)
17. R. Nevanlinna, Über beschränkte Funktionen die in gegebenen Punkten vorgeschriebene Werte annehmen. Ann. Acad. Sci. Fenn. **13**(1), 1–71 (1919)
18. G. Pick, Über die Beschränkungen analytischer Funktionen, welche durch vorgegebene Funktionswerte bewirkt werden. Math. Ann. **77**(1), 7–23 (1916)
19. S. Ruscheweyh, W.B. Jones, Blaschke product interpolation and its application to the design of digital filters. Constr. Approx. **3**, 405–409 (1987)
20. D. Sarason, Nevanlinna–Pick interpolation with boundary data. Integr. Equ. Oper. Theory **30**, 231–250 (1998)
21. G. Semmler, E. Wegert, Boundary interpolation with Blaschke products of minimal degree. Comput. Methods Funct. Theory **6**(2), 493–511 (2006)
22. R. Younis, Interpolation by a finite Blaschke product. Proc. Am. Math. Soc. **78**(3), 451–452 (1980)

Support Points and the Bieberbach Conjecture in Higher Dimension

Filippo Bracci and Oliver Roth

Abstract We describe some open questions related to support points in the class S^0 and introduce some useful techniques toward a higher dimensional Bieberbach conjecture.

Keywords Loewner theory • Support points • Bieberbach conjecture in higher dimension

2010 Mathematics Subject Classification 32A19, 32A30, 30C80

1 Introduction

Let $\mathbb{B}^n \subset \mathbb{C}^n$ denote the unit ball for the standard Hermitian product in $\mathbb{C}^n$, $n \geq 1$. For the sake of simplicity, we consider only the case $n = 1$ and $n = 2$, but, in fact, all results which we are going to discuss for $\mathbb{B}^2$ hold in any dimension. We denote $\mathbb{D} := \mathbb{B}^1$.

Let $S(\mathbb{B}^n)$ denote the class of univalent maps $f : \mathbb{B}^n \to \mathbb{C}^n$ normalized so that $f(0) = 0, df_0 = \text{id}$. We consider $S(\mathbb{B}^n)$ as a subspace of the Frechét space of holomorphic maps from $\mathbb{B}^n$ to $\mathbb{C}^n$ with the topology of uniform convergence on compacta.

F. Bracci (✉)
Dipartimento Di Matematica, Università di Roma "Tor Vergata", Via Della Ricerca Scientifica 1, 00133 Roma, Italy
e-mail: fbracci@mat.uniroma2.it

O. Roth
Department of Mathematics, University of Würzburg, Emil Fischer Straße 40, 97074 Würzburg, Germany
e-mail: roth@mathematik.uni-wuerzburg.de

© Springer International Publishing AG 2018
M. Agranovsky et al. (eds.), *Complex Analysis and Dynamical Systems*, Trends in Mathematics, https://doi.org/10.1007/978-3-319-70154-7_4

For $n = 1$, the set $S(\mathbb{D})$ is compact (see, e.g., [12]). For $n > 1$, the set $S(\mathbb{D})$ is not compact: a simple example is given by considering the sequence

$$\{(z_1, z_2) \mapsto (z_1 + mz_2^2, z_2)\}_{m \in \mathbb{N}},$$

which belongs to $S(\mathbb{B}^2)$ but for $m \to \infty$ does not converge.

Since the set $S(\mathbb{D})$ is compact, for every continuous linear operator $L :$ $\mathsf{Hol}(\mathbb{D}, \mathbb{C}) \to \mathbb{C}$ there exists $f \in S(\mathbb{D})$ such that $\mathsf{Re}\, L(f) \geq \mathsf{Re}\, L(g)$ for all $g \in S(\mathbb{D})$. If L is not constant, such an f is called a *support point* for L.

It is known that every support point in $S(\mathbb{D})$ is unbounded and is, in fact, a slit map (see [14]). The most interesting linear functionals to be considered are perhaps those defined as $L_m(f) = b_m$, where $f(z) = \sum_{j \geq 0} b_j z^j \in S(\mathbb{D})$, $m \in \mathbb{N}$. The Bieberbach conjecture, proved in the 1980s by L. de Branges, states that $|b_m| \leq m$ for $f \in S(\mathbb{D})$ and that equality is reached only by rotations of the Koebe function.

In higher dimensions, since the class $S(\mathbb{B}^2)$ is not compact, one is forced to consider suitable compact subclasses. Convex maps and starlike maps form compact subclasses, but, for many purposes, these classes are too small. In [8] (see also [7]), it was introduced a compact subclass, denoted by $S^0(\mathbb{B}^2)$ (or simply S^0), for which the membership depends on the existence of a parametric representation, a condition that is always satisfied in dimension one thanks to the classical Loewner theory (see Sect. 2).

The class S^0 is strictly contained in $S := S(\mathbb{B}^2)$, but evidences are that every map in S might be factorized as the composition of an element of S^0 and a normalized entire univalent map of $\mathbb{C}^2$ (this is known to be true for univalent maps f on $\mathbb{B}^2$ which extend C^∞ up to the boundary, $f(\mathbb{B}^2)$ is strongly pseudoconvex and $\overline{f(\mathbb{B}^2)}$ is polynomially convex[1]; see [2]). Were this the case, one could somehow split the difficulties in understanding univalent maps on $\mathbb{B}^2$ into two pieces: understanding the compact class S^0 and automorphisms/Fatou-Bieberbach maps in $\mathbb{C}^2$. In light of this, the class S^0 seems to be a natural candidate to study in higher dimensions.

The present note focuses on support points on S^0. Our aim is, on the one hand, to state some natural open questions originating in the recent works [3, 5, 11, 13], and, on the other hand, to develop some new techniques to handle such problems (in particular, slice reduction and decoupling harmonic terms tricks). The paper [10] contains other open questions in this direction and an extensive bibliography on the subject, to which we refer the reader. Here we mainly focus our attention on those questions which relate the class S^0 to the (huge) group of automorphisms of $\mathbb{C}^2$, highlighting the deep differences between dimension 1 and dimension 2 (see Sect. 4).

We also develop the ideas in [3] (see Sect. 6), which allowed to construct an example of a bounded support point in S^0. With these tools in hand, we state a

---

[1]In [11], the result was extended to univalent maps on $\mathbb{B}^2$ which extend C^1 up to $\partial \mathbb{B}^2$ and whose image is Runge in $\mathbb{C}^2$, but, unfortunately, there is a gap in the proof.

Bieberbach-type conjecture in S^0 for coefficients of pure terms in z_1 and z_2 in the expansion at the origin.

We thank the referee for the comments which improved the original manuscript.

2 The Class S^0

In what follows we denote by $\mathbb{R}^+$ the semigroup of nonnegative real numbers and by $\mathbb{N}$ the semigroup of nonnegative integers.

Let

$$\mathcal{M} := \{h \in \mathsf{Hol}(\mathbb{B}^2, \mathbb{C}^2) : h(0) = 0, dh_0 = \mathsf{id}, \mathsf{Re}\,\langle h(z), z \rangle > 0, \forall z \in \mathbb{B}^2 \setminus \{0\}\},$$

where $\langle \cdot, \cdot \rangle$ denotes the Euclidean inner product in $\mathbb{C}^2$.

The set $\mathcal{M}$ is compact in $\mathsf{Hol}(\mathbb{B}^2, \mathbb{C}^2)$ endowed with the topology of uniform convergence on compacta (see [7]).

Definition 2.1 A *Herglotz vector field associated with the class* $\mathcal{M}$ *on* $\mathbb{B}^2$ is a mapping $G : \mathbb{B}^2 \times \mathbb{R}^+ \to \mathbb{C}^2$ with the following properties:

(i) the mapping $G(z, \cdot)$ is measurable on $\mathbb{R}^+$ for all $z \in \mathbb{B}^2$.
(ii) $-G(\cdot, t) \in \mathcal{M}$ for a.e. $t \in [0, +\infty)$.

Remark 2.2 Due to the estimates for the class $\mathcal{M}$ (see [7]), a Herglotz vector field associated with the class $\mathcal{M}$ on $\mathbb{B}^2$ is an L^∞-Herglotz vector field on $\mathbb{B}^2$ in the sense of [4].

Definition 2.3 A family $(f_t)_{t \geq 0}$ of holomorphic mappings from $\mathbb{B}^2$ to $\mathbb{C}^2$ such that $f_t(0) = 0$ and $d(f_t)_0 = e^t \mathsf{id}$ for all $t \geq 0$, is called a *normalized regular family* if

(i) the mapping $t \mapsto f_t$ is continuous with respect to the topology in $\mathsf{Hol}(\mathbb{B}^2, \mathbb{C}^2)$ induced by the uniform convergence on compacta in $\mathbb{B}^2$,
(ii) there exists a set of zero measure $N \subset [0, +\infty)$ such that for all $t \in [0, +\infty) \setminus N$ and all $z \in \mathbb{B}^2$ the partial derivative $\frac{\partial f_t}{\partial t}(z)$ exists and is holomorphic.

For a given Herglotz vector field $G(z, t)$ associated with the class $\mathcal{M}$ on $\mathbb{B}^2$, a *normalized solution* to the *Loewner-Kufarev PDE* associated to $G(z, t)$ consists of a normalized regular family $(f_t)_{t \geq 0}$ such that the following equation is satisfied for a.e. $t \geq 0$ and for all $z \in \mathbb{B}^2$

$$\frac{\partial f_t}{\partial t}(z) = -d(f_t)_z \cdot G(z, t). \tag{2.1}$$

Definition 2.4 A *normalized subordination chain* $(f_t)_{t \geq 0}$ is a family of holomorphic mappings $f_t : \mathbb{B}^2 \to \mathbb{C}^2$, such that $f_t(0) = 0$, $d(f_t)_0 = e^t \mathsf{id}$ for all $t \geq 0$, and for every $0 \leq s \leq t$ there exists $\varphi_{s,t} : \mathbb{B}^2 \to \mathbb{B}^2$ holomorphic such that $f_s = f_t \circ \varphi_{s,t}$.

A normalized subordination chain $(f_t)_{t\geq 0}$ is called a *normalized Loewner chain* if for all $t \geq 0$ the mapping f_t is univalent.

Definition 2.5 A normalized Loewner chain $(f_t)_{t\geq 0}$ on $\mathbb{B}^2$ is called a *normal Loewner chain* if the family $\{e^{-t}f_t(\cdot)\}_{t\geq 0}$ is normal.

From [7, Chapter 8], [1, Prop. 2.6] and [9], we have the following:

Theorem 2.6

(1) *If $(f_t)_{t\geq 0}$ is a normalized Loewner chain on $\mathbb{B}^2$, then it is a normalized solution to a Loewner-Kufarev PDE (2.1) for some Herglotz vector field $G(z,t)$ associated with the class $\mathcal{M}$ in $\mathbb{B}^2$.*

(2) *Let $G(z,t)$ be a Herglotz vector field associated with the class $\mathcal{M}$ on $\mathbb{B}^2$. Then there exists a unique normal Loewner chain $(g_t)_{t\geq 0}$—called the canonical solution—which is a normalized solution to (2.1). Moreover, $\bigcup_{t\geq 0} g_t(\mathbb{B}^2) = \mathbb{C}^2$.*

(3) *If $(f_t)_{t\geq 0}$ is a normalized solution to (2.1), then $(f_t)_{t\geq 0}$ is a normalized subordination chain on $\mathbb{B}^2$. Moreover, there exists a holomorphic mapping $\Phi : \mathbb{C}^2 \to \bigcup_{t\geq 0} f_t(\mathbb{B}^2)$, with $\Phi(0) = 0$ and $d\Phi_0 = \text{id}$ such that $f_t = \Phi \circ g_t$, where $(g_t)_{t\geq 0}$ is the canonical solution to (2.1). In particular, $(f_t)_{t\geq 0}$ is a normalized Loewner chain if and only if Φ is univalent.*

Remark 2.7 Let $(f_t)_{t\geq 0}$ be a family of holomorphic mappings such that $f_t(0) = 0$, $d(f_t)_0 = e^t\text{id}$ for all $t \geq 0$. Assume that $(f_t)_{t\geq 0}$ satisfies (i) of Definition 2.3 and for all fixed $z \in \mathbb{B}^2$ the mapping $t \mapsto f_t(z)$ is absolutely continuous. If for all fixed $z \in \mathbb{B}^2$ the family $(f_t)_{t\geq 0}$ satisfies (2.1) for a.e. $t \geq 0$, then it a normalized solution to the Loewner-Kufarev PDE and thus is a regular family.

Definition 2.8 Let $f \in S$. We say that f admits *parametric representation* if

$$f(z) = \lim_{t\to\infty} e^t \varphi(z,t)$$

locally uniformly on $\mathbb{B}^2$, where $\varphi(z,0) = z$ and

$$\frac{\partial \varphi}{\partial t}(z,t) = G(\varphi(z,t),t), \quad \text{a.e. } t \geq 0, \quad \forall z \in \mathbb{B}^2, \tag{2.2}$$

for some Herglotz vector field G associated with the class $\mathcal{M}$ on $\mathbb{B}^2$.

We denote by S^0 the set consisting of univalent mappings which admit parametric representation.

The following result is in [7, Chapter 8] (see also [8]):

Theorem 2.9

(1) *A normalized univalent map $f : \mathbb{B}^2 \to \mathbb{C}^2$ has parametric representation if and only if there exists a normal Loewner chain $(f_t)_{t\geq 0}$ on $\mathbb{B}^2$ such that $f_0 = f$.*

(2) *The class S^0 is compact in the topology of uniform convergence on compacta.*

3 Support Points and Extreme Points

Definition 3.1

(i) Let K be a compact subset of $\mathsf{Hol}(\mathbb{B}^2, \mathbb{C}^2)$ endowed with the topology of uniform convergence on compacta. A mapping $f \in K$ is called a *support point* if there exists a continuous linear operator $L : \mathsf{Hol}(\mathbb{B}^2, \mathbb{C}^2) \to \mathbb{C}$ not constant on K such that $\max_{g \in K} \mathsf{Re}\, L(g) = \mathsf{Re}\, L(f)$. We denote by $\mathsf{Supp}(K)$ the set of support points of K.

(ii) A mapping $f \in K$ is called an *extreme point* if $f = tg + (1 - t)h$, where $t \in (0, 1), g, h \in K$, implies $f = g = h$. We denote by $\mathsf{Ex}(K)$ the set of extreme points of K.

Note that the notion of extreme points is not related to topology, but only on the geometry of the set. If K is a compact subset of $\mathsf{Hol}(\mathbb{B}^2, \mathbb{C}^2)$ and $a \in K$ is a support point which maximizes the continuous linear operator L, then $\mathcal{L} := \{b \in \mathsf{Hol}(\mathbb{B}^2, \mathbb{C}^2) : \mathsf{Re}\, L(b) = \mathsf{Re}\, L(a)\}$ is a real hyperplane and $\mathcal{L} \cap K$ contains extreme points. Therefore, for any continuous linear operator L which is not constant on K there exists a point $a \in K$ which is both a support point (for L) and an extreme point for K.

In dimension one it is known that all support points for S^0 are slit mappings (see [14]). In higher dimension, the situation is considerably more complicated.

Proposition 3.2 ([13]) *Let $f \in S^0$ be a support point. Let $G(z, t)$ be a Herglotz vector field associated with the class $\mathcal{M}$ which generates a normal Loewner chain (f_t) such that $f_0 = f$. Then $G(z, t)$ is a support point of $-\mathcal{M}$ for a.e. $t \geq 0$.*

Question 3.3 Let $f \in S^0$ be a support point.

(1) Does there exist only one normal Loewner chain (f_t) such that $f_0 = f$?
(2) Does there exist a Herglotz vector field associated with the class $\mathcal{M}$ which generates a normal Loewner chain (f_t) with $f_0 = f$ such that $t \mapsto G(\cdot, t)$ is continuous and $G(\cdot, t) \in supp(-\mathcal{M})$ for all $t \geq 0$?

Question 3.4 Let $G(z, t)$ be a Herglotz vector field associated with the class $\mathcal{M}$ which generates a normal Loewner chain (f_t).

(1) If f_0 is extreme in S^0, is it true that $G(z, t)$ is extreme in $-\mathcal{M}$ for a.e. $t \geq 0$?
(2) If f_0 is extreme in S^0, is $G(z, t)$ uniquely determined?

Proposition 3.5 *[15] Let (f_t) be a normal Loewner chain. Then for all $t \geq 0$, $e^{-t} f_t \in S^0$. Moreover, if f_0 is a support/extreme point for S^0, so is $e^{-t} f_t$ for all $t \geq 0$.*

Definition 3.6 Let $(f_t)_{t \geq 0}$ be a normalized Loewner chain in $\mathbb{B}^2$ and $G(z, t)$ be the associated Herglotz vector field. We say that $(f_t)_{t \geq 0}$ is *exponentially squeezing in*

$[T_1, T_2)$, for $0 \leq T_1 < T_2 \leq +\infty$ (with squeezing ratio $a \in (0, 1)$) if for a.e. $t \in [T_1, T_2)$ and for all $z \in \mathbb{B}^2 \setminus \{0\}$,

$$\operatorname{Re} \left\langle G(z, t), \frac{z}{\|z\|^2} \right\rangle \leq -a. \tag{3.1}$$

In [5] it is proved that (3.1) is equivalent to: for all $T_1 \leq s < t < T_2$,

$$\|f_t^{-1}(f_s(z))\| \leq e^{a(s-t)}\|z\|, \quad \text{for all } z \in \mathbb{B}^2. \tag{3.2}$$

Hence, if (f_t) is exponentially squeezing in $[T_1, T_2)$, then f_t is bounded for all $t \in [0, T_2)$ and $\overline{f_s(\mathbb{B}^2)} \subset f_t(\mathbb{B}^2)$ for all $T_1 \leq s < t < T_2$.

Using the results of [5], or of [13] for support points, one can prove

Proposition 3.7 *Let (f_t) be a normal Loewner chain which is exponentially squeezing in $[T_1, T_2)$. Then $f_0 \notin \operatorname{Supp}(S^0) \cup \operatorname{Ex}(S^0)$.*

Example 3.8 Let $f \in S^0$. Let (f_t) be one parametric representation of f. Let $r \in (0, 1)$. Consider $f_{r,t}(z) := r^{-1}f_t(rz)$. Then $(f_{r,t})$ is an exponentially squeezing normal Loewner chain and in particular, $f_r \in S^0 \setminus (\operatorname{Supp}(S^0) \cup \operatorname{Ex}(S^0))$.

Question 3.9 Let $f \in S^0 \setminus (\operatorname{Supp}(S^0) \cup \operatorname{Ex}(S^0))$ be a bounded function. Is it true that f can be embedded into an exponentially squeezing Loewner chain?

Question 3.10 Let $G(z, t)$ be a Herglotz vector field associated with the class $\mathcal{M}$ which generates a normal Loewner chain (f_t). Assume that

$$\limsup_{z \to A} \operatorname{Re} \left\langle G(z, t), \frac{z}{\|z\|^2} \right\rangle \leq -a,$$

for some $A \subset \partial \mathbb{B}^2$ and a.e. $t \geq 0$. Is it true that $f_0 \notin \operatorname{Supp}(S^0) \cup \operatorname{Ex}(S^0)$?

4   Automorphisms of $\mathbb{C}^2$ and Support Points

Let

$$\operatorname{Aut}_0(\mathbb{C}^2) := \{f \in \operatorname{Aut}(\mathbb{C}^2) : f(0) = 0, df_0 = \operatorname{id}\}.$$

Given $f \in \operatorname{Aut}_0(\mathbb{C}^2)$, for every $r > 0$, the map $f^r : \mathbb{B}^2 \to \mathbb{C}^2$ defined by

$$f^r(z) = \frac{1}{r}f(rz),$$

is normalized and univalent. For $r << 1$, the image $f^r(\mathbb{B}^2)$ is convex, hence $f^r \in S^0$. For $f \in \operatorname{Aut}_0(\mathbb{C}^2)$, let

$$r(f) := \sup\{t > 0 : f^t \in S^0\}.$$

Since S^0 is compact and $\{f^t\}_{t>0}$ is not normal except for $f = \mathsf{id}$, it follows that for $f \in \mathsf{Aut}_0(\mathbb{C}^2) \setminus \{\mathsf{id}\}$

$$0 < r(f) < +\infty, \quad f^{r(f)} \in S^0.$$

Question 4.1 Let $f \in \mathsf{Aut}_0(\mathbb{C}^2) \setminus \{\mathsf{id}\}$. Is it true that $f^{r(f)} \in \mathsf{Supp}(S^0)$?

The previous question has a positive solution in the case $f(z_1, z_2) = (z_1 + a z_2^2, z_2)$, as we discuss later (or see [3]).

Let

$$\mathcal{A} := \{f \in S^0 : \text{there exists } \Psi \in \mathsf{Aut}(\mathbb{C}^2) : \Psi|_{\mathbb{B}^2} = f\}$$

Note that in dimension one the analogue of $\mathcal{A}$ contains only the identity mapping. In higher dimension we have

Theorem 4.2 ([11]) $\overline{\mathcal{A}} = S^0$.

Take $f \in S^0$, and expand f as

$$f(z_1, z_2) = (z_1 + \sum_{\alpha \in \mathbb{N}^2, |\alpha| \geq 2} b_\alpha^1 z^\alpha, z_2 + \sum_{\alpha \in \mathbb{N}^2, |\alpha| \geq 2} b_\alpha^2 z^\alpha).$$

By a result of F. Forstnerič [6], for any $M \in \mathbb{N}$ there exists $g \in \mathsf{Aut}_0(\mathbb{C}^2)$ such that $f - g = O(\|z\|^{M+1})$ (that is f, g have the same jets up to order M). However, such a g does not belong to S^0 in general.

Question 4.3 For which $\alpha \in \mathbb{N}^2$ is it true that for any $f \in S^0$ there exists $g \in \mathcal{A}$ having the same coefficients b_α^1 as f?

5 Coefficient Bounds in $\mathbb{B}^2$

We use the following notation: $f \in S^0$,

$$f(z_1, z_2) = (z_1 + \sum_{\alpha \in \mathbb{N}^2, |\alpha| \geq 2} b_\alpha^1 z^\alpha, z_2 + \sum_{\alpha \in \mathbb{N}^2, |\alpha| \geq 2} b_\alpha^2 z^\alpha).$$

If (f_t) is a normal Loewner chain, we denote by $b_\alpha^j(t)$ the corresponding coefficients of f_t.

For $G(z, t)$ a Herglotz vector field associated with the class $\mathcal{M}$

$$G(z, t) = (-z_1 + \sum_{\alpha \in \mathbb{N}^2, |\alpha| \geq 2} q_\alpha^1(t) z^\alpha, -z_2 + \sum_{\alpha \in \mathbb{N}^2, |\alpha| \geq 2} q_\alpha^2(t) z^\alpha).$$

For an evolution family $\varphi_{s,t} := f_t^{-1} \circ f_s$,

$$\varphi_{s,t}(z_1, z_2) = (e^{s-t}z_1 + \sum_{\alpha \in \mathbb{N}^2, |\alpha| \geq 2} a_\alpha^1(s,t)z^\alpha, e^{s-t}z_2 + \sum_{\alpha \in \mathbb{N}^2, |\alpha| \geq 2} a_\alpha^2(s,t)z^\alpha).$$

For $f \in S^0, \alpha \in \mathbb{N}^2, |\alpha| \geq 2, j = 1, 2$, let

$$L_\alpha^j(f) := b_\alpha^j.$$

The problem of finding the maximal possible sharp bound for coefficients of mappings in the class S^0, consists in fact in finding the support points in S^0 for the linear functionals L_α^j.

If $(f_1(z_1, z_2), f_2(z_1, z_2)) \in S^0$, then $(f_2(z_2, z_1), f_1(z_2, z_1)) \in S^0$. Therefore, it is enough to solve the problem for L_α^1. More generally, if U is a 2×2 unitary matrix, given a normal Loewner chain (f_t), $(U^* f_t(Uz))$ is again a normal Loewner chain. This enables us to assume that a given coefficient $b_\alpha^1 > 0$, so that, in fact, $\max_{g \in S^0} \operatorname{Re} L_\alpha^1(g) = \max_{g \in S^0} |b_\alpha^1(g)|$.

Let (f_t) be a normal Loewner chain, $G(z,t)$ the associated Herglotz vector field and $(\varphi_{s,t})$ the associated evolution equation. Expanding the Loewner ODE one gets

$$\frac{\partial a_\alpha^1(s,t)}{\partial t} = -a_\alpha^1(s,t) + q_\alpha^1(t)e^{|\alpha|(s-t)} + R_\alpha, \tag{5.1}$$

where R_α is the coefficient of z^α in the expansion of

$$\sum_{2 \leq |\gamma| \leq |\alpha|-1} q_\gamma^1(t)(e^{s-t}z_1 + \sum_{2 \leq |\beta| \leq |\alpha|-1} a_\beta^1(s,t)z^\beta)^{\gamma_1}(e^{s-t}z_1 + \sum_{2 \leq |\beta| \leq |\alpha|-1} a_\beta^2(s,t)z^\beta)^{\gamma_2}.$$

Since $f_0 = \lim_{t \to \infty} e^t \varphi_{0,t}$ uniformly on compacta, we have

$$b_\alpha^1 = \lim_{t \to \infty} e^t a_\alpha^1(0,t). \tag{5.2}$$

Therefore, in order to get a sharp bound on the coefficients, one should try first to reduce the problem (if possible) to a simple problem involving the least possible number of coefficients of G, then find a sharp bound for such coefficients and solve the associated ODE.

Below, we describe some methods which can be used to simplify the problem and then turn to some applications.

6 Operations in the Class $\mathcal{M}$

6.1 Decoupling Harmonic Terms

Let $G(z)$ be an autonomous Herglotz vector field associated with the class $\mathcal{M}$. Then

$$\text{Re} \langle G(z), z \rangle \leq 0.$$

Such an inequality translates in terms of expansion as

$$- |z_1|^2 - |z_2|^2 + \sum_{|\alpha| \geq 2} \text{Re } q_\alpha^1 z_1^{\alpha_1} \overline{z_1} z_2^{\alpha_2} + \sum_{|\alpha| \geq 2} \text{Re } q_\alpha^2 z_1^{\alpha_1} z_2^{\alpha_2} \overline{z_2} \leq 0. \tag{6.1}$$

Replacing (z_1, z_2) by $(e^{i\theta k_1} z_1, e^{i\theta k_2} z_2)$, with $\theta \in \mathbb{R}$ and $k_1, k_2 \in \mathbb{Z}$, we obtain the expression

$$
\begin{aligned}
-|z_1|^2 - |z_2|^2 + &\sum_{|\alpha| \geq 2, (\alpha_1 - 1)k_1 + \alpha_2 k_2 = 0} \text{Re } q_\alpha^1 z_1^{\alpha_1} \overline{z_1} z_2^{\alpha_2} \\
&+ \sum_{|\alpha| \geq 2, \alpha_1 k_1 + (\alpha_2 - 1)k_2 = 0} \text{Re } q_\alpha^2 z_1^{\alpha_1} z_2^{\alpha_2} \overline{z_2} + R(e^{i\theta}) \leq 0,
\end{aligned}
\tag{6.2}
$$

where $R(e^{i\theta})$ are harmonic terms with some common period. Integrating (6.2) in θ over such a period causes the term $R(e^{i\theta})$ to disappear, and we get a new expression

$$
\begin{aligned}
-|z_1|^2 - |z_2|^2 + &\sum_{|\alpha| \geq 2, (\alpha_1 - 1)k_1 + \alpha_2 k_2 = 0} \text{Re } q_\alpha^1 z_1^{\alpha_1} \overline{z_1} z_2^{\alpha_2} \\
&+ \sum_{|\alpha| \geq 2, \alpha_1 k_1 + (\alpha_2 - 1)k_2 = 0} \text{Re } q_\alpha^2 z_1^{\alpha_1} z_2^{\alpha_2} \overline{z_2} \leq 0.
\end{aligned}
\tag{6.3}
$$

This means that the vector field

$$G^{(k_1, k_2)}(z) = (-z_1 + \sum_{|\alpha| \geq 2, (\alpha_1 - 1)k_1 + \alpha_2 k_2 = 0} q_\alpha^1 z^\alpha, -z_2 + \sum_{|\alpha| \geq 2, \alpha_1 k_1 + (\alpha_2 - 1)k_2 = 0} q_\alpha^2 z^\alpha),$$

is again a Herglotz vector field associated with the class $\mathcal{M}$.

6.2 Slice Reduction

Let $\|v\| = 1$. Let $G(z)$ be an autonomous Herglotz vector field associated with the class $\mathcal{M}$. For $\zeta \in \mathbb{D}$, let

$$-\zeta p_v(\zeta) = \langle G(\zeta v), v \rangle.$$

It is easy to see that $p_v(\zeta) = 1 + \tilde{p}_v(\zeta)$ belongs to the Carathéodory class; in particular, (see, e.g., [12]), its coefficients are bounded by 2. A direct computation gives

$$p_v(\zeta) = 1 - \sum_{m=1}^{\infty} \left(\sum_{|\alpha|=m+1} (q_\alpha^1 \overline{v_1} + q_\alpha^2 \overline{v_2}) v^\alpha \right) \zeta^m.$$

In particular, for all $m \in \mathbb{N}$, $m \geq 1$,

$$\sup_{\|v\|=1} \left| \sum_{|\alpha|=m+1} (q_\alpha^1 \overline{v_1} + q_\alpha^2 \overline{v_2}) v^\alpha \right| \leq 2. \tag{6.4}$$

This condition is necessary but not sufficient for p_v to belong to the Carathéodory class. Observe that by [12, Corollary 2.3], if for some $\|v\| = 1$ and $m \geq 1$,

$$\left| \sum_{|\alpha|=m+1} (q_\alpha^1 \overline{v_1} + q_\alpha^2 \overline{v_2}) v^\alpha \right| = 2,$$

then

$$p_v(\zeta) = \sum_{l=1}^{m} t_m \frac{e^{i\theta + 2\pi i l/m} + z}{e^{i\theta + 2\pi i l/m} - z}$$

for some $\theta \in \mathbb{R}$ and $t_j \geq 0$ with $\sum t_j = 1$.

A necessary and sufficient condition for p_v to belong to the Carathéodory class is the following ([12, Thm. 2.4]). For all $m \geq 1$,

$$\sum_{k=0}^{m} \sum_{l=0}^{m} \left(\sum_{|\alpha|=k-l+1} (q_\alpha^1 \overline{v_1} + q_\alpha^2 \overline{v_2}) v^\alpha \right) \lambda_k \overline{\lambda_l} \geq 0, \tag{6.5}$$

for all $\lambda_0, \ldots, \lambda_m \in \mathbb{C}$, with the convention that for $k - l + 1 \leq -2$,

$$\left(\sum_{|\alpha|=k-l+1} (q_\alpha^1 \overline{v_1} + q_\alpha^2 \overline{v_2}) v^\alpha \right) = \overline{\left(\sum_{|\alpha|=l-k-1} (q_\alpha^1 \overline{v_1} + q_\alpha^2 \overline{v_2}) v^\alpha \right)}$$

and $\sum_{|\alpha| \leq 1} (q_\alpha^1 \overline{v_1} + q_\alpha^2 \overline{v_2}) v^\alpha = 2$.

7 Coefficient Bounds: $q_{m,0}^1$ and $b_{m,0}^1$

Let $G(z)$ be an autonomous Herglotz vector field associated with the class $\mathcal{M}$ and fix $m \in \mathbb{N}$, $m \geq 2$. Using the trick of harmonic decoupling, consider $G^{(0,1)}(z)$. Then

$$G^{(0,1)}(z_1, z_2) = \left(-z_1 + \sum_{m \geq 2} q_{m,0}^1 z_1^m, -z_2 + \sum_{m \geq 2} q_{m,1}^2 z_1^m z_2\right).$$

From (6.4) we obtain for all $m \geq 2$,

$$\sup_{\|v\|=1} \left| q_{m,0}^1 |v_1|^2 + q_{m-1,1}^2 |v_2|^2 \right| |v_1^{m-1}| \leq 2 \tag{7.1}$$

Taking $v_1 = 1$, $v_2 = 0$, we obtain

$$|q_{m,0}^1| \leq 2. \tag{7.2}$$

This bound is sharp, as can be seen by considering the autonomous Herglotz vector field $G(z_1, z_2) = (-z_1(1 + z_1)(1 - z_1)^{-1}, -z_2)$.

Now, for $m = 2$, from (5.1), we obtain

$$a_{2,0}^1(t) = e^{-t} \int_0^t e^{-\tau} q_{2,0}^1(\tau) d\tau.$$

By (7.2) and (5.2), we then have for all $f \in S^0$,

$$|b_{2,0}^1| \leq 2.$$

The bound is sharp, as one sees by considering the map $(k(z_1), z_2)$, where $k(z_1)$ is the Koebe function in $\mathbb{D}$.

A similar bound for $|b_{m,0}^1|$ is not known.

Question 7.1 Is it true that $|b_{m,0}^1| \leq m$ for all $f \in S^0$ and $m \in \mathbb{N}$, $m \geq 3$?
Note that if the bound is correct, it is sharp as one sees from the function $(k(z_1), z_2)$, where $k(z_1)$ is the Koebe function in $\mathbb{D}$.

8 Coefficient Bounds: $q_{0,m}^1$ and $b_{0,m}^1$

Let $G(z)$ be an autonomous Herglotz vector field associated with the class $\mathcal{M}$. Fix $m \in \mathbb{N}$, $m \geq 2$. Using the decoupling harmonic terms trick, consider the vector field $G^{(m,1)}$, given by

$$G^{(m,1)}(z_1, z_2) = \left(-z_1 + q_{0,m}^1(t) z_2^m, -z_2\right).$$

Since $-G^{(m,1)} \in \mathcal{M}$, imposing the condition $\mathsf{Re}\,\langle G^{(m,1)}(z), z \rangle \leq 0$, we get

$$-|z_1|^2 - |z_2|^2 + \mathsf{Re}\,q_{0,m}^1 \overline{z_1} z_2^m \leq 0.$$

Setting $z_1 = xe^{i(\theta+\eta)}$, $z_2 = ye^{i\theta/m}$ with $x, y \geq 0$ and $q_{0,m}^1 e^{-i\eta} = |q_{0,m}^1|$, we obtain the equivalent equation

$$-x^2 - y^2 + |q_{0,m}^1| xy^m \leq 0, \quad x, y \geq 0, x^2 + y^2 \leq 1.$$

Using the method of Lagrange multipliers, one checks easily that the maximum for the function $(x, y) \mapsto -x^2 - y^2 + |q_{0,m}^1| xy^m$ under the constraint $x, y \geq 0, x^2 + y^2 \leq 1$ is attained at the point $x = \frac{1}{\sqrt{(1+m)}}, y = \sqrt{\frac{m}{(1+m)}}$. Hence the previous inequality is satisfied if and only if

$$|q_{0,m}^1| \leq \frac{(1+m)^{\frac{m+1}{2}}}{m^{\frac{m}{2}}}. \tag{8.1}$$

Note that (8.1) gives the sharp bound for the coefficients $q_{0,m}^1$ in the class $\mathcal{M}$.

As before, for $m = 2$, from (5.1), (7.2) and (5.2) we then have for all $f \in S^0$,

$$|b_{0,2}^1| \leq \frac{3\sqrt{3}}{2}.$$

In particular, the map

$$\Phi : (z_1, z_2) \mapsto (z_1 + \frac{3\sqrt{3}}{2} z_2^2, z_2) \in S^0,$$

is a *bounded* support point. Note also that Φ is an automorphism of $\mathbb{C}^2$ and provides an affirmative answer to Question 4.1 for this automorphism.

This result, together with the germinal idea of decoupling harmonic terms, was proved in [3].

Question 8.1 Is it true that

$$|b_{0,m}^1| \leq u_m := \frac{(1+m)^{\frac{m+1}{2}}}{m^{\frac{m}{2}}}$$

for all $f \in S^0$ and $m \in \mathbb{N}, m \geq 3$?

Note that if the bound is correct, it is sharp (consider the function $(z_1 + u_m z_2^m, z_2)$).

Acknowledgement Filippo Bracci was supported by the ERC grant "HEVO - Holomorphic Evolution Equations" no. 277691.

References

1. L. Arosio, F. Bracci, E.F. Wold, Solving the Loewner PDE in complete hyperbolic starlike domains of $\mathbb{C}^N$. Adv. Math. **242**, 209–216 (2013)
2. L. Arosio, F. Bracci, E.F. Wold, Embedding univalent functions in filtering Loewner chains in higher dimension. Proc. Am. Math. Soc. **143**, 1627–1634 (2015)
3. F. Bracci, Shearing process and an example of a bounded support function in $S^0(\mathbb{B}^2)$. Comput. Methods Funct. Theory **15**, 151–157 (2015)
4. F. Bracci, M.D. Contreras, S. Díaz-Madrigal, Evolution families and the Loewner equation II: complex hyperbolic manifolds. Math. Ann. **344**, 947–962 (2009)
5. F. Bracci, I. Graham, H. Hamada, G. Kohr, Variation of Loewner chains, extreme and support points in the class S^0 in higher dimensions. Constr. Approx. **43**, 231–251 (2016)
6. F. Forstnerič, Interpolation by holomorphic automorphisms and embeddings in $\mathbb{C}^n$. J. Geom. Anal. **9**, 93–117 (1999)
7. I. Graham, G. Kohr, *Geometric Function Theory in One and Higher Dimensions* (Marcel Dekker, Inc., New York, 2003)
8. I. Graham, H. Hamada, G. Kohr, Parametric representation of univalent mappings in several complex variables. Can. J. Math. **54**, 324–351 (2002)
9. I. Graham, G. Kohr, J.A. Pfaltzgraff, The general solution of the Loewner differential equation on the unit ball of $\mathbb{C}^n$, in *Complex Analysis and Dynamical Systems II*. Contemporary Mathematics, vol. 382 (American Mathematical Society, Providence, RI, 2005), pp. 191–203
10. I. Graham, H. Hamada, G. Kohr, M. Kohr, Extreme points, support points, and the Loewner variation in several complex variables. Sci. China Math. **55**, 1353–1366 (2012)
11. M. Iancu, Some applications of variation of Loewner chains in several complex variables. J. Math. Anal. Appl. **421**, 1469–1478 (2015)
12. C. Pommerenke, Univalent functions. Studia Mathematica/Mathematische Lehrbücher, Band XXV. Vandenhoeck and Ruprecht, Göttingen (1975)
13. O. Roth, Pontryagin's maximum principle for the Loewner equation in higher dimensions. Can. J. Math. **67**, 942–960 (2015)
14. A.C. Schaeffer, D.C. Spencer, *Coefficient Regions for Schlicht Functions*. American Mathematical Society Colloquium Series (American Mathematical Society, New York, NY, 1950)
15. S. Schleissinger, On support points of the class $S^0(B^n)$. Proc. Am. Math. Soc. **142**, 3881–3887 (2014)

Some Unsolved Problems About Condenser Capacities on the Plane

Vladimir N. Dubinin

Dedicated to the memory of my friend Sasha Vasil'ev

Abstract In the paper we pose fourteen open problems of Potential Theory involved the conformal capacity of condensers with three and more plates, the logarithmic capacity, the relative capacity and extremal decompositions of the unit disk or the Riemann sphere. All problems are closely related to various applications in Geometric Function Theory of a complex variable.

Keywords Capacity of condenser • Logarithmic capacity • Relative capacity • Symmetrization • Asymptotic formulae • Extremal decompositions

2010 Mathematics Subject Classification Primary: 31A15, 31C15, 31C20; Secondary: 30C85, 30C70

1 Introduction

The effectiveness of the potential-theoretic approach to the solution of the extremal problems in Geometric Function Theory is well-known. In that regard, the geometric transformations of sets and condensers, including symmetrization [6], are of special importance. In the survey paper [7] the symmetrization method in the complex plane was considered, and some open problems were formulated. Fifteen years later we supplemented this list of open problems in the book [9] (see, also [14]). Three relatively simple problems 7.1, 8.1, and 9.1 from [7] were soon there after solved [25, 27, 35]. All other problems remain unresolved. Most problems presented

V.N. Dubinin (✉)
Far Eastern Federal University, Vladivostok, Russia

Institute of Applied Mathematics, Vladivostok, Russia
e-mail: dubinin@iam.dvo.ru

© Springer International Publishing AG 2018   81
M. Agranovsky et al. (eds.), *Complex Analysis and Dynamical Systems*,
Trends in Mathematics, https://doi.org/10.1007/978-3-319-70154-7_5

in this paper are of more recent origin. All problems are closely associated with applications in the geometric theory of functions of a complex variable. In particular, considerable attention is paid to the problems about condensers with three and more plates. A first version of such condenser showed up in [8]. Subsequently, these condensers proved useful for studying of the metric properties of sets (see, for example, [14, Ch. 5]), in addressing the extremal decomposition problems [14, Ch. 6], for establishing the distortion theorems [18, 20], in obtaining the geometric estimates of the Schwarzian derivative of a holomorphic function [14, 15, Ch. 7.6], the inequalities for harmonic measures [16], and the inequalities for polynomials and rational functions [11, 17]. We hope that solution of our problems stated here will lead to a progress in the application of condenser capacities in the above areas.

2 Generalized Condensers

The notion of capacity of a condenser with two plates and the problem of the behavior of the capacity of a condenser under various transformations of its plates go back to the papers of Pólya and Szegő (see [32] as well as [14]). The study of the behavior of the capacity of condensers with three or more plates is motivated by numerous applications of such condensers in geometric function theory (see, for example, [14]). Let us pass to formal statements. Suppose that B is a domain on the complex sphere $\overline{\mathbb{C}}_z$. In order to distinguish the boundary points of B and not to overcomplicate the exposition of the main ideas, we confine ourselves to finitely connected domains. Let $\overline{B}$ denote the compactification of B by Carathéodory's prime ends and let the boundary ∂B be a collection of the prime ends. A neighborhood is any open set in $\overline{B}$. When this cannot lead to a confusion, we will make no distinction between the elements of $\overline{B}$ corresponding to the inner points of B and these points. We also use the same notation for the support of an accessible boundary point and the boundary point itself. If B is a Jordan domain, $\overline{B}$ and ∂B defined above agree with the usual closure and boundary and a neighborhood in $\overline{B}$ is the intersection of $\overline{B}$ with an open set in $\overline{\mathbb{C}}_z$. A *generalized condenser* (or, simply, *condenser* in what follows) is a triple $C = (B, \, \mathcal{E}, \, \Delta)$, where $\mathcal{E} = \{E_k\}_{k=1}^n$ is a collection of closed pairwise disjoint sets E_k, $k = 1, \ldots, n$, in $\overline{B}$ and $\Delta = \{\delta_k\}_{k=1}^n$ is collection of real numbers. The sets E_k are called the plates of the condenser C. The capacity cap C of C is defined as the infimum of the Dirichlet integral

$$\int \int_B |\nabla v|^2 dx dy, \quad z = x + iy,$$

taken over all admissible functions $v : \overline{B} \to \mathbb{R}$, i.e. continuous real-valued functions in $\overline{B}$ satisfying the Lipschitz condition in a neighborhood of every finite point of B possibly excluding a finite number of such points and taking the value δ_k in a neighborhood of the plate E_k, $k = 1, \ldots, n$. Condenser capacity enjoys some nice properties such as monotonicity, the conformal invariance, the composition principles and other (see details in [14]). In the classical case, we have

$$\operatorname{cap}(B, \{E_1, E_2\}, \{0, 1\}) = 1/\lambda(\Gamma),$$

where B is a domain and $\lambda(\Gamma)$ is the extremal length of the family Γ of curves in B which join E_1 and E_2. This equality is almost certainly due to Beurling [1, pp. 65, 81].

Problem 2.1 For a condenser C with three or more plates express $\operatorname{cap} C$ through the extremal length of some family (or families) of curves.

It is possible that Jenkins' approach [23, 29] may be useful when addressing Problem 2.1.

Assume that a condenser $C = (\overline{\mathbb{C}}_z, \{E_1, E_2\}, \{0, 1\})$ has a connected field $\overline{\mathbb{C}}_z \setminus (E_1 \cup E_2)$ and its plates E_1 and E_2 are infinite sets. For an arbitrary ordered quadruple z_1, z_2, z_3, z_4 on the Riemann sphere $\overline{\mathbb{C}}_z$, where $z_1 \neq z_3$ and $z_2 \neq z_4$, we set

$$|z_1, z_2, z_3, z_4| = |\varphi(z_1)/\varphi(z_3)|,$$

where φ is a Möbius transformation of the sphere $\overline{\mathbb{C}}$ such that $\varphi(z_2) = 0$ and $\varphi(z_4) = \infty$. Let $a_1, \ldots, a_n$ be some points in the plate E_1 and $b_1, \ldots, b_n$ some points in E_2. We set

$$W_n = \inf \binom{n}{2}^{-1} \sum_{1 \le i < j \le n} \log |a_i, b_j, b_i, a_j|.$$

The sequence $\{W_n\}$ is nondecreasing. Denote by $\operatorname{md} C$ the (finite or infinite) limit of the sequence $\{W_n\}$. In 1967 Bagby [3] proved the following equality

$$\operatorname{cap} C = \frac{2\pi}{\operatorname{md} C}.$$

Problem 2.2 Give a geometric interpretation of the capacity of an arbitrary condenser

$$C = (\overline{\mathbb{C}}_z, \{E_k\}_{k=1}^n, \{\delta_k\}_{k=1}^n)$$

which coincides with Bagby's equality in the case when C has two plates.

There are many geometric transformations of a condenser with two plates under which its capacity does not increase [14]. What can be said about a condenser with three plates? We have some very special transformations which decrease the capacity of such condenser [8, 10]. For example, let us consider the Steiner symmetrization. Denote by $l(x)$ the vertical line $\operatorname{Re} z = x$, $-\infty < x < +\infty$. Let B be an open subset of $\overline{\mathbb{C}}_z$. By the *Steiner symmetrization of B with respect to the real axis* we mean the transformation of this set into the mirror-symmetric set $\operatorname{St} B$

defined by

$$\mathrm{St}\,B = \{x + iy : \ B \cap l(x) \neq \emptyset, \ 2|y| < m(B \cap l(x))\},$$

where $m(\cdot)$ is the linear Lebesgue measure. For a closed set $E \subset \mathbb{C}_z$ its Steiner symmetrization is

$$\mathrm{St}\,E = \{x + iy : \ E \cap l(x) \neq \emptyset, \ 2|y| \leq m(E \cap l(x))\}.$$

Consider the condenser

$$C = (\overline{\mathbb{C}}_z, \ \{E_1, E_2, E_3\}, \ \{1, -1, 0\})$$

such that $E_1 \subset \{z : \ \mathrm{Re}\,z < 0\}$, $E_2 = \{z : \ -\overline{z} \in E_1\}$ and let E_3 be symmetric with respect to the imaginary axis. Put

$$\mathrm{St}\,C = (\overline{\mathbb{C}}_z, \ \{\mathrm{St}\,E_1, \ \mathrm{St}\,E_2, \ \overline{\mathbb{C}}_z \setminus \mathrm{St}\,(\overline{\mathbb{C}}_z \setminus E_3)\}, \ \{1, -1, 0\}).$$

Then it is easy to see that

$$\mathrm{cap}\,C \geq \mathrm{cap}\,\mathrm{St}\,C \tag{2.1}$$

(see a special case of (2.1) in [8]).

Problem 2.3 Is it true that (2.1) still holds if the plate E_3 is not symmetric?

If the answer to the above question were in the affirmative, we would have a geometric estimates of the Schwarzian derivative with a weaker restriction on the holomorphic functions [14, Theorem 7.35]. We shall notice, that the symmetry of the plates E_1 and E_2 is an essential condition for Problem 2.3.

3 Asymptotic Formulae

Let $g_B(z, z_0)$ denote (the classical) *Green function* of a connected component of an open set $B \subset \overline{\mathbb{C}}_z$ with pole at $z_0 \neq \infty$, and let

$$r(B, z_0) = \exp\{\lim_{z \to z_0} [g_B(z, z_0) + \log|z - z_0|]\}$$

be the *inner radius* of B with respect to the point z_0. In [12] the following statement is proved. Let B be a finitely connected domain on $\overline{\mathbb{C}}_z$ with nondegenerate boundary components, $B \neq \overline{\mathbb{C}}_z$, $\{z_k\}_{k=1}^n$ be a set of distinct finite points of the domain B, $\Delta = \{\delta_k\}_{k=0}^n$ be a set of real numbers such that $\delta_0 = 0$, $\delta_k \neq 0$, $k = 1, \ldots, n$, and let

μ_k, ν_k, $k = 1, \ldots, n$, be arbitrary positive numbers. Suppose that the continua

$$E_k(r), \quad r > 0, \quad k = 1, \ldots, n,$$

have the logarithmic capacity

$$\operatorname{cap} E_k(r) = \mu_k r^{\nu_k} (1 + o(1)) \text{ as } r \to 0$$

and

$$E_k(r) \to z_k, \quad r \to 0, \quad k = 1, \ldots, n.$$

Then the following asymptotic formula holds:

$$\operatorname{cap}\left(B, \{\partial B, E_1(r), \ldots, E_n(r)\}, \Delta\right) = -2\pi \sum_{k=1}^{n} \frac{\delta_k^2}{\nu_k \log r} - \tag{3.1}$$

$$-2\pi \left\{ \sum_{k=1}^{n} \frac{\delta_k^2}{\nu_k^2} \log \frac{r(B, z_k)}{\mu_k} + \sum_{k=1}^{n} \sum_{\substack{l=1 \\ l \neq k}}^{n} \frac{\delta_k \delta_l}{\nu_k \nu_l} g_B(z_k, z_l) \right\} \left(\frac{1}{\log r}\right)^2$$

$$+ o\left(\left(\frac{1}{\log r}\right)^2\right), \quad r \to 0.$$

Problem 3.1 Prove formula (3.1) without the connectivity condition on the closed sets $E_k(r)$, $k = 1, \ldots, n$, $n \geq 2$.

For $n = 1$ Problem 3.1 was solved by Aseev and Lazareva (see [31]).

Problem 3.2 Show that formula (3.1) is valid for the condenser

$$(B(r), \{\partial B(r), E_1(r), \ldots, E_n(r)\}, \Delta)$$

where $E_k(r)$, $k = 1, \ldots, n$, are not necessarily connected, and the kernel convergence $B_r \to B$ takes place as $r \to 0$.

Assume that $\{z_k\}_{k=1}^{n}$ is a collection of different finite points in $\overline{\mathbb{C}}_z$, $\Delta = \{\delta_k\}_{k=1}^{n}$ is a collection of real numbers such that at least two of them are different, and μ_k, ν_k, $k = 1, \ldots, n$, are positive numbers. Let $\{E_k(r)\}_{r>0}$ be a family of continua satisfying the following conditions: the logarithmic capacity $\operatorname{cap} E_k(r)$ is representable in the form $\mu_k r^{\nu_k}(1 + o(1))$ as $r \to 0$ and $E_k(r) \to z_k$ as $r \to 0$, $k = 1, \ldots, n$. Introduce the notation:

$$\delta = \left(\sum_{k=1}^{n} \frac{\delta_k}{\nu_k}\right) \bigg/ \sum_{k=1}^{n} \frac{1}{\nu_k}.$$

The following asymptotic formula for the condenser capacity holds:

$$\text{cap}\,(\overline{\mathbb{C}},\,\{E_k(r)\}_{k=1}^n,\,\Delta) = -2\pi \sum_{k=1}^n \frac{(\delta_k - \delta)^2}{\nu_k \log r} + \tag{3.2}$$

$$+R\left(\frac{1}{\log r}\right)^2 + o\left(\left(\frac{1}{\log r}\right)^2\right) \quad \text{as } r \to 0,$$

where

$$R = 2\pi \left\{ \sum_{k=1}^n \frac{(\delta_k - \delta)^2}{\nu_k^2} \log \mu_k + \sum_{k=1}^n \sum_{\substack{l=1 \\ l \neq k}}^n \frac{(\delta_k - \delta)(\delta_l - \delta)}{\nu_k \nu_l} \log |z_k - z_l| \right\}$$

(see [13]).

Problem 3.3 Prove formula (3.2) for the closed sets $E_k(r)$, $k = 1, \ldots, n$, which are not necessarily connected.

4 Extremal Decompositions

The extremal decomposition problems have a long history and now form an important part of the geometric function theory (see, for example, [14, 29, 36, 37]). We propose some special cases of such problems concerned with the estimates of the condenser capacities using (3.1).

Problem 4.1 (9.2 in [7] and 16 in [14]) Consider the product

$$r^\alpha(B_0, a_0) \prod_{k=1}^n r(B_k, a_k),$$

where $B_0, B_1, \ldots, B_n$ $(n \geq 2)$ are pairwise disjoint domains in $\overline{\mathbb{C}}_z$, $a_0 = 0$, $|a_k| = 1$, $k = 1, \ldots, n$, and $0 < \alpha \leq n$. Show that it attains its maximum at a configuration of domains B_k and points a_k possessing rotational n-symmetry.

The proof is due to this author for $\alpha = 1$ [5] and to Kuz'mina [28] for $0 < \alpha < 1$ and simply connected domains B_k, $k = 1, \ldots, n$. Kovalev [26] solved this problem under the additional assumption that the angles between neighboring line segments $[0, a_k]$ do not exceed $2\pi/\sqrt{\alpha}$. Subsequently, the considerable progress in the solution of Problem 4.1 has been achieved by Ukrainian mathematicians (see the article [4] and the bibliography in there). However, in its full generality Problem 4.1 is still open.

Problem 4.2 (17 in [14]) Find the maximum value of the product

$$\prod_{k=1}^{n} r(B_k, a_k)$$

over various systems of nonoverlapping domains $B_k \subset \overline{\mathbb{C}}_z$ and points $a_k \in B_k \cap [-1, 1]$, $k = 1, \ldots, n$, $n > 4$.

Let us introduce the notation:

$$U = \{z : |z| < 1\}, \; C(\lambda) = \text{cap}\,(\overline{\mathbb{C}}_z, \{[0, \lambda]\}, \{z : |z| \geq 1\}, \{0, 1\}\}), \; 0 < \lambda < 1.$$

Problem 4.3 For two distinct fixed points a_1, $a_2 \in U$ and a constant c, $C(|a_2 - a_1|/|1 - \overline{a}_1 a_2|) < c < \infty$, find the maximum value of the product

$$r(B_1, a_1) r(B_2, a_2)$$

over all couples of nonoverlapping domains $B_k \subset U$, $a_k \in B_k$, $k = 1, 2$, for which there exists a condenser $C = (\overline{\mathbb{C}}_z, \{E_0, E_1\}, \{0, 1\})$ such that $\text{cap}\, C \leq c$ and $E_0 \supset B_1 \cup B_2$, $E_1 \supset \overline{\mathbb{C}}_z \setminus U$.

Problem 4.3 has been solved for $c = 2\pi/\log(1/\rho)$ where $\rho + 1/\rho = 2|1 - \overline{a}_1 a_2|/|a_2 - a_1|$, $0 < \rho < 1$ [14, corollary 6.1].

5 Set Capacities

Let E be a closed infinite set in the plane $\mathbb{C}_z$, and $\text{cap}\, E$ be the logarithmic capacity of this set. If B is a connected component of the complement $\overline{\mathbb{C}}_z \setminus E$, containing the point $z = \infty$, then

$$\text{cap}\,(B, \{\partial B, \{z : |z| \geq 1/r\}\}, \{0, 1\})$$

$$= -\frac{2\pi}{\log r} + \frac{2\pi \log \text{cap}\, E}{(\log r)^2} + o\left(\left(\frac{1}{\log r}\right)^2\right), \; r \to 0$$

[14, Sec. 2.4]. We mention the important properties of the logarithmic capacity such as monotonicity and contractibility. The latter means that if a function T maps a compact set E to a compact set $T(E)$ so that for any points z, $\zeta \in E$ we have

$$|T(z) - T(\zeta)| \leq |z - \zeta|,$$

then

$$\text{cap}\, E \geq \text{cap}\, T(E).$$

It is interesting to compare the inner radius of a bounded domain D to the logarithmic capacity of the closure $\overline{D}$. Both values do not decrease under extension of the domain D but when we carry out a symmetrization (for instance Steiner symmetrization) of the domain D the inner radius $r(D, \cdot)$ does not decrease, while $\text{cap}\,\overline{D}$ does not increase. In [2], among other things, the *linear radial transformation* was introduced as the mapping from a compact set $E \subset \mathbb{C}_z$ to the set

$$LRE = \{re^{i\theta} : 0 \le r \le m(\theta, E), \ 0 \le \theta \le 2\pi\},$$

where $m(\theta, E)$ is the linear measure of the intersection of E with the ray $\arg z = \theta$. In some special cases the contractibility property yields the inequality

$$\text{cap}\,E \ge \text{cap}\,LRE. \tag{5.1}$$

For example, inequality (5.1) is true if one of the following conditions holds:

   (i) The set E lies on several rays with initial point at the origin which form the angles of at least $\pi/2$;
  (ii) The set E lies on the rays $\arg z = \pm\theta_1$ and $\arg z = \pm\theta_2$, where $0 \le \theta_1 \le \theta_2 - \pi/2 \le \pi/2$, or E is mirror-symmetric relative to the real axis;
 (iii) The set

$$E = \bigcup_{k=1}^{n} \{re^{i\theta_k} : r \in F\},$$

where $\theta_1, \theta_2, \ldots, \theta_n$ are real numbers and F is a compact subset of the nonnegative half-axis.

Problem 5.1 (1 in [9]) Is it true that (5.1) holds for any compact set E?
Note that

$$m(\theta, E) \ge M(\theta, E) := \lim_{\rho \to 0} \rho \exp\left(\int_{H(\rho, \theta, E)} \frac{dr}{r}\right),$$

where $H(\rho, \theta, E) = \{r : re^{i\theta} \in E, \ \rho \le r < \infty\}$, and for $ME := \{re^{i\theta} : 0 \le r \le M(\theta, E), \ 0 \le \theta \le 2\pi\}$ we have

$$\text{cap}\,E \ge \text{cap}\,ME$$

for any compact set $E \subset \mathbb{C}_z$ [9, corollary 3.4].
Let E be an arbitrary closed subset of the unit circle $|z| = 1$ and let l_k be the linear measure of the intersection of E with the arc $\{z = e^{i\theta} : |\theta - 2\pi k/n| \le \pi/n\}$, $k = 1, \ldots, n$. Then

$$\text{cap}\,E \ge \prod_{k=1}^{n} (\sin(nl_k/4))^{2/n^2}.$$

The equality holds only in the case of even n and E consisting of $n/2$ equal arcs with centers either at the points $\exp(i(\pi/n + 4\pi k/n))$, $k = 1, \ldots, n/2$, or at the points $\exp(i(-\pi/n + 4\pi k/n))$, $k = 1, \ldots, n/2$, [7, Theorem 2.9].

Problem 5.2 (7.2 in [7], 14 in [14]) Under the above conditions find a sharp lower bound for the logarithmic capacity in the case of odd n.

The next result [33, p. 348] (cf. [22, 24]) is an analogue of Löwner's lemma. Let f map U conformally into U with $f(0) = 0$. If E is a Borel set on ∂U such that the angular limit of $f(z)$ exists and $f(z) \in \partial U$ for $z \in E$, then

$$\operatorname{cap} f(E) \geq \frac{\operatorname{cap} E}{\sqrt{|f'(0)|}}. \tag{5.2}$$

Let us take a look at a discrete analogue of the inequality (5.2). We recall that n-diameter of a compact set E is defined as

$$d_n(E) = \max_{z_k, z_l \in E} \left\{ \prod_{k=1}^{n} \prod_{\substack{l=1 \\ l \neq k}}^{n} |z_k - z_l| \right\}^{\frac{1}{n(n-1)}}, \quad n \geq 0.$$

It's a known fact that

$$\lim_{n \to \infty} d_n(E) = \operatorname{cap} E.$$

If E is a compact set on ∂U, f is defined above, and there exists a constant M such that $|f'(z)| \leq M$ for $z \in E$ (f' is the angular derivative), then the following discrete analogue of (5.2) holds:

$$d_n(f(E)) \geq \frac{d_n(E)}{|f'(0)|^{\frac{n}{2(n-1)}} M^{\frac{1}{(n-1)}}}$$

(see [19, Sec. 2]).

Problem 5.3 Find a discrete analogue of the inequality [34, Theorem 9.53]:

$$\operatorname{cap} F(E) \geq (\operatorname{cap} E)^2.$$

Here E is a compact set on ∂U, and F is univalent function of the form $F(z) = z + b_0 + b_1 z^{-1} + \ldots$, for $|z| > 1$, such that the angular limit $F(z)$ exists for $z \in E$.

Let D be a domain in the extended complex plane $\overline{\mathbb{C}}_z$, and let z_0 be a finite accessible boundary point of the domain D. Suppose that in some neighborhood of z_0 the boundary ∂D is represented by an analytic arc γ. Then for every set E relatively closed with respect to D and such that $\operatorname{dist}(E, z_0) > 0$, the following expansion holds:

$$\frac{r(D \setminus E, z)}{r(D, z)} = 1 - c|z - z_0|^2 + o(|z - z_0|^2), \quad z \in D \setminus E, \ z \to z_0, \tag{5.3}$$

where the convergence of z to the point z_0 takes place along a path, perpendicular to the arc γ at z_0 and $c \geq 0$ is a constant, depending only on the set E, the domain D and the point z_0 [21, theorem 2.1]. The asymptotic expansion in the case $z_0 = \infty$ is contained in (5.3) under the change of variable $z - z_0 \mapsto 1/z$. We say that relcap $E := c/2$ is *the relative capacity of the set E* with respect to the domain D at the point z_0. In the case when the domain D is the upper half plane $\operatorname{Im} z > 0$ and the point $z_0 = \infty$ we have

$$\text{relcap}\, E = \text{hcap}\, E,$$

where hcap E is *the half-plane capacity* of the set E [30]. The concept of the relative capacity can be useful for obtaining the geometric estimates of the Schwarzian derivative [21].

From the proof of Theorem 2.1 in [21] it is clear that the requirement of the analyticity of the arc γ in (5.3) can be weakened (see [21, Remark 2.5]).

Problem 5.4 Find necessary and sufficient conditions for the boundary ∂D near the point z_0 (instead of analyticity of γ) under which the expansion (5.3) holds.

6 A General Question

In [6] the generalized capacity of a set and condenser in n-space $\mathbb{R}^n$ is introduced as a lower bound of the functional $I(v)$, furnished with some symmetry, where the function v satisfies certain normalizing conditions. It is established that, depending on the type of symmetry of the functional $I(v)$, the generalized capacity does not increase under such geometric transformations of sets and condensers as polarization, Gonchar standardization, Steiner k-dimensional symmetrization, k-dimensional spherical symmetrization and dissymmetrization. These results strengthen many known statements of this kind for the particular cases of capacities. The functional $I(v)$ in [6] depends on the values of the function v and it's first partial derivatives.

Problem 6.1 Build upon approach from [6] to establish the comparison theorems for the capacities associated with the functionals involving the partial derivatives of the second and higher other.

Acknowledgement Vladimir N. Dubinin was supported by the Russian Science Foundation (Grant 14-11-00022).

References

1. L.V. Ahlfors, *Conformal Invariants. Topics in Geometric Function Theory* (McGraw-Hill Book Co., New York, 1973)
2. E.G. Akhmedzyanova, V.N. Dubinin, Radial transformations of sets, and inequalities for the transfinite diameter. Russ. Math. (Izvestiya VUZ. Matematika) **43**(4), 1–6 (1999)
3. T. Bagby, The modulus of a plane condenser. J. Math. Mech. **17**(4), 315–329 (1967)
4. A.K. Bakhtin, I.Y. Dvorak, I.V. Denega, A separating transformation in the problems of extremal decomposition of the complex plane. Dop. NAS Ukraine (12), 7–12 (2015) [in Russian]
5. V.N. Dubinin, The separating transformation of domains and problems on the extremal partition. Analytical theory of numbers and theory of functions. Part 9. Zap. Nauchn. Sem. LOMI **168**, 48–66 (1988) [in Russian]
6. V.N. Dubinin, Capacities and geometric transformations of subsets in n-space. Geom. Funct. Anal. **3**(4), 342–369 (1993)
7. V.N. Dubinin, Symmetrization in the geometric theory of functions of a complex variable. Russ. Math. Surv. **49**(1), 1–79 (1994)
8. V.N. Dubinin, Symmetrization, Green's function, and conformal mappings. J. Math. Sci. (New York) **89**(1), 967–975 (1998)
9. V.N. Dubinin, *Condenser Capacities and Symmetrization in the Geometric Theory of Functions of a Complex Variable* (Dal'nauka, Vladivostok, 2009) [in Russian]
10. V.N. Dubinin, Steiner symmetrization and the initial coefficients of univalent functions. Izv. Math. **74**(4), 735–742 (2010)
11. V.N. Dubinin, Methods of geometric function theory in classical and modern problems for polynomials. Russ. Math. Surv. **67**(4), 599–684 (2012)
12. V.N. Dubinin, Asymptotic behavior of the capacity of a condenser as some of its plates contract to points, in *Mathematical Notes*, vol. 96(2) (Springer, Berlin, 2014), pp. 187–198
13. V.N. Dubinin, On the reduced modulus of the complex sphere. Sib. Math. J. **55**(5), 882–892 (2014)
14. V.N. Dubinin, *Condenser Capacities and Symmetrization in Geometric Function Theory*. Translated from the Russian by Nikolai G. Kruzhilin (Birkhäuser/Springer, Basel, 2014)
15. V.N. Dubinin, Schwarzian derivative and covering arcs of a pencil of circles by holomorphic functions. Math. Not. **98**(6), 920–925 (2015)
16. V.N. Dubinin, Inequalities for harmonic measures with respect to non-overlapping domains. Izv. Math. **79**(5), 902–918 (2015)
17. V.N. Dubinin, The logarithmic energy of zeros and poles of a rational function. Sib. Math. J. **57**(6), 981–986 (2016)
18. V.N. Dubinin, D.B. Karp, Generalized condensers and distortion theorems for conformal mappings of planar domains, in *The Interaction of Analysis and Geometry*, ed. by V.I. Burenkov, T. Iwaniec, S.K.Vodopyanov. Contemporary Mathematics, vol. 424 (American Mathematical Society, Providence, RI, 2007), pp. 33–51
19. V.N. Dubinin, V.Y. Kim, Generalized condensers and boundary distortion theorems for conformal mappings. Dal'nevost. Mat. Zh. **13**(2), 196–208 (2013) [in Russian]
20. V.N. Dubinin, M. Vuorinen, Robin functions and distortion theorems for the regular mapping. Math. Nach. **283**(11), 1589–1602 (2010)
21. V.N. Dubinin, M. Vuorinen, Ahlfors-Beurling conformal invariant and relative capacity of compact sets. Proc. Am. Math. Soc. **142**(11), 3865–3879 (2014)
22. J.A. Jenkins, Some theorems on boundary distortion. Trans. Am. Math. Soc. **81**, 477–500 (1956)
23. J.A. Jenkins, On the existence of certain general extremal metrics. Ann. Math. **66**(3), 440–453 (1957); Tohoku Math. J. **45**(2), 249–257 (1993)
24. Y. Komatu, Uber eine Verscharfung des Lownerschen Hilfssatzes. Proc. Imp. Acad. Jpn. **18**, 354–359 (1942)

25. E.V. Kostyuchenko, The solution of one problem of extremal decomposition. Dal'nevost. Mat. Zh. **2**(1), 13–15 (2001) [in Russian]
26. L.V. Kovalev, On the problem of extremal decomposition with free poles on a circle. Dal'nevost. Mat. Sb. **2**, 96–98 (1996) [in Russian]
27. G.V. Kuz'mina, Connections between different problems on extremal decomposition. J. Math. Sci. (New York) **105**(4), 2197–2209 (2001)
28. G.V. Kuz'mina, The method of extremal metric in extremal decomposition problems with free parameters. J. Math. Sci. (New York) **129**(3), 3843–3851 (2005)
29. G.V. Kuz'mina, Geometric function theory. Jenkins results. The method of modules of curve families. Analytical theory of numbers and theory of functions. Part 31. Zap. Nauchn. Sem. POMI **445**, 181–249 (2016)
30. G.F. Lawler, *Conformally Invariant Processes in the Plane*. Mathematical Surveys and Monographs, vol. 114 (American Mathematical Society, Providence, RI, 2005)
31. O.A. Lazareva, *Capacity Properties of Uniformly Perfect Sets and Condensers: Classical Results and New Studies* (Lambert Academic Publishing, Saarbrücken, 2012)
32. G. Polya, G. Szego, *Isoperimetric Inequalities in Mathematical Physics* (Princeton University Press, Princeton, NJ, 1951)
33. C. Pommerenke, *Univalent Functions, with a Chapter on Quadratic Differentials by Gerd Jensen* (Vandenhoeck & Ruprecht, Gottingen, 1975)
34. C. Pommerenke, *Boundary Behaviour of Conformal Maps* (Springer, Berlin, 1992)
35. A.Y. Solynin, Extremal configurations of certain problems on the capacity and harmonic measure. J. Math. Sci. (New York) **89**(1), 1031–1049 (1998)
36. A.Y. Solynin, Modules and extremal metric problems. St. Petersbg. Math. J. **11**(1), 1–65 (2000)
37. A. Vasil'ev, *Moduli of Families of Curves for Conformal and Quasiconformal Mappings*. Lecture Notes in Mathematics, vol. 1788 (Springer, Berlin, 2002)

Filtration of Semi-complete Vector Fields Revisited

Mark Elin, David Shoikhet, and Toshiyuki Sugawa

We dedicate it to the memory of Sasha Vasil'ev who was involved to the problems described here

Abstract In this note we use a technique based on subordination theory in order to continue the study of filtration theory of semi-complete vector fields.

Keywords Semi-complete vector field • Subordination theory • Holomorphic function

2010 MSC 30C35, 37C10, 30C80

The most significant issue in the generation theory of semigroups (groups) of holomorphic mappings in complex spaces is the study of semi-complete (complete) vector fields. In the study of this object various criteria for a vector field f on a complex manifolds to be semi-complete have been found during last 30 years; see [3]. In this note we focus on the one-dimensional case and discuss some sufficient conditions that provide additional properties of generated semigroups (a detailed discussion concerning these properties can be found in [12] and [1]). We deal with some special subclasses of semi-complete vector fields that have very transparent geometric and dynamical properties. This leads us to the so-called filtration theory of semi-complete vector fields which was introduced and partially studied by using

M. Elin (✉)
ORT Braude College, Karmiel 21982, Israel
e-mail: mark_elin@braude.ac.il

D. Shoikhet
Department of Mathematics, Holon Institute of Technology, Holon 5810201, Israel
e-mail: davidsho@hit.ac.il

T. Sugawa
GSIS, Tohoku University, Sendai 9808579, Japan
e-mail: sugawa@math.is.tohoku.ac.jp

© Springer International Publishing AG 2018
M. Agranovsky et al. (eds.), *Complex Analysis and Dynamical Systems*,
Trends in Mathematics, https://doi.org/10.1007/978-3-319-70154-7_6

some parametrization in [12] and developed in more details in [1]. Also note that we are interested in parametrization methods which determinate subclasses ordered by the inclusions.

Our purpose here is to provide another approach to study some relations which is based on the subordination theory and gives shorter proofs for some results of the above-mentioned works.

The use of a new technique enable us to concretize some previous results as well as to formulate new open problems.

The following notations will be used. We denote by Δ the open unit disk in the complex plane $\mathbb{C}$. The set of holomorphic functions on Δ is denoted by $\mathrm{Hol}(\Delta, \mathbb{C})$, and the set of holomorphic self-maps of Δ by $\mathrm{Hol}(\Delta)$. A function $f \in \mathrm{Hol}(\Delta, \mathbb{C})$ is called a *semi-complete vector field* on Δ if for any $z \in \Delta$ the Cauchy problem:

$$\frac{\partial u}{\partial t} + f(u) = 0 \quad \text{and} \quad u(0) = z \tag{1}$$

has the unique solution $u = u(t)$, $0 \le t < +\infty$. The set of all semi-complete vector fields on Δ will be denoted by $\mathcal{G}$.

Definition 1 Let $\widetilde{\mathcal{G}}$ be a subclass of $\mathcal{G}$. A *filtration* of $\widetilde{\mathcal{G}}$ is a family $\mathfrak{F} = \{\mathfrak{F}_s\}_{s \in [a,b]}$ of subsets of $\widetilde{\mathcal{G}}$, where $a, b \in [-\infty, +\infty]$ with $a < b$, such that $\mathfrak{F}_s \subseteq \mathfrak{F}_t$ whenever $a \le s \le t \le b$. Moreover, we say that

- the filtration $\{\mathfrak{F}_s\}_{s \in [a,b]}$ is *strict* if $\mathfrak{F}_s \subsetneq \mathfrak{F}_t$ for $s < t$; and
- the filtration $\{\mathfrak{F}_s\}_{s \in [a,b]}$ is *exhaustive on* $\widetilde{\mathcal{G}}$ if $\mathfrak{F}_b = \widetilde{\mathcal{G}}$.

We set

$$\partial \mathfrak{F}_\alpha = \mathfrak{F}_\alpha \setminus \bigcup_{\beta < \alpha} \mathfrak{F}_\beta.$$

Note that the filtration $\mathfrak{F}$ is strict if $\partial \mathfrak{F}_\alpha \ne \emptyset$ for $a < \alpha \le b$.

It is known (see [11]) that a function $f \in \mathrm{Hol}(\Delta, \mathbb{C})$ is a semi-complete vector field if and only if it admits the following representation:

$$f(z) = f(0) - \overline{f(0)} z^2 + z p(z), \tag{2}$$

where $p \in \mathrm{Hol}(\Delta, \mathbb{C})$ satisfies $\mathrm{Re}\, p(z) \ge 0$ for all $z \in \Delta$. For simplicity, we shall study filtrations of subclasses of $\mathcal{G}$ determined by a given value at zero, say, $f(0) = 0$. In this case $f(z) = z p(z)$, where p has positive real part by (2).

The class of functions with positive real part and its subclasses play an important role in various fields of Complex Analysis, in particular in Geometric Function Theory and Theory of Dynamical Systems. One of the most important subclasses of it is the Carathéodory class $\mathcal{P}$ consisting of holomorphic functions p of positive real part and normalized by $p(0) = 1$. It is impossible to list all of the subjects where appearance of such functions is crucial, we recall only that classical criteria for starlikeness, spirallikeness and convexity are given in terms of this class. In addition, the well-known Noshiro-Warschawski condition $\operatorname{Re} f'(z) > 0$ is sufficient for the univalence of $f \in \operatorname{Hol}(\Delta, \mathbb{C})$ [2, 5].

As we have already mentioned, if $f(0) = 0$, then $f \in \mathcal{G}$ if and only if $\operatorname{Re} \frac{f(z)}{z} > 0$ for $f \in \operatorname{Hol}(\Delta, \mathbb{C})$. In this notes we concentrate on the subclass $\mathcal{G}_0$ of $\mathcal{G}$ consisting of functions vanishing at zero and normalized by $f'(0) = 1$. To study the asymptotic behavior of dynamical systems, it is very important to examine whether the last inequality can be strengthened. For example, it is of strong interest to find (sharp) $k, \beta \geq 0$ such that $\operatorname{Re} \frac{f(z)}{z} \geq k$ and $|\arg[f(z)/z]| \leq \pi\beta/2$. Recall in this connection that a classical result of Marx and Strohhäcker (see, for example, [7]), asserts that for $f \in \mathcal{G}_0$,

$$f \quad \text{is convex} \quad \Rightarrow \quad \operatorname{Re} \frac{zf'(z)}{f(z)} > \frac{1}{2} \quad \Rightarrow \quad \operatorname{Re} \frac{f(z)}{z} > \frac{1}{2}. \tag{3}$$

To clarify the problem, we introduce for a subclass $\mathcal{G}_1$ of $\mathcal{G}_0$ the following two quantities

$$K(\mathcal{G}_1) = \inf_{f \in \mathcal{G}_1} \inf_{z \in \Delta} \operatorname{Re} \frac{f(z)}{z}$$

and

$$B(\mathcal{G}_1) = \frac{2}{\pi} \inf_{f \in \mathcal{G}_1} \inf_{z \in \Delta} \left| \arg \frac{f(z)}{z} \right|.$$

It can be easily seen that if f satisfies the Noshiro-Warschawski condition, then

$$\operatorname{Re} \frac{f(z)}{z} = \int_0^1 \operatorname{Re} f'(tz)dt > 0,$$

hence, f is a semi-complete vector field. In this case, for any $\alpha \in [0, 1]$, we have

$$\operatorname{Re} \left[\alpha \frac{f(z)}{z} + (1-\alpha)f'(z) \right] \geq 0, \quad z \in \Delta. \tag{4}$$

So, it is natural to consider this convex combination and consider subclasses $\mathfrak{A}_\alpha \subset \mathcal{G}_0$, $\alpha \in [0, 1]$, of all functions $f \in \mathcal{G}_0$ satisfying (4).

Question 2 Does the Noshiro-Warschawski condition (or, more generally, the inclusion $f \in \mathfrak{A}_\alpha$) imply that $\operatorname{Re}(f(z)\bar{z}) \geq k|z|^2$ for some $k > 0$? More precisely, can one compute the quantity $K(\mathfrak{A}_\alpha)$?

To study this and other questions related to the classes $\mathfrak{A}_\alpha$, recall some known results concerning the subordination theory.

A function $f \in \operatorname{Hol}(\Delta, \mathbb{C})$ is said to be subordinate to another $g \in \operatorname{Hol}(\Delta, \mathbb{C})$ and written as $f \prec g$ or $f(z) \prec g(z)$ if $f = g \circ \omega$ for an $\omega \in \operatorname{Hol}(\Delta)$ with $\omega(0) = 0$. Note that the inequality $|\omega'(0)| \leq 1$ holds with equality if and only if $\omega(z) = e^{i\theta}z$ for some $\theta \in \mathbb{R}$ by Schwarz's lemma. Since $f'(0) = g'(0)\omega'(0)$, we see the inequality $|f'(0)| \leq |g'(0)|$. For instance, a function $q \in \mathrm{Hol}(\Delta, \mathbb{C})$ with $q(0) = 1$ has positive real part if and only if $q \prec p_1$, where p_1 is defined by

$$p_1(z) = \frac{1+z}{1-z} = 1 + 2z + 2z^2 + 2z^3 + \cdots$$

and maps the unit disk univalently onto the right half-plane $\mathrm{Re}\, w > 0$.

Let γ be a non-zero complex number with $\mathrm{Re}\,\gamma \geq 0$ and $q(z)$ be an analytic function on the unit disk Δ with $q(0) = 1$. Then $q(z)$ can be expressed as a power series of the form

$$q(z) = 1 + \sum_{n=1}^{\infty} b_n z^n, \quad |z| < 1.$$

We observe that the analytic solution on Δ to the differential equation

$$p(z) + \frac{zp'(z)}{\gamma} = q(z)$$

is given by $p = I_\gamma[q]$, where I_γ is a sort of the Bernardi integral operator (see [9]) defined by

$$I_\gamma[q](z) = \gamma z^{-\gamma} \int_0^z \tau^{\gamma-1} q(\tau) d\tau = \gamma \int_0^1 t^{\gamma-1} q(tz) dt$$

$$= \int_0^1 q(s^{1/\gamma} z) ds$$

$$= 1 + \sum_{n=1}^{\infty} \frac{\gamma b_n}{n+\gamma} z^n.$$

Then a theorem of Hallenbeck and Ruscheweyh [6, Theorem 1] can be rephrased as follows.

Theorem 3 *Let $Q(z)$ be a convex univalent function on Δ with $Q(0) = 1$. If $q \prec Q$, then $I_\gamma[q] \prec I_\gamma[Q]$ for $\mathrm{Re}\,\gamma > 0$.*

As an immediate consequence, we obtain the following result for the class $\mathfrak{A}_\alpha$.

Theorem 4 *Let $0 \leq \alpha < 1$. A function f in the class $\mathfrak{A}_\alpha$ satisfies $f(z)/z \prec F_{1/(1-\alpha)}(z)$, where*

$$F_\gamma(z) = I_\gamma[p_1](z) = 1 + 2 \sum_{n=1}^{\infty} \frac{\gamma}{n+\gamma} z^n = 2\,{}_2F_1(1, \gamma; \gamma+1; z) - 1$$

Here, we recall that the Gauss hypergeometric function is defined by

$$_2F_1(a, b; c; z) = \sum_{n=0}^{\infty} \frac{(a)_n (b)_n}{(c)_n n!} z^n, \quad |z| < 1,$$

for $a, b, c \in \mathbb{C}$ with $c \neq 0, -1, -2, \ldots$.

Proof If we put $p(z) = f(z)/z$, then

$$\alpha \frac{f(z)}{z} + (1 - \alpha) f'(z) = p(z) + (1 - \alpha) z p'(z).$$

Thus the condition $f \in \mathfrak{A}_\alpha$ is equivalent to $q(z) := p(z) + (1 - \alpha) z p'(z) \prec p_1(z) = (1 + z)/(1 - z)$. Letting $\gamma = 1/(1 - \alpha)$, the Hallenbeck-Ruscheweyh theorem now implies that $p = I_\gamma[q] \prec I_\gamma[p_1] = F_\gamma$. $\qquad\square$

Note that $F_1(z) = -(2/z) \log(1 - z) - 1$ when $\alpha = 0$ and $F_\infty(z) = p_1(z) = (1 + z)/(1 - z)$ when $\alpha = 1$. We remark that the function F_γ appears in [1] as p_α with $\gamma = 1/(1 - \alpha)$. We also have the following formulae for the quantity $\varkappa(\alpha) = p_\alpha(-1) = \int_0^1 \frac{1 - t^{1-\alpha}}{1 + t^{1-\alpha}} dt$ which appears frequently in [1]:

$$\varkappa(\alpha) = F_\gamma(-1) = 2\,_2F_1(1, \gamma; \gamma + 1; -1) - 1$$

$$= -1 + 2 \sum_{n=0}^{\infty} \frac{(-1)^n \gamma}{n + \gamma} = -1 + \gamma \sum_{m=0}^{\infty} \left(\frac{2}{2m + \gamma} - \frac{2}{2m + 1 + \gamma} \right)$$

$$= \gamma \left[\psi \left(\frac{\gamma + 1}{2} \right) - \psi \left(\frac{\gamma}{2} \right) \right] - 1$$

$$= \frac{1}{1 - \alpha} \left[\psi \left(\frac{2 - \alpha}{2(1 - \alpha)} \right) - \psi \left(\frac{1}{2(1 - \alpha)} \right) \right] - 1$$

with $\gamma = 1/(1 - \alpha)$. Here $\psi(x)$ denotes the digamma function $\Gamma'(x)/\Gamma(x)$ and the following well-known formula for $x \neq 0, -1, -2, \ldots$ was used:

$$\psi(x) = \psi(1) - \frac{1}{x} + \sum_{m=1}^{\infty} \left(\frac{1}{m} - \frac{1}{m + x} \right).$$

Moreover, part (i) of Proposition 4.11 in [1] immediately follows from the next result due to Ruscheweyh [10] (Theorem 5 with $n = 1$ therein, see also [13, §4]).

Theorem 5 *The function $F_\gamma(z)$ is convex univalent on $|z| < 1$ for $\gamma \neq 0$ with* Re $\gamma \geq 0$.

Also, since for each $f \in \mathfrak{A}_\alpha$ the function $f(z)/z$ is subordinate to the convex univalent function $F_\gamma(z)$ for $\gamma = 1/(1 - \alpha)$, by the comment right after Definition 1.5

in [1], the function $f_\alpha = zF_\gamma$ is totally extremal for $\mathfrak{A}_\alpha$ in the sense that

$$\min_{|z|=r} \mathrm{Re}\left(\lambda\frac{f(z)}{z}\right) \geq \min_{|z|=r} \mathrm{Re}\left(\lambda\frac{f_\alpha(z)}{z}\right)$$

for every $\lambda \in \mathbb{C}, 0 < r < 1$ and $f \in \mathfrak{A}_\alpha$. In particular, we obtain the following answer to Question 2.

Corollary 6 *The quantities $K(\mathfrak{A}_\alpha)$ and $B(\mathfrak{A}_\alpha)$ are given by*

$$K(\mathfrak{A}_\alpha) = F_\gamma(-1) = \varkappa(\alpha)$$

and

$$B(\mathfrak{A}_\alpha) = \frac{2}{\pi} \cdot \max_{0<\theta<\pi} \arg F_\gamma(e^{i\theta}),$$

where F_γ is given in Theorem 4 with $\gamma = 1/(1-\alpha)$.

By using Mathematica, we can produce graphs of these quantities. See Figs. 1 and 2.

We can also show the following known result [1] in a simple manner.

Corollary 7 *The family $\mathfrak{A}_\alpha$, $0 \leq \alpha \leq 1$, is strict and exhaustive.*

Proof The strictness of the family $\mathfrak{A}$ is seen by the fact that $f_\alpha = zF_\gamma \in \partial\mathfrak{A}_\alpha$, where $\gamma = 1/(1-\alpha)$ as before. Indeed, $f_\alpha \in \mathfrak{A}_\alpha$ is clear by definition of F_γ. If $f_\alpha \in \mathfrak{A}_\beta$ for some $\beta < \alpha$, then by Theorem 4, we obtain $F_\gamma \prec F_\delta$ for $\delta = 1/(1-\beta) < \gamma$. In particular, $|F'_\gamma(0)| \leq |F'_\delta(0)|$; namely, $2\gamma/(1+\gamma) \leq 2\delta/(1+\delta)$. However, this is impossible because $2x/(1+x)$ is strictly increasing in $x > 0$. The exhaustivity is obvious. $\square$

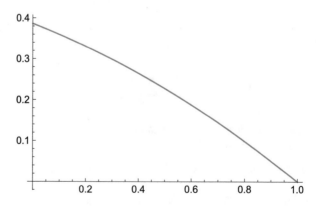

Fig. 1 The graph of $\varkappa(\alpha)$

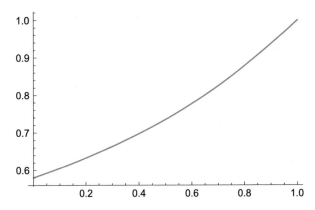

Fig. 2 The graph of $B(\mathfrak{A}_\alpha)$

We next consider the class $\mathfrak{H}_\alpha$ consisting of functions $f \in \mathcal{G}_0$ with

$$\mathrm{Re}\left[\alpha |z|^2 \frac{f(z)}{z} + (1-\alpha)(1-|z|^2)f'(z)\right] \geq 0.$$

It is known that the filtration determined by these classes is strict and exhaustive. More precisely, it was proven in [1] that $\mathfrak{H}_\alpha \subsetneq \mathfrak{H}_\beta \subsetneq \mathfrak{H}_{2/3}$ for $0 \leq \alpha < \beta < \frac{2}{3}$ and that $\mathfrak{H}_\alpha = \mathfrak{H}_1$ for $2/3 \leq \alpha \leq 1$. At the same time, the following problem seems to be of general importance.

Question 8 How to describe the set $\partial\mathfrak{H}_\alpha = \mathfrak{H}_\alpha \setminus (\cup_{\alpha' < \alpha}\mathfrak{H}_{\alpha'})$?
   As for this question, we can show the following.

Theorem 9 *Let $\alpha \in \left(0, \frac{2}{3}\right]$, and let $f_\alpha(z) = z(1-z)^{\frac{\alpha}{2(1-\alpha)}}$. Then $f_\alpha \in \partial\mathfrak{H}_\alpha$.*

Proof We write $z \in \Delta$ in the form $z = 1 - re^{i\phi}$ with $|\phi| < \frac{\pi}{2}$ and $r \in (0, 2\cos\phi)$ and let $\lambda = \frac{\alpha}{2(1-\alpha)}$. We note that $0 < \lambda \leq 1$. Then, we calculate

$$2\mu|z|^2 \frac{f_\alpha(z)}{z} + (1-|z|^2)f'_\alpha(z) \qquad (5)$$

$$= 2\mu\left(1 + r^2 - 2r\cos\phi\right)r^\lambda e^{i\phi\lambda} + r(2\cos\phi - r) \cdot$$

$$\cdot \left(r^\lambda e^{i\phi\lambda} - \lambda\left(1 - re^{i\phi}\right)r^{\lambda-1}e^{i\phi(\lambda-1)}\right)$$

$$= r^\lambda\left[2\mu\left(1 + r^2 - 2r\cos\phi\right)e^{i\phi\lambda} + (2\cos\phi - r)\left(r(1+\lambda)e^{i\phi\lambda} - \lambda e^{i\phi(\lambda-1)}\right)\right]$$

If $\mu < \lambda$, we put $\phi = 0$ in (5) to get

$$2\mu|z|^2 \frac{f_\alpha(z)}{z} + (1-|z|^2)f'_\alpha(z)$$

$$= r^\lambda\left[2(\mu-\lambda) + r\{2\mu(r-2) + (2-r)(1+\lambda) + \lambda\}\right].$$

Clearly, this expression is negative when r is small enough. Therefore, $f_\alpha \notin \mathfrak{H}_{\alpha'}$ when $\mu = \frac{\alpha'}{2(1-\alpha')}$ for $\alpha' < \alpha$.

If $\mu = \lambda$, we get from (5),

$$r^{-\lambda}\mathrm{Re}\left[2\lambda|z|^2\frac{f_\alpha(z)}{z} + (1-|z|^2)f'_\alpha(z)\right] \tag{6}$$

$$= 2\lambda\cos(\phi\lambda) + 2r\lambda(r - 2\cos\phi)\cos(\phi\lambda)$$

$$+(2\cos\phi - r)r(1 + \lambda)\cos(\phi\lambda) - (2\cos\phi - r)\lambda\cos(\phi(\lambda - 1))$$

$$= (2\cos\phi - r)r(1 - \lambda)\cos(\phi\lambda) + r\lambda\cos(\phi(\lambda - 1))$$

$$+2\lambda\left(\cos(\phi\lambda) - \cos\phi \cdot \cos(\phi(\lambda - 1))\right).$$

Since obviously $(2\cos\phi - r)r(1 - \lambda)\cos(\phi\lambda) \geq 0$ and $r\lambda\cos(\phi(\lambda - 1)) \geq 0$, we have to examine whether the last summand is positive. Indeed, applying an addition theorem for $\cos(\phi\lambda) = \cos(\phi + \phi(\lambda - 1))$, we get

$$\cos(\phi\lambda) - \cos\phi \cdot \cos(\phi(\lambda - 1)) = \sin\phi \cdot \sin(\phi(1 - \lambda)) \geq 0.$$

Therefore, $f_\alpha \in \mathfrak{H}_\alpha$. $\qquad\square$

This theorem leads us to the following conjecture.

Conjecture 10 Let $f(z) = z\left(\dfrac{1 + z}{1 - z}\right)^{\frac{\alpha}{2(1-\alpha)}}$ for $0 \leq \alpha < 2/3$. Then $f \in \partial\mathfrak{H}_\alpha$.

To explain this better, we have the following observation. Let

$$p_\beta(z) = \left(\frac{1 + z}{1 - z}\right)^\beta$$

for $0 < \beta$. We remark that p_β maps the unit disk univalently onto the convex sector $|\arg w| < \pi\beta/2$ for $0 < \beta \leq 1$. By the Alexander correspondence, we also observe that

$$zp'_\beta(z) = \frac{2\beta z}{1 - z^2}\left(\frac{1 + z}{1 - z}\right)^\beta = \frac{2\beta z}{1 - z^2}p_\beta(z)$$

is starlike univalent with respect to the origin. Put

$$P_{\alpha,\beta}(z) = \left\{\alpha r^2 + (1 - \alpha)(1 - r^2)\right\}p_\beta(z) + (1 - \alpha)(1 - r^2)zp'_\beta(z)$$

$$= p_\beta(z)\left\{\alpha r^2 + (1 - \alpha)(1 - r^2) + (1 - \alpha)(1 - r^2) \cdot \frac{2\beta z}{1 - z^2}\right\}$$

for $r = |z| < 1$ (cf. [4]). Fix $0 < \alpha < 2/3$. We now observe that

$$P_{\alpha,\beta}(-r) = \{\alpha r^2 + (1-\alpha)(1-r^2) - 2(1-\alpha)\beta r\} p_\beta(-r).$$

Question 11 Does the following implication hold: if $P_{\alpha,\beta}(-1) = \alpha - 2(1-\alpha)\beta \geq 0$, then $\operatorname{Re} P_{\alpha,\beta}(z) \geq 0$ on Δ?

It might be helpful to see that $\operatorname{Re} P_{\alpha,\beta}(z) \geq 0$ precisely when

$$-\operatorname{Re}\left[z p'_\beta(z)\right] \leq \left(1 + \frac{2\beta r^2}{1-r^2}\right) \operatorname{Re} p_\beta(z), \quad r = |z| < 1.$$

If the answer to the last question is affirmative, we could conclude that

$$z p_\beta \in \mathfrak{H}_\alpha \Leftrightarrow \beta \leq \frac{\alpha}{2(1-\alpha)},$$

and then this would prove the conjecture above.

We remark that $K(\mathfrak{H}_0) = K(\mathfrak{A}_0) = \varkappa(0) = 2\log 2 - 1 = 0.38629\ldots$ and $B(\mathfrak{H}_0) = B(\mathfrak{A}_0) = 0.58035\ldots$. Using results in [4], one can easily conclude that $K(\mathfrak{H}_\alpha) = 0$ for every $\alpha \in (0,1]$ (also our Theorem 9 may be used to show it). However, we do not know even whether $B(\mathfrak{H}_\alpha) < 1$ for a general $\alpha \in (0, 2/3)$.

Question 12 Compute the quantity $B(\mathfrak{H}_\alpha)$.

As we have already mentioned, the Noshiro-Warschawski condition $\operatorname{Re} f'(z) > 0$ implies that f is a semi-complete vector field, hence for any $z \in \Delta$ both the points $w_0 = f'(z)$ and $w_1 = \frac{f(z)}{z}$ lie in the right half-plane. In the filtration $\mathfrak{A}_\alpha$, we considered the Euclidean convex combination $\alpha w_1 + (1-\alpha)w_0$ in (4). Since w_0, w_1 are usually non-vanishing, we may consider the quasihyperbolic geometry on the punctured plane $\mathbb{C}^* = \mathbb{C} \setminus \{0\}$. Then the quasihyperbolic geodesic of w_0 and w_1 can be expressed by

$$w_0^{1-\alpha} w_1^\alpha = w_0 \left(\frac{w_1}{w_0}\right)^\alpha = f'(z) \left(\frac{f(z)}{z f'(z)}\right)^\alpha$$

(see [8] for example). Here, we note that one can take a single-valued analytic branch of $h(z) = [z f'(z)/f(z)]^{-\alpha}$ on Δ with $h(0) = 1$ whenever f is non-vanishing except for $z = 0$ and locally univalent; namely, $f(z) \neq 0$ and $f'(z) \neq 0$ for $0 < |z| < 1$. Let $\mathfrak{F}_\alpha$ the class of such functions f in $\mathcal{G}_0$ with

$$\operatorname{Re}\left[f'(z)\left(\frac{f(z)}{z f'(z)}\right)^\alpha\right] > 0$$

on Δ. We have now the following fundamental problem.

Question 13 Is the family $\mathfrak{F}_\alpha$, $0 \leq \alpha \leq 1$, a filtration? If the answer is affirmative, is this filtration strict and exhaustive?

Acknowledgements The work of the first and the second authors was partially supported by the European Commission under the project STREVCOMS PIRSES-2013-612669. The second author was financially supported by the Ministry of Education and Science of the Russian Federation (the Agreement number 02.A03.21.0008). The third author is very grateful to the Organizing Committee of the Third Workshop on Complex and Harmonic Analysis held at Holon and Haifa in 2017 (especially, to Prof. A. Golberg) for the opportunity to visit Israel and to discuss the problems considered in the paper.

References

1. F. Bracci, M.D. Contreras, S. Díaz-Madrigal, M. Elin, D. Shoikhet, Filtrations of semigroup generators (2016). Preprint
2. P. Duren, *Univalent Functions* (Springer, New York, 1983)
3. M. Elin, D. Shoikhet, *Linearization Models for Complex Dynamical Systems*. Topics in Univalent Functions, Functional Equations and Semigroup Theory (Birkhäuser, Basel, 2010)
4. M. Elin, D. Shoikhet, N. Tuneski, Parametric embedding of starlike functions. Comp. Anal. Oper. Theory **11**, 1543–1556 (2017).
5. A.W. Goodman, *Univalent Functions*, vol. 1 (Mariner Publishing Company, Tampa, 1983)
6. D.J. Hallenbeck, St. Ruscheweyh, Subordination by convex functions. Proc. Am. Math. Soc. **52**, 191–195 (1975)
7. Y. Komatu, On starlike and convex mappings of a circle. Kodai Math. Semin. Rep. **13**, 123–126 (1961)
8. G.J. Martin, B.G. Osgood, The quasihyperbolic metric and associated estimates on the hyperbolic metric. J. Anal. Math. **47**, 37–53 (1986)
9. S.S. Miller, P.T. Mocanu, *Differential Subordinations. Theory and Applications* (Dekker, New York, 2000)
10. St. Ruscheweyh, New criteria for univalent functions. Proc. Am. Math. Soc. **49**, 109–115 (1975)
11. D. Shoikhet, *Semigroups in Geometrical Function Theory* (Kluwer Academic, Dordrecht, 2001)
12. D. Shoikhet, Rigidity and parametric embedding of semi-complete vector fields on the unit disk. Milan J. Math. **77**, 127–150 (2010)
13. T. Sugawa, L.-M. Wang, Spirallikeness of shifted hypergeometric functions. Ann. Acad. Sci. Fenn. Math. **42**, 963–977 (2017)

Polynomial Lemniscates and Their Fingerprints: From Geometry to Topology

Anastasia Frolova, Dmitry Khavinson, and Alexander Vasil'ev

Abstract We study shapes given by polynomial lemniscates, and their fingerprints. We focus on the inflection points of fingerprints, their number and geometric meaning. Furthermore, we study dynamics of zeros of lemniscate-generic polynomials and their 'explosions' that occur by planting additional zeros into a defining polynomial at a certain moment, and then studying the resulting deformation. We call this dynamics *polynomial fireworks* and show that it can be realized by a construction of a non-unitary operad.

Keywords Lemniscates • Conformal welding • Blaschke products •
Versal deformations • Operads • Quadratic differentials • Braids

2010 Mathematics Subject Classification Primary: 30C10; Secondary: 37E10

1 Introduction

A polynomial lemniscate is a plane algebraic curve of degree $2n$, defined as a level curve of the modulus of a polynomial $p(z) = \prod_{k=1}^{n}(z - z_k)$ of degree n with its roots z_k called the nodes of the lemniscate. Lemniscates have been objects of interest since 1680 when they were first studied by French-Italian astronomer Giovanni Domenico Cassini see, e.g., [4], and later christened as "ovals of Cassini". 14 years later Jacob Bernoulli, unaware of Cassini's work, described a curve forming 'a figure 8 on its side', which is defined as a solution of the equation

$$(x^2 + y^2)^2 = 2c^2(x^2 - y^2), \quad \text{or } |z - c|^2 |z + c|^2 = c^4, \tag{1}$$

A. Frolova (✉) • A. Vasil'ev
Department of Mathematics, University of Bergen, P.O. Box 7800, Bergen N-5020, Norway
e-mail: Anastasia.Frolova@math.uib.no; alexander.vasiliev@math.uib.no

D. Khavinson
Department of Mathematics & Statistics, University of South Florida, Tampa, FL 33620-5700, USA
e-mail: dkhavins@cas.usf.edu

© Springer International Publishing AG 2018
M. Agranovsky et al. (eds.), *Complex Analysis and Dynamical Systems*,
Trends in Mathematics, https://doi.org/10.1007/978-3-319-70154-7_7

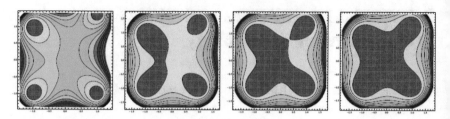

Fig. 1 Evolution of a lemniscate of a polynomial of degree 4

where $z = x + iy$. (Curiously, his brother Johann independently discovered lemniscate in 1694 in a different context.) Observe that the level curves $|p(z)| = R$ can form a closed Jordan curve for sufficiently large R, and split into n disconnected components when $R \to 0^+$. Great interest in lemniscates in eighteenth century was inspired by their links to elliptic integrals. The dynamics of a lemniscate as R grows is illustrated in Fig. 1.

The recent revival of lemniscates owes to the newly emerging field of vision and pattern recognition. The key idea is to consider the space of 2D 'shapes', domains in the complex plane $\mathbb{C}$ bounded by smooth C^∞ Jordan curves Γ dividing $\mathbb{C}$ into two simply connected domains one of which Ω_- contains infinity and the other one, Ω_+, is bounded. The study of an enormous space of 2D shapes was one of the central problems in a program outlined by Mumford at the ICM 2002 in Beijing [15]. We will also call the boundary curves Γ shapes for convenience.

Here we focus on 'fingerprints' of shapes obtained by means of conformal welding. Let Γ be a curve defining a shape Ω_+, and let ϕ_+ be a conformal mapping of the unit disk $\mathbb{D} = \mathbb{D}_+$ onto the domain Ω_+ bounded by Γ. The matching function ϕ_- maps the exterior $\mathbb{D}_-$ of $\mathbb{D}$ onto Ω_-. One can either normalize the interior maps by shifting and scaling Γ so that $0 \in \Omega_+$ and the conformal radius of Ω_+ with respect to the origin is 1, or the exterior one by the 'hydrodynamical' normalization

$$\phi_-(z) = z + \sum_{k=1}^\infty \frac{c_k}{z^k} \in \mathcal{G},$$

where by $\mathcal{G}$ we denote the class of all analytic functions in $\mathbb{D}_-$ normalized as above and C^∞ smooth up to the boundary. The conformal welding produces the function $k : [0, 2\pi] \to [0, 2\pi]$ defined by $e^{ik(\theta)} = (\phi_-)^{-1} \circ \phi_+(e^{i\theta})$, which is monotone, smooth and '2π-periodic' in the sense that $k(\theta + 2\pi) = k(\theta) + 2\pi$. In the first case, the function $e^{ik(\theta)}$ is a representative of an element of the smooth Teichmüller space $\mathrm{Diff}S^1/\mathrm{Rot}$, i.e., the connected component of the identity of the quotient Lie-Fréchet group $\mathrm{Diff}S^1$ of orientation preserving diffeomorphisms of the unit circle S^1 over the subgroup of rotations Rot. In the second case, $e^{ik(\theta)}$ is an element of the smooth Teichmüller space $\mathrm{Diff}S^1/\mathrm{Möb}$, where Möb is the group $\mathrm{PSL}(2,\mathbb{C})$ restricted to S^1. The fibration $\pi : \mathrm{Diff}S^1/\mathrm{Rot} \to \mathrm{Diff}S^1/\mathrm{Möb}$ has the typical fiber $\mathrm{Möb}/S^1 \simeq \mathbb{D}_-$. The homogeneous spaces $\mathrm{Diff}S^1/\mathrm{Möb}$ and $\mathrm{Diff}S^1/\mathrm{Rot}$ carry the structure of

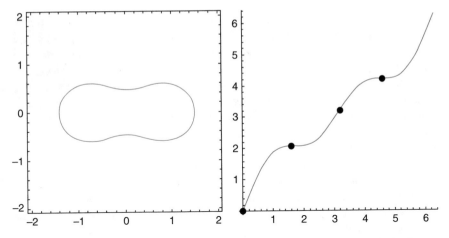

Fig. 2 Cassini oval (Bernoulli's lemniscate of degree 2) and its fingerprint (the marked points are the points of inflection)

infinite-dimensional, homogeneous, complex analytic Kählerian manifolds which appeared in the classification of orbits in the coadjoint representation of the Virasoro-Bott group see, e.g., [7, 8]. There is a biholomorphic isomorphism between $\mathcal{G}$ and $\mathrm{Diff}S^1/\mathrm{M\ddot{o}b}$, see [8].

The construction of fingerprints is straightforward whereas the reconstruction of shapes from their fingerprints is a non-trivial task. Let us mention two more recent reconstructing algorithms: one provided by Mumford and Sharon [20], the second is the 'zipper' algorithm by Marshall [11, 12].

By Hilbert's theorem, polynomial lemniscates approximate any smooth shape with respect to the Hausdorff distance in the plane, see e.g., [6] and [21, Ch. 4]. An advantage of this approach is that the fingerprint of a polynomial lemniscate is given by the nth root of a Blaschke product of degree n, which was proved in [5], see also [22] for a simplified proof and extensions to rational lemniscates. The reciprocal statement also holds: a fingerprint given by an n-th root of a Blaschke product of degree n represents the shape given by a polynomial lemniscate of the same degree [5, 19, 22]. Also, cf. [19] for some generalizations.

The proof of Hilbert's theorem was based on approximation of the Riemann integral by the Riemann sums and the intermediate points of the partitions were chosen as the nodes of the approximating lemniscates. This approach is algorithmically poor. Indeed, the more precise approximation we need, the higher degree of the polynomial we must choose. So, even starting with a shape already defined by a lemniscate of a low degree, the approximating lemniscate will be of much higher degree. This problem was addressed by Rakcheeva in [17, 18]. She proposed a focal algorithm which starts with a just a polynomial of degree 1, and then, 'budding' the zeros of the polynomial iteratively according to the shape, a better approximation is achieved. Based on this idea we also start with a lemniscate for the polynomial of degree 1, just a circle. Then we place a power $z^{n_k^j}$ at a simple zero z_k during the j-th step of the iteration, and then, perform a deformation, i.e., we move the new zeros without significant change in the structure of lemniscates. From the algebraic viewpoint, as a result, we arrive at a braid operad which we call the *polynomial fireworks operad*. This way, smooth shapes encode the polynomial fireworks operad.

The structure of the paper is as follows. First, we observe some simple properties of real analytic shapes and shapes with corners. Then, we study the role of inflection points of a fingerprint and their relation to the corresponding shape. Finally, after presenting some necessary definitions and some background we outline the construction of the polynomial fireworks operad.

Note. Professor Vasil'ev (Sasha to his many friends and colleagues) conceived the idea of this work and enthusiastically worked towards its completion. The untimely death didn't let him to see the final version. We shall all miss him, his friendship, insights and his kindness.

2 On the Geometry of Fingerprints

In this section we describe some simple relations between shapes and their fingerprints, which we couldn't find in the literature.

2.1 Real-Analytic Fingerprints

As is shown in [7, 8], any smooth increasing function $k: [0, 2\pi] \to [0, 2\pi]$, satisfying $k(\theta + 2\pi) = k(\theta) + 2\pi$, defines a smooth shape Γ. However, if Γ is real analytic, it restricts the fingerprints severely. Indeed, if an analytic shape Γ contains a circular arc, then Γ is a circle. Furthermore, Γ is bounded, and therefore, a real analytic Γ can not contain a line segment. At the same time, periodicity of $k(\theta)$ implies that the fingerprint in the square $[0, 2\pi] \times [0, 2\pi]$ can not contain a segment of a straight line unless $k(\theta) = \theta + const$.

Theorem 1 *Let $p(z)$ be a polynomial of degree at least two. If the fingerprint $k(\theta)$ of a shape Γ is given by the relation $e^{ik(\theta)} = p(e^{i\theta})$ in some closed interval $\sigma \subseteq [0, 2\pi]$, where $p(z)$ is a polynomial, then Γ is not an analytic curve.*

Proof Indeed, if Γ were analytic, then its fingerprint $k \in \text{Diff} S^1 / \text{Möb}$ would be given by the relation $e^{ik(\theta)} = (\phi_-)^{-1} \circ \phi_+(e^{i\theta})$, where

$$\phi_+(\zeta) = \sum_{n=0}^{\infty} a_n \zeta^n, \quad \text{and} \quad \phi_-(\zeta) = \zeta + \sum_{n=1}^{\infty} \frac{b_n}{\zeta^n}.$$

Both functions have continuous extension to S^1. Moreover, $\phi_- \circ p(e^{i\theta})$ and $\phi_+(e^{i\theta})$ have analytic extensions to a neighbourhood of the arc $\{z: z = e^{i\theta}, \, \theta \in \sigma\}$. Thus, by Morera's theorem, $\phi_- \circ p(e^{i\theta})$ and $\phi_+(e^{i\theta})$ are analytic extensions of each other, hence, equating the coefficients, we conclude that all coefficients a_n and b_n must vanish except for the case $k(\theta) = \theta$. A simple adjustment of the normalization of $\phi_+(e^{i\theta})$ yields the proof in the case $k \in \mathrm{DiffS}^1/\mathrm{Rot}$. □

2.2 Shapes with Corners

Let us now remark on the case when the boundary curve Γ has a corner of opening $\pi\alpha$ $(0 \leq \alpha \leq 2)$ at a point $z_\alpha \in \Gamma$. Then the interior mapping ϕ_+ satisfies the condition

$$\arg[\phi_+(e^{it}) - \phi_+(e^{i\theta_\alpha})] \to \begin{cases} \beta & \text{as } t \to \theta_\alpha + 0, \\ \beta + \pi\alpha & \text{as } t \to \theta_\alpha - 0, \end{cases}$$

where $z_\alpha = \phi_+(\theta_\alpha)$. If $\alpha = 1$, then Γ has a tangent at z_α, if $\alpha = 0$ or $\alpha = 2$, then Γ has an outward-pointing or an inward-pointing cusp respectively. If Γ is smooth except at the point z_α, $\alpha \in (0, 1)$, then the derivative $(\phi_+)'(\zeta)$ has a continuous extension to $S^1 \setminus \{e^{i\theta_\alpha}\}$, and the functions $(\zeta - e^{i\theta_\alpha})^{1-\alpha}(\phi_+)'(\zeta)$ and $(\zeta - e^{i\theta_\alpha})^{-\alpha}(\phi_+(\zeta) - \phi_+(e^{i\theta_\alpha}))$ are continuous in some neighbourhood of $e^{i\theta_\alpha}$ in $\hat{\mathbb{D}}$, see e.g., [16, Theorem 3.9]. Similar conclusions hold for the exterior mapping ϕ_-. The fingerprint $k(\theta)$ does not longer represent an element of $\mathrm{DiffS}^1/\mathrm{Rot}$ or $\mathrm{DiffS}^1/\mathrm{M\ddot{o}b}$ but it belongs to $\mathrm{Hom}\, S^1/\mathrm{Rot}$ or $\mathrm{Hom}\, S^1/\mathrm{M\ddot{o}b}$, where Hom denotes a group of quasisymmetric homeomorphisms of S^1. Since $\phi_-(e^{ik(\theta)}) = \phi_+(e^{i\theta})$, we have

$$\phi_-(\zeta) = z_\alpha + b_1(\zeta - e^{ik(\theta_\alpha)})^{2-\alpha} + o(|\zeta - e^{ik(\theta_\alpha)}|^{2-\alpha})$$

$$\phi_+(\zeta) = z_\alpha + d_1(\zeta - e^{i\theta_\alpha})^{\alpha} + o(|\zeta - e^{i\theta_\alpha}|^{\alpha})$$

in the corresponding neighbourhoods of the points $e^{ik(\theta_\alpha)}$ and $e^{i\theta_\alpha}$ in $\hat{\mathbb{D}}^-$ and $\hat{\mathbb{D}}$ respectively. After performing the conformal welding it is clear that the original fingerprint $k(\theta)$ has the singular point $\theta_\alpha \in [0, 2\pi)$, i.e., the graph has a singularity $(k'(\theta_\alpha) = \infty$ when $\alpha = 2)$ of order $2(\alpha - 1)/(2 - \alpha)$, $k'(\theta) \sim |\theta - \theta_\alpha|^{\frac{2(\alpha-1)}{2-\alpha}}$.

2.3 Dynamics of Proper Lemniscates When They Are Approaching Critical Points

Let $\Gamma(R)$ be a polynomial lemniscate of degree n, i.e.,

$$\Gamma(R) = \{z \in \mathbb{C} : |p(z)| = R\}, \quad p(z) = \prod_{k=1}^{n}(z - z_k), \quad R > 0.$$

Without loss of generality let us, whenever possible, assume $R = 1$. Let us denote the Riemann sphere by $\hat{\mathbb{C}} = \mathbb{C} \cup \{\infty\}$. The lemniscate $\Gamma(1)$ is called *proper* if the region $\Omega^+ = \{z \in \hat{\mathbb{C}} : |p(z)| < 1\}$, is connected. Let us denote by Ω^- the unbounded component of $\hat{\mathbb{C}} \setminus \Gamma(1)$, i.e. $\Omega^- = \{z \in \hat{\mathbb{C}} : |p(z)| > 1\}$. If $\Gamma(1)$ is proper, the Riemann map $\phi_- : \mathbb{D}_- \to \Omega_-$ has a simple inverse $(\phi_-)^{-1}(z) = \sqrt[n]{p(z)}$. Let $\{y_j\}_{j=1}^{n-1}$ denote the *critical points* of p, i.e. the zeros of the derivative $p'(z)$. Then $\Gamma(1)$ is proper if and only if all the critical points $\{y_j\}_{j=1}^{n-1}$ of p lie in Ω^+, or, equivalently, all the critical values $p(y_j)$ lie in $\mathbb{D}_+$, cf. [5].

Let $B(\zeta)$ stand for the Blaschke product

$$B(\zeta) = e^{i\alpha} \prod_{k=1}^{n} \frac{\zeta - a_k}{1 - \bar{a}_k \zeta},$$

for some real α and $|a_k| < 1$. Then Theorem 2.2 from [5] states that the fingerprint $k : [0, 2\pi] \to [0, 2\pi]$ of the proper lemniscate $\Gamma(1)$ is given by

$$e^{ik(\theta)} = \sqrt[n]{B(e^{i\theta})}, \tag{2}$$

where the branch of $\sqrt[n]{\cdot}$ is appropriately chosen, e.g., by fixing the branch $\sqrt[n]{1} = 1$, and the zeros a_k of $B(\zeta)$ are the pre-images of the nodes z_k under ϕ_+, repeated according to the multiplicity.

Let us discuss the dynamics of proper lemniscates $\Gamma(R)$, $R > |p(y_{n-1})|$, as $R \searrow |p(y_{n-1})|$, where we first assume that $|p(y_k)|$, $k = 1, \ldots, n-1$ are the modules of the critical values of p at the points y_k, $y_{n-1} \neq y_k$ and $|p(y_{n-1})| > |p(y_k)|$, $k = 1, \ldots, n-2$. When the proper lemniscate approaches the first critical point, and the domain Ω_+ splits up into exactly two domains with the multiple boundary point of valence 4. Let $k(\theta)$ represent an element of $\mathrm{Diff}S^1/\mathrm{M\ddot{o}b}$. If $R \searrow |p(y_{n-1})| + 0$, then the exterior mapping ϕ_- still exists at the limit point. In order to give any reasonable sense to what happens with the interior conformal map let us use the Carathéodory theorem, see e.g., [16, Theorem 1.8] fixing a point in one of the parts Ω_1 or Ω_2 of Ω_+ bounded by $\Gamma(|p(y_{n-1})|)$. For example, we can specify one of the nodes of the lemniscate $z_1 \in \Omega_1$ as the image of the origin by ϕ_+, $\phi_+(0) = z_1$, for all $R > |p(y_{n-1})|$. Then, the preimages $a_1, \ldots, a_m$ of those nodes $z_1, \ldots, z_m$ that remain in Ω_1 will remain in $\mathbb{D}$, while all other preimages $a_{m+1}, \ldots, a_n$ of the nodes $z_{m+1}, \ldots, z_n \in \Omega_2$ will tend to S^1 as $R \searrow |p(y_{n-1})|$. At the same time, the preimage of Ω_2 will collapse and the Carathéodory convergence theorem guarantees that the limiting map ϕ_+ will be well-defined in $\mathbb{D}$ (the kernel), $\phi_+ : \mathbb{D} \to \Omega_1$, as $R \searrow |p(y_{n-1})|$. The inverse $(\phi_-)^{-1}$ of the exterior map can be continuously (non-bijectively) extended to $\partial\Omega_1$, where the bifurcation point is understood to be two different points over the same support. Then the fingerprint $k(\theta)$ can be made sense of only between two points corresponding to the images of the bifurcation point. That is, as in the previous section, the graph of $k(\theta)$ will have the vertical tangent at the points $\theta_{1/2}^1$ and $\theta_{1/2}^2$, $\theta \in [\theta_{1/2}^1, \theta_{1/2}^2]$, and the order of the singularity is $(-3/2)$, as the lemniscate has a corner of opening $\pi/2$ at the singular point.

This argument can be generalized to the case $|p(y_{n-1})| \geq |p(y_k)|$, $k = 1, \ldots, n-2$ and y_{n-1} is allowed to coincide with other critical points. In this case, the domain Ω_1 containing a fixed node can have several angular points of different angles and the fingerprint will have several singularities of different orders.

Let us mention in passing that similar arguments apply to $\mathrm{Diff}S^1/\mathrm{Rot}$ as long as we specify which domain bounded by the critical lemniscate we consider as the Carathéodory kernel. This can be achieved, e.g., by considering equivalent polynomials. (Recall that two polynomials p_1 and p_2 are said to be from the same conjugacy class $[p]$ if there is an affine map A such that $p_2 = A^{-1} \circ p_1 \circ A$. In this case the geometry of the lemniscates of p_1 and p_2 is the same up to scaling, translation and rotation, i.e., as 'shapes' those lemniscates are indistinguishable.)

3 Nodes of Lemniscates and Inflection Points of Fingerprints

The first feature one observes looking at a fingerprint of a smooth shape is that it possesses a number of inflection points. It turns out that lemniscates' fingerprints must have at least two inflection points. More precisely, the following is true, cf. Fig. 2.

Theorem 2 *The fingerprint $k(\theta)$ given by (2) has an even number of inflection points, at least two and at most $4n - 2$.*

Proof If we write $a_k = |a_k|e^{i\theta_k}$, then

$$k'(\theta) = -\frac{i}{n}\frac{\partial}{\partial\theta}\log B(e^{i\theta}) = \frac{1}{n}\sum_{k=1}^{n}\frac{1 - |a_k|^2}{1 + |a_k|^2 - 2|a_k|\cos(\theta_k - \theta)}$$

or, in terms of the Poisson kernel,

$$k'(\theta) = \frac{1}{n}\mathrm{Re}\sum_{k=1}^{n}\frac{\zeta + a_k}{\zeta - a_k}, \quad \zeta = e^{i\theta}.$$

Respectively,

$$k''(\theta) = \frac{1}{n}\mathrm{Re}\sum_{k=1}^{n}\frac{-2a_k i\zeta}{(\zeta - a_k)^2} = -\frac{1}{2n}\sum_{k=1}^{n}\left(\frac{2a_k i\zeta}{(\zeta - a_k)^2} - \frac{2\bar{a}_k i\zeta}{(1 - \bar{a}_k\zeta)^2}\right).$$

First, observe that the rational function $\sum_{k=1}^{n}\frac{\zeta + a_k}{\zeta - a_k}$ maps the unit circle onto a smooth closed curve with possible self-intersections. Hence, its real part $k'(\theta)$

attains at least one maximum and one minimum. Therefore, the function $k''(\theta)$ has at least two zeros in $[0, 2\pi)$ at which $k''(\theta)$ changes the sign. An elementary calculus theorem states that for the graph of a differentiable function of one variable $f(x)$, the number of points c where f has a local extremum in the interval $[a, b]$ is even if $f'(a)$ and $f'(b)$ have the same sign and this number is odd if the signs are different, assuming that there is a finite number of critical points in the interval, and that the sign of the derivative changes as we go from left to right passing through a zero of the first derivative. Consider finite alternating sequences of, let say, $(+)$ and $(-)$. If one has an equal number of alternating $(+)$'s and $(-)$'s, then the sign changes odd number of times, and if the number of alternating $(+)$'s and $(-)$'s differs by 1, then the sign changes even number of times.

Hence, the periodic function k' has an even number of critical points corresponding to the inflection points of k.

At the same time, the rational function

$$Z(\zeta) = \zeta \sum_{k=1}^{n} \left(\frac{2ia_k}{(\zeta - a_k)^2} - \frac{2i\bar{a}_k}{(1 - \bar{a}_k\zeta)^2} \right)$$

has degree $4n$, has simple zeros at the origin and ∞, and satisfies the relation $\bar{Z}(\zeta) = Z(1/\bar{\zeta})$. Therefore, if $b_1, \ldots, b_m$ are zeros of the function $Z(\zeta)$ in $\mathbb{D} \setminus \{0\}$, then $1/\bar{b}_1, \ldots, 1/\bar{b}_m$ are also its zeros in $\mathbb{D}_- \setminus \{\infty\}$, so $0 \leq m \leq 2n - 1$. The rest of $4n - 2 - 2m$ zeros of Z lie on the unit circle and are precisely the zeros of the function k''. □

However, the maximal number $4n - 2$ of the inflection points need not be achieved. The following theorem provides a more detailed explanation of this phenomenon.

Theorem 3 *If all nodes of the n-Blaschke product B lie on the same radius of $\mathbb{D}$, then the number of the inflection points of the fingerprint k, $e^{ik(\theta)} = \sqrt[n]{B(e^{i\theta})}$ is at most $4n - 4$. In the particular case when $n = 2$, the number of the inflection points is at most 4 (not 6!) for arbitrary position of the nodes of B.*

Proof Set

$$\Psi(\zeta) = \prod_{k=1}^{n} (1 - \bar{a}_k\zeta)^2 (\zeta - a_k)^2,$$

and define the polynomial P of the form

$$P(\zeta) = \Psi(\zeta) \sum_{j=1}^{n} \left(\frac{a_j}{(\zeta - a_j)^2} - \frac{\bar{a}_j}{(1 - \bar{a}_j\zeta)^2} \right).$$

Then

$$
\frac{P'(0)}{P(0)} = \frac{\Psi'(0)}{\Psi(0)} + 2\frac{\sum_{k=1}^{n}\left(\frac{1}{a_k^2} - \bar{a}_k^2\right)}{\sum_{k=1}^{n}\left(\frac{1}{a_k} - \bar{a}_k\right)} =
$$

$$
= -2\sum_{k=1}^{n}\left(\frac{1}{a_k} + \bar{a}_k\right) + 2\frac{\sum_{k=1}^{n}\left(\frac{1}{a_k^2} - \bar{a}_k^2\right)}{\sum_{k=1}^{n}\left(\frac{1}{a_k} - \bar{a}_k\right)} =
$$

$$
= 2\frac{\sum_{k=1}^{n}\left(\frac{1}{a_k^2} - \bar{a}_k^2\right) - \sum_{k=1}^{n}\left(\frac{1}{a_k} + \bar{a}_k\right)\sum_{k=1}^{n}\left(\frac{1}{a_k} - \bar{a}_k\right)}{\sum_{k=1}^{n}\left(\frac{1}{a_k} - \bar{a}_k\right)} =
$$

$$
= -2\frac{\sum_{1\le k\ne j\le n}\left(\frac{1}{a_k} + \bar{a}_k\right)\left(\frac{1}{a_j} - \bar{a}_j\right)}{\sum_{k=1}^{n}\left(\frac{1}{a_k} - \bar{a}_k\right)} =
$$

$$
= -4(n-1)\ \frac{\sum_{1\le k<j\le n}\dfrac{1 - |a_k a_j|^2}{a_k a_j}}{\sum_{1\le k<j\le n}\dfrac{a_j(1 - |a_k|^2) + a_k(1 - |a_j|^2)}{a_k a_j}}.
$$

If all a_k lie on the same radius, then we have

$$
\left|\frac{P'(0)}{P(0)}\right| > 4(n-1); \tag{3}
$$

and ± 1 are among the roots. Therefore, we have $|\frac{P'(0)}{P(0)}| = 2|\sum_{k=1}^{2n-2}\cos\theta_k| \le 4n-4$, which contradicts (3) and finishes the proof of the first statement of the theorem.

In the case $n = 2$,

$$
\frac{P'(0)}{P(0)} = \frac{4(1 - |a_1|^2|a_2|^2)}{a_1(1 - |a_2|^2) + a_2(1 - |a_1|^2)},
$$

so we have

$$
\left|\frac{P'(0)}{P(0)}\right| \ge \frac{4(1 - |a_1|^2|a_2|^2)}{|a_1|(1 - |a_2|^2) + |a_2|(1 - |a_1|^2)} = \frac{4(1 + |a_1||a_2|)}{|a_1| + |a_2|} > 4. \tag{4}
$$

The polynomial $P(\zeta)$ is self-inversive because $\zeta^6 \bar{P}(1/\zeta) = -P(\zeta)$. Therefore,

- If b is a root, then $1/\bar{b}$ is also a root;
- If $e^{i\theta}$ is a root, then $e^{-i\theta}$ is also a root.

If $b_1, \ldots, b_6$ are the roots of $P(\zeta)$, then, by Vieta's theorem, and recalling that $P(\zeta)$ is self-inversive,

$$\sum_{k=1}^{6} b_k = \overline{\frac{P'(0)}{P(0)}}.$$

Let us assume that there are exactly six zeros $e^{\pm i\theta_k}$, $k = 1, 2, 3$, in S^1. Then

$$\overline{\frac{P'(0)}{P(0)}} = 2(\cos\theta_1 + \cos\theta_2 + \cos\theta_3),$$

and so $P'(0)/P(0)$ is real. There are two possibilities:

(1) $|a_1| = |a_2|$ and $a := a_1 = \bar{a}_2$;
(2) $|a_1| \neq |a_2|$ and a_1, a_2 are real.

In the first case,

$$\begin{aligned}
P(\zeta) = {} & a(1 - \bar{a}\zeta)^2(1 - a\zeta)^2(\zeta - \bar{a})^2 \\
& + \bar{a}(1 - \bar{a}\zeta)^2(1 - a\zeta)^2(\zeta - a)^2 \\
& - \bar{a}(1 - a\zeta)^2(\zeta - a)^2(\zeta - \bar{a})^2 \\
& - a(1 - \bar{a}\zeta)^2(\zeta - a)^2(\zeta - \bar{a})^2,
\end{aligned}$$

and this polynomial has ± 1 among the roots. Therefore, $\frac{P'(0)}{P(0)} = 2\cos\theta_3$, which contradicts (4).

In the second case,

$$\begin{aligned}
P(\zeta) = {} & a_1(1 - a_1\zeta)^2(1 - a_2\zeta)^2(\zeta - a_2)^2 \\
& + a_2(1 - a_1\zeta)^2(1 - a_2\zeta)^2(\zeta - a_1)^2 \\
& - a_1(1 - a_2\zeta)^2(\zeta - a_1)^2(\zeta - a_2)^2 \\
& - a_2(1 - a_1\zeta)^2(\zeta - a_1)^2(\zeta - a_2)^2,
\end{aligned}$$

and this polynomial has again ± 1 among the roots, which contradicts (4) for the same reason.

Summarizing, the polynomial $P(\zeta)$ has at least one root in $\mathbb{D}_+$ and hence, another in $\mathbb{D}_-$, and the maximal number of the inflection points in the fingerprint of a Bernoulli lemniscate is 4. $\qquad\square$

Remark 1 The upper bound 4 of zeros of the function $Z(\zeta)$ for $n = 2$ is achieved. For example, if $a_1 = -1/2$, and $a_2 = 1/2$, then the function has a double zero at the origin and at infinity, and four zeros $1, i, -1, -i$ on the circle S^1.

The following statement clarifies the geometric meaning of the fingerprints' inflection points in general setting.

Theorem 4 *The inflection points of the fingerprint $k(\theta)$ divide the unit circle S^1 into m arcs $\gamma_j = \{e^{i\theta} : \theta \in [\theta_j, \theta_{j+1})\}, \theta_{m+1} = \theta_1 + 2\pi, \text{where } j = 1, \ldots, m, \text{ so that the ratio of the rates of change of the harmonic measures of the arc } \alpha \subset \Gamma, \alpha = \{\phi_+(s) : s \in [\theta_1, \theta)\} \text{ with respect to } (\Omega^+, 0) \text{ and } (\Omega^-, \infty) \text{ respectively, alternates its monotonicity.}*

Proof The fingerprint $k(\theta)$ of a curve α is defined by

$$e^{ik(\theta)} = \phi_+^{-1} \circ \phi_-(e^{i\theta}).$$

We rewrite the last expression as

$$\phi_+(e^{ik(\theta)}) = \phi_-(e^{i\theta}),$$

and differentiate it

$$k'(\theta)e^{ik(\theta)}\phi_+'(e^{ik(\theta)}) = \phi_-'(e^{i\theta})e^{i\theta}.$$

Without loss of generality we can assume that $\theta_1 = 0$, and

$$\phi_+(1) = \phi_-(1).$$

We consider an arc α on $\partial\Omega_+$ starting at the point $\phi_+(1)$.

Let γ_+ and γ_- denote the images of α by ϕ_+^{-1} and ϕ_-^{-1} correspondingly, which can be parametrized as follows: $\gamma_-(\theta) = e^{i\theta}, \gamma_+(\theta) = e^{ik(\theta)}$. Let us determine the harmonic measure

$$\omega_-(\alpha, \infty) = \int_\alpha \frac{\partial g(z, \infty)}{\partial n} |ds|$$

where $g(z, \infty)$ is Green's function, $g(z, \infty) = \mathrm{Re}\, G(z, \infty)$, and $G(z, \infty)$ is complex Green's function, that satisfies

$$G(z, \infty) = \log \phi_-^{-1}(z).$$

The normal derivative of $g(z, \infty)$ has form

$$\frac{\partial g(z, \infty)}{\partial n} = \mathrm{Re}\left(G'(z, \infty) \frac{e^{i\theta}\phi_-'}{|\phi_-'|} \right),$$

and thus the harmonic measure satisfies

$$\omega_-(\alpha, \infty) = \operatorname{Re} \int_\alpha G'(z, \infty) \frac{e^{i\theta} \phi'_-}{|\phi'_-|} |dz| =$$

$$\int_{\gamma_-} G'(\phi_-(\zeta), \infty) \zeta \phi'_-(\zeta) |d\zeta| = \int_{\gamma_-} |d\zeta| = \theta,$$

where $\zeta = e^{i\theta}$. We differentiate the resulting harmonic measure $\omega_-(\alpha, \infty)$

$$\frac{d}{d\theta} \omega_-(\alpha, \infty) = G'(\phi_-(e^{i\theta}), \infty) e^{i\theta} \phi'_-(e^{i\theta}).$$

Analogously, we obtain

$$\frac{d}{d\theta} \omega_+(\alpha, 0) = G'(\phi_+(e^{ik(\theta)}), 0) e^{ik(\theta)} \phi'_+(e^{ik(\theta)}) k'(\theta).$$

Thus

$$k'(\theta) \frac{\partial}{\partial\theta} \omega_-(\alpha, \infty) = \frac{\partial}{\partial\theta} \omega_+(\alpha, 0).$$

Therefore, $k'(\theta)$ shows the ratio of the rates of change of the respective harmonic measures.   □

Remark 2 We can now rephrase Theorem 2 as follows. In the case of a lemniscate of degree n, the number m of points where the rates of change of interior and exterior harmonic measures change roles in dominating one another, is even and is at least 2 and at most $4n - 2$.

4 Polynomial Fireworks

As it was observed in Introduction, the focal algorithm [17, 18] suggests a process of construction of lemniscates approximating smooth shapes by budding the new nodes, i.e., blowing up the old ones iteratively. So any smooth shape encodes a tree of evolution of lemniscate's nodes. In this section we study dynamics of zeros of lemniscate-generic polynomials and their explosion planting singularities at certain moment, and then, performing their deformation and evolution. We call this dynamics *polynomial fireworks* and it is realized by a construction of a non-unitary operad. The term is chosen because of the similarity to the real fireworks, cf. Fig. 3.

Fig. 3 Polynomial fireworks

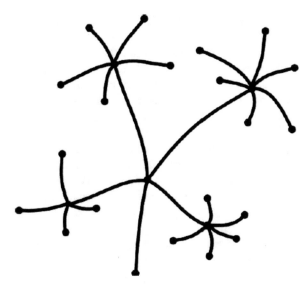

4.1   Trees

Following [2], we call a polynomial $p(z)$ *lemniscate-generic* if the zeros $y_1, \ldots, y_{n-1}$ of $p'(z)$ are distinct, $w_k = p(y_k) \neq 0$ for $k = 1, \ldots, n-1$ and $|w_i| < |w_j|$ for $i < j$. Then, only finitely many level sets $\Gamma(R)$, $R > 0$ of $|p(z)|$ are not unions of 1-manifolds. Such a singular level set for a fixed R is called a *big lemniscate*. Each big lemniscate contains one singular connected component, i.e. a 'figure-eight', which is called a *small lemniscate*.

A lemniscate-generic polynomial $p(z)$ of degree n has exactly $n - 1$ big and, correspondingly, small lemniscates. If p and p^* are lemniscate-generic polynomials of degree n, the unions Λ and Λ^* of all big lemniscates of p and p^* belong to the same *lemniscate configuration* $(\Lambda, \mathbb{C})$ if there exists a homeomorphism $h : \mathbb{C} \to \mathbb{C}$ mapping Λ onto Λ^*. Thus, we identify lemniscate configurations and the isotopy classes of lemniscate-generic polynomials.

Consider the set $\mathcal{P}_n$ of all polynomials of degree exactly n and the subset $\mathcal{L}_n$ of lemniscate-generic polynomials. Then $\mathcal{L}_n$ is open and $\mathcal{P}_n \setminus \mathcal{L}_n$ is a union of real hypersurfaces as it will be shown in Sect. 4.3. The lemniscate configuration does not change if p varies in a connected component E of $\mathcal{L}_n$.

If $[R_1, R_2]$ is an interval in $\mathbb{R}^+$ containing none of the points $|w_k| = |p(y_k)|$, where w_k are the critical values of p, then $\Gamma(R_1)$ is diffeomorphic to $\Gamma(R_2)$ by a gradient flow as it follows from Morse theory, see [14]. That is $|p|^2$ becomes a local Morse function whose Hessian matrix is non-degenerate at the critical points and whose gradient generates a flow between them.

Catanese and Paluszny [2] established a bijection between the connected components of $\mathcal{L}_n$ and the lemniscate configurations of polynomials of degree n. They also showed bijection between lemniscate configurations and simple central balanced trees of length $n - 1$. Let us recall some terminology.

By a tree we understand a connected graph without cycles. A valence of a vertex is the number of edges adjacent to it. A vertex of valence 3 is called a node, a vertex of valence 1 is called an end. Any two vertices a and b of a tree can be connected by a path, i.e., a sequence of edges that connects the vertices. Any such path is in fact a sub-tree. The distance between a and b is the number of edges in a shortest path connecting a and b. A chain of edges is a tree that consists of subsequent edges $e_1, \ldots, e_n$, such that e_k shares a vertex with e_{k+1}, $1 \leq k \leq n - 1$.

A root radius of a vertex a is the maximal of distances from a to the leaves of the tree. A tree has vertex v as a center if v is a vertex with a minimal root radius. A tree is central if it has only one center. The length $|T|$ of a central tree T is the distance from the center of T to the ends of T. We call a tree binary if it only has vertices of valence no bigger than 3.

A tree T is called a simple central balanced tree of length $n - 1$ if it has $n - 1$ leaves, it is central, the root radius of the center v is $n - 1$, the minimal of distances from v to the leaves is $n - 1$; the valence of the center is 2, there is exactly one node at distance j from the center ($1 \leq j \leq n - 2$), and the tree does not have vertices of valence ≥ 3.

To each polynomial $p(z) \in \mathcal{L}_n$ we can assign a simple central balanced tree T of length $n - 1$. The vertices of T of valence ≥ 2 represent connected components of big lemniscates. The leaves of T represent the zeros of p. Vertices of T at distance k from the center ($0 \leq k \leq n - 2$) represent connected components of a big lemniscate L_k: the nodes represent the figure-eight-like branchings and the vertices of valence 2 represent circumferences. T has $(n - 1)(n - 2)/2$ vertices corresponding to circumferences, $n - 1$ vertices corresponding to the figure-eights, n vertices corresponding to zeros. The edges of T can be interpreted as the doubly connected domains situated between the connected components of critical level sets.

4.2 Operad

The notion of an operad first appeared and was coined in May's book [13] in 1972, and in the original work in algebraic topology by Boardman and Vogt [1], in the study of iterated loop spaces formalizing the idea of an abstract space of operations, see also more modern review in [9]. Also, cf. [10]. Thus, it is not surprising that operad appears in our case of shapes, i.e., 1-D loop space. Here is a precise definition.

Definition 1 An operad is a sequence $\{\mathcal{O}(n)\}_{n=1}^{\infty}$ of sets (topological spaces, vector spaces, complexes, etc.), an identity element $e \in \mathcal{O}(1)$, and composition map $\circ$ defined for all positive integers $n; k_1, k_2, \ldots, k_n$

$$\circ: \mathcal{O}(n) \times \mathcal{O}(k_1) \times \cdots \times \mathcal{O}(k_n) \to \mathcal{O}(k_1 + \cdots + k_n)$$

$$\theta, \theta_1, \ldots, \theta_n \to \theta \circ (\theta_1, \ldots, \theta_n) := (\theta; \theta_1, \ldots, \theta_n);$$

satisfying the following axioms:

- Associativity:

$$\theta \circ (\theta_1 \circ (\theta_{1,1}, \ldots, \theta_{1,k_1}), \ldots, \theta_n \circ (\theta_{n,1}, \ldots, \theta_{n,k_n}))$$
$$= (\theta \circ (\theta_1, \ldots, \theta_n)) \circ (\theta_{1,1}, \ldots, \theta_{1,k_1}, \ldots, \theta_{n,1}, \ldots, \theta_{n,k_n}).$$

- Identity: $\theta \circ (1, \ldots, 1) = \theta = 1 \circ \theta$.

Important examples of operads are the endomorphism operad, Lie operad, tree operad, 'little something' operad, etc.

A *non-unitary* operad is an operad without the identity axiom.

4.3 Construction of Polynomial Fireworks

The idea of the construction is as follows. We work with the space $\mathcal{M}_n$ of complex conjugacy classes $[p]$ of polynomials p of degree n where affine maps appear as a precomposition from the right and multiplication by a complex constant acts as a postcomposition from the left. Since any $p \in [p]$ belongs to the same connected component of $\mathcal{L}_n$, we will write simply p as a representative of $[p]$. The operation of composition of lemniscates (which will be used in the operad construction) consists of planting a zero of higher order in the place of the original zero and deforming it into simple zeros at the first moment.

Now the question is what happens analytically?

Take a polynomial $p \in [p] \in \mathcal{M}_n \subset \mathcal{L}_n$, and look at one of its zeros z_k. Let us consider a polynomial $q \in [q] \in \mathcal{M}_m \subset \mathcal{L}_m$. We want to define the operation $[p] \circ_k [q]$. Take another polynomial $\tilde{q} \in [q]$ such that all big lemniscates of $\tilde{q}$ are found inside the disk $U_r(z_k) = \{z \in \mathbb{C}: |z - z_k| < r\}$ for a sufficient small r such that U_r is inside the circular domain of the lemniscate configuration centered at z_k. Construct $\tilde{p} = (z - z_k)^{-1} p(z) \tilde{q}(z)$. Components of $\mathcal{L}_n$ are invariant under pre-composition with affine maps so we can assume without loss of generality that the polynomial p_n has one zero at 0 instead of z_k. If the polynomial $\tilde{p}$ is lemniscate generic, then the class $[\tilde{p}] \in \mathcal{M}_{n+m-1}$ will be the result of the superposition $[\tilde{p}] = [p] \circ_k [q]$. It is, of course, not true in general however it is always possible to find a path from the boundary point of $\mathcal{L}_{m+n-1}$ containing the non-lemniscate generic polynomial $z^{m-1} p(z)$ inside every component of $\mathcal{L}_{m+n-1}$ performing a deformation of z^{m-1} to $\tilde{q} \in [q]$, which will be shown in what follows.

Let us first show that given p and $\tilde{q}$ the deformation of z^{m-1} to $\tilde{q}$ keeps the roots and critical points of $\tilde{p} = z^{-1} p(z) \tilde{q}(z)$ in the same neighbourhood as those of $\tilde{q}$.

Lemma 1 *Let $z_1, \ldots, z_n \in \mathbb{D} = \{z \in \mathbb{C} : |z| < 1\}$ and $p(z) = \prod_{k=1}^{n}(z - z_k)$. Then*

$$\left| \frac{p'(z)}{p(z)} \right| = \left| \sum_{k=1}^{n} \frac{1}{z - z_k} \right| > \frac{n}{2}, \quad on \ \mathbb{T} = \partial \mathbb{D}.$$

Proof Indeed, consider the mapping $w = \frac{1}{1-z}$ of $\mathbb{D}$ onto the right half-plane $\{w \in \mathbb{C} : \mathrm{Re}\, w > \frac{1}{2}\}$. Then assume $\zeta_k \in \mathbb{D}$ we have

$$\left| \frac{1}{1 - \zeta_k} \right| \geq \mathrm{Re} \frac{1}{1 - \zeta_k} > \frac{1}{2} \quad \text{and} \quad \left| \sum_{k=1}^{n} \frac{1}{1 - \zeta_k} \right| \geq \mathrm{Re} \sum_{k=1}^{n} \frac{1}{1 - \zeta_k} > \frac{n}{2}.$$

If $|z| = 1$, then

$$\left| \sum_{k=1}^{n} \frac{1}{z - z_k} \right| = \left| \sum_{k=1}^{n} \frac{\bar{z}}{1 - \bar{z} z_k} \right| = \left| \sum_{k=1}^{n} \frac{1}{1 - \bar{z} z_k} \right|.$$

Substituting $\bar{z} z_k = \zeta_k$ we finish the proof. $\qquad\square$

Corollary 1 *Let $z_1, \ldots, z_n \in \{z \in \mathbb{C} : |z| < \varepsilon\}$ and $p(z) = \prod_{k=1}^{n}(z - z_k)$. Then*

$$\left| \frac{p'(z)}{p(z)} \right| = \left| \sum_{k=1}^{n} \frac{1}{z - z_k} \right| > \frac{n}{2\varepsilon}, \quad on \ \{z \in \mathbb{C} : |z| = \varepsilon\}.$$

Theorem 5 *Given lemniscate generic polynomials $p_n(z) = z \prod_{k=1}^{n-1}(z - z_k)$ and $q_m(z) = z \prod_{j=1}^{m-1}(z - w_j)$ with*

$$|w_j| < \varepsilon = \frac{m}{2(n-1) + m} \min_k |z_k|,$$

the critical points of q_m and $m - 1$ critical points of $P(z) = z^{-1} q_m p_n$ 'inherited' from q_m lie within the same disk $|z| < \varepsilon$.

Proof Indeed,

$$\left| \frac{p_n'(z)}{p_n(z)} - \frac{1}{z} \right| = \left| \sum_{k=1}^{n-1} \frac{1}{z - z_k} \right| \leq \sum_{k=1}^{n-1} \frac{1}{|z - z_k|} \leq \frac{n-1}{\min_k \min_{|z|=\varepsilon}\{|z - z_k|\}}.$$

Using the simple identities $\min_{|z|=\varepsilon}\{|z - z_k|\} = |z_k| - \varepsilon$ and

$$\frac{n-1}{\min_k\{|z_k| - \varepsilon\}} = \frac{m}{2\varepsilon}$$

for $\varepsilon = \frac{m}{2(n-1)+m} \min_k |z_k|$ we conclude by Corollary 1 that

$$\left| \frac{p'_n(z)}{p_n(z)} - \frac{1}{z} \right| \le \frac{m}{2\varepsilon} < \left| \frac{1}{z} + \sum_{j=1}^{m-1} (z - w_j) \right| = \left| \frac{q'_m(z)}{q_m(z)} \right|$$

on the circle $\{|z| = \varepsilon\}$. So,

$$\left| \frac{1}{z} q_m p'_n - \frac{1}{z^2} q_m p_n \right| < \left| \frac{1}{z} q'_m p_n \right|,$$

and Rouchè's theorem implies that the holomorphic functions $P'(z)$ and $\frac{1}{z} q'_m p_n$ have the same number of zeros inside the disk $\{|z| < \varepsilon\}$. Since the function $\frac{1}{z} p_n$ has no zeros in this disk then $P'(z)$ and q'_m have the same number of zeros there which proves the theorem.   □

Next, we note a conic-like structure of the sets $\mathcal{L}_n$, $\mathcal{L}_m$ and $\mathcal{L}_{n+m-1}$.

Theorem 6 *Let a lemniscate generic polynomial* $p_n(z) = z \prod_{k=1}^{n-1} (z - z_k)$ *belong to a connected component* $E' \subset \mathcal{L}_n$, *and let* $q_m(z) = z \prod_{j=1}^{m-1} (z - w_j)$ *belong to a connected component* $E'' \subset \mathcal{L}_m$. *There is a small deformation of the polynomial* $z^{m-1} p_n(z)$ *such that the resulting polynomial* $P(z)$ *belongs to a connected component* $E''' \subset \mathcal{L}_{n+m-1}$ *such that its projection to* $\mathcal{L}_n$ *is from* E' *and its projection to* $\mathcal{L}_m$ *is from* E''.

Proof Following [3] define a map $\psi \colon \mathbb{C}^* \times \mathbb{C} \times \mathbb{C}^{n-1} \to V_n$ by

$$\psi(a_n, a_0, y) = n a_n \left(\int \prod_{k=1}^{n-1} (z - y_k) dz \right) + a_0, \quad a_n \in \mathbb{C}^*, a_0 \in \mathbb{C}, y \in \mathbb{C}^{n-1},$$

where $\mathbb{C}^* = \mathbb{C} \setminus \{0\}$, V_n is the space of polynomials of degree n, and y_k, $k = 1, \ldots, n-1$ are the critical points of the polynomial p_n. The set $\psi^{-1}(V_n \setminus \mathcal{L}_n)$ is a real hypersurface whose equation is $\prod_{i<j} |P_y(y_i)| - |P_y(y_j)| = 0$, where

$$P_y(z) = n \left(\int \prod_{k=1}^{n-1} (z - y_k) dz \right).$$

All connected components of $\mathcal{L}_n$ have a unique common point at their boundaries that corresponds to the polynomial z^n. So starting from this point we can enter any of them. The polynomial $z^k p_n(z)$, $p_n \in \mathcal{L}_n$, is in the boundary of $\mathcal{L}_{n+k}$ and $\psi^{-1}(z^k p_n(z)) \in \partial \psi^{-1}(\mathcal{L}_{n+k})$, $\psi^{-1}(\mathcal{L}_{n+k}) = \psi^{-1}(\mathcal{L}_n) \times \psi^{-1}(\mathcal{L}_k)$, and there is a projection to $\psi^{-1}(\mathcal{L}_n)$ defined by $z^k p_n(z) \to p_n(z)$. Since z^k is a common boundary point for all connected components of $\mathcal{L}_k$, there exists a path connecting the boundary point $z^k \in \partial \mathcal{L}_k$ and an arbitrary point in every connected component of $\mathcal{L}_k$.

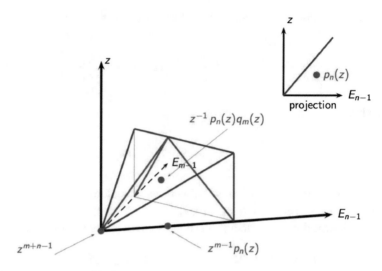

Fig. 4 Projections of $\psi^{-1}(z^{-1}p_n q_m)$

Applying now these arguments to the polynomials $P(z)$ and $z^{m-1}p_n(z)$ yields the conclusion of the theorem. (Figure 4 gives a schematic idea how these projections are realized for $\psi^{-1}(\mathcal{L}_{n+m-1})$.)   □

4.4 Operad Construction

We construct a non-unitary operad on central binary trees with labelled ends with numbers assigned to them. Let us note that central balanced trees representing lemniscate generic polynomials are in particular central and binary.

A generic element of $\mathcal{O}(k)$ is a central binary tree T with k labelled ends, together with a sequence of admissible pairs of numbers $((l_1, v_1), \ldots, (l_k, v_k))$. The tuple $(v_1, \ldots, v_k)$ consists of labels of the ends of T. Let s be the distance from the end v_j to the closest vertex with valence greater than one, which is either the center or one of the nodes. If $l_j \geq -s + 1$, we call the number l_j and the pair (l_j, v_j) admissible.

Example 1 The tree T on Fig. 5 is a central balanced tree of length 2. The center of the tree (marked by white) is the closest to the end 3 vertex of valence larger than one. The distance s between the end 3 and the center is 2. This means that an admissible number assigned to the vertex 3 must be be greater or equal -1. Vertices 1 and 2 can come in pair with some non-negative integers. For example, T together with $((0, 1), (0, 2), (1, 3))$ is an element of $\mathcal{O}(3)$.

We define the identity $1 \in \mathcal{O}(1)$ to be one labelled vertex with 0 assigned to it.

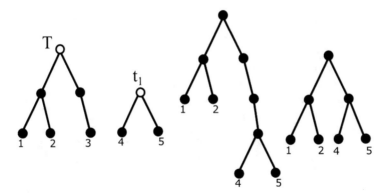

Fig. 5 Attaching a tree to an end: example

In what follows we define the non-unitary operad composition operation. Given an element of $\mathcal{O}(k)$ which is a central binary tree T with k pairs

$$\left((l_1^0, v_1^0), \ldots, (l_k^0, v_k^0)\right)$$

and k elements $t_1 \in \mathcal{O}(j_1)$ with pairs $\left((l_1, v_1), \ldots, (l_{j_1}, v_{j_1})\right), \ldots, t_k \in \mathcal{O}(j_k)$ with pairs $\left((l_{n-j_k+1}, v_{n-j_k+1}), \ldots, (l_n, v_n)\right)$, where $n = j_1 + \cdots + j_k$. The result $\widetilde{T} = (T; t_1, \ldots, t_k)$ of the composition of T and $t_1, \ldots, t_k$ must be an element of $\mathcal{O}(n)$, i.e. a central binary tree with n ends together with n admissible pairs.

Let us first describe the construction of the tree $\widetilde{T}$. As index p varies from 1 to k, we attach the center of the tree t_p to the end v_p^0 "at distance" l_p^0 from the end. Namely, if $l_p^0 \geq 0$, we attach the center of tree t_p to a chain of l_p^0 edges, and then we attach the resulting tree to the end v_p^0. If, in turn, the number l_p^0 is negative, we erase a chain of $|l_p^0|$ edges containing v_p^0 and attach the center of the tree t_p to the vertex which connected the deleted chain and the rest of the tree.

Example 2 Let T on Fig. 5 have pairs $((1, 3), (0, 1), (0, 2))$. The second from the right tree on Fig. 5 is the result of attaching the tree t_1 to the end 3 "at distance" 1. Suppose now T has pairs $((-1, 3), (0, 1), (0, 2))$, the result of attaching the tree t_1 is the tree on the right hand side of the figure.

Remark 3 Let β be the closest to v_j^0 vertex of valence larger than one in T. An admissible number l_j^0 is defined so that we never erase β and have a freedom to attach a tree closer or further from β.

When all $t_1, \ldots, t_k$ are attached to T, the resulting tree $\overline{T}$ is not necessarily central. We transform the resulting tree into a central tree $\widetilde{T}$.

The former center v of T will be the center of the tree $\widetilde{T}$, i.e. the distance from the center v of $\widetilde{T}$ to all the ends of $\widetilde{T}$ must be the same. The possible distances from v to the ends of $\overline{T}$ are $S_p = |T| + l_p^0 + |t_p|$, where $|T|$ and $|t_p|$ are lengths of T and t_p respectively where p varies from 1 to k. The maximum S among S_p, $1 \leq p \leq k$, will

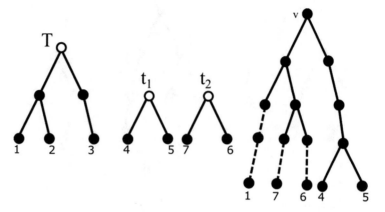

Fig. 6 Composition of trees: example

be the length of $\widetilde{T}$, i.e.

$$|\widetilde{T}| = |(T; t_1, \ldots, t_k)| = S = |T| + \max_{1 \le j \le k} \left(l_j^0 + |t_j| \right).$$ (5)

If an end of $\overline{T}$ is at distance S_p from v, we add a chain of

$$\delta_p := S - S_p = \max_{1 \le j \le k} \left(l_j^0 + |t_j| \right) - \left(l_p^0 + |t_p| \right)$$

edges to the end, the new end inherits the label.

Example 3 Let T on Fig. 6 have pairs $((1, 3), (0, 2), (0, 1))$. Let us construct the composition of the trees T and $(t_1, t_2, 1)$. We will not specify the pairs assigned to the trees t_1 and t_2 for now and focus on the construction of the tree. We attach the tree t_1 to the vertex 3 "at distance" 1, the tree t_2 to the vertex 2 and the identity element 1 to the vertex 1. The tree $\overline{T}$ is shown on Fig. 6 with solid lines. The maximal distance from the ends of $\overline{T}$ to the former center v is 4. We attach a chain of 2 edges to the vertex 1 and chains of 1 edge to the vertices 7 and 6, the added edges are shown with dashed lines. The new ends inherit the labels of the old ends. The resulting tree $\widetilde{T}$ is a central tree of length 4 with 5 ends.

The resulting tree $\widetilde{T}$ has $j_1 + \cdots + j_k = n$ vertices

$$v_1, \ldots, v_{j_1}, v_{j_1+1}, \ldots, v_{j_1+j_2}, \ldots, v_{n-j_k+1}, \ldots, v_n.$$

Given an integer m between 1 and n, assume $j_1 + \cdots + j_{p-1} + 1 \le m \le j_1 + \cdots + j_p$, for some $1 \le p \le k$. To the vertex v_m, which is inherited from t_p, we assign the number $l_m - \delta_p$, where δ_p is defined above and represents the difference between the length S of the tree $\widetilde{T}$ and the distance S_p from the vertex v_m to v in $\overline{T}$. The pairs assigned to the tree $\widetilde{T}$ are as follows:

$$((l_1 - \delta_1, v_1), \ldots, (l_{j_1} - \delta_1, v_{j_1}), (l_{j_1+1} - \delta_2, v_{j_1+1}), \ldots, (l_{j_1+j_2} - \delta_2, v_{j_1+j_2}), \ldots,$$

$$(l_{n-j_k+1} - \delta_k, v_{n-j_k+1}), \ldots, (l_n - \delta_k, v_n)).$$

This concludes the construction of the composition $(T; t_1, \ldots, t_k) \in \mathcal{O}(n)$.

Example 4 In the previous example we have $\delta_1 = 0$, $\delta_2 = 1$, $\delta_3 = 2$. Let t_1 and t_2 have sequences $((0, 4), (1, 5))$ and $((1, 7), (2, 6))$ correspondingly. The identity element is a vertex with a pair $(0, 1)$.

The tree $\widetilde{T}$ is assigned the following sequence:

$$((0, 4), (1, 5), (0, 7), (1, 6), (-2, 1)).$$

Let us verify that the number $l_m - \delta_p$ assigned to the end v_m in $\widetilde{T} = (T; t_1, \ldots, t_k)$ is admissible. The integer l_m assigned to the vertex v_m in t_p is admissible, i.e.,

$$l_m \geq -s + 1, \tag{6}$$

where s is the distance from v_m to the closest vertex of valence greater than one in t_p. Note that the former center of t_p becomes a node in $\widetilde{T}$. The end v_m of $\widetilde{T}$ is at distance $s + \delta_p$ from the closest node of $\widetilde{T}$, and this distance is shorter than the distance to the center v of $\widetilde{T}$. An admissible number assigned to v_m in $\widetilde{T}$ should be larger or equal to $-(s + \delta_p) + 1$. By construction we assign to v_m a number $l_m - \delta_p$, from (6) we obtain that $l_m \geq -s + 1 - \delta_p$. Therefore, the new number l_m for the vertex v_m of the composition $\widetilde{T}$ is admissible and thus the operation of composition of elements of $\mathcal{O}(k)$ is well-defined.

Theorem 7 *The sequence* $\{\mathcal{O}(n)\}_{n=1}^{\infty}$ *of central binary trees with labelled vertices and admissible numbers assigned to them, together with operation of composition defined above, forms a non-unitary operad.*

The operation of composition is defined in accordance with Definition 1 of an operad. We only need to show that the associativity axiom holds. In order to do that consider an element $T \in \mathcal{O}(k)$, with a sequence

$$((l_1^0, v_1^0), \ldots, (l_k^0, v_k^0)),$$

k elements $t_J \in \mathcal{O}(j_J)$ $1 \leq J \leq k$ with pairs

$$((l_{j_1+\cdots+j_{J-1}+1}, v_{j_1+\cdots+j_{J-1}+1}), \ldots, (l_{j_1+\cdots+j_J}, v_{j_1+\cdots+j_J})),$$

$\sum_1^k j_J = n$, and n elements $\tau_L \in \mathcal{O}(m_L)$, $1 \leq L \leq n$, with sequences

$$((\lambda_{m_1+\cdots+m_{L-1}+1}, \beta_{m_1+\cdots+m_{L-1}+1}), \ldots, (\lambda_{m_1+\cdots+m_L}, \beta_{m_1+\cdots+m_L})).$$

We need to prove that

$$((T; t_1, \ldots, t_k); \tau_1, \ldots, \tau_n) = \left(T; (t_1; \tau_1, \ldots, \tau_{j_1}), \ldots, (t_k; \tau_{n-j_k+1}, \ldots, \tau_n)\right). \quad (7)$$

We denote the left hand side element by T_1 and the right hand side element by T_2.

Let us first show that the trees T_1 and T_2 coincide. Then we conclude the proof by showing that the elements T_1 and T_2 have the same sets of pairs.

Lemma 2 *The trees T_1 and T_2 coincide.*

Proof Let us show that the trees T_1 and T_2 have the same length

First, we calculate $|T_1| = |((T; t_1, \ldots, t_k); \tau_1, \ldots \tau_n)|$.

By (5) the length $|(T; t_1, \ldots, t_k)|$ is given by

$$|(T; t_1, \ldots, t_k)| = |T| + \max_{1 \le j \le k}\{l_j^0 + |t_j|\}.$$

The tree $(T; t_1, \ldots, t_k)$ has vertices

$$v_1, \ldots, v_{j_1}, \ldots, v_{n-j_k+1}, \ldots, v_n$$

with numbers

$$l_1 - (\max_{1 \le j \le k}\{l_j^0 + |t_j|\} - (l_1^0 + |t_1|)), \ldots,$$

$$l_{j_1} - (\max_{1 \le j \le k}\{l_j^0 + |t_j|\} - (l_1^0 + |t_1|)), \ldots,$$

$$l_{n-j_k+1} - (\max_{1 \le j \le k}\{l_j^0 + |t_j|\} - (l_k^0 + |t_k|)), \ldots,$$

$$l_n - (\max_{1 \le j \le k}\{l_j^0 + |t_j|\} - (l_k^0 + |t_k|)).$$

Thus the length of $((T; t_1, \ldots, t_k); \tau_1, \ldots \tau_n)$ is

$$|T| + \max_{1 \le j \le k}\{l_j^0 + |t_j|\} + \max_{1 \le m \le n}\{l_m + |\tau_m|\} =$$
$$= |T| + \max_{1 \le m \le n}\{|\tau_m| + l_m + l_{j(m)}^0 + |t_{j(m)}|\}. \quad (8)$$

Here $j(m) = 1$ if $m = 1, \ldots, j_1$, $j(m) = 2$ if $m = j_1 + 1, \ldots, j_1 + j_2$, and so on. Given m, the tree τ_m is glued to an end of some tree t_j, so $j = j(m)$ is the index of this tree, which is uniquely determined by m.

Let us calculate now $|T_2| = |(T; (t_1; \tau_1, \ldots \tau_{j_1}), \ldots, (t_k; \tau_{n-j_k+1}, \ldots, \tau_n))|$.

$$|T_2| = |T| + \max_{1 \le p \le k}\{l_p^0 + |t_p| + \max_{j_1 + \cdots + j_{p-1}+1 \le m \le j_1 + \cdots + j_p}\{l_m + |\tau_m|\}\}. \quad (9)$$

Given $1 \le p \le k$, we define $m(p)$, $1 \le m \le n$, to be indices of τ_m that are glued to the tree t_p. Given p, the values of $m(p)$ are integers between $j_1 + \cdots + j_{p-1} + 1$

and $j_1 + \cdots + j_p$. The expression (9) can be rewritten as

$$|T_2| = |T| + \max_{1 \leq p \leq k} \max_{m(p)} \{l_p^0 + |t_p| + l_m + |\tau_m|\}. \tag{10}$$

The expressions (10) and (8) coincide. Therefore, the trees T_1 and T_2 have the same length.

Let us show now that the trees T_1 and T_2 coincide.

Let us fix $1 \leq p \leq k$ and let v be the center of t_p. The center of t_p is glued to the end v_p^0 of T "at distance" l_p^0 during the construction of both T_1 and T_2.

We denote by β a vertex in T of valence greater than one closest to v_p^0. The vertex β is either the center, or a node of T. The end v_p^0 is attached to β by a chain of edges. During construction of T_1 and T_2 we replace this chain of edges with a new one and attach the center v of t_p to it. It is clear that the distance from v to the center of T in both cases is the same and equal to $|T| + l_p^0$. We can conclude that the trees t_p, $1 \leq p \leq k$ are attached to the same positions in T_1 and T_2.

Let us now denote by b the center of τ_L, $1 \leq L \leq n$. Suppose τ_L is attached to the end v_L of t_J for some J between 0 and k. We denote by V the vertex of t_J of valence larger than one, that is closest to v_L. The end v_L is attached to V by a chain of edges. We replace this chain of edges with another one and attach the center b of τ_L to the new chain. The distance from b to V is $|t_J| + l_L$ both, in $\left(T; (t_1; \tau_1, \ldots \tau_{j_1}), \ldots, (t_k; \tau_{n-j_k+1}, \ldots, \tau_n)\right)$, and $((T; t_1, \ldots, t_k); \tau_1, \ldots \tau_n)$. We can conclude that the trees τ_L, $1 \leq L \leq n$ are attached to the same positions in T_1 and T_2.

In addition, the lengths of T_1 and T_2 coincide, thus the trees T_1 and T_2 coincide.

$\square$

The following lemma concludes the proof of the theorem.

Lemma 3 *The pairs assigned to T_1 and T_2 are identical.*

Proof First we calculate the sequences for the left hand side of (7).

The composition $(T; t_1, \ldots, t_k)$ is an element in $\mathcal{O}(n)$ with the sequence

$$\left((l_1 - (S - S_1), v_1), \ldots, (l_{j_1} - (S - S_1), v_{j_1}),\right.$$
$$(l_{j_1+1} - (S - S_2), v_{j_1+1}), \ldots, (l_{j_1+j_2} - (S - S_2), v_{j_1+j_2}), \ldots,$$
$$\left.(l_{n-j_k+1} - (S - S_k), v_{n-j_k+1}), \ldots, (l_n - (S - S_k), v_n)\right),$$

where $S_J = |T| + l_J^0 + |t_J|$, $1 \leq J \leq k$; S is the maximum of S_J. S is the length of the tree $(T; t_1, \ldots, t_k)$.

Composition $((T; t_1, \ldots, t_k); \tau_1, \ldots \tau_n)$ is an element in $\mathcal{O}(m_1 + \cdots + m_n)$. Let L be an integer between 1 and n. Recall that $n = j_1 + \cdots + j_k$ and suppose $j_1 + \cdots + j_{p-1} + 1 \leq L \leq j_1 + \cdots + j_p$ for some $1 \leq p \leq k$. Note that $1 \leq L - (j_1 + \cdots + j_{p-1}) \leq j_p$. We attach the tree τ_L to the end v_L "at distance" $l_L - (S - S_p)$. We define

$$\Delta_L = |(T; t_1, \ldots, t_k)| + l_L - (S - S_p) + |\tau_L|,$$

which can be rewritten as

$$\Delta_L = l_L + S_p + |\tau_L|,$$

and let Δ denote the maximum of Δ_L. We obtain

$$\Delta = |\ ((T;t_1,\ldots,t_k);\tau_1,\ldots\tau_n)\ |.$$

The composition $((T;t_1,\ldots,t_k);\tau_1,\ldots\tau_n)$ has the following sequence:

$$((\lambda_1 - (\Delta - \Delta_1), \beta_1),\ldots, (\lambda_{m_1} - (\Delta - \Delta_1), \beta_{m_1}),$$

$$(\lambda_{m_1+1} - (\Delta - \Delta_2), \beta_{m_1+1}),\ldots, (\lambda_{m_1+m_2} - (\Delta - \Delta_2), \beta^2_{m_1+m_2}),\ldots,$$

$$(\lambda_{m_1+\cdots+m_{n-1}+1} - (\Delta - \Delta_n), \beta_{m_1+\cdots+m_{n-1}+1}),\ldots,$$

$$(\lambda_{m_1+\cdots+m_n} - (\Delta - \Delta_n), \beta_{m_1+\cdots+m_n}))\,.$$

Let R be an integer between 1 and $m_1 + \cdots + m_n$. Suppose

$$m_1 + \cdots + m_{L-1} + 1 \le R \le m_1 + \cdots + m_L$$

for some $1 \le L \le n$ and, also, assume

$$j_1 + \cdots + j_{p-1} + 1 \le L \le j_1 + \cdots + j_p$$

for some $1 \le p \le k$. The R-th pair in the sequence assigned to T_1 has form $(\lambda_R - (\Delta - \Delta_L), \beta_R)$, where

$$\Delta - \Delta_L = |\ ((T;t_1,\ldots,t_k);\tau_1,\ldots\tau_n)\ | - (l_L + S_p + |\tau_L|) =$$

$$= |\ ((T;t_1,\ldots,t_k);\tau_1,\ldots\tau_n)\ | - (l_L + |T| + l_p^0 + |t_p| + |\tau_L|)\,.$$

Let us now write down the sequence of pairs for T_2.

Let $1 \le L \le n$ and also assume that $j_1 + \cdots + j_{p-1} + 1 \le L \le j_1 + \cdots + j_p$ for some $1 \le p \le k$. We define $\sigma_L = |t_p| + l_L + |\tau_L|$, Σ_p is the maximum of σ_L, where $j_1 + \cdots + j_{p-1} + 1 \le L \le j_1 + \cdots + j_p$, it is the length of the tree $(t_p; \tau_{j_1+\cdots+j_{p-1}+1},\ldots,\tau_{j_1+\cdots+j_p})$.

Given p between 1 and k, the element $(t_p; \tau_{j_1+\cdots+j_{p-1}+1},\ldots,\tau_{j_1+\cdots+j_p})$ has the sequence

$$\left(\left(\lambda_{m_1+\cdots+m_{j_1+\cdots+j_{p-1}}+1} - (\Sigma_p - \sigma_{j_1+\cdots+j_{p-1}+1}), \beta_{m_1+\cdots+m_{j_1+\cdots+j_{p-1}}+1}\right),\ldots,\right.$$

$$\left(\lambda_{m_1+\cdots+m_{j_1+\cdots+j_{p-1}+1}} - (\Sigma_p - \sigma_{j_1+\cdots+j_{p-1}+1}), \beta_{m_1+\cdots+m_{j_1+\cdots+j_{p-1}+1}}\right),\ldots,$$

$$\Big(\lambda_{m_1+\cdots+m_{j_1}+\cdots+j_p-1+1} - (\Sigma_p - \sigma_{j_1+\cdots+j_p}), \beta_{m_1+\cdots+m_{j_1}+\cdots+j_p-1+1}\Big), \ldots,$$

$$\Big(\lambda_{m_1+\cdots+m_{j_1}+\cdots+j_p} - (\Sigma_p - \sigma_{j_1+\cdots+j_p}), \beta_{m_1+\cdots+m_{j_1}+\cdots+j_p}\Big)\Big).$$

We attach the center of the tree $(t_p; \tau_{j_1+\cdots+j_{p-1}+1}, \ldots, \tau_{j_1+\cdots+j_p})$ to the vertex v_p^0 "at distance" l_p^0, as p varies between 1 and k.

We define $Q_p = |T| + l_p^0 + |(t_p; \tau_{j_1+\cdots+j_{p-1}+1}, \ldots, \tau_{j_1+\cdots+j_p})| = |T| + l_p^0 + \Sigma_p$. We define Q to be the maximum among Q_p, $1 \le p \le k$; Q is the length of the tree T_2.

Let us choose R, where $1 \le R \le m_1 + \cdots + m_n$, and write down the R-th pair of T_2. Suppose $m_1 + \cdots + m_{L-1} + 1 \le R \le m_1 + \cdots + m_L$ for some $1 \le L \le n$ and suppose $j_1 + \cdots + j_{p-1} + 1 \le L \le j_1 + \cdots + j_p$ for some $1 \le p \le k$. The R-th pair in the sequence for T_2 has the form

$$\big(\lambda_R - (\Sigma_p - \sigma_L) - (Q - Q_p), \beta_R\big).$$

Let us rewrite

$$(\Sigma_p - \sigma_L) + (Q - Q_p) = \Sigma_p - (|t_p| + l_L + |\tau_L|) + |T_2| - (|T| + l_p^0 + \Sigma_p) =$$

$$= |T_2| - \big(|t_p| + l_L + |\tau_L| + |T| + l_p^0\big).$$

We can see that the R-th pair for $((T; t_1, \ldots, t_k); \tau_1, \ldots \tau_n)$ coincides with the R-th pair for $\big(T; (t_1; \tau_1, \ldots \tau_{j_1}), \ldots, (t_k; \tau_{n-j_k+1}, \ldots, \tau_n)\big)$. □

Remark 4 In Sect. 4.3 we discussed the composition $[p] \circ_k [q]$ of conjugacy classes of lemniscate generic polynomials $p \in \mathcal{L}_n$ and $q \in \mathcal{L}_m$.

Let T and t be the central balanced trees of length $n - 1$ and $m - 1$ which correspond to p and q respectively. Let us label the zeros of p, and respectively ends of T, by numbers from 1 to n.

Geometrically, composition $[p] \circ_k [q]$ can be described as follows. We take a neighbourhood containing the big lemniscates of q, shrink it and replace with it a small neighbourhood of the zero k of p. In terms of trees this is analogous to gluing the tree t to the vertex k of T "at distance" 0. To obtain the lemniscate configuration of $[p] \circ_k [q]$ we add extra circumferences around the zeros of p, or, in terms of trees, extend the tree (as described in Sect. 4.4) so that it becomes a central balanced tree of length $n + m - 1$.

The composition of p and q can be realized as composition of an element $T \in \mathcal{O}(n)$ with the tuple $(1, \ldots, 1, t, 1, \ldots, 1)$ of length n, there $t \in \mathcal{O}(m)$ in at the k-th place in the tuple, the trees T and t represent p and q, 1 is the identity in $\mathcal{O}(1)$. The polynomial fireworks can be realized through the non-unitary operad constructed in Sect. 4.4. Thus, the goal we set in the Introduction has been achieved.

Acknowledgements All authors were supported by the Norwegian Research Council #213440/BG. The third author was also supported by the grants of the Norwegian Research Council #239033/F20 and by EU FP7 IRSES program STREVCOMS, grant no. PIRSES-GA-2013-612669.

References

1. J.M. Boardman, R.M. Vogt, *Homotopy Invariant Algebraic Structures on Topological Spaces*. Lecture Notes in Mathematics, vol. 347 (Springer, Berlin, 1973)
2. F. Catanese, M. Paluszny, Polynomial-lemniscates, trees and braids. Topology **30**(4), 623–640 (1991)
3. F. Catanese, B. Wajnryb, The fundamental group of generic polynomials. Topology **30**(4), 641–651 (1991)
4. D.A. Cox, *Galois Theory* (Wiley, Hoboken, NJ, 2004), pp. 457–508
5. P. Ebenfelt, D. Khavinson, H.S. Shapiro, Two-dimensional shapes and lemniscates, in *Complex Analysis and Dynamical Systems IV. Part 1*. Contemporary Mathematics, vol. 553 (American Mathematical Society, Providence, RI, 2011), pp. 45–59
6. D. Hilbert, Über die Entwickelung einer beliebigen analytischen Function einer Variabeln in eine unendliche, nach ganzen rationalen Functionen fortschreitende Reihe. Gött. Nachr. **1897**, 63–70 (1897)
7. A.A. Kirillov, Geometric approach to discrete series of unirreps for Vir. J. Math. Pures Appl. **77**, 735–746 (1998)
8. A.A. Kirillov, D.V. Yur'ev, Kähler geometry and the infinite-dimensional homogenous space $M = \mathrm{Diff}_+(S^1)/\mathrm{Rot}(S^1)$. Funct. Anal. Appl. **21**(4), 284–294 (1987)
9. M. Markl, S. Shnider, J. Stasheff, *Operads in Algebra, Topology and Physics*. Mathematical Surveys and Monographs, vol. 96 (American Mathematical Society, Providence, RI, 2002)
10. M. Markl, S. Shnider, J. Stasheff, *Operads in Algebra, Topology and Physics (English Summary)*. Mathematical Surveys and Monographs, vol. 96 (American Mathematical Society, Providence, RI, 2002)
11. D.E. Marshall, Zipper, fortran programs for numerical computation of conformal maps, and C programs for X-11 graphics display of the maps. Sample pictures, Fortran, and C code available online at http://www.math.washington.edu/marshall/personal.html
12. D.E. Marshall, S. Rohde, Convergence of a variant of the zipper algorithm for conformal mapping. SIAM J. Numer. Anal. **45**(6), 2577–2609 (2007)
13. J.P. May, *The Geometry of Iterated Loop Spaces*. Lecture Notes in Mathematics, vol. 271 (Springer, Berlin, 1972)
14. J. Milnor, *Morse Theory*. Annals of Mathematics Studies (Princeton University Press, Princeton, NJ, 1963)
15. D. Mumford, Pattern theory: the mathematics of perception, in *Proceedings ICM 2002*, vol. 1, pp. 401–422
16. Ch. Pommerenke, *Boundary Behaviour of Conformal Maps* (Springer, Berlin, 1992)
17. T.A. Rakcheeva, Multifocus lemniscates: approximation of curves. Zh. Vychisl. Mat. Mat. Fiz. **50**(11), 2060–2072 (2010). Translation in Comput. Math. Math. Phys. **51**(11), 1956–1967 (2011)
18. T.A. Rakcheeva, Focal approximation on the complex plane. Zh. Vychisl. Mat. Mat. Fiz. **51**(11), 1963–1972 (2011). Translation in Comput. Math. Math. Phys. **51**(11), 1847–1855 (2011)
19. T. Richards, M. Younsi, Conformal models and fingerprints of pseudo-lemniscates. Constr. Approx. **45**(1), 129–141 (2017)
20. E. Sharon, D. Mumford, 2d-shape analysis using conformal mapping. Int. J. Comput. Vis. **70**(1), 55–75 (2006)
21. J.L. Walsh, *Interpolation and Approximation by Rational Functions in the Complex Domain*, 4th edn. American Mathematical Society Colloquium Publications, vol. XX (American Mathematical Society, Providence, RI, 1965)
22. M. Younsi, Shapes, fingerprints and rational lemniscates. Proc. Am. Math. Soc. **144**(3) 1087–1093 (2016)

Regularity of Mappings with Integrally Restricted Moduli

Anatoly Golberg and Ruslan Salimov

Dedicated to the memory of Sasha Vasil'ev

Abstract We consider certain classes of homeomorphisms of domains in $\mathbb{R}^n$ with integrally bounded p-moduli of the families of curves and surfaces, which essentially extend the well-known classes of mappings such as quasiconformal, quasiisometric, Lipschitzian, etc. In the paper we survey the known results in this field regarded to the differential properties of such homeomorphisms, but mainly present a wide range of open related problems.

Keywords Weighted p-module • Q-homeomorphisms • Lipschitz mappings • Quasiconformal mappings • Hölder continuity • Lusin's (N) • (N^{-1})-properties

2010 Mathematics Subject Classification Primary: 30C65, 26B05; Secondary: 26B30, 26B35

1 Introduction

It is well known that the class of conformal mappings of the multidimensional Euclidean spaces exhausts only Möbius transformations. This strong rigidity has forced to introduce various classes of mappings which preserve in some sense

A. Golberg (✉)
Department of Mathematics, Holon Institute of Technology, 52 Golomb St., P.O.B. 305, Holon 5810201, Israel
e-mail: golberga@hit.ac.il

R. Salimov
Institute of Mathematics, National Academy of Sciences of Ukraine, 3 Tereschenkivska St., Kiev-4 01601, Ukraine
e-mail: ruslan623@yandex.ru

© Springer International Publishing AG 2018
M. Agranovsky et al. (eds.), *Complex Analysis and Dynamical Systems*, Trends in Mathematics, https://doi.org/10.1007/978-3-319-70154-7_8

the main features of analyticity in the plane. Then the class of quasiconformal mappings (homeomorphisms of the Sobolev space $W^{1,n}$ with uniformly bounded distortion coefficient $K(x)$) has been defined first in the plane and later in $\mathbb{R}^n$. It is well known that such mappings admit rich analytic and geometric properties like differentiability almost everywhere, preservation of sets of zero measure (Lusin's (N) and (N^{-1})-properties), etc. Nonhomeomorphic quasiconformal mappings are called quasiregular or mappings with bounded distortion (following Reshetnyak). These mappings are always open and discrete and their multiplicity function is bounded on compact sets. The next essential extension of quasiconformality and quasiregularity leads to the class of mappings of finite distortion. Here the uniform boundedness of the distortion function is relaxed by finiteness almost everywhere. Many important features of quasiregular mappings are preserved by assuming appropriate conditions on $K(x)$ like local integrability, etc. On the other hand, the mappings of finite distortion have interesting applications in elasticity theory. For the main properties of quasiconformal, quasiregular mappings and mappings of finite distortion we refer to the monographs [20, 22, 23, 26, 32, 33, 41, 42].

One of the most fruitful tools for studying various properties of mappings is given by the module method which goes back to the classical papers by Ahlfors-Beurling [1] and Fuglede [7]. A crucial point here is that in contrast to other methods (like variational methods, analytic methods based on studying the Beltrami differential equations, etc.), the module method is easily applied for higher dimensions and allows one to derive many deep analytic properties of mappings; see, e.g. [2–6, 9–12, 14–18, 21, 24, 25, 35–38, 40].

In this paper we restrict ourselves by discussing various open questions on regularity of homeomorphisms with integrally controlled p-moduli in multidimensional domains.

2   Regularity Properties for General Homeomorphisms

We intend to provide here a short sketch of the regularity properties of homeomorphic (or more general continuous) mappings in higher dimensions outlined mainly the book [22]. For recent results regarding differentiability we also refer [30, 40].

First we note that every mapping of the Sobolev class $W^{1,p}$ is differentiable almost everywhere (a.e.) when $p > n$ and $n \geq 2$. The restriction $p > n$ cannot be omitted because even in the case of $p = n$ there exist mappings from $W^{1,p}$ which are not continuous at any point and, hence, fail to be differentiable a.e.

A sufficient condition for a mapping $f : G \to \mathbb{R}^n$ to be differentiable a.e. is the classical Stepanov's result, which states that the finiteness of the limit

$$\limsup_{y \to x} \frac{|f(x) - f(y)|}{|x - y|} < \infty \quad \text{a.e.} \quad x \in G,$$

ensures the differentiability a.e. in G.

In the case of homeomorphisms $f \in W^{1,p}$, the weaker restrictions with $p > n - 1$ for $n > 2$ and $p \geq 1$ for $n = 2$ guaranty differentiability a.e.

We also remind that a continuous mapping f satisfies (N)-property with respect to k-dimensional Hausdorff area if $\mathcal{H}^k_{\mathcal{S}}(f(B)) = 0$ whenever $\mathcal{H}^k_{\mathcal{S}}(B) = 0$. Similarly, f has (N^{-1})-property if $\mathcal{H}^k_{\mathcal{S}}(B) = 0$ whenever $\mathcal{H}^k_{\mathcal{S}}(f(B)) = 0$.

From the mathematical point of view, these properties are crucial for the change variable formulas.

A continuous mapping $f \in W^{1,p}$ always satisfies the Lusin (N)-property with respect to the n-dimensional Lebesgue measure when $p > n$, and there exist continuous mappings $f \in W^{1,n}$, which fail to have the (N)-property. For homeomorphisms, the restriction for p can be reduced to $p \geq n$.

On equivalence of nonvanishing of Jacobian and (N^{-1})-property see [31].

3 Homeomorphisms with Integrally Restricted Moduli

Let $\mathcal{S}_k$ be a family of k-dimensional surfaces $\mathcal{S}$ in $\mathbb{R}^n$, $1 \leq k \leq n - 1$ (curves for $k = 1$). The *conformal module* of $\mathcal{S}_k$ is defined as

$$\mathcal{M}(\mathcal{S}_k) = \inf \int_{\mathbb{R}^n} \rho^n \, dm(x), \tag{1}$$

where the infimum is taken over all admissible functions for $\mathcal{S}_k$ (abr. $\rho \in \operatorname{adm} \mathcal{S}_k$), i.e. Borel measurable functions $\rho \geq 0$ such that $\int_{\mathcal{S}} \rho^k \, d\sigma_k \geq 1$ for every $\mathcal{S} \in \mathcal{S}_k$.

The quasiinvariance of the conformal module, i.e.

$$K^{\frac{k-n}{n-1}} \mathcal{M}(\mathcal{S}_k) \leq \mathcal{M}(f(\mathcal{S}_k)) \leq K^{\frac{n-k}{n-1}} \mathcal{M}(\mathcal{S}_k),$$

completely characterizes quasiconformality [39], while the quasiinvariance of p-module ($p \neq n$), where the exponent n in (1) is replaced by p, describes quasiisometric (bilipschitz) mappings; see, e.g. [13, 24].

Let $f : G \to G^*$, $G, G^* \subset \mathbb{R}^n$, be a homeomorphism such that f and its inverse f^{-1} are ACL and differentiable almost everywhere with nonzero Jacobians in G and G^*, respectively. It was shown in [12], that if, in addition, f preserves (N) and (N^{-1})-properties with respect to k-dimensional Hausdorff areas and $K^{-1}_{O,p}, K_{I,p} \in L^1_{\text{loc}}(G)$, then the following bounds for the p-module of k-dimensional surface families

$$\inf_{\varrho \in \operatorname{ext}_p \operatorname{adm} \mathcal{S}_k} \int_G \frac{\varrho^p(x)}{K_{O,p}(x,f)} \, dm(x) \leq \mathcal{M}_p(f(\mathcal{S}_k)) \leq \inf_{\rho \in \operatorname{adm} \mathcal{S}_k} \int_G \rho^p(x) K_{I,p}(x,f) \, dm(x)$$

$$\tag{2}$$

are fulfilled. Here $K_{I,p}(x,f)$ and $K_{O,p}(x,f)$ stand for the p-inner and p-outer dilatations of f at $x \in G$ (see, e.g., [9]). The notation $\varrho \in \text{ext}_p\text{adm }\mathcal{S}_k$ means that ϱ is p-extensively admissible for the family $\mathcal{S}_k$, that is $\varrho \in \text{adm }(\mathcal{S}_k \backslash \widetilde{\mathcal{S}}_k)$, provided that $\mathcal{M}_p(f(\widetilde{\mathcal{S}}_k)) = 0$.

The basic inequality (2) gives raise to investigate more general module type relation for mappings which satisfy

$$\mathcal{M}_p(f(\mathcal{S}_k)) \leq \inf_{\rho \in \text{adm }\mathcal{S}_k} \int_G \rho^p(x) Q(x) \, dm(x), \tag{3}$$

or

$$\mathcal{M}_p(f(\mathcal{S}_k)) \geq \inf_{\varrho \in \text{ext}_p\text{adm }\mathcal{S}_k} \int_G \frac{\varrho^p(x)}{Q(x)} \, dm(x), \tag{4}$$

with a given measurable function $Q : G \to [0, \infty]$. Note that the behavior of mappings satisfying (4) is completely unknown except for a special case when the family $\mathcal{S}_k$ contains only spherical segments and $p > n$. The situation with (3) for $k > 1$ is also absolutely unknown. In this paper we collect all known regularity results regarding the case $k = 1$ for the inequality (3) and its weaker form and formulate several open questions.

4 Q-Homeomorphisms

Inequalities of type (3) naturally lead to important extensions of quasiconformality and quasiisometry called (p, Q)-*homeomorphisms* or Q-*homeomorphisms with respect to p-module*. The mappings subject to (4) are called *lower Q-homeomorphisms with respect to p-module*. An essential difference between (3) and (4) is based on the following property: the infimum in (3) can be omitted assuming that ρ is an arbitrary admissible metric for the appropriate family $\mathcal{S}_k$. However, such dropping infimum is not allowable in the inequality (4).

As was mentioned above, we shall restrict ourselves by considering only the inequalities of kind (3) related to the case $k = 1$. More precisely, given a measurable function $Q : G \to [0, \infty]$, we call a homeomorphism $f : G \to \mathbb{R}^n$ the Q-*homeomorphism with respect to p-module* if

$$\mathcal{M}_p(f(\Gamma)) \leq \int_G Q(x) \rho^p(x) \, dm(x) \tag{5}$$

for every family of curves Γ in G and for every admissible function ρ for Γ.

For $p = n$, the class of Q-homeomorphisms was introduced in [27]. It admits various rich properties; see, e.g. [28] and [38] (the latter for open and discrete mappings). We are focused here to the case $p \neq n$. The ranges $1 \leq p < n$ and $p > n$ for p are needed to apply, similar to the classical case $p = n$, the "continuum" analysis, since for $p < n$ and $p > n$ there appear some essential differences from the case $p = n$.

The class of Q-homeomorphisms with respect to p-module provides a nice extension of Lipschitz homeomorphisms. It easily follows from Gehring's result [8] assuming $Q \in L^\infty$. Recall that $f : G \to G^*$ belongs to $\text{Lip}(K)$ if there is K, $0 < K < \infty$, satisfying

$$\limsup_{y \to x} \frac{|f(y) - f(x)|}{|y - x|} \leq K$$

for each $x \in G$.

Theorem 4.1 *If $f : G \to G^*$ is a Q-homeomorphism with respect to p-module such that $Q(x) \leq K$ a.e. in G, then*

(1) $f \in \text{Lip}(K_0)$ for $n - 1 < p < n$;
(2) $f^{-1} \in \text{Lip}(K_0)$ for $p > n$.

In both cases K_0 depends only on p, n, and K.

The class of Q-homeomorphisms with respect to p-moduli have been introduced in [10]. The regularity properties like differentiability almost everywhere and absolute continuity on lines are established for $p > n - 1$ and locally integrable majorant Q. For $p = n$ see [34].

Theorem 4.2 *Let $f : G \to G^*$ be a Q-homeomorphism with respect to p-module with $Q \in L^1_{\text{loc}}(G)$ and $p > n - 1$. Then*

(1) f is ACL-homeomorphism;
(2) f is differentiable a.e. in G;
(3) $f \in W^{1,1}_{\text{loc}}(G)$.

For the proofs of the above results see also [10, 14, 34]. Omitting the restrictions on p and Q, it is naturally to ask

Question 4.1 Let $f : G \to G^*$ be a Q-homeomorphism with respect to p-module.

(1) Should $Q \in L^1_{\text{loc}}(G)$ be a necessary condition in Theorem 4.2?
(2) Do the assertions of Theorem 4.2 hold whereas $1 < p \leq n - 1$?

Question 4.2 Given a Q-homeomorphism f with respect to p-module, when do we have that also its inverse f^{-1} is differentiable a.e., ACL, belongs to $W^{1,q}_{\text{loc}}(G^*)$?

Assuming a higher integrability of the majorant Q, we established in [14] the preservation of the n-dimensional Lebesque null-measure under Q-homeomorphisms with respect to p-module.

Theorem 4.3 *Let $f : G \to G^*$ be a Q-homeomorphism with respect to p-module such that $Q(x) \in L_{\text{loc}}^{\frac{n}{n-p}}$, $n - 1 < p < n$. Then f satisfies (N)-property and, moreover,*

$$mf(E) \leq \gamma_{n,p} \int_E Q^{\frac{n}{n-p}}(x) \, dm(x)$$

for any Lebesgue measurable set E. Here $\gamma_{n,p}$ is a constant depending only on n and p.

Question 4.3 Does (N)-property hold for Q-homeomorphisms f with respect to p-module under weaker conditions than $Q \in L_{\text{loc}}^{\frac{n}{n-p}}(G)$ and $n - 1 < p < n$?

A preservation the n-dimensional Lebesgue null-measure under f^{-1} ((N^{-1})-property) is equivalent to $J(x,f) \neq 0$ almost everywhere; see, e.g. [31]. The following result was established in [18].

Theorem 4.4 *Let G and G^* be domains in $\mathbb{R}^n$, $n \geq 2$, and $f : G \to G^*$ be a Q-homeomorphism with respect to p-module and $Q \in L_{\text{loc}}^1$, $p \geq n$. Then f has Lusin's (N^{-1})-property and $J(x,f) \neq 0$ a.e.*

Question 4.4 Must the Q-homeomorphisms f admit Lusin's (N^{-1})-property under appropriate conditions on the majorant Q, when $1 < p < n$?

The following question on regularity of Q-homeomorphisms concerns the optimal exponent of integrability of J and $1/J$. We first mention the results of [18].

Theorem 4.5 *Let G and G^* be domains in $\mathbb{R}^n$, $n \geq 2$, and $f : G \to G^*$ be a Q-homeomorphism with respect to p-module.*

(1) If $n - 1 < p < n$ and $Q \in L_{\text{loc}}^{\frac{\alpha n}{n-p}}$, $\alpha > 1$, then $J \in L_{\text{loc}}^{\alpha}$;
(2) if $p > n$ and $Q \in L_{\text{loc}}^{\frac{\alpha n}{p-n}}$, $\alpha > 1$, then $1/J \in L_{\text{loc}}^{\alpha}$.

Question 4.5 What happens with both J and $1/J$ when $1 < p \leq n - 1$? Moreover, what are the optimal degrees for $1/J$ in the case $n - 1 < p < n$, and for J when $p > n$?

5 Ring Q-Homeomorphisms

Now we consider a weakened module inequality related to the spherical rings. We call homeomorphisms $f : G \to \mathbb{R}^n$ to be the *ring Q-homeomorphism with respect to p-module* at a point $x_0 \in G$, $1 < p \leq n$, if the inequality

$$\mathcal{M}_p\left(\Delta\left(f(S_1), f(S_2), f(G)\right)\right) \leq \int_{A(x_0, r_1, r_2)} Q(x) \, \eta^p(|x - x_0|) \, dm(x) \tag{6}$$

holds for any spherical ring $A(x_0, r_1, r_2)$, $0 < r_1 < r_2 < d_0$ with the boundaries S_1, S_2, and for any measurable function $\eta : (r_1, r_2) \to [0, \infty]$ such that

$$\int_{r_1}^{r_2} \eta(r)\, dr \geq 1.$$

Here $\Delta(E_1, E_2, D)$ denotes a family of all curves joining E_1 and E_2 in D. A homeomorphism $f : G \to \mathbb{R}^n$ is called *ring Q-homeomorphism with respect to p-module in the domain G*, if (6) holds for all points $x_0 \in G$. The ring Q-homeomorphisms with respect to p-module are defined with respect to a priory fixed point and contain as a proper subclass all Q-homeomorphisms with respect to p-module. It was established in [21] that for $p = n$ the majorant Q in (6) can be replaced by the angular dilatation. For $p \neq n$ see [16].

As it was mentioned above, there exist the homeomorphisms from Sobolev class $W^{1,p}_{loc}$, $p \leq n - 1$, which are not differentiable almost everywhere. To ensure differentiability, one has to restrict the class of measurable functions Q in (5) and (6) either assuming their local integrability with appropriate degree or requiring a controlled behavior of their spherical means

$$q_{x_0}(r) = \frac{1}{\omega_{n-1} r^{n-1}} \int_{S(x_0, r)} Q(x)\, d\mathcal{A},$$

which are well defined for almost all $r > 0$. Here $S(x_0, r) = \{x \in \mathbb{R}^n : |x - x_0| = r\}$, and $d\mathcal{A}$ denotes the element of surface area on $S(x_0, r)$. It was shown in [17], that any ring Q-homeomorphism f with respect to p-module $(n - 1 < p < n)$ having $q_{x_0}(r) \leq C_{x_0} r^{p-n}$ is logarithmically Hölder continuous.

In the recent paper [19], we established the differentiability of ring Q-homeomorphisms with respect to p-module under a wider condition on Q than local integrability. Namely, we say that the function Q satisfies the *(p, λ)-condition at $x_0 \in G$*, if there exists a constant $\lambda = \lambda(x_0) > 1$ such that

$$\liminf_{\varepsilon \to 0} \varepsilon^{\frac{n-p}{p-1}} \int_{\varepsilon}^{\lambda \varepsilon} \frac{dt}{t^{\frac{n-1}{p-1}} q_{x_0}^{\frac{1}{p-1}}(t)} > 0. \tag{7}$$

A close type of restrictions has been used in [29].

Theorem 5.1 *Let G and G^* be two domains in $\mathbb{R}^n$ $(n \geq 2)$ and $f : G \to G^*$ be a ring Q-homeomorphism with respect to p-module, $p > n - 1$. Suppose that Q satisfies (p, λ)-condition a.e. in G. Then f is differentiable a.e. in G.*

Question 5.1 Is this theorem sharp or it fails if one requires something less restrictive than (7)? Can the restriction on p, namely $p > n - 1$, be reduced?

Remark 5.1 The condition $Q \in L^1_{loc}$ is stronger than (7). So, the local integrability of Q guaranties the differentiability of ring Q-homeomorphisms with respect to p-module almost everywhere for $p > n - 1$.

The following results provide an explicit type of continuity, moreover in the stronger Lipschitz or Hölder sense.

Theorem 5.2 *Let G and G^* be domains in $\mathbb{R}^n$, $n \geq 2$, and let $f : G \to G^*$ be a ring Q-homeomorphism with respect to p-module, $n - 1 < p < n$, at a point $x_0 \in G$. Then the following estimate*

$$|f(x) - f(x_0)| \leq C_{n,p} \|Q\|_{\frac{n}{n-p}}^{\frac{1}{n-p}} \left(\log \frac{1}{|x - x_0|} \right)^{-\frac{p(n-1)}{n(n-p)}} \tag{8}$$

holds for any $x \in G$ satisfying $|x - x_0| = r < \delta = \min\{1, \text{dist}^4(x_0, \partial G)\}$. Here $C_{n,p}$ is a positive constant depending only on n and p, and

$$\|Q\|_{\frac{n}{n-p}} = \left(\int\limits_{B(x_0,\delta)} Q^{\frac{n}{n-p}}(x) \, dm(x) \right)^{\frac{n-p}{n}} . \tag{9}$$

Theorem 5.2 was proved in [17] under much stronger conditions: the weight function Q is itself integrable over the balls centered at x_0. In [16], we require the integrability of p-angular dilatations over some rings. It was also shown that the integrability degree $n/(n - p)$ cannot be decreased, and it is impossible to replace the logarithmic Hölder continuity by the usual Hölder continuity; in addition, the exponent $p(n - 1)/n(n - p)$ in (8) is sharp.

The next statement provides the Hölder continuity condition for ring Q-homeomorphisms if $\alpha > n/(n - p)$ and p ranges between $n - 1$ and n.

Theorem 5.3 *Let G and G^* be domains in $\mathbb{R}^n$, $n \geq 2$, and let $f : G \to G^*$ be a ring Q-homeomorphism with respect to p-module at point $x_0 \in G$ with $n - 1 < p < n$ and $\alpha > \frac{n}{n-p}$. Then*

$$|f(x) - f(x_0)| \leq C_{n,p} \|Q\|_{\alpha}^{\frac{1}{n-p}} |x - x_0|^{1 - \frac{n}{\alpha(n-p)}},$$

for any x satisfying $|x - x_0| < \delta = \frac{1}{4}\text{dist}(x_0, \partial G)$. Here $\|Q\|_{\alpha}$ is α-norm defined by (9) over the ball $B(x_0, \delta)$ and $C_{n,p}$ is a positive constant depending only on n and p.

The sufficient conditions for ring Q-homeomorphisms to be finite Lipschitzian at a prescribed point x_0 are given in [16] and [35]. These conditions involve an integral mean of the form

$$Q_{x_0} = \limsup_{r \to 0} \frac{1}{mB(x_0, r)} \int\limits_{B(x_0,r)} Q(x) \, dm(x).$$

Theorem 5.4 *Let G and G^* be domains in $\mathbb{R}^n$, $n \geq 2$, and let $f : G \to G^*$ be a ring Q-homeomorphism with respect to p-module, $n-1 < p < n$, at $x_0 \in G$. If $Q_{x_0} < \infty$, then*

$$L(x_0, f) = \limsup_{x \to x_0} \frac{|f(x) - f(x_0)|}{|x - x_0|} \leq C_{n,p} Q_{x_0}^{\frac{1}{n-p}},$$

where $C_{n,p}$ is a positive constant that depends only on n and p.

Question 5.2 What are sharp bounds for the constants $C_{n,p}$ in Theorems 5.2–5.4?

Question 5.3 Could the restrictions on the parameter p in Theorems 5.2–5.4 be weakened?

Question 5.4 Are the inverse mappings f^{-1} Lipschitz/Hölder/logarithmic Hölder continuous?

The following theorem given in [15] implies an upper bound for Q-homeomorphisms with respect to p-module $(1 < p < n)$, which can be regarded as a generalization of Schwarz's lemma.

Theorem 5.5 *For any Q-homeomorphism $f : \mathbb{B}^n \to \mathbb{B}^n$ with respect to p-module, $1 < p < n$, with $Q \in L^\alpha_{\text{loc}}(\mathbb{B}^n)$, $\alpha > \frac{n}{n-p}$, and normalized by $f(0) = 0$, we have*

$$\liminf_{x \to 0} \frac{|f(x)|}{|x|^\gamma} \leq c_0 \, ||Q||_\alpha^{\frac{1}{n-p}}, \tag{10}$$

where $\gamma = 1 - \frac{n}{\alpha(n-p)}$ and c_0 is a constant depending only on α, p and n.

The following theorem from [15] estimates the growth of Q-homeomorphisms of the unit ball $\mathbb{B}^n$, in the case when $p = n$, again in the spirit of Schwarz's lemma.

Theorem 5.6 *Let $f : \mathbb{B}^n \to \mathbb{B}^n$ be a Q-homeomorphism preserving the origin. If for some real β,*

$$\mathcal{I} = \int_0^1 \left(\beta - \frac{1}{q^{\frac{1}{n-1}}(t)} \right) \frac{dt}{t} < \infty,$$

then

$$\liminf_{x \to 0} \frac{|f(x)|}{|x|^\beta} \leq e^{\mathcal{I}}.$$

It seems likely that the assumption $Q \in L^\alpha$ with $\alpha > \frac{n}{n-p}$ in Theorem 5.5 can not be improved. In fact, if $\alpha \leq n/(n-p)$, the estimate (10) becomes trivial because in this case the assumptions of this theorem yields $\gamma \leq 0$, and, obviously,

$$\liminf_{x \to 0} \frac{|f(x)|}{|x|^\gamma} = 0.$$

The same situation occurs when $||Q||_\alpha = \infty$. This shows that the assumptions on the growth of Q are sharp in some sense.

The sharpness of the bound (10) was illustrated in [15]. The behavior of Q-homeomorphisms with weighted $|f(x)|$ located in the closed unit ball has been established in [37].

Question 5.5 Does there exist an analogue of (10) for $p > n$?

Question 5.6 Does an analogue of the Schwarz lemma hold for the inverse f^{-1}?

6 Lower Q-Homeomorphisms

Let G and G^* be two bounded domains in $\mathbb{R}^n$, $n \geq 2$ and $x_0 \in \overline{G}$. Given a Lebesgue measurable function $Q : G \to [0, \infty]$, a homeomorphism $f : G \to G^*$ is called the *lower Q-homeomorphism with respect to p-module at x_0* if

$$\mathcal{M}_p(f(\Sigma_\varepsilon)) \geq \inf_{\rho \in \text{ext}_p \text{adm } \Sigma_\varepsilon} \int_{G_\varepsilon(x_0)} \frac{\rho^p(x)}{Q(x)} \, dm(x),$$

where

$$G_\varepsilon(x_0) = G \cap \{x \in \mathbb{R}^n : \varepsilon < |x - x_0| < \varepsilon_0\}, \quad 0 < \varepsilon < \varepsilon_0, \quad 0 < \varepsilon_0 < \sup_{x \in G} |x - x_0|,$$

and Σ_ε denotes the family of all pieces of spheres centered at x_0 of radii r, $\varepsilon < r < \varepsilon_0$, located in G.

The following result given in [14] provides a necessary and sufficient condition for the ring homeomorphisms to be lower homeomorphisms ones with respect to p-module.

Theorem 6.1 *Every lower Q-homeomorphism with respect to p-module $f : G \to G^*$ at $x_0 \in G$, with $p > n - 1$ and $Q \in L_{\text{loc}}^{\frac{n-1}{p-n+1}}$, is a ring $\widetilde{Q}$-homeomorphism with respect to α-module at x_0 with $\widetilde{Q} = Q^{\frac{n-1}{p-n+1}}$ and $\alpha = \frac{p}{p-n+1}$.*

Question 6.1 Is Theorem 6.1 true also for $1 < p \leq n - 1$?
Note that the condition $n - 1 < p/(p - n + 1) < n$ is equivalent to

$$n < p < (n-1)^2/(n-2). \tag{11}$$

So, in the case $n > 2$ there appears naturally the following interesting question:

Question 6.2 Can be the upper bound for p in (11) replaced by the infinity similarly to the planar case?

Acknowledgement The first author was supported by EU FP7 IRSES program STREVCOMS, grant no. PIRSES-2013-612669.

References

1. L. Ahlfors, A. Beurling, Conformal invariants and function-theoretic null-sets. Acta Math. **83**, 101–129 (1950)
2. C. Andreian Cazacu, Some formulae on the extremal length in n-dimensional case, in *Proceedings of the Romanian-Finnish Seminar on Teichmüller Spaces and Quasiconformal Mappings* (Braşov, 1969) (Publishing House of the Academy of the Socialist Republic of Romania, Bucharest, 1971), pp. 87–102
3. C. Andreian Cazacu, Module inequalities for quasiregular mappings. Ann. Acad. Sci. Fenn. Ser. A I Math. **2**, 17–28 (1976)
4. C. Bishop, V. Gutlyanskiĭ, O. Martio, M. Vuorinen, On conformal dilatation in space. Int. J. Math. Math. Sci. **22**, 1397–1420 (2003)
5. M. Cristea, Local homeomorphisms satisfying generalized modular inequalities. Complex Var. Elliptic Equ. **59**(10), 1363–1387 (2014)
6. M. Cristea, On generalized quasiconformal mappings. Complex Var. Elliptic Equ. **59**(2), 232–246 (2014)
7. B. Fuglede, Extremal length and functional completion. Acta Math. **98**, 171–219 (1957)
8. F.W. Gehring, Lipschitz mappings and the p-capacity of rings in n-space, in *Advances in the Theory of Riemann Surfaces. Proceedings of the Conference*, Stony Brook, NY, 1969. Annals of Mathematical Studies, vol. 66 (Princeton University Press, Princeton, NJ, 1971), pp. 175–193
9. A. Golberg, Homeomorphisms with finite mean dilatations. Contemp. Math. **382**, 177–186 (2005)
10. A. Golberg, *Differential Properties of (α, Q)-Homeomorphisms*. Further Progress in Analysis (World Scientific, Hackensack, NJ, 2009), pp. 218–228
11. A. Golberg, Integrally quasiconformal mappings in space. Trans. Inst. Math. Natl. Acad. Sci. Ukr. **7**(2), 53–64 (2010)
12. A. Golberg, Homeomorphisms with integrally restricted moduli. Contemp. Math. **553**, 83–98 (2011)
13. A. Golberg, Quasiisometry from different points of view. J. Math. Sci. **196**, 617–631 (2014)
14. A. Golberg, R. Salimov, Topological mappings of integrally bounded p-moduli. Ann. Univ. Buchar. Math. Ser. **3(LXI)**(1), 49–66 (2012)
15. A. Golberg, R. Salimov, Extension of the Schwarz Lemma to homeomorphisms with controlled p-module. Georgian Math. J. **21**(3), 273–279 (2014)
16. A. Golberg, R. Salimov, Homeomorphisms Lipschitzian in the mean, in *Complex Analysis and Potential Theory with Applications* (Cambridge Scientific Publishers, Cambridge, 2014), pp. 95–111
17. A. Golberg, R. Salimov, Logarithmic Hölder continuity of ring homeomorphisms with controlled p-moduli. Complex Var. Elliptic Equ. **59**(1), 91–98 (2014)
18. A. Golberg, R. Salimov, Mappings with upper integral bounds for p-moduli. Contemp. Math. **659**, 91–113 (2016)
19. A. Golberg, R. Salimov, Differentiability of ring homeomorphisms with controlled p-module. Contemp. Math. **699**, 121–217 (2017)
20. V. Gol'dshtein, Yu.G. Reshetnyak, *Quasiconformal Mappings and Sobolev Spaces* (Kluwer Academic Publishers Group, Dordrecht, 1990)
21. V.Ya. Gutlyanskiĭ, A. Golberg, On Lipschitz continuity of quasiconformal mappings in space. J. Anal. Math. **109**, 233–251 (2009)

22. S. Hencl, P. Koskela, *Lectures on Mappings of Finite Distortion*. Lecture Notes in Mathematics, vol. 2096 (Springer, Cham, 2014)
23. T. Iwaniec, G. Martin, *Geometric Function Theory and Non-linear Analysis*. Oxford Mathematical Monographs (The Clarendon Press, Oxford University Press, New York, 2001)
24. A.P. Kopylov, Quasi-isometric mappings and the p-moduli of path families. Probl. Anal. Issues Anal. **5**(23)(2), 33–37 (2016)
25. D. Kovtonyk, V. Ryazanov, On the theory of mappings with finite area distortion. J. Anal. Math. **104**, 291–306 (2008)
26. S.L. Krushkal, *Quasiconformal Mappings and Riemann Surfaces* (V.H. Winston & Sons/Wiley, Washington, DC/New York-Toronto, Ontorio-London, 1979)
27. O. Martio, V. Ryazanov, U. Srebro, E. Yakubov, Q-homeomorphisms. Contemp. Math. **364**, 193–203 (2004)
28. O. Martio, V. Ryazanov, U. Srebro, E. Yakubov, *Moduli in Modern Mapping Theory*. Springer Monographs in Mathematics (Springer, New York, 2009)
29. V.M. Miklyukov, On some criteria for the existence of the total differential. Sib. Math. J. **51**(4), 639–647 (2010)
30. J. Onninen, V. Tengvall, Mappings of L^p-integrable distortion: regularity of the inverse. Proc. R. Soc. Edinb. Sect. A **146**(3), 647–663 (2016)
31. S.P. Ponomarev, The N^{-1}-property of mappings, and Luzin's (N) condition. Math. Notes **58**(3–4), 960–965 (1995)
32. Yu.G. Reshetnyak, *Space Mappings with Bounded Distortion* (American Mathematical Society, Providence, 1989)
33. S. Rickman, *Quasiregular Mappings*. Results in Mathematics and Related Areas, vol. 26(3) (Springer, Berlin, 1993)
34. R. Salimov, ACL and differentiability of Q-homeomorphisms. Ann. Acad. Sci. Fenn. Math. **33**(1), 295–301 (2008)
35. R. Salimov, On finitely Lipschitz space mappings. Sib. Elektron. Mat. Izv. **8**, 284–295 (2011)
36. R. Salimov, On Q-homeomorphisms with respect to p-modulus. Ann. Univ. Buchar. Math. Ser. **2(LX)**(1), 207–213 (2011)
37. R. Salimov, Estimation of the measure of the image of the ball. Sib. Mat. Zh. **53**(4), 920–930 (2012)
38. R. Salimov, E. Sevost'yanov, Theory of ring Q-mappings and geometric function theory. Sb. Math. **201**(5–6), 909–934 (2010)
39. B.V. Shabat, The modulus method in space. Sov. Math. Dokl. **130**, 1210–1213 (1960)
40. V. Tengvall, Absolute continuity of mappings with finite geometric distortion. Ann. Acad. Sci. Fenn. Math. **40**(1), 3–15 (2015)
41. J. Väisälä, *Lectures on n-Dimensional Quasiconformal Mappings* (Springer, Berlin, 1971)
42. A. Vasil'ev, *Moduli of Families of Curves for Conformal and Quasiconformal Mappings*. Lecture Notes in Mathematics, vol. 1788 (Springer, Berlin, 2002)

Extremal Problems for Mappings with g-Parametric Representation on the Unit Polydisc in $\mathbb{C}^n$

Ian Graham, Hidetaka Hamada, and Gabriela Kohr

This paper is dedicated to the memory of Prof. Alexander Vasil'ev

Abstract In this paper we are concerned with extremal problems for mappings with g-parametric representation on the unit polydisc $\mathbb{U}^2$ of $\mathbb{C}^2$, where g is a univalent holomorphic function on the unit disc $\mathbb{U}$ such that $g(0) = 1$, and which satisfies some natural conditions. In the first part of the paper, we obtain certain results related to extreme points and support points associated with the Carathéodory family $\mathcal{M}_g(\mathbb{U}^n)$ and the family $S_g^*(\mathbb{U}^n)$ of g-starlike mappings on $\mathbb{U}^n$. In particular, if g is a convex function on $\mathbb{U}$, we use an analogue of the shearing process due to F. Bracci, to obtain sharp coefficient bounds for the family $\mathcal{M}_g(\mathbb{U}^2)$. In the last part of the paper, we are concerned with support points for the family $S_g^0(\mathbb{U}^2)$ of mappings with g-parametric representation on $\mathbb{U}^2$, where g is a convex function on $\mathbb{U}$ with $g(0) = 1$, $\Re g(\zeta) > 0$, $\zeta \in \mathbb{U}$, and which satisfies certain natural conditions. Sharp coefficient bounds for the family $S_g^0(\mathbb{U}^2)$, and various consequences and examples are obtained. Certain questions and conjectures are also formulated. This work complements recent work on extremal problems on the Euclidean unit ball $\mathbb{B}^2$ in $\mathbb{C}^2$.

I. Graham
Department of Mathematics, University of Toronto, Toronto, ON, Canada M5S 2E4
e-mail: graham@math.toronto.edu

H. Hamada
Faculty of Science and Engineering, Kyushu Sangyo University, 3-1 Matsukadai 2-Chome, Higashi-ku, Fukuoka 813-8503, Japan
e-mail: h.hamada@ip.kyusan-u.ac.jp

G. Kohr (✉)
Faculty of Mathematics and Computer Science, Babeş-Bolyai University, 1 M. Kogălniceanu Str., 400084 Cluj-Napoca, Romania
e-mail: gkohr@math.ubbcluj.ro

© Springer International Publishing AG 2018
M. Agranovsky et al. (eds.), *Complex Analysis and Dynamical Systems*, Trends in Mathematics, https://doi.org/10.1007/978-3-319-70154-7_9

Keywords Carathéodory family • Extreme point • Herglotz vector field • Loewner chain • Loewner differential equation • Support point

2000 Mathematics Subject Classification Primary 32H02; Secondary 30C45, 46G20

1 Introduction

In this paper, we are concerned with extremal problems for mappings with parametric representation on the unit polydisc $\mathbb{U}^n$ in $\mathbb{C}^n$. We also study extreme points associated with the family $S_g^*(\mathbb{U}^n)$, consisting of g-starlike mappings on $\mathbb{U}^n$, where $g : \mathbb{U} \to \mathbb{C}$ is a univalent holomorphic function such that $g(0) = 1$ and $\Re g(\zeta) > 0$, $\zeta \in \mathbb{U}$ (see [41], for $g(\zeta) = \frac{1-\zeta}{1+\zeta}$, $\zeta \in \mathbb{U}$). Finally, we include certain conjectures and questions related to extremal problems in $\mathbb{C}^n$. The main results complement recent extremal results for mappings with parametric representation on the Euclidean unit ball $\mathbb{B}^2$, obtained in [4, 5, 20, 21, 27].

The main results of this paper are given below. The notations will be explained in the next sections.

Theorem 1.1 *Let* $g : \mathbb{U} \to \mathbb{C}$ *be a convex (univalent) function on* $\mathbb{U}$ *such that* $g(0) = 1$. *Also, let* $h : \mathbb{U}^n \to \mathbb{C}^n$ *be given by*

$$h(z) = (z_1 p_1(z), \ldots, z_n p_n(z)), \quad z = (z_1, \ldots, z_n) \in \mathbb{U}^n,$$

where $p_j : \mathbb{U}^n \to \mathbb{C}$ $(j = 1, \ldots, n)$ *are holomorphic functions which satisfy necessary conditions for* $h \in \mathcal{M}_g(\mathbb{U}^n)$. *If* $p_j(\cdot z) \in \mathrm{ex}\, s(g)$, *for all* $z \in \partial_\infty \mathbb{U}^n$, $j = 1, \ldots, n$, *then* $h \in \mathrm{ex}\, \mathcal{M}_g(\mathbb{U}^n)$. *In particular, if* $p_j(\cdot z) \in \mathrm{ex}\mathcal{P}$, *for all* $z \in \partial_\infty \mathbb{U}^n$, $j = 1, \ldots, n$, *then* $h \in \mathrm{ex}\, \mathcal{M}(\mathbb{U}^n)$.

Theorem 1.2 *Let* $g : \mathbb{U} \to \mathbb{C}$ *be a convex (univalent) function on* $\mathbb{U}$ *such that* $g(0) = 1$. *Also, let* $h : \mathbb{U}^n \to \mathbb{C}^n$ *be given by*

$$h(z) = (z_1 p_1(z_{i_1}), \ldots, z_n p_n(z_{i_n})), \quad z = (z_1, \ldots, z_n) \in \mathbb{U}^n,$$

where $i_1, \ldots, i_n \in \{1, \ldots, n\}$ *and* $p_j \in s(g)$, $j = 1, \ldots, n$. *Then* $h \in \mathrm{supp}\, \mathcal{M}_g(\mathbb{U}^n)$ *if and only if there exists* $j \in \{1, \ldots, n\}$ *such that* $p_j \in \mathrm{supp}\, s(g)$. *In particular, if* $p_j \in \mathcal{P}$, $j = 1, \ldots, n$, *then* $h \in \mathrm{supp}\, \mathcal{M}(\mathbb{U}^n)$ *if and only if there exists* $j \in \{1, \ldots, n\}$ *such that* $p_j \in \mathrm{supp}\, \mathcal{P}$.

Theorem 1.3 *Let* $g : \mathbb{U} \to \mathbb{C}$ *be a univalent holomorphic function such that* $g(0) = 1$ *and* $\Re g(\zeta) > 0$, $\zeta \in \mathbb{U}$. *Also, let* $f \in S_g^*(\mathbb{U}^n)$ *and let* $h(z) = [Df(z)]^{-1}f(z)$, $z \in \mathbb{U}^n$. *Assume that* $h(z) = (z_1 p_1(z), \ldots, z_n p_n(z))$, $z = (z_1, \ldots, z_n) \in \mathbb{U}^n$. *If* $p_j(\cdot z) \in \{g_x : |x| = 1\}$, *for all* $z \in \partial_\infty \mathbb{U}^n$ *and* $j = 1, \ldots, n$, *where* $g_x(\zeta) = g(x\zeta)$, $|\zeta| < 1$, $|x| = 1$, *then* $f \in \mathrm{ex}\, S_g^*(\mathbb{U}^n)$.

Theorem 1.4 *Let $g : \mathbb{U} \to \mathbb{C}$ be a convex (univalent) function such that $g(0) = 1$. Assume that $\Re g(\zeta) > 0$, $g(\bar{\zeta}) = \overline{g(\zeta)}$, $\zeta \in \mathbb{U}$, and*

$$\min_{|\zeta|=r} \Re g(\zeta) = g(r) \quad and \quad \max_{|\zeta|=r} \Re g(\zeta) = g(-r), \ \forall \, r \in (0,1).$$

If $h = (h_1, h_2) \in \mathcal{M}_g(\mathbb{U}^2)$, then $\max\{|q_{0,2}^1|, |q_{2,0}^2|\} \leq a_0$, where a_0 is given by

$$a_0 = \inf_{r \in (0,1)} \left\{ \frac{\min\{1 - g(r), g(-r) - 1\}}{r} \right\},$$

$q_{0,2}^1 = \frac{1}{2} \frac{\partial^2 h_1}{\partial z_2^2}(0)$ *and* $q_{2,0}^2 = \frac{1}{2} \frac{\partial^2 h_2}{\partial z_1^2}(0)$. *If, in addition, $\mathbb{U}(1, a_0) \subseteq g(\mathbb{U})$, then $h^{[c]}, \widetilde{h}^{[c]} \in \mathcal{M}_g(\mathbb{U}^2)$, where $h^{[c]}(z) = \left(z_1 + q_{0,2}^1 z_2^2, z_2\right)$ and $\widetilde{h}^{[c]}(z) = \left(z_1, z_2 + q_{2,0}^2 z_1^2\right)$, $z = (z_1, z_2) \in \mathbb{U}^2$. In particular, if $h \in \mathcal{M}(\mathbb{U}^2)$, then $\max\{|q_{0,2}^1|, |q_{2,0}^2|\} \leq 1$, and $h^{[c]}, \widetilde{h}^{[c]} \in \mathcal{M}(\mathbb{U}^2)$.*

Theorem 1.5 *Let $g : \mathbb{U} \to \mathbb{C}$ be a convex (univalent) function such that $g(0) = 1$. Assume that g satisfies the conditions of Theorem 1.4. Also, assume that $\mathbb{U}(1, a_0) \subseteq g(\mathbb{U})$, where a_0 is given in the above result. Let $f = (f_1, f_2) \in S_g^0(\mathbb{U}^2)$. Then $\max\{|a_{0,2}^1|, |a_{2,0}^2|\} \leq a_0$, where $a_{0,2}^1 = \frac{1}{2} \frac{\partial^2 f_1}{\partial z_2^2}(0)$ and $a_{2,0}^2 = \frac{1}{2} \frac{\partial^2 f_2}{\partial z_1^2}(0)$. These estimates are sharp. In particular, if $f \in S^0(\mathbb{U}^2)$, then $\max\{|a_{0,2}^1|, |a_{2,0}^2|\} \leq 1$, and these estimates are sharp.*

2 Preliminaries

Let $\mathbb{C}^n$ be the space of n complex variables $z = (z_1, \ldots, z_n)$ with the maximum norm $\|z\|_\infty = \max\{|z_1|, \ldots, |z_n|\}$. Also, let $\mathbb{U}^n$ be the unit polydisc in $\mathbb{C}^n$ and let $\mathbb{U}_1 = \mathbb{U}$ be the unit disc. Let $\partial_\infty \mathbb{U}^n = \prod_{k=1}^n \partial \mathbb{U}$ be the distinguished boundary of $\mathbb{U}^n$. The disc $\{\zeta \in \mathbb{C} : |\zeta - a| < r\}$ is denoted by $\mathbb{U}(a, r)$.

Let $L(\mathbb{C}^n)$ be the space of linear operators from $\mathbb{C}^n$ into $\mathbb{C}^n$ with the standard operator norm, and let I_n be the identity in $L(\mathbb{C}^n)$.

We denote by $H(\mathbb{U}^n)$ the family of holomorphic mappings from $\mathbb{U}^n$ into $\mathbb{C}^n$ with the standard topology of locally uniform convergence. If $f \in H(\mathbb{U}^n)$, we say that f is normalized if $f(0) = 0$ and $Df(0) = I_n$. Let $S(\mathbb{U}^n)$ be the subset of $H(\mathbb{U}^n)$ consisting of all normalized univalent (biholomorphic) mappings on $\mathbb{U}^n$, and let $S^*(\mathbb{U}^n)$ (respectively $K(\mathbb{U}^n)$) be the subset of $S(\mathbb{U}^n)$ consisting of starlike mappings with respect to the origin (respectively convex mappings on $\mathbb{U}^n$). The family $S(\mathbb{U}^1)$ is denoted by S. Also, let $\mathcal{LS}(\mathbb{U}^n)$ be the family of normalized locally biholomorphic mappings on $\mathbb{U}^n$.

Definition 2.1 A mapping $f : \mathbb{U}^n \times [0, \infty) \to \mathbb{C}^n$ is called a Loewner chain (or a normalized univalent subordination chain) if $e^{-t} f(\cdot, t) \in S(\mathbb{U}^n)$ for $t \geq 0$, and

$f_s(\mathbb{U}^n) \subseteq f_t(\mathbb{U}^n)$, for all $0 \le s \le t < \infty$, where $f_t(z) = f(z,t)$. The Loewner chain $f(z,t)$ is said to be normal if $\{e^{-t}f(\cdot,t)\}_{t\ge 0}$ is a normal family on $\mathbb{U}^n$.

The following subsets of $H(\mathbb{U}^n)$ are generalizations to $\mathbb{C}^n$ of the Carathéodory family on $\mathbb{U}$ (see [39]):

$$\mathcal{N}(\mathbb{U}^n) = \left\{ h = (h_1, \ldots, h_n) \in H(\mathbb{U}^n) : h(0) = 0, \Re\left[\frac{h_j(z)}{z_j}\right] > 0, \right.$$

$$\left. |z_j| = \|z\|_\infty \in (0,1), j = 1, \ldots, n \right\},$$

$$\mathcal{M}(\mathbb{U}^n) = \{ h \in \mathcal{N}(\mathbb{U}^n) : Dh(0) = I_n \}.$$

In the case $n = 1$, $f \in \mathcal{M}(\mathbb{U})$ if and only if $q \in \mathcal{P}$, where $q(z) = f(z)/z$ for $z \ne 0$, and $q(0) = 1$, and $\mathcal{P}$ is the Carathéodory family on the unit disc $\mathbb{U}$ given by

$$\mathcal{P} = \{ p \in H(\mathbb{U}) : p(0) = 1, \Re p(\zeta) > 0, |\zeta| < 1 \}.$$

It is known that the family $\mathcal{M}(\mathbb{U}^n)$ is compact in the topology of $H(\mathbb{U}^n)$ (see e.g. [13] and [32]). For various applications of this family in the Loewner theory in higher dimensions, the reader may consult [2, 3, 7, 9, 12–15, 19, 24, 25, 27, 30, 32, 33, 36, 37, 39, 40]. See also, [10, 11, 35] and [39], for other applications of the family $\mathcal{M}(\mathbb{U}^n)$ in the study of univalent mappings in $\mathbb{C}^n$.

Let $\alpha \in [0,1)$ and let

$$\mathcal{P}(\alpha) = \{ p \in H(\mathbb{U}) : p(0) = 1, \Re p(\zeta) > \alpha, \zeta \in \mathbb{U} \}.$$

Then $\mathcal{P}(\alpha)$ is a compact and convex subset of $H(\mathbb{U})$, $\mathcal{P}(\alpha) \subseteq \mathcal{P}$, and $\mathcal{P}(0) = \mathcal{P}$.

For every $z \in \mathbb{C}^n \setminus \{0\}$, let

$$T(z) = \{ \ell_z \in L(\mathbb{C}^n, \mathbb{C}) : \|\ell_z\| = 1, \ell_z(z) = \|z\|_\infty \}.$$

The following subset of $H(\mathbb{U}^n)$ plays a special role in our treatment (see [13]).

Definition 2.2 Let $g : \mathbb{U} \to \mathbb{C}$ be a univalent holomorphic function such that $g(0) = 1$. Also, let $h \in H(\mathbb{U}^n)$ be such that $h(0) = 0$ and $Dh(0) = I_n$. We say that h belongs to the family $\mathcal{M}_g(\mathbb{U}^n)$ if

$$\frac{1}{\|z\|_\infty} \ell_z(h(z)) \in g(\mathbb{U}), \quad z \in \mathbb{U}^n \setminus \{0\}, \quad \ell_z \in T(z).$$

Remark 2.3

(i) Let $z \in \mathbb{C}^n \setminus \{0\}$ and $\ell_z \in T(z)$. Then (see e.g. [39])

$$\ell_z(w) = \sum_{\{j:|z_j|=\|z\|_\infty\}} t_j \frac{w_j \bar{z}_j}{\|z\|_\infty}, \quad \text{where } t_j \ge 0, \sum_{\{j:|z_j|=\|z\|_\infty\}} t_j = 1. \tag{2.1}$$

Hence it is not difficult to deduce that if $h \in H(\mathbb{U}^n)$ is a normalized mapping, and g is a convex (univalent) function on $\mathbb{U}$ with $g(0) = 1$, then $h \in \mathcal{M}_g(\mathbb{U}^n)$ if and only if

$$\frac{h_j(z)}{z_j} \in g(\mathbb{U}), \quad 0 < |z_j| = \|z\|_\infty < 1, \quad j \in \{1, \dots, n\}. \tag{2.2}$$

(ii) If $g : \mathbb{U} \to \mathbb{C}$ satisfies the conditions in Definition 2.2 and $\Re g(\zeta) > 0, \zeta \in \mathbb{U}$, then $\mathcal{M}_g(\mathbb{U}^n) \subseteq \mathcal{M}(\mathbb{U}^n)$, and if $g(\zeta) = \frac{1-\zeta}{1+\zeta}, \zeta \in \mathbb{U}$, then $\mathcal{M}_g(\mathbb{U}^n) = \mathcal{M}(\mathbb{U}^n)$.

(iii) If $g \in H(\mathbb{U})$ is a univalent holomorphic function such that $g(0) = 1$, then $\mathcal{M}_g(\mathbb{U}^n)$ is a compact family, by Graham et al. [17, Corollary 3.10]. If, in addition, g is a convex (univalent) function on $\mathbb{U}$ such that $g(0) = 1$, then it is easily seen that $\mathcal{M}_g(\mathbb{U}^n)$ is a convex family.

Definition 2.4 Let $g : \mathbb{U} \to \mathbb{C}$ be a univalent holomorphic function such that $g(0) = 1$ and $\Re g(\zeta) > 0, |\zeta| < 1$. Also, let $f \in \mathcal{LS}(\mathbb{U}^n)$. The mapping f is said to be g-starlike (denoted by $f \in S_g^*(\mathbb{U}^n)$) if (see [26])

$$\frac{1}{\|z\|_\infty} \ell_z(h(z)) \in g(\mathbb{U}), \quad z \in \mathbb{U}^n \setminus \{0\}, \quad \ell_z \in T(z),$$

where $h(z) = [Df(z)]^{-1} f(z), z \in \mathbb{U}^n$.

Remark 2.5

(i) In view of (2.1) and (2.2), we deduce that if g is a convex (univalent) function on $\mathbb{U}$ such that $g(0) = 1$, and if $f \in \mathcal{LS}(\mathbb{U}^n)$, then $f \in S_g^*(\mathbb{U}^n)$ if and only if $\frac{h_j(z)}{z_j} \in g(\mathbb{U}), 0 < |z_j| = \|z\|_\infty < 1, j \in \{1, \dots, n\}$, where $h(z) = [Df(z)]^{-1} f(z)$ for $z \in \mathbb{U}^n$.

(ii) Let $g : \mathbb{U} \to \mathbb{C}$ be a univalent holomorphic function such that $g(0) = 1$. Then the family $S_g^*(\mathbb{U}^n)$ is compact, by using arguments similar to those in the proof of [8, Lemma 2.12] (see also [18]).

Next, we recall the notion of parametric representation for mappings in $H(\mathbb{U}^n)$ (see [32] and [13]).

Definition 2.6 A normalized mapping $f \in H(\mathbb{U}^n)$ is said to have parametric representation (denoted by $f \in S^0(\mathbb{U}^n)$) if there exists a mapping $h : \mathbb{U}^n \times [0, \infty) \to \mathbb{C}^n$ such that

(i) $h(\cdot, t) \in \mathcal{M}(\mathbb{U}^n)$ for almost all $t \geq 0$,
(ii) $h(z, \cdot)$ is measurable on $[0, \infty)$ for $z \in \mathbb{U}^n$,

and $f(z) = \lim_{t \to \infty} e^t v(z, t)$ locally uniformly on $\mathbb{U}^n$, where $v = v(z, t)$ is the unique locally absolutely continuous solution on $[0, \infty)$ of the initial value problem

$$\frac{\partial v}{\partial t} = -h(v, t) \quad \text{a.e.} \quad t \geq 0, \quad v(z, 0) = z, \tag{2.3}$$

for all $z \in \mathbb{U}^n$.

Remark 2.7

(i) Let $f \in H(\mathbb{U}^n)$ be a normalized mapping. Then $f \in S^0(\mathbb{U}^n)$ if and only if there is a normal Loewner chain $f(z,t)$ such that $f = f(\cdot, 0)$ (cf. [13, 32]).

(ii) It is clear that $S^0(\mathbb{U}^n) \subseteq S(\mathbb{U}^n)$. Then $S^0(\mathbb{U}^n) \subsetneq S(\mathbb{U}^n)$ for $n \geq 2$ (see [13] and [32]), while $S^0(\mathbb{U}) = S$ (see [31]).

(iii) Geometric interpretations of the notion of parametric representation were obtained in [14] and [33].

Definition 2.8 A mapping $h = h(z,t) : \mathbb{U}^n \times [0, \infty) \to \mathbb{C}^n$ which satisfies the conditions (i) and (ii) in Definition 2.6 is called a Herglotz vector field (cf. [6]).

Remark 2.9 Let $f = f(z,t) : \mathbb{U}^n \times [0, \infty) \to \mathbb{C}^n$ be a Loewner chain. Then there exists a Herglotz vector field h such that $f(z,t)$ is a solution of the Loewner differential equation (see [12, 13, 32])

$$\frac{\partial f}{\partial t}(z,t) = Df(z,t)h(z,t), \quad \text{a.e.} \quad t \geq 0, \quad \forall z \in \mathbb{U}^n. \tag{2.4}$$

Now, we consider the notions of a g-Loewner chain and g-parametric representation on $\mathbb{U}^n$ (see [13, 18]).

Definition 2.10 Let $f = f(z,t) : \mathbb{U}^n \times [0, \infty) \to \mathbb{C}^n$ be a normal Loewner chain, and let $g : \mathbb{U} \to \mathbb{C}$ be a univalent holomorphic function such that $g(0) = 1$ and $\Re g(\zeta) > 0$, $|\zeta| < 1$. We say that $f(z,t)$ is a g-Loewner chain if $h(\cdot, t) \in \mathcal{M}_g(\mathbb{U}^n)$, for almost all $t \geq 0$, where $h = h(z,t)$ is the Herglotz vector field given by (2.4).

Definition 2.11 Let $g : \mathbb{U} \to \mathbb{C}$ be a univalent holomorphic function such that $g(0) = 1$ and $\Re g(\zeta) > 0$, $\zeta \in \mathbb{U}$. Let $f \in H(\mathbb{U}^n)$ be a normalized mapping. The mapping f is said to have g-parametric representation (denoted by $f \in S_g^0(\mathbb{U}^n)$) if there exists a Herglotz vector field $h : \mathbb{U}^n \times [0, \infty) \to \mathbb{C}^n$ such that $h(\cdot, t) \in \mathcal{M}_g(\mathbb{U}^n)$ for a.e. $t \geq 0$, and $f(z) = \lim_{t \to \infty} e^t v(z,t)$ locally uniformly on $\mathbb{U}^n$, where $v = v(z,t)$ is the unique locally Lipschitz continuous solution on $[0, \infty)$ of the initial value problem (2.3).

Remark 2.12 Let $g : \mathbb{U} \to \mathbb{C}$ be a univalent holomorphic function such that $g(0) = 1$ and $\Re g(\zeta) > 0$, $\zeta \in \mathbb{U}$.

(i) Clearly, $S_g^0(\mathbb{U}^n) \subseteq S^0(\mathbb{U}^n)$, and if $g(\zeta) = \frac{1-\zeta}{1+\zeta}$, $|\zeta| < 1$, then $S_g^0(\mathbb{U}^n) = S^0(\mathbb{U}^n)$.

(ii) Let $f \in H(\mathbb{U}^n)$ be a normalized mapping. Then $f \in S_g^0(\mathbb{U}^n)$ if and only if there exists a g-Loewner chain $f(z,t)$ such that $f = f(\cdot, 0)$, by Definition 2.11 and Remark 2.9 (cf. [13]).

(iii) It is easy to see that $f \in S_g^*(\mathbb{U}^n)$ if and only if $f(z,t) = e^t f(z)$ is a g-Loewner chain, and thus $S_g^*(\mathbb{U}^n) \subseteq S_g^0(\mathbb{U}^n)$ ([13]; see [32], in the case $g(\zeta) = \frac{1-\zeta}{1+\zeta}$, $\zeta \in \mathbb{U}$).

(iv) If $g \in H(\mathbb{U})$ is a convex (univalent) function with $g(0) = 1$, and $\Re g(\zeta) > 0$, $\zeta \in \mathbb{U}$, then as in the proof of [21, Proposition 3.3] (cf. [27, Theorem 3.11]), we may deduce that the family $S_g^0(\mathbb{U}^n)$ is a compact subset of $H(\mathbb{U}^n)$.

Remark 2.13

(i) Let $\alpha \in [0, 1)$ and $g(\zeta) = \frac{1-\zeta}{1+(1-2\alpha)\zeta}$, $|\zeta| < 1$. In this case, the family $\mathcal{M}_g(\mathbb{U}^n)$ will be denoted by $\mathcal{M}_\alpha(\mathbb{U}^n)$, while the family $S_g^0(\mathbb{U}^n)$ will be denoted by $S_\alpha^0(\mathbb{U}^n)$. On the other hand, the family $S_g^*(\mathbb{U}^n)$ is the usual family $S_\alpha^*(\mathbb{U}^n)$ of starlike mappings of order α on $\mathbb{U}^n$ (see e.g. [12, Chapter 6]).

Let $h = (h_1, \ldots, h_n) \in \mathcal{M}_\alpha(\mathbb{U}^n)$. Then $h \in H(\mathbb{U}^n)$ is a normalized mapping, and for $\alpha \in (0, 1)$, we have

$$\left| \frac{h_j(z)}{z_j} - \frac{1}{2\alpha} \right| < \frac{1}{2\alpha}, \quad 0 < |z_j| = \|z\|_\infty < 1, \quad j \in \{1, \ldots, n\}.$$

We remark that the family $\mathcal{M}_\alpha(\mathbb{U}^n)$ is related to the family $S_\alpha^*(\mathbb{U}^n)$. Indeed, $f \in S_\alpha^*(\mathbb{U}^n)$ if $h \in \mathcal{M}_\alpha(\mathbb{U}^n)$, where $h(z) = [Df(z)]^{-1}f(z)$, $z \in \mathbb{U}^n$ (see e.g. [12, Chapter 6]). Then $S_\alpha^*(\mathbb{U}^n) \subseteq S_\alpha^0(\mathbb{U}^n) \subseteq S^0(\mathbb{U}^n)$, $\alpha \in [0, 1)$. Also, if $f \in K(\mathbb{U}^n)$, then $f \in S_{1/2}^*(\mathbb{U}^n)$, $z \in \mathbb{U}^n$ (see [39]; e.g. [12, Chapter 6]; cf. [13]).

(ii) Let $g(\zeta) = \frac{1-(1-2\alpha)\zeta}{1+\zeta}$, $\zeta \in \mathbb{U}$, where $\alpha \in [0, 1)$. If $h = (h_1, \ldots, h_n) \in H(\mathbb{U}^n)$ is a normalized mapping, then $h \in \mathcal{M}_g(\mathbb{U}^n)$ if and only if

$$\Re \left[\frac{h_j(z)}{z_j} \right] > \alpha, \quad 0 < |z_j| = \|z\|_\infty < 1, \quad j \in \{1, \ldots, n\}.$$

The family $\mathcal{M}_g(\mathbb{U}^n)$ is related to the family of almost starlike mappings of order $\alpha \in [0, 1)$ on $\mathbb{U}^n$ (see [42]). Indeed, a normalized mapping $f \in H(\mathbb{U}^n)$ is almost starlike of order α on $\mathbb{U}^n$ if $h \in \mathcal{M}_g(\mathbb{U}^n)$, where $h(z) \equiv [Df(z)]^{-1}f(z)$ (see [42]).

Definition 2.14 Let $\mathcal{G} \subseteq H(\mathbb{U}^n)$ be a nonempty set.

(i) A point $f \in \mathcal{G}$ is called a support point of $\mathcal{G}$ (denoted by $f \in \text{supp}\,\mathcal{G}$) if there is a continuous linear functional $L : H(\mathbb{U}^n) \to \mathbb{C}$ such that $\Re L|_\mathcal{G} \neq$ constant and $\Re L(f) = \max\{\Re L(q) : q \in \mathcal{G}\}$.

(ii) A point $f \in \mathcal{G}$ is called an extreme point of $\mathcal{G}$ (denoted by $f \in \text{ex}\,\mathcal{G}$) if $f \equiv \lambda g + (1 - \lambda)h$, for some $\lambda \in (0, 1)$, $g, h \in \mathcal{G}$, implies that $f \equiv g \equiv h$.

Note that if $\mathcal{G} \neq \emptyset$ is a compact subset of $H(\mathbb{U}^n)$, then $\text{ex}\,\mathcal{G} \neq \emptyset$, and if $\mathcal{G}$ has at least two points, then $\text{supp}\,\mathcal{G} \neq \emptyset$ (see e.g. [22]).

Let $h : \mathbb{U} \to \mathbb{C}$ be a univalent holomorphic function, and let

$$s(h) = \{p \in H(\mathbb{U}) : p(0) = h(0), p(\mathbb{U}) \subseteq h(\mathbb{U})\}.$$

Remark 2.15

(i) Obviously, if $p \in H(\mathbb{U})$, then $p \in s(h)$ if and only if $p \prec h$, where "$\prec$" is the usual symbol of subordination.

(ii) If h is a univalent holomorphic function on $\mathbb{U}$, then $s(h)$ is a compact subset of $H(\mathbb{U})$, in view of [22, Lemma 5.19] (see also [28, p. 374]). Hence $\text{ex}\,s(h) \neq \emptyset$

and supp $s(h) \neq \emptyset$. If, in addition, h is a convex function on $\mathbb{U}$, then it is easily seen that $s(h)$ is a convex subset of $H(\mathbb{U})$.

(iii) Let $h : \mathbb{U} \to \mathbb{C}$ be a univalent holomorphic function. Then

$$\{h_x : |x| = 1\} \subseteq \mathrm{ex}\,\overline{\mathrm{co}}s(h),$$

where $\overline{\mathrm{co}}s(h)$ is the closed convex hull of $s(h)$, $h_x(\zeta) = h(x\zeta)$ for $|\zeta| < 1$ and $|x| = 1$ (see [22, Theorem 5.5], [28, p. 379]). If, in addition, $\mathbb{C} \setminus h(\mathbb{U})$ is a convex set, then

$$\mathrm{ex}\,\overline{\mathrm{co}}s(h) = \{h_x : |x| = 1\},$$

by Hallenbeck et al. [23] and MacGregor and Wilken [28, Theorem 9].

(iv) Let $h : \mathbb{U} \to \mathbb{C}$ be a univalent holomorphic function. Then

$$\{h_x : |x| = 1\} \subseteq \mathrm{supp}\,s(h),$$

by Hallenbeck and MacGregor [22, Theorem 7.13] (see also [28, p. 381]).

The following condition will be useful in the next sections (see [13]):

Assumption 2.16 Let $g : \mathbb{U} \to \mathbb{C}$ be a univalent holomorphic function such that $g(0) = 1$, $g(\bar{\zeta}) = \overline{g(\zeta)}$ for $\zeta \in \mathbb{U}$, and $\Re g(\zeta) > 0$ on $\mathbb{U}$. We assume that g satisfies the following conditions:

$$\min\{\Re g(\zeta) : |\zeta| = r\} = g(r), \quad \max\{\Re g(\zeta) : |\zeta| = r\} = g(-r), \ \forall\, r \in (0, 1). \tag{2.5}$$

In this paper we consider extreme points and support points associated with the Carathéodory family $\mathcal{M}_g(\mathbb{U}^n)$, where $g : \mathbb{U} \to \mathbb{C}$ is a convex (univalent) holomorphic function with $g(0) = 1$. We also consider extreme points associated with the family $S_g^*(\mathbb{U}^n)$ of g-starlike mappings on $\mathbb{U}^n$ in $\mathbb{C}^n$, where $g : \mathbb{U} \to \mathbb{C}$ is a univalent holomorphic function with $g(0) = 1$ and $\Re g(\zeta) > 0$, $\zeta \in \mathbb{U}$. In particular, if g is a convex function on $\mathbb{U}$ which satisfies the conditions of Assumption 2.16, we are concerned with bounded support points for the family $S_g^0(\mathbb{U}^2)$ of mappings with g-parametric representation on $\mathbb{U}^2$. Note that interesting extremal problems for convex mappings on the Euclidean unit ball $\mathbb{B}^n$ in $\mathbb{C}^n$ were investigated in [29].

3 Extreme Points and Support Points for $\mathcal{M}_g(\mathbb{U}^n)$ and $S_g^*(\mathbb{U}^n)$

In this section, we are concerned with extreme/support points for the compact family $\mathcal{M}_g(\mathbb{U}^n)$, where $g : \mathbb{U} \to \mathbb{C}$ is a convex (univalent) holomorphic function such that $g(0) = 1$. We also obtain a characterization of extreme points associated with the family $S_g^*(\mathbb{U}^n)$, where $g : \mathbb{U} \to \mathbb{C}$ is a univalent holomorphic function such that $g(0) = 1$ and $\Re g(\zeta) > 0, \zeta \in \mathbb{U}$.

First, we start with the following extremal result for the family $\mathcal{M}_g(\mathbb{U}^n)$ (see [41, Proposition 2.2.1], in the case $g(\zeta) = \frac{1-\zeta}{1+\zeta}, \zeta \in \mathbb{U}$).

Proposition 3.1 *Let $g : \mathbb{U} \to \mathbb{C}$ be a convex (univalent) function on $\mathbb{U}$ such that $g(0) = 1$. Also, let $h : \mathbb{U}^n \to \mathbb{C}^n$ be given by*

$$h(z) = \big(z_1 p_1(z), \ldots, z_n p_n(z)\big), \quad z = (z_1, \ldots, z_n) \in \mathbb{U}^n,$$

where $p_j : \mathbb{U}^n \to \mathbb{C}$ $(j = 1, \ldots, n)$ are holomorphic functions which satisfy necessary conditions for $h \in \mathcal{M}_g(\mathbb{U}^n)$. If $p_j(\cdot z) \in \operatorname{ex} s(g)$, for all $z \in \partial_\infty \mathbb{U}^n$, $j = 1, \ldots, n$, then $h \in \operatorname{ex} \mathcal{M}_g(\mathbb{U}^n)$.

Proof We use arguments similar to those in the proof of [41, Proposition 2.2.1]. First, note that $\mathcal{M}_g(\mathbb{U}^n)$ is a compact family, in view of Remark 2.3 (iii), and thus $\operatorname{ex} \mathcal{M}_g(\mathbb{U}^n) \neq \emptyset$. The same is true related to the family $s(g)$, by Remark 2.15 (ii). Assume that $p_j(\cdot z) \in \operatorname{ex} s(g)$, $z \in \partial_\infty \mathbb{U}^n$, $j = 1, \ldots, n$. Let $q \in H(\mathbb{U}^n)$ be such that $h \pm q \in \mathcal{M}_g(\mathbb{U}^n)$. Then $q(0) = 0$ and $Dq(0) = 0$. We show that $q \equiv 0$. To this end, fix $z \in \partial_\infty \mathbb{U}^n$ and $j \in \{1, \ldots, n\}$. Since $h \pm q \in \mathcal{M}_g(\mathbb{U}^n)$, it follows that

$$\frac{h_j(\zeta z)}{\zeta z_j} \pm \frac{q_j(\zeta z)}{\zeta z_j} \in g(\mathbb{U}), \quad \zeta \in \mathbb{U}, \quad |z_j| = 1,$$

i.e. $p_j(\zeta z) \pm \frac{q_j(\zeta z)}{\zeta z_j} \in g(\mathbb{U})$ for $\zeta \in \mathbb{U}$. Hence $p_j(\cdot z) \pm \frac{q_j(\cdot z)}{(\cdot z_j)} \in s(g)$. On the other hand, since $p_j(\cdot z) \in \operatorname{ex} s(g)$, we deduce in view of the previous relation that $q_j(\cdot z)/(\cdot z_j) = 0$. Since $z \in \partial_\infty \mathbb{U}^n$ and $j \in \{1, \ldots, n\}$ are arbitrary, we deduce that $q \equiv 0$, as desired.

Next, we suppose that $h = (1 - \lambda)h^1 + \lambda h^2$ for some $\lambda \in (0, 1)$ and $h^1, h^2 \in \mathcal{M}_g(\mathbb{U}^n)$. Without loss of generality, we may assume that $\lambda \in (0, 1/2]$. Let $h^3 = (1 - 2\lambda)h^1 + 2\lambda h^2$. Since $\mathcal{M}_g(\mathbb{U}^n)$ is a convex family, it follows that $h^3 \in \mathcal{M}_g(\mathbb{U}^n)$. Also, it is easily seen that if $q = (h^1 - h^3)/2$, then $h + q = h^1 \in \mathcal{M}_g(\mathbb{U}^n)$ and $h - q = h^3 \in \mathcal{M}_g(\mathbb{U}^n)$. Taking into account the first part of the proof, we deduce that $q \equiv 0$, and thus $h^1 = h^3$, i.e. $h^1 = h^2$, as desired. This completes the proof. $\square$

Remark 3.2

(i) In the above proposition, we should take $z \in \partial_\infty \mathbb{U}^n$. For example, if $z \in \partial \mathbb{U}^n$ and $z_n = 0$ and some p_j is a function only in z_n, i.e. $p_j(z) = p_j(z_n)$, then $p_j(\cdot z) = p_j(0)$ cannot be an extreme point of $s(g)$.

(ii) Let $g : \mathbb{U} \to \mathbb{C}$ be a univalent holomorphic function such that $g(0) = 1$, and let $h : \mathbb{U}^n \to \mathbb{C}^n$ be given by

$$h(z) = \big(z_1 p_1(z), \ldots, z_n p_n(z)\big), \quad z = (z_1, \ldots, z_n) \in \mathbb{U}^n,$$

where $p_j : \mathbb{U}^n \to \mathbb{C}$ $(j = 1, \ldots, n)$ are holomorphic functions which satisfy necessary conditions for $h \in \overline{\mathrm{co}}\mathcal{M}_g(\mathbb{U}^n)$. If $p_j(\cdot z) \in \mathrm{ex}\,\overline{\mathrm{co}}s(g)$, for all $z \in \partial_\infty \mathbb{U}^n$, $j = 1, \ldots, n$, then $h \in \mathrm{ex}\,\overline{\mathrm{co}}\mathcal{M}_g(\mathbb{U}^n)$.

In view of Proposition 3.1, we obtain the following consequence (see [41, Proposition 2.2.1], in the case $\alpha = 0$).

Corollary 3.3 *Let $\alpha \in [0, 1)$ and let $h : \mathbb{U}^n \to \mathbb{C}^n$ be defined as in Proposition 3.1. If $p_j(\cdot z) \in \mathrm{ex}\,\mathcal{P}(\alpha)$, for all $z \in \partial_\infty \mathbb{U}^n$, $j = 1, \ldots, n$, then $h \in \mathrm{ex}\,\mathcal{M}_g(\mathbb{U}^n)$, where $g(\zeta) = \frac{1-(1-2\alpha)\zeta}{1+\zeta}$, $\zeta \in \mathbb{U}$.*

If we restrict the form of the mapping h, then we can obtain the converse of Proposition 3.1 (cf. [41, Proposition 2.2.1] in the case $g(\zeta) = \frac{1-\zeta}{1+\zeta}$, $\zeta \in \mathbb{U}$).

Proposition 3.4 *Let $g : \mathbb{U} \to \mathbb{C}$ be a convex (univalent) function on $\mathbb{U}$ such that $g(0) = 1$. Also, let $h : \mathbb{U}^n \to \mathbb{C}^n$ be given by*

$$h(z) = \big(z_1 p_1(z_{i_1}), \ldots, z_n p_n(z_{i_n})\big), \quad z = (z_1, \ldots, z_n) \in \mathbb{U}^n,$$

where $i_1, \ldots, i_n \in \{1, \ldots, n\}$ and $p_j \in s(g)$, $j = 1, \ldots, n$. Then $h \in \mathrm{ex}\,\mathcal{M}_g(\mathbb{U}^n)$ if and only if $p_j \in \mathrm{ex}\,s(g)$, $j = 1, \ldots, n$.

Question 3.5 Let $g : \mathbb{U} \to \mathbb{C}$ be a convex (univalent) function on $\mathbb{U}$ such that $g(0) = 1$. Also, let $h : \mathbb{U}^n \to \mathbb{C}^n$ be defined as in Proposition 3.1. If $p_j(\cdot z) \in \mathrm{supp}\,s(g)$, for all $z \in \partial_\infty \mathbb{U}^n$, $j = 1, \ldots, n$, then is it true that $h \in \mathrm{supp}\,\mathcal{M}_g(\mathbb{U}^n)$?

Related to Question 3.5, we obtain the following characterizations of support points for the family $\mathcal{M}_g(\mathbb{U}^n)$.

Proposition 3.6 *Let $g : \mathbb{U} \to \mathbb{C}$ be a convex (univalent) function on $\mathbb{U}$ such that $g(0) = 1$. Also, let $h : \mathbb{U}^n \to \mathbb{C}^n$ be given by*

$$h(z) = (z_1 p_1(z), \tilde{h}(z)), \quad z = (z_1, \ldots, z_n) \in \mathbb{U}^n,$$

where $p_1 : \mathbb{U}^n \to \mathbb{C}$ and $\tilde{h} : \mathbb{U}^n \to \mathbb{C}^{n-1}$ are holomorphic functions which satisfy necessary conditions for $h \in \mathcal{M}_g(\mathbb{U}^n)$. If $p_1(\cdot z) \in \mathrm{supp}\,s(g)$, for some $z \in \partial_\infty \mathbb{U}^n$, then $h \in \mathrm{supp}\,\mathcal{M}_g(\mathbb{U}^n)$.

Proof Since $p_1(\cdot z) \in \mathrm{supp}\,s(g)$, there exists a continuous linear functional $L_1 : H(\mathbb{U}) \to \mathbb{C}$ such that $\Re L_1|_{s(g)} \neq$ constant and $\Re L_1(p_1(\cdot z)) = \max\{\Re L_1(q) : q \in s(g)\}$. Let $L : H(\mathbb{U}^n) \to \mathbb{C}$ be given by

$$L(f) = L_1(q_f), \quad f = (f_1, \ldots, f_n) \in H(\mathbb{U}^n),$$

where

$$q_f(\zeta) = \frac{f_1(\zeta z) - f_1(0)}{\zeta z_1}, \quad \zeta \in \mathbb{U}.$$

Then L is a continuous linear functional on $H(\mathbb{U}^n)$, $\Re L|_{\mathcal{M}_g(\mathbb{U}^n)} \neq$ constant, and

$$\Re L(h) = \Re L_1(p_1(\cdot z)) = \max\{\Re L_1(q) : q \in s(g)\} = \max\{\Re L(f) : f \in \mathcal{M}_g(\mathbb{U}^n)\}.$$

Thus, $h \in \operatorname{supp} \mathcal{M}_g(\mathbb{U}^n)$, as desired. $\qquad\qquad\qquad\qquad\qquad\qquad\qquad\square$

Corollary 3.7 *Let $g : \mathbb{U} \to \mathbb{C}$ be a convex (univalent) function such that $g(0) = 1$, and let $h : \mathbb{U}^n \to \mathbb{C}^n$ be defined as in Proposition 3.6. If $p_1(\cdot z) \in \{g_x : |x| = 1\}$, for some $z \in \partial_\infty \mathbb{U}^n$, then $h \in \operatorname{supp} \mathcal{M}_g(\mathbb{U}^n)$. In particular, if $h \in \mathcal{M}(\mathbb{U}^n)$ and there exist $x \in \partial \mathbb{U}$ and $z \in \partial_\infty \mathbb{U}^n$ such that $p_1(\zeta z) = \frac{1-x\zeta}{1+x\zeta}$, $\zeta \in \mathbb{U}$, then $h \in \operatorname{supp} \mathcal{M}(\mathbb{U}^n)$.*

Proof It suffices to apply Proposition 3.6 and Remark 2.15 (iv). $\qquad\qquad\qquad\square$

Proposition 3.8 *Let $g : \mathbb{U} \to \mathbb{C}$ be a convex (univalent) function on $\mathbb{U}$ such that $g(0) = 1$. Also, let $h : \mathbb{U}^n \to \mathbb{C}^n$ be given by*

$$h(z) = (z_1 p_1(z_{i_1}), \ldots, z_n p_n(z_{i_n})), \quad z = (z_1, \ldots, z_n) \in \mathbb{U}^n,$$

where $i_1, \ldots, i_n \in \{1, \ldots, n\}$ and $p_j \in s(g), j = 1, \ldots, n$. Then $h \in \operatorname{supp} \mathcal{M}_g(\mathbb{U}^n)$ if and only if there exists $j \in \{1, \ldots, n\}$ such that $p_j \in \operatorname{supp} s(g)$.

Proof If there is $j \in \{1, \ldots, n\}$ such that $p_j \in \operatorname{supp} s(g)$, then $h \in \operatorname{supp} \mathcal{M}_g(\mathbb{U}^n)$, in view of Proposition 3.6.

Conversely, assume that $h \in \operatorname{supp} \mathcal{M}_g(\mathbb{U}^n)$. There exists a continuous linear functional $L : H(\mathbb{U}^n) \to \mathbb{C}$ such that $\Re L|_{\mathcal{M}_g(\mathbb{U}^n)} \neq$ constant and

$$\Re L(h) = \max\{\Re L(f) : f \in \mathcal{M}_g(\mathbb{U}^n)\}. \qquad\qquad (3.1)$$

There exists $j \in \{1, \ldots, n\}$ such that $\Re L_j|_{\mathcal{M}_g(\mathbb{U}^n)} \neq$ constant, where $L_j(f) = L((0, \ldots, f_j, \ldots, 0))$. We may assume that $j = 1$. There exists $f = (f_1, \ldots, f_n) \in \mathcal{M}_g(\mathbb{U}^n)$ such that $\Re L_1((z_1, \ldots, z_n)) < \Re L_1(f)$. From (3.1), we have

$$\begin{aligned}
\Re L_1((z_1, z_2 p_2(z_{i_2}) \ldots, z_n p_n(z_{i_n}))) &= \Re L_1((z_1, \ldots, z_n)) \\
&< \Re L_1(f) \\
&= \Re L_1((f_1, z_2 p_2(z_{i_2}) \ldots, z_n p_n(z_{i_n}))) \\
&\leq \Re L_1(h).
\end{aligned}$$

Let $\tilde{L}_1 : H(\mathbb{U}) \to \mathbb{C}$ be given by $\tilde{L}_1(q) = L_1(h_q)$, $q \in H(\mathbb{U})$, where

$$h_q(z) = (z_1 q(z_{i_1}), z_2 p_2(z_{i_2}) \ldots, z_n p_n(z_{i_n})), \quad z = (z_1, \ldots, z_n) \in \mathbb{U}^n.$$

Then $\tilde{L}_1$ is a continuous linear functional such that $\Re\tilde{L}_1|_{s(g)} \neq$ constant and

$$\Re\tilde{L}_1(p_1) = \Re L(h) - \Re L((0, z_2 p_2(z_{i_2}), \ldots, z_n p_n(z_{i_n})))$$
$$\geq \max\{\Re L(h_q) : q \in s(g)\} - \Re L((0, z_2 p_2(z_{i_2}), \ldots, z_n p_n(z_{i_n})))$$
$$= \max\{\Re\tilde{L}_1(q) : q \in s(g)\}.$$

Thus, $p_1 \in \operatorname{supp} s(g)$, as desired. This completes the proof. $\qquad\square$

In view of Proposition 3.8, we obtain the following consequence.

Corollary 3.9 *Let* $\alpha \in [0, 1)$ *and let* $g(\zeta) = \frac{1-(1-2\alpha)\zeta}{1+\zeta}$, $\zeta \in \mathbb{U}$. *Also, let* $h : \mathbb{U}^n \to \mathbb{C}^n$ *be defined as in Proposition 3.8. Then* $h \in \operatorname{supp}\mathcal{M}_g(\mathbb{U}^n)$ *if and only if there exists* $j \in \{1, \ldots, n\}$ *such that* $p_j \in \operatorname{supp}\mathcal{P}(\alpha)$. *In particular, if* $p_j \in \mathcal{P}$, $j = 1, \ldots, n$, *then* $h \in \operatorname{supp}\mathcal{M}(\mathbb{U}^n)$ *if and only if there exists* $j \in \{1, \ldots, n\}$ *such that* $p_j \in \operatorname{supp}\mathcal{P}$.

Next, we prove the following result related to extreme points for the compact family $S_g^*(\mathbb{U}^n)$ (see [41, Proposition 2.2.2]) in the case $g(\zeta) = \frac{1-\zeta}{1+\zeta}$, $\zeta \in \mathbb{U}$).

Theorem 3.10 *Let* $g : \mathbb{U} \to \mathbb{C}$ *be a univalent holomorphic function such that* $g(0) = 1$ *and* $\Re g(\zeta) > 0$, $\zeta \in \mathbb{U}$. *Also, let* $f \in S_g^*(\mathbb{U}^n)$ *and let* $h(z) = [Df(z)]^{-1}f(z)$, $z \in \mathbb{U}^n$. *Assume that*

$$h(z) = \big(z_1 p_1(z), \ldots, z_n p_n(z)\big), \quad z = (z_1, \ldots, z_n) \in \mathbb{U}^n, \tag{3.2}$$

and

$$\{p_j(\cdot z) : j = 1, \ldots, n\} \subseteq \{g_x : |x| = 1\}, \quad \forall z \in \partial_\infty \mathbb{U}^n, \tag{3.3}$$

where $g_x(\zeta) = g(x\zeta)$, $\zeta \in \mathbb{U}$, $|x| = 1$. *Then* $f \in \operatorname{ex} S_g^*(\mathbb{U}^n)$.

Proof We use arguments similar to those in the proof of [41, Proposition 2.2.2]. Suppose that $f = (1 - \lambda)f^1 + \lambda f^2$, for some $\lambda \in (0, 1)$ and $f^1, f^2 \in S_g^*(\mathbb{U}^n)$. We prove that $f^1 \equiv f^2$. To this end, let $h^1, h^2 \in \mathcal{M}_g(\mathbb{U}^n)$ be such that $h^i(z) = [Df^i(z)]^{-1}f^i(z)$, $z \in \mathbb{U}^n$. Also, let $Q_2(z) = \frac{1}{2}D^2h(0)(z^2)$, $Q_2^i(z) = \frac{1}{2}D^2h^i(0)(z^2)$, $P_2(z) = \frac{1}{2}D^2f(0)(z^2)$ and $P_2^i(z) = \frac{1}{2}D^2f^i(0)(z^2)$, for $z \in \mathbb{U}^n$ and $i = 1, 2$. Then it is easily seen that $P_2 = -Q_2$ and $P_2^i = -Q_2^i$, $i = 1, 2$. Hence, we obtain that

$$Q(z) = (1 - \lambda)Q_2^1(z) + \lambda Q_2^2(z), \quad z \in \mathbb{U}^n. \tag{3.4}$$

Next, fix $z \in \partial_\infty \mathbb{U}^n$ and let $j \in \{1, \ldots, n\}$. Also, let $p_j^i(\zeta) = \frac{h_j^i(\zeta z)}{\zeta z_j}$ for $0 < |\zeta| < 1$, and $p_j^i(0) = 1$, $i = 1, 2$. Since $h^i \in \mathcal{M}_g(\mathbb{U}^n)$, it follows that $p_j^i \prec g$, $i = 1, 2$. Then $|\frac{dp_j^i}{d\zeta}(0)| \leq |g'(0)|$, $i = 1, 2$, by the subordination principle. It is easy to obtain that $\frac{dp_j^i}{d\zeta}(0) = \frac{Q_{2,j}^i(z)}{z_j}$, where $Q_{2,j}^i$ is the j-th component of Q_2^i, $i = 1, 2$. Hence

$$|Q_{2,j}^i(z)| \leq |g'(0)|, \quad i = 1, 2. \tag{3.5}$$

On the other hand, in view of the relation (3.3), we deduce that there exists $x_j \in \mathbb{C}$, $|x_j| = 1$, such that $p_j(\cdot z) = g(\cdot x_j)$, and thus

$$\frac{h_j(\zeta z)}{\zeta z_j} = p_j(\zeta z) = g(\zeta x_j), \quad \zeta \in \mathbb{U}.$$

Consequently, the j-th component $Q_{2,j}$ of Q_2 is given by $Q_{2,j}(z) = g'(0)x_j z_j$. Taking into account the relation (3.4), we obtain that

$$Q_{2,j}(z) = (1 - \lambda)Q^1_{2,j}(z) + \lambda Q^2_{2,j}(z),$$

and since $|Q_{2,j}(z)| = |g'(0)|$, we deduce in view of (3.5) and the above equality that $Q_{2,j}(z) = Q^1_{2,j}(z) = Q^2_{2,j}(z) = g'(0)x_j z_j$. Further, since $j \in \{1, \ldots, n\}$ and $z \in \partial_\infty \mathbb{U}^n$ are arbitrary, we conclude that $Q_2 \equiv Q^1_2 \equiv Q^2_2$.

Now, since $\frac{dp^i_j}{d\zeta}(0)z_j = Q^i_{2,j}(z) = g'(0)x_j z_j$, the subordination principle yields that $p^i_j(\zeta) = g(\zeta x_j)$, $i = 1, 2$, i.e. $\frac{h^i_j(\zeta z)}{\zeta z_j} = g(\zeta x_j)$, $i = 1, 2$. Since $z \in \partial_\infty \mathbb{U}^n$ and $j \in \{1, \ldots, n\}$ are arbitrary, it follows that $h^1 \equiv h^2 \equiv h$. Finally, the previous relation yields that $f^1 \equiv f^2 \equiv f$, as desired. This completes the proof. □

From Theorem 3.10 we obtain the following result (cf. [41, Proposition 2.2.2]).

Corollary 3.11 *Let* $g : \mathbb{U} \to \mathbb{C}$ *be a univalent holomorphic function such that* $g(0) = 1$ *and* $\Re g(\zeta) > 0$, $\zeta \in \mathbb{U}$. *Also, let* $f \in S^*_g(\mathbb{U}^n)$ *and let* $h(z) = [Df(z)]^{-1}f(z)$, $z \in \mathbb{U}^n$. *Assume that*

$$\text{ex} \, \overline{\text{cos}}(g) = \{g_x : |x| = 1\}, \tag{3.6}$$

where $g_x(\zeta) = g(x\zeta)$, $|\zeta| < 1$, $|x| = 1$. *If* $p_1, \ldots, p_n$ *are given by (3.2) and* $p_1(\cdot z), \ldots, p_n(\cdot z) \in \text{ex} \, \overline{\text{cos}}(g)$, *for all* $z \in \partial_\infty \mathbb{U}^n$, *then* $f \in \text{ex} \, S^*_g(\mathbb{U}^n)$.

In view of Corollary 3.11 and Remark 2.15 (iii), we obtain the following result.

Corollary 3.12 *Let* $g : \mathbb{U} \to \mathbb{C}$ *be a univalent holomorphic function such that* $g(0) = 1$ *and* $\Re g(\zeta) > 0$, $\zeta \in \mathbb{U}$. *Assume that* $\mathbb{C} \setminus g(\mathbb{U})$ *is a convex set. Also, let* $f \in S^*_g(\mathbb{U}^n)$ *and let* $h(z) = [Df(z)]^{-1}f(z)$, $z \in \mathbb{U}^n$. *If* $p_1, \ldots, p_n$ *are given by (3.2) and* $p_1(\cdot z), \ldots, p_n(\cdot z) \in \text{ex} \, \overline{\text{cos}}(g)$, *for all* $z \in \partial_\infty \mathbb{U}^n$, *then* $f \in \text{ex} \, S^*_g(\mathbb{U}^n)$.

Proof Since $\mathbb{C} \setminus g(\mathbb{U})$ is convex, it follows that the relation (3.6) holds, by Remark 2.15 (iii). Then the result follows in view of Corollary 3.11. □

The following result is a direct consequence of Proposition 3.4 and Corollary 3.11 (see [41, Proposition 2.2.2], in the case $g(\zeta) = \frac{1-\zeta}{1+\zeta}$, $\zeta \in \mathbb{U}$).

Corollary 3.13 *Let* $g : \mathbb{U} \to \mathbb{C}$ *be a convex (univalent) function such that* $g(0) = 1$ *and* $\Re g(\zeta) > 0$, $\zeta \in \mathbb{U}$. *Also, let* $f \in S^*_g(\mathbb{U}^n)$ *and let* $h(z) = [Df(z)]^{-1}f(z)$, $z \in \mathbb{U}^n$. *Assume that* $\text{ex} \, s(g) = \{g_x : |x| = 1\}$. *If*

$$h(z) = \left(z_1 p_1(z_{i_1}), \ldots, z_n p_n(z_{i_n})\right), \quad z = (z_1, \ldots, z_n) \in \mathbb{U}^n,$$

where $i_1, \ldots, i_n \in \{1, \ldots, n\}$ *and* $h \in \text{ex} \, \mathcal{M}_g(\mathbb{U}^n)$, *then* $f \in \text{ex} \, S^*_g(\mathbb{U}^n)$.

In view of Corollaries 3.3 and 3.12, we obtain the following consequence (see [41, Proposition 2.2.2], in the case $\alpha = 0$).

Corollary 3.14 *Let $\alpha \in [0, 1)$ and let $g(\zeta) = \frac{1-(1-2\alpha)\zeta}{1+\zeta}$, $\zeta \in \mathbb{U}$. Also, let $f \in S_g^*(\mathbb{U}^n)$ and let $h(z) = [Df(z)]^{-1}f(z)$, for all $z \in \mathbb{U}^n$. If $p_1, \ldots, p_n$ are given by (3.2) and $p_j(\cdot z) \in \mathrm{ex}\, \mathcal{P}(\alpha)$, for all $z \in \partial_\infty \mathbb{U}^n$ and $j = 1, \ldots, n$, then $h \in \mathrm{ex}\, \mathcal{M}_g(\mathbb{U}^n)$ and $f \in \mathrm{ex}\, S_g^*(\mathbb{U}^n)$.*

Proof Since $s(g) = \mathcal{P}(\alpha)$ and $\mathbb{C} \setminus g(\mathbb{U})$ is a convex set, the result follows from Corollaries 3.3, 3.11 and 3.12, as desired. □

4 Coefficient Bounds for the Family $\mathcal{M}_g(\mathbb{U}^2)$

We begin this section with the following notion on the unit polydisc $\mathbb{U}^2$, which is an analogue on $\mathbb{U}^2$ of the shearing process due to Bracci [4, Definition 1.3]. Then we obtain some coefficient bounds for the family $\mathcal{M}_g(\mathbb{U}^2)$, where g is a convex (univalent) function on $\mathbb{U}$ such that $g(0) = 1$, and which satisfies the conditions of Assumption 2.16. We also obtain various particular cases. We shall apply Proposition 4.2, to obtain examples of support points for the family $S_g^0(\mathbb{U}^2)$.

Definition 4.1 Let $\rho, \sigma \in \mathbb{C} \setminus \{0\}$ and let $h \in H(\mathbb{U}^2)$ be such that $h(0) = 0$ and

$$h(z) = \left(\rho z_1 + q_{0,2}^1 z_2^2 + O(|z_1|^2, |z_1 z_2|, \|z\|^3), \sigma z_2 + O(\|z\|^2)\right), \quad z = (z_1, z_2) \in \mathbb{U}^2.$$

Then the shearing $h^{[c]}$ of h is given by

$$h^{[c]}(z) = \left(\rho z_1 + q_{0,2}^1 z_2^2, \sigma z_2\right), \quad z = (z_1, z_2) \in \mathbb{U}^2.$$

Next, we prove the following result, by using arguments similar to those in the proofs of [4, Proposition 2.1] and [21, Proposition 4.2]. To this end, if $g : \mathbb{U} \to \mathbb{C}$ is a convex (univalent) function with $g(0) = 1$, and which satisfies the conditions of Assumption 2.16, let a_0 be given by

$$a_0 = \inf_{r \in (0,1)} \left\{ \frac{\min\{1 - g(r), g(-r) - 1\}}{r} \right\}. \tag{4.1}$$

It is easily seen that $a_0 \geq |g'(0)|/2$, by the well known growth result for the family K of normalized convex functions on $\mathbb{U}$. Hence $a_0 > 0$.

Proposition 4.2 *Let $g : \mathbb{U} \to \mathbb{C}$ be a convex (univalent) function such that $g(0) = 1$. Assume that g satisfies the conditions of Assumption 2.16. If $h = (h_1, h_2) \in \mathcal{M}_g(\mathbb{U}^2)$, then*

$$\max\{|q_{0,2}^1|, |q_{2,0}^2|\} \leq a_0, \tag{4.2}$$

where a_0 is given by (4.1), $q_{0,2}^1 = \frac{1}{2}\frac{\partial^2 h_1}{\partial z_2^2}(0)$ and $q_{2,0}^2 = \frac{1}{2}\frac{\partial^2 h_2}{\partial z_1^2}(0)$. If, in addition, $\mathbb{U}(1, a_0) \subseteq g(\mathbb{U})$, then $h^{[c]} \in \mathcal{M}_g(\mathbb{U}^2)$.

Proof We shall use some arguments similar to those in the proof of [4, Proposition 2.1]. Since $h(0) = 0$ and $Dh(0) = I_2$, we deduce that h has the following power series expansion on $\mathbb{U}^2$:

$$h(z) = \left(z_1 + \sum_{\alpha \in \mathbb{N}^2, |\alpha| \geq 2} q_\alpha^1 z^\alpha, z_2 + \sum_{\alpha \in \mathbb{N}^2, |\alpha| \geq 2} q_\alpha^2 z^\alpha \right), \quad z = (z_1, z_2) \in \mathbb{U}^2.$$

Since $h \in \mathcal{M}_g(\mathbb{U}^2)$, it follows in view of [13, eq.(2.3)] (see also [17, Lemma 3.1]) and the relation (2.5) that

$$\|z\|_\infty g(\|z\|_\infty) \leq \Re \ell_z(h(z)) \leq \|z\|_\infty g(-\|z\|_\infty), \quad z \in \mathbb{U}^2 \setminus \{0\}, \quad \ell_z \in T(z).$$

Taking into account the relation (2.1), we obtain that

$$g(r) \leq \Re \left[\frac{h_j(z)}{z_j} \right] \leq g(-r), \quad |z_j| = \|z\|_\infty = r \in (0, 1), \quad j = 1, 2.$$

Hence

$$g(r) \leq \Re \left[\frac{h_j(z)}{z_j} \right] = 1 + \sum_{\alpha \in \mathbb{N}^2, |\alpha| \geq 2} \Re \left[q_\alpha^j \frac{z^\alpha}{z_j} \right] \leq g(-r), \tag{4.3}$$

for $|z_j| = \|z\|_\infty = r \in (0, 1), j = 1, 2$. Next, fix $z = (z_1, z_2) \in \mathbb{C}^2$ such that $|z_1| = |z_2| = r \in (0, 1)$, and let $\eta \in [0, 2\pi)$ be such that $q_{0,2}^1 = |q_{0,2}^1| e^{i(\pi + \eta)}$, let $\theta \in [0, 2\pi)$ be such that $z_1 = re^{i(\eta + \theta)}$ and $z_2 = re^{i\theta/2}$. Using arguments similar to those in the proof of [4, Proposition 2.1] and taking into account the relation (4.3), we deduce that

$$1 - |q_{0,2}^1| r \geq g(r),$$

and thus

$$|q_{0,2}^1| \leq \frac{1 - g(r)}{r}, \quad \forall r \in (0, 1).$$

Letting

$$a_1 = \inf_{r \in (0,1)} \left\{ \frac{1 - g(r)}{r} \right\}, \tag{4.4}$$

we obtain that $|q_{0,2}^1| \leq a_1$. Since $\Re[h_2(z)/z_2] \geq g(r)$ for $|z_2| = \|z\|_\infty = r \in (0,1)$, we obtain by an argument similar to above that $|q_{2,0}^2| \leq a_1$, and thus

$$\max\{|q_{0,2}^1|, |q_{2,0}^2|\} \leq a_1. \tag{4.5}$$

Now, fix $z = (z_1, z_2) \in \mathbb{U}^2 \setminus \{0\}$ such that $|z_1| = |z_2| = r$. Also, let $\delta \in [0, 2\pi)$ be such that $q_{0,2}^1 = |q_{0,2}^1|e^{i\delta}$, and let $\theta \in [0, 2\pi)$ be such that $z_1 = re^{i(\delta+\theta)}$ and $z_2 = re^{i\theta/2}$. Using arguments similar to above and the relation (4.3), we obtain that

$$1 + |q_{0,2}^1|r \leq g(-r),$$

and thus

$$|q_{0,2}^1| \leq \frac{g(-r) - 1}{r}, \quad \forall r \in (0,1).$$

Similarly, since $\Re[h_2(z)/z_2] \leq g(-r)$ for $|z_2| = \|z\|_\infty = r \in (0,1)$, we obtain that $|q_{2,0}^2| \leq \frac{g(-r)-1}{r}$. Letting

$$a_2 = \inf_{r \in (0,1)} \left\{ \frac{g(-r) - 1}{r} \right\}, \tag{4.6}$$

we obtain that

$$\max\{|q_{0,2}^1|, |q_{2,0}^2|\} \leq a_2. \tag{4.7}$$

Hence, the relation (4.2) follows from (4.5) and (4.7), and since $a_0 = \min\{a_1, a_2\}$.
Next, we assume that $\mathbb{U}(1, a_0) \subseteq g(\mathbb{U})$. In view of (4.2), we obtain that

$$\left| \frac{h_1^{[c]}(z)}{z_1} - 1 \right| = \left| \frac{q_{0,2}^1 z_2^2}{z_1} \right| \leq a_0, \quad 0 < |z_1| = \|z\|_\infty < 1,$$

and thus $h_1^{[c]}(z)/z_1 \in \mathbb{U}(1, a_0)$ for $0 < |z_1| = \|z\|_\infty < 1$. Since $\mathbb{U}(1, a_0) \subseteq g(\mathbb{U})$, it follows that $h_1^{[c]}(z)/z_1 \in g(\mathbb{U})$ for $0 < |z_1| = \|z\|_\infty < 1$. Obviously, $h_2^{[c]}(z)/z_2 = 1 \in g(\mathbb{U})$ for $0 < |z_2| = \|z\|_\infty < 1$. Thus, taking into account the relation (2.2), we conclude that $h^{[c]} \in \mathcal{M}_g(\mathbb{U}^2)$, as desired. This completes the proof. □

Remark 4.3 Let $h = (h_1, h_2) \in H(\mathbb{U}^2)$ be a normalized mapping, and let

$$\widetilde{h}^{[c]}(z) = \left(z_1, z_2 + q_{2,0}^2 z_1^2\right), \quad z = (z_1, z_2) \in \mathbb{U}^2. \tag{4.8}$$

Also, let $g : \mathbb{U} \to \mathbb{C}$ be a convex (univalent) function such that $g(0) = 1$, and g satisfies the conditions of Assumption 2.16. Assume that $\mathbb{U}(1, a_0) \subseteq g(\mathbb{U})$, where a_0 is given by (4.1). Taking into account (4.2), we deduce as in the proof of Proposition 4.2 that if $h \in \mathcal{M}_g(\mathbb{U}^2)$, then $\widetilde{h}^{[c]} \in \mathcal{M}_g(\mathbb{U}^2)$ and $|q_{2,0}^2| \leq a_0$.

In connection with the above results, it would be interesting to see if the assumption $\mathbb{U}(1, a_0) \subseteq g(\mathbb{U})$ may be omitted in the statement of Proposition 4.2.

Question 4.4 Let $g : \mathbb{U} \to \mathbb{C}$ be a convex (univalent) function such that $g(0) = 1$. Assume that g satisfies the conditions of Assumption 2.16. If $h = (h_1, h_2) \in \mathcal{M}_g(\mathbb{U}^2)$, is it true that $h^{[c]} \in \mathcal{M}_g(\mathbb{U}^2)$?

Remark 4.5 In [21, Remark 4.4], the authors provided an example of a convex (univalent) function g on $\mathbb{U}$ such that $g(0) = 1$ and which satisfies the conditions of Assumption 2.16, but for which $\mathbb{U}(1, a_0) \not\subseteq g(\mathbb{U})$, where a_0 is given by (4.1).

The following result yields that the estimate (4.2) is sharp (cf. [4] and [21], in the case of the Euclidean unit ball $\mathbb{B}^2$ in $\mathbb{C}^2$).

Proposition 4.6 *Let $g : \mathbb{U} \to \mathbb{C}$ be a convex (univalent) function such that $g(0) = 1$. Assume that g satisfies the conditions of Assumption 2.16 and that $\mathbb{U}(1, a_0) \subseteq g(\mathbb{U})$, where a_0 is given by (4.1). Let $h : \mathbb{U}^2 \to \mathbb{C}^2$ be given by*

$$h(z) = \left(z_1 \pm a_0 z_2^2, z_2\right), \quad z = (z_1, z_2) \in \mathbb{U}^2.$$

Then $h \in \mathcal{M}_g(\mathbb{U}^2)$ and $f = f(z, t) : \mathbb{U}^2 \times [0, \infty) \to \mathbb{C}^2$ given by

$$f(z, t) = e^t\left(z_1 \pm a_0 z_2^2, z_2\right), \quad z = (z_1, z_2) \in \mathbb{U}^2, \quad t \geq 0,$$

is a g-Loewner chain. In particular, $F \in S_g^0(\mathbb{U}^2)$ and F is g-starlike on $\mathbb{U}^2$, where

$$F(z) = \left(z_1 \pm a_0 z_2^2, z_2\right), \quad z = (z_1, z_2) \in \mathbb{U}^2. \tag{4.9}$$

Proof The fact that $h \in \mathcal{M}_g(\mathbb{U}^2)$ follows from the proof of Proposition 4.2. Also, it is easy to see that $f(z, t)$ is a normal Loewner chain which satisfies the Loewner differential equation

$$\frac{\partial f}{\partial t}(z, t) = Df(z, t)h(z), \quad \forall (z, t) \in \mathbb{U}^2 \times [0, \infty).$$

Since $h \in \mathcal{M}_g$, it follows that $f(z, t)$ is a g-Loewner chain, and thus $F = f(\cdot, 0) \in S_g^0(\mathbb{U}^2)$. Finally, since $f(z, t) = e^t F(z)$ is a g-Loewner chain, it follows that $F \in S_g^*(\mathbb{U}^2)$, as desired. This completes the proof. $\qquad\square$

Remark 4.7 Let $g : \mathbb{U} \to \mathbb{C}$ be a convex (univalent) function such that $g(0) = 1$. Assume that g satisfies the conditions of Assumption 2.16 and that $\mathbb{U}(1, a_0) \subseteq g(\mathbb{U})$, where a_0 is given by (4.1). Using arguments similar to those in the proof of Proposition 4.6, we deduce that $G \in S_g^0(\mathbb{U}^2)$, where

$$G(z) = \left(z_1, z_2 + a_0 z_1^2\right), \quad z = (z_1, z_2) \in \mathbb{U}^2. \tag{4.10}$$

Let $g(\zeta) = \frac{1-\zeta}{1+\zeta}$, $|\zeta| < 1$. Then $\mathcal{M}_g(\mathbb{U}^2) = \mathcal{M}(\mathbb{U}^2)$ and $a_0 = 1$, where a_0 is given by (4.1). In view of Propositions 4.2 and 4.6, we obtain the following consequence (compare [4, Proposition 2.1 and Corollary 2.2] and [21], in the case of the Euclidean unit ball $\mathbb{B}^2$; compare [32, Lemma 5] and [38]).

Corollary 4.8 *Let* $h = (h_1, h_2) \in \mathcal{M}(\mathbb{U}^2)$ *and let* $q_{0,2}^1 = \frac{1}{2}\frac{\partial^2 h_1}{\partial z_2^2}(0)$ *and* $q_{2,0}^2 = \frac{1}{2}\frac{\partial^2 h_2}{\partial z_1^2}(0)$. *Then* $h^{[c]}, \widetilde{h}^{[c]} \in \mathcal{M}(\mathbb{U}^2)$ *and* $\max\{|q_{0,2}^1|, |q_{2,0}^2|\} \leq 1$. *These estimates are sharp.*

Let $\alpha \in [0, 1)$ and $g(\zeta) = \frac{1-\zeta}{1+(1-2\alpha)\zeta}$, $|\zeta| < 1$. Then $\mathcal{M}_g(\mathbb{U}^2) = \mathcal{M}_\alpha(\mathbb{U}^2)$. Also, let $h = (h_1, h_2) \in \mathcal{M}_\alpha(\mathbb{U}^2)$. Then $h \in H(\mathbb{U}^2)$, $h(0) = 0$, $Dh(0) = I_2$, and for $\alpha \in (0, 1)$, we obtain that

$$\left| \frac{h_j(z)}{z_j} - \frac{1}{2\alpha} \right| < \frac{1}{2\alpha}, \quad 0 < |z_j| = \|z\|_\infty < 1.$$

Let a_0 be given by (4.1). Then $a_0 = a_0(\alpha)$, where (see [21])

$$a_0(\alpha) = \begin{cases} 1, & \alpha \in [0, \frac{1}{2}] \\ \frac{1-\alpha}{\alpha}, & \alpha \in (\frac{1}{2}, 1). \end{cases} \tag{4.11}$$

It is not difficult to deduce that if $\alpha \in [0, 1)$, then $\mathbb{U}(1, a_0(\alpha)) \subseteq g(\mathbb{U})$ (see [21]). In this case, we obtain the following sharp estimate for the family $\mathcal{M}_\alpha(\mathbb{U}^2)$ (compare [4] and [21], in the case of the unit ball $\mathbb{B}^2$).

Corollary 4.9 *Let* $\alpha \in [0, 1)$, *let* $h = (h_1, h_2) \in \mathcal{M}_\alpha(\mathbb{U}^2)$, *and let* $q_{0,2}^1 = \frac{1}{2}\frac{\partial^2 h_1}{\partial z_2^2}(0)$ *and* $q_{2,0}^2 = \frac{1}{2}\frac{\partial^2 h_2}{\partial z_1^2}(0)$. *Then* $h^{[c]}, \widetilde{h}^{[c]} \in \mathcal{M}_\alpha(\mathbb{U}^2)$ *and* $\max\{|q_{0,2}^1|, |q_{2,0}^2|\} \leq a_0(\alpha)$, *where* $a_0(\alpha)$ *is given by* (4.11). *These estimates are sharp.*

Next, let $g(\zeta) = \frac{1-(1-2\alpha)\zeta}{1+\zeta}$, $\zeta \in \mathbb{U}$. In this case, it is easily seen that if a_0 is given by (4.1), then $a_0 = 1 - \alpha$. In view of Propositions 4.2 and 4.6, we obtain (compare [4] and [21], in the case of the unit ball $\mathbb{B}^2$).

Corollary 4.10 *Let* $\alpha \in [0, 1)$ *and let* $h = (h_1, h_2) \in \mathcal{M}_g(\mathbb{U}^2)$, *where* $g(\zeta) = \frac{1-(1-2\alpha)\zeta}{1+\zeta}$, $\zeta \in \mathbb{U}$. *Also, let* $q_{0,2}^1 = \frac{1}{2}\frac{\partial^2 h_1}{\partial z_2^2}(0)$ *and* $q_{2,0}^2 = \frac{1}{2}\frac{\partial^2 h_2}{\partial z_1^2}(0)$. *Then* $h^{[c]}, \widetilde{h}^{[c]} \in \mathcal{M}_g(\mathbb{U}^2)$ *and* $\max\{|q_{0,2}^1|, |q_{2,0}^2|\} \leq 1 - \alpha$. *These estimates are sharp.*

From Proposition 4.6 and Corollary 4.9, we deduce the following result (cf. [4, Lemma 2.1] and [21], in the case of the Euclidean unit ball $\mathbb{B}^2$).

Corollary 4.11 *Let* $\alpha \in [0, 1)$ *and* $\Phi_\alpha : \mathbb{U}^2 \to \mathbb{C}^2$ *be given by*

$$\Phi_\alpha(z) = (z_1 + a_0(\alpha)z_2^2, z_2), \quad z = (z_1, z_2) \in \mathbb{U}^2, \tag{4.12}$$

where $a_0(\alpha)$ *is given by* (4.11). *Then* $\Phi_\alpha \in S_\alpha^*(\mathbb{U}^2)$, *and thus* $\Phi_\alpha \in S_\alpha^0(\mathbb{U}^2)$.

From Proposition 4.6 and Corollary 4.10, we obtain (cf. [4, Lemma 2.1] and [21], in the case of the Euclidean unit ball $\mathbb{B}^2$).

Corollary 4.12 *Let $\alpha \in [0, 1)$ and $\Psi_\alpha : \mathbb{U}^2 \to \mathbb{C}^2$ be given by*

$$\Psi_\alpha(z) = \left(z_1 + (1 - \alpha)z_2^2, z_2\right), \quad z = (z_1, z_2) \in \mathbb{U}^2. \tag{4.13}$$

Then $\Psi_\alpha \in S_g^*(\mathbb{U}^2)$, and thus $\Psi_\alpha \in S_g^0(\mathbb{U}^2)$, where $g(\zeta) = \frac{1-(1-2\alpha)\zeta}{1+\zeta}$, $\zeta \in \mathbb{U}$.

We close this section with the following coefficient bounds for the family $\mathcal{M}_g(\mathbb{U}^2)$ (cf. [17, Theorem 3.6]; compare [5], [32, Lemma 5], for $g(\zeta) = \frac{1-\zeta}{1+\zeta}$, $\zeta \in \mathbb{U}$).

Proposition 4.13 *Let $g : \mathbb{U} \to \mathbb{C}$ be a univalent holomorphic function with $g(0) = 1$, and let $h = (h_1, h_2) \in \mathcal{M}_g(\mathbb{U}^2)$. Then*

$$\max\left\{|q_{2,0}^j|, |q_{0,2}^j| : j = 1, 2\right\} \leq |g'(0)|, \tag{4.14}$$

where $q_{2,0}^j = \frac{1}{2}\frac{\partial^2 h_j}{\partial z_1^2}(0)$ and $q_{0,2}^j = \frac{1}{2}\frac{\partial^2 h_j}{\partial z_2^2}(0)$, $j = 1, 2$. These estimates are sharp.

Proof Fix $u \in \mathbb{C}^2$ such that $\|u\|_\infty = 1$ and $\ell_u \in T(u)$. Also, let $p : \mathbb{U} \to \mathbb{C}$ given by $p(\zeta) = \frac{1}{\zeta}\ell_u(h(\zeta u))$, $0 < |\zeta| < 1$, and $p(0) = 1$. Then $p \in H(\mathbb{U})$, and since $h \in \mathcal{M}_g(\mathbb{U}^2)$, it is easily seen that $p \prec g$. Hence $|p'(0)| \leq |g'(0)|$. An elementary computation yields that $p'(0) = \frac{1}{2}\ell_u(D^2h(0)(u^2))$, and thus

$$\left|\frac{1}{2}\ell_u(D^2h(0)(u^2))\right| \leq |g'(0)|.$$

Since $u \in \mathbb{C}^2$, $\|u\|_\infty = 1$, and $\ell_u \in T(u)$ are arbitrary, we deduce that (see [17, Lemma 1.1])

$$\left\|\frac{1}{2}D^2h(0)(u^2)\right\|_\infty \leq |g'(0)|.$$

On the other hand, since $D^2h(0)(u, \cdot) = \left(\sum_{m=1}^2 \frac{\partial^2 h_j}{\partial z_k \partial z_m}(0)u_m\right)_{1 \leq j,k \leq 2}$, it follows in view of the above relation that

$$\left|\frac{1}{2}\frac{\partial^2 h_j}{\partial z_1^2}(0)u_1^2 + \frac{\partial^2 h_j}{\partial z_1 \partial z_2}(0)u_1 u_2 + \frac{1}{2}\frac{\partial^2 h_j}{\partial z_2^2}(0)u_2^2\right| \leq |g'(0)|, \tag{4.15}$$

for all $u = (u_1, u_2) \in \mathbb{C}^2$, $\|u\|_\infty = 1$ and $j = 1, 2$. Hence, the relation (4.14) easily follows from (4.15), as desired.

To deduce sharpness of the relation (4.14), let $u = (u_1, u_2) \in \mathbb{C}^2$ be such that $\|u\|_\infty = 1$, and let $\ell_u \in T(u)$. Also, let $h_u(z) = g(\ell_u(z))z$, $z \in \mathbb{U}^2$. Then $h_u \in \mathcal{M}_g(\mathbb{U}^2)$ and it is elementary to deduce that $\frac{1}{2}D^2h_u(0)(u^2) = g'(0)u$. Now, if $u = (1, 0)$, then $q_{2,0}^1 = g'(0)$, while if $u = (0, 1)$, then $q_{0,2}^2 = g'(0)$. This completes the proof. $\qquad\square$

Remark 4.14 Let $g : \mathbb{U} \to \mathbb{C}$ be a univalent holomorphic function such that $g(0) = 1$. Let $\theta = -\arg g'(0)$. Also, let $L : H(\mathbb{U}^2) \to \mathbb{C}$ be given by $L(h) = \frac{1}{2} e^{i\theta} \frac{\partial^2 h_1}{\partial z_1^2}(0)$, for all $h = (h_1, h_2) \in H(\mathbb{U}^2)$. Then $\Re L$ is a continuous linear functional on $H(\mathbb{U}^2)$, and in view of Proposition 4.13, we obtain that $\Re L(h) \leq |g'(0)| = \Re L(q)$, for all $h \in M_g(\mathbb{U}^2)$, where $q \in M_g(\mathbb{U}^2)$ is given by $q(z) = g(z_1)z$, for all $z = (z_1, z_2) \in \mathbb{U}^2$. Since $\Re L(\mathrm{id}_{\mathbb{U}^2}) = 0 < |g'(0)| = \Re L(q)$ and $\mathrm{id}_{\mathbb{U}^2} \in M_g(\mathbb{U}^2)$, it follows that $\Re L|_{M_g(\mathbb{U}^2)}$ is not constant. Hence q is a support point for the family $M_g(\mathbb{U}^2)$ with respect to the functional L. Similarly, if $r(z) = g(z_2)z$, $z = (z_1, z_2) \in \mathbb{U}^2$, then r is also a support point for $M_g(\mathbb{U}^2)$.

Remark 4.15 In connection with the above result, note that recently the authors in [17, Theorem 3.6] (see also [18]) proved that if g is a convex (univalent) function on $\mathbb{U}$ with $g(0) = 1$, and if $h \in M_g(\mathbb{U}^n)$, then $\| P_m(z) \|_\infty \leq |g'(0)|$, for $\|z\|_\infty = 1$ and $m \geq 2$, where $P_m(z) = \frac{1}{m!} D^m h(0)(z^m)$. These estimates are sharp.

5   Bounded Support Points for the Family $S_g^0(\mathbb{U}^2)$

In this section we obtain some examples of bounded support points for the family $S_g^0(\mathbb{U}^2)$, where g is a convex (univalent) function on $\mathbb{U}$ such that $g(0) = 1$, and which satisfies the conditions of Assumption 2.16. This part is a continuation of a recent work in [21], on bounded support points for $S_g^0(\mathbb{B}^2)$.

The following result is a generalization of [4, Theorem 1.4] and [21, Theorem 4.8] to the case of g-Loewner chains on $\mathbb{U}^2 \times [0, \infty)$, where $g : \mathbb{U} \to \mathbb{C}$ is a convex (univalent) function such that $g(0) = 1$ and which satisfies the conditions of Assumption 2.16, such that $\mathbb{U}(1, a_0) \subseteq g(\mathbb{U})$. We omit the proof of Theorem 5.1, since it suffices to use arguments similar to those in the proof of [21, Theorem 4.8].

Theorem 5.1 *Let $g : \mathbb{U} \to \mathbb{C}$ be a convex (univalent) function such that $g(0) = 1$. Assume that g satisfies the conditions of Assumption 2.16. Also, assume that $\mathbb{U}(1, a_0) \subseteq g(\mathbb{U})$, where a_0 is given by (4.1). Also, let $f(z, t) : \mathbb{U}^2 \times [0, \infty) \to \mathbb{C}^2$ be a g-Loewner chain. Then $f^{[c]}(z, t)$ is also a g-Loewner chain. In particular, if $f \in S_g^0(\mathbb{U}^2)$, then $f^{[c]} \in S_g^0(\mathbb{U}^2)$.*

Next, we prove the following sharp coefficient estimate for the family $S_g^0(\mathbb{U}^2)$, where $g : \mathbb{U} \to \mathbb{C}$ is a convex (univalent) function such that $g(0) = 1$, and which satisfies the conditions of Assumption 2.16, and $\mathbb{U}(1, a_0) \subseteq g(\mathbb{U})$ (compare [4, Theorem 3.1] and [21], in the case of the unit ball $\mathbb{B}^2$ in $\mathbb{C}^2$).

Theorem 5.2 *Let $g : \mathbb{U} \to \mathbb{C}$ be a convex (univalent) function such that $g(0) = 1$. Assume that g satisfies the conditions of Assumption 2.16. Also, assume that $\mathbb{U}(1, a_0) \subseteq g(\mathbb{U})$, where a_0 is given by (4.1). Also, let $f = (f_1, f_2) \in S_g^0(\mathbb{U}^2)$. Then $\max\{|a_{0,2}^1|, |a_{2,0}^2|\} \leq a_0$, where $a_{0,2}^1 = \frac{1}{2} \frac{\partial^2 f_1}{\partial z_2^2}(0)$ and $a_{2,0}^2 = \frac{1}{2} \frac{\partial^2 f_2}{\partial z_1^2}(0)$. These estimates are sharp.*

Proof First, we prove that $|a_{0,2}^1| \leq a_0$. To this end, we use some arguments similar to those in the proofs of [4, Theorem 3.1] and [21, Theorem 4.10]. Since $f \in S_g^0(\mathbb{U}^2)$, there exists a g-Loewner chain $f(z, t)$ such that $f = f(\cdot, 0)$. Let $h_t(z) = h(z, t)$ be the Herglotz vector field associated with $f(z, t)$. In view of Theorem 5.1, we deduce that the shearing $f^{[c]}(z, t)$ of $f(z, t)$ is a g-Loewner chain. Then it is not difficult to deduce that $h_t^{[c]}(z)$ is the associated Herglotz vector field of $f_t^{[c]}(z)$. Let

$$f_t(z) = e^t z + \cdots = \left(e^t z_1 + \beta(t) z_2^2 + \cdots, e^t z_2 + \cdots \right), \quad z \in \mathbb{U}^2.$$

Then it is clear that $a_{0,2}^1 = \beta(0)$. Also, let

$$h_t(z) = z + \cdots = \left(z_1 + q(t) z_2^2 + \cdots, z_2 + \cdots \right)$$

be the power series expansion of h_t on $\mathbb{U}^2$. Then

$$f_t^{[c]}(z) = (e^t z_1 + \beta(t) z_2^2, e^t z_2), \quad z = (z_1, z_2) \in \mathbb{U}^2, \quad t \geq 0,$$
$$h_t^{[c]}(z) = (z_1 + q(t) z_2^2, z_2), \quad z = (z_1, z_2) \in \mathbb{U}^2, \quad t \geq 0.$$

Since $h_t^{[c]}$ is the Herglotz vector field of $f_t^{[c]}(z)$, we deduce that

$$\frac{\partial f_t^{[c]}}{\partial t}(z) = Df_t^{[c]}(z) h_t^{[c]}(z), \quad \text{a.e. } t \geq 0, \quad \forall z \in \mathbb{U}^2.$$

Identifying the coefficients in both sides of the above equality, we obtain that

$$\beta'(t) - 2\beta(t) = e^t q(t), \quad \text{a.e. } t \geq 0, \quad \forall z \in \mathbb{U}^2.$$

Therefore, we obtain

$$\frac{d}{dt} \left(e^{-2t} \beta(t) \right) = e^{-t} q(t), \quad \text{a.e.} \quad t \geq 0.$$

Integrating both sides of the above equality from 0 to t, and using the fact that $\beta(0) = a_{0,2}^1$, we deduce that

$$e^{-2t} \beta(t) - a_{0,2}^1 = \int_0^t e^{-\tau} q(\tau) d\tau, \quad t \geq 0. \tag{5.1}$$

On the other hand, in view of Proposition 4.2, we deduce that $|q(t)| \leq a_0$, for a.e. $t \geq 0$, where a_0 is given by (4.1). Hence, in view of (5.1), we deduce that

$$\left| e^{-2t} \beta(t) - a_{0,2}^1 \right| \leq a_0(1 - e^{-t}), \quad t \geq 0. \tag{5.2}$$

Let $f_t^{[c]}(z) = (f_1^{[c]}(z, t), f_2^{[c]}(z, t))$, $z \in \mathbb{U}^2$ and $t \geq 0$. Since $\beta(t) = \frac{1}{2} \frac{\partial^2 f_1^{[c]}}{\partial z_2^2}(0, t)$ and $\{e^{-t} f^{[c]}(\cdot, t)\}_{t \geq 0}$ is a normal family on $\mathbb{U}^2$, we obtain that $\lim_{t \to \infty} e^{-2t} \beta(t) = 0$, by using an argument similar to that in the proof of [21, Theorem 4.10]. Letting $t \to \infty$ in (5.2), we deduce that $|a_{0,2}^1| \leq a_0$, as desired. Sharpness of this relation is provided by the mapping $F \in S_g^0(\mathbb{U}^2)$ given by (4.9).

Finally, we deduce that $|a_{2,0}^2| \leq a_0$. To this end, it suffices to use Remark 4.3 and arguments similar to those in the above. Sharpness of this relation is provided by the mapping G given by (4.10). This completes the proof. □

From Theorem 5.2, we obtain the following consequences (cf. [4, 5, 21, 38]).

Corollary 5.3 *Let $\alpha \in [0, 1)$ and let $f = (f_1, f_2) \in S_\alpha^0(\mathbb{U}^2)$. Also, let $a_0(\alpha)$ be given by (4.11). Then $\max\{|a_{0,2}^1|, |a_{2,0}^2|\} \leq a_0(\alpha)$, where $a_{0,2}^1 = \frac{1}{2} \frac{\partial^2 f_1}{\partial z_2^2}(0)$ and $a_{2,0}^2 = \frac{1}{2} \frac{\partial^2 f_2}{\partial z_1^2}(0)$. These estimates are sharp.*

Corollary 5.4 *Let $\alpha \in [0, 1)$ and let $g(\zeta) = \frac{1-(1-2\alpha)\zeta}{1+\zeta}$, $\zeta \in \mathbb{U}$. If $f = (f_1, f_2) \in S_g^0(\mathbb{U}^2)$, then $\max\{|a_{0,2}^1|, |a_{2,0}^2|\} \leq 1 - \alpha$, where $a_{0,2}^1 = \frac{1}{2} \frac{\partial^2 f_1}{\partial z_2^2}(0)$ and $a_{2,0}^2 = \frac{1}{2} \frac{\partial^2 f_2}{\partial z_1^2}(0)$. These estimates are sharp.*

Remark 5.5 Let $g : \mathbb{U} \to \mathbb{C}$ be a convex (univalent) function such that $g(0) = 1$, and which satisfies the conditions of Assumption 2.16. Assume that $\mathbb{U}(1, a_0) \subseteq g(\mathbb{U})$, where a_0 is given by (4.1). Let $L : H(\mathbb{U}^2) \to \mathbb{C}$ be given by $L(f) = \frac{1}{2} \frac{\partial^2 f_1}{\partial z_2^2}(0)$, for $f = (f_1, f_2) \in H(\mathbb{U}^2)$. Then $\Re L$ is a continuous linear functional on $H(\mathbb{U}^2)$, and in view of Theorem 5.2, we obtain that $\Re L(q) \leq a_0 = \Re L(F)$, for all $q \in S_g^0(\mathbb{U}^2)$, where $F \in S_g^0(\mathbb{U}^2)$ is given by

$$F(z) = (z_1 + a_0 z_2^2, z_2), \quad z = (z_1, z_2) \in \mathbb{U}^2. \tag{5.3}$$

Since $\Re L(\mathrm{id}_{\mathbb{U}^2}) = 0 < a_0 = \Re L(F)$ and $\mathrm{id}_{\mathbb{U}^2} \in S_g^0(\mathbb{U}^2)$, it follows that $\Re L|_{S_g^0(\mathbb{U}^2)}$ is not constant. Hence F is a support point for the family $S_g^0(\mathbb{U}^2)$, which is bounded on $\mathbb{U}^2$ (compare [4, 20, 21], in the case of the unit ball $\mathbb{B}^2$).

Taking into account Remark 4.7, we deduce by arguments similar to those in the above that $G \in \mathrm{supp}\, S_g^0(\mathbb{U}^2)$, where G is given by (4.10).

Remark 5.6 Let $\alpha \in [0, 1)$. In view of Remark 5.5, we deduce that the mapping Φ_α given by (4.12) is a bounded support point for the family $S_\alpha^0(\mathbb{U}^2)$. On the other hand, if $g(\zeta) = \frac{1-(1-2\alpha)\zeta}{1+\zeta}$, $\zeta \in \mathbb{U}$, where $\alpha \in [0, 1)$, then the mapping Ψ_α given by (4.13) is a bounded support point for $S_g^0(\mathbb{U}^2)$ (cf. [4] and [21], in the case of the Euclidean unit ball $\mathbb{B}^2$ in $\mathbb{C}^2$).

We close this section with other coefficient bounds for the family $S_g^0(\mathbb{U}^2)$ (cf. [13, Theorem 2.14], [18, Theorem 10]; cf. [5], [32, Theorem 3], for $g(\zeta) = \frac{1-\zeta}{1+\zeta}$, $\zeta \in \mathbb{U}$).

Theorem 5.7 *Let $g : \mathbb{U} \to \mathbb{C}$ be a univalent holomorphic function such that $g(0) = 1$ and $\Re g(\zeta) > 0$, $\zeta \in \mathbb{U}$. Also, let $f \in S_g^0(\mathbb{U}^2)$. Then*

$$\max \{|a_{2,0}^j|, |a_{0,2}^j| : j = 1, 2\} \leq |g'(0)|, \tag{5.4}$$

where $a_{2,0}^j = \frac{1}{2}\frac{\partial^2 f_j}{\partial z_1^2}(0)$ and $a_{0,2}^j = \frac{1}{2}\frac{\partial^2 f_j}{\partial z_2^2}(0)$, $j = 1, 2$. These estimates are sharp.

Proof Since $f \in S_g^0(\mathbb{U}^2)$, there exists a g-Loewner chain $f(z, t)$ such that $f = f(\cdot, 0)$. Let $h_t(z) = h(z, t)$ be the Herglotz vector field associated with $f(z, t)$. Then $h(\cdot, t) \in \mathcal{M}_g(\mathbb{U}^2)$ for a.e. $t \geq 0$, and

$$\frac{\partial f}{\partial t}(z, t) = Df(z, t)h(z, t), \quad \text{a.e.} \quad t \geq 0, \quad \forall\, z \in \mathbb{U}^2.$$

Integrating both sides of the above equality, we deduce as in the proof of [32, Theorem 3] (see also the proof of [13, Theorem 2.14]) that

$$e^{-2T}D^2f(0, T)(w^2) - D^2f(0, 0)(w^2) = \int_0^T e^{-t}D^2h(0, t)(w^2)dt, \tag{5.5}$$

for all $T \in (0, \infty)$ and $w \in \mathbb{C}^2$, $\|w\|_\infty = 1$. Taking into account the fact that $\{e^{-t}f(\cdot, t)\}_{t \geq 0}$ is a normal family on $\mathbb{U}^2$, we deduce in view of the Cauchy integral formula for holomorphic mappings on $\mathbb{U}^2$ that $\lim_{T \to \infty} e^{-2T}D^2f(0, T)(w^2) = 0$. Finally, letting $T \to \infty$ in (5.5) and using (4.14), we obtain after elementary computations the relation (5.4), as desired.

Next, we prove sharpness of (5.4). To this end, let $h : \mathbb{U}^2 \to \mathbb{C}^2$ be given by $h(z) = g(z_1)z$. Then $h \in \mathcal{M}_g(\mathbb{U}^2)$. Since $\Re g(\zeta) > 0$, $\zeta \in \mathbb{U}$, it follows that $\mathcal{M}_g(\mathbb{U}^2) \subseteq \mathcal{M}(\mathbb{U}^2)$ and $S_g^0(\mathbb{U}^2) \subseteq S^0(\mathbb{U}^2)$. On the other hand, there exists $f \in S_g^*(\mathbb{U}^2)$ such $[Df(z)]^{-1}f(z) = h(z)$, $z \in \mathbb{U}^2$ (see [18]). It is not difficult to deduce that $\frac{1}{2}D^2f(0)(u^2) = -\frac{1}{2}D^2h(0)(u^2)$, $u \in \mathbb{C}^2$, and thus $\frac{1}{2}D^2f(0)(e_1^2) = -g'(0)e_1$, where $e_1 = (1, 0)$. Hence $a_{2,0}^1 = -g'(0)$, which yields that the relation (5.4) is sharp, as desired. This completes the proof. $\qquad\square$

Remark 5.8 Let $g : \mathbb{U} \to \mathbb{C}$ be a univalent holomorphic function such that $g(0) = 1$ and $\Re g(\zeta) > 0$, $\zeta \in \mathbb{U}$. Let $\theta = -\arg g'(0)$. Also, let $\Lambda : H(\mathbb{U}^2) \to \mathbb{C}$ be given by $\Lambda(f) = \frac{1}{2}\frac{\partial^2 f_1}{\partial z_1^2}(0)$ for $f = (f_1, f_2) \in H(\mathbb{U}^2)$. Then $\Re\Lambda$ is a continuous linear functional on $H(\mathbb{U}^2)$, and from (5.4) we deduce that $\Re\Lambda(f) \leq |g'(0)| = \Re\Lambda(F_\theta)$, where $F_\theta \in S_g^*(\mathbb{U}^2)$ is given by $[DF_\theta(z)]^{-1}F_\theta(z) = g(e^{i\theta}z_1)z$, $z = (z_1, z_2) \in H(\mathbb{U}^2)$. Since $\Re\Lambda(\mathrm{id}_{\mathbb{U}^2}) = 0 < \Re\Lambda(F_\theta)$ and $\mathrm{id}_{\mathbb{U}^2} \in S_g^0(\mathbb{U}^2)$, it follows that $\Re\Lambda$ is not constant on $S_g^0(\mathbb{U}^2)$. Hence $F_\theta \in \mathrm{supp}\, S_g^0(\mathbb{U}^2)$ with respect to the functional Λ.

6   Questions and Conjectures

In connection with the above results, we formulate the following conjectures and questions. In the case of the Euclidean unit ball $\mathbb{B}^n$, $n \geq 2$, see [5, 15, 21, 36]. Recent surveys on Loewner theory in $\mathbb{C}^n$ may be found in [1, 16] and [18].

Question 6.1 Let $g : \mathbb{U} \to \mathbb{C}$ be a univalent holomorphic function such that $g(0) = 1$ and $\Re g(\zeta) > 0$, $\zeta \in \mathbb{U}$. Also, assume that g satisfies the relation *(3.6)*. Let $f \in S_g^*(\mathbb{U}^n)$ and let $h(z) = [Df(z)]^{-1}f(z)$, $z \in \mathbb{U}^n$. If $h \in \operatorname{ex} \mathcal{M}_g(\mathbb{U}^n)$, then is it true that $f \in \operatorname{ex} S_g^*(\mathbb{U}^n)$?

Question 6.2 Let $g : \mathbb{U} \to \mathbb{C}$ be a convex (univalent) function such that $g(0) = 1$ and $\Re g(\zeta) > 0$, $\zeta \in \mathbb{U}$. Let $f \in S_g^*(\mathbb{U}^n)$ and let $h(z) = [Df(z)]^{-1}f(z)$, $z \in \mathbb{U}^n$. Assume that there exist $p_1, \dots, p_n \in s(g)$ such that h is given by *(3.2)*. If $h \in \operatorname{supp} \mathcal{M}_g(\mathbb{U}^n)$, then is it true that $f \in \operatorname{supp} S_g^*(\mathbb{U}^n)$?

Conjecture 6.3 Let g be a convex (univalent) function on $\mathbb{U}$ such that $g(0) = 1$. Assume that g satisfies the conditions of Assumption *2.16*. Also, assume that $\mathbb{U}(1, a_0) \subseteq g(\mathbb{U})$, where a_0 is given by *(4.1)*. Also, let $F : \mathbb{U}^2 \to \mathbb{C}^2$ be given by *(5.3)*. Then $F \in \operatorname{ex} S_g^0(\mathbb{U}^2)$.

Question 6.4 Let g be a convex (univalent) function on $\mathbb{U}$ such that $g(0) = 1$. Assume that g satisfies the conditions of Assumption *2.16*. Also, assume that $\mathbb{U}(1, a_0) \subseteq g(\mathbb{U})$, where a_0 is given by *(4.1)*. Let $f \in S_g^0(\mathbb{U}^n)$ and let (f_t) be a g-Loewner chain such that $f = f_0$. Also, let $h = h(z, t)$ be the associated Herglotz vector field of (f_t).

   (i) Is it true that $h(\cdot, t) \in \operatorname{ex} \mathcal{M}_g(\mathbb{U}^n)$ for a.e. $t \in [0, \infty)$ if and only if $f \in \operatorname{ex} S_g^0(\mathbb{U}^n)$?
  (ii) If $h(\cdot, t) \in \operatorname{supp} \mathcal{M}_g(\mathbb{U}^n)$ for a.e. $t \geq 0$, is it true that $f \in \operatorname{supp} S_g^0(\mathbb{U}^n)$?
 (iii) Assume that $f \in \operatorname{supp} S_g^0(\mathbb{U}^n)$. Is it true that $h(\cdot, t) \in \operatorname{supp} \mathcal{M}_g(\mathbb{U}^n)$ for a.e. $t \geq 0$?

   Roth [36, Theorem 1.5] proved that Question 6.4 (iii) is true in the case of the family $S^0(\mathbb{B}^n)$ (see also [5, Question 3.4]).

   Recently, the authors in [15] proved that if $f \in \operatorname{ex} S^0(\mathbb{B}^n)$ and (f_t) is a normal Loewner chain such that $f = f_0$, then $e^{-t}f_t(\cdot) \in \operatorname{ex} S^0(\mathbb{B}^n)$, for all $t \geq 0$. On the other hand, Schleissinger [37] proved that if $f \in \operatorname{supp} S^0(\mathbb{B}^n)$ and (f_t) is a normal chain such that $f = f_0$, then $e^{-t}f_t(\cdot) \in \operatorname{supp} S^0(\mathbb{B}^n)$, for all $t \geq 0$. This result provides a positive answer to [15, Conjecture 2.6]. Extensions of these results to the case of mappings with A-parametric representation on $\mathbb{B}^n$ (see [14, Definition 1.5]), where $A \in L(\mathbb{C}^n)$ is such that $k_+(A) < 2m(A)$, may be found in [19] and [20] (see also [27], in the case of time-dependent linear operators in $\mathbb{C}^n$). Here $k_+(A)$ is the Lyapunov index of A (see e.g. [10] and [34]) and $m(A) = \min\{\Re\langle A(z), z\rangle : \|z\| = 1\}$. In connection with these results, it would be interesting to give an answer to the following conjecture. We remark that partial results related to Conjecture 6.5 were obtained in [8] and [18], in the case of the family $S_g^0(\mathbb{B}^n)$, where g is a convex (univalent) function on $\mathbb{U}$ such that $g(0) = 1$ and $\Re g(\zeta) > 0$, $\zeta \in \mathbb{U}$. Recently, Schleissinger [38] proved that Conjecture 6.5 is true in the case of the family $S^0(\mathbb{U}^n)$.

Conjecture 6.5 Let g be a convex (univalent) function on $\mathbb{U}$ such that $g(0) = 1$ and $\Re g(\zeta) > 0$, $\zeta \in \mathbb{U}$. Assume that g satisfies the conditions of Assumption 2.16. Let $f \in S^0_g(\mathbb{U}^n)$ and let (f_t) be a g-Loewner chain such that $f = f_0$. If $f \in \text{supp } S^0_g(\mathbb{U}^n)$ (respectively $f \in \text{ex } S^0_g(\mathbb{U}^n)$), then $e^{-t}f_t(\cdot) \in \text{supp } S^0_g(\mathbb{U}^n)$ (respectively $e^{-t}f_t(\cdot) \in \text{ex } S^0_g(\mathbb{U}^n)$), for all $t \geq 0$.

Let $M \in (1, \infty)$ and let $g : \mathbb{U} \to \mathbb{C}$ be a univalent function such that $g(0) = 1$ and $\Re g(\zeta) > 0$, $\zeta \in \mathbb{U}$. Also, let $S^0_g(\mathbb{U}^2, M) = \{f \in S^0_g(\mathbb{U}^2) : \|f(z)\| \leq M, z \in \mathbb{U}^2\}$, and let $S^0(\mathbb{U}^2, M) = S^0_{g_0}(\mathbb{U}^2, M)$, where $g_0(\zeta) = \frac{1-\zeta}{1+\zeta}$, $\zeta \in \mathbb{U}$.

Next, we formulate the following question related to mappings in $S^0(\mathbb{U}^2, M)$. In the case of the Euclidean unit ball $\mathbb{B}^2$, see [5, 7], and [15].

Question 6.6 Let $M \in (1, \infty)$ and $f \in S^0(\mathbb{U}^2, M)$. Also, let $a^1_{0,2} = \frac{1}{2}\frac{\partial^2 f_1}{\partial z_2^2}(0)$ and $a^2_{2,0} = \frac{1}{2}\frac{\partial^2 f_2}{\partial z_1^2}(0)$. Is it true that $\max\{|a^1_{0,2}|, |a^2_{2,0}|\} \leq 1 - \frac{1}{M}$?

In connection with Question 6.6, it is natural to ask the following question. We remark that sharp coefficient estimates for a proper subset of $S^0_g(\mathbb{B}^2)$ consisting of bounded mappings on $\mathbb{B}^2$ were obtained in [21] (see [5, 7], for $g(\zeta) = \frac{1-\zeta}{1+\zeta}$, $\zeta \in \mathbb{U}$).

Question 6.7 Let $M \in (1, \infty)$ and let g be a convex (univalent) function on $\mathbb{U}$ such that $g(0) = 1$ and $\Re g(\zeta) > 0$, $\zeta \in \mathbb{U}$. Assume that g satisfies the conditions of Assumption 2.16. Let $f \in S^0_g(\mathbb{U}^2, M)$ and let a_0 be given by (4.1). Assume that $\mathbb{U}(1, a_0) \subseteq g(\mathbb{U})$. Is it true that $\max\{|a^1_{0,2}|, |a^2_{2,0}|\} \leq a_0(1 - \frac{1}{M})$?

The following question is related to Questions 6.6 and 6.7. In the case of the Euclidean unit ball $\mathbb{B}^2$, see [7, 15], and [21].

Question 6.8 Let $M \in (1, \infty)$ and let g be a convex (univalent) function on $\mathbb{U}$ such that $g(0) = 1$ and $\Re g(\zeta) > 0$, $\zeta \in \mathbb{U}$. Assume that g satisfies the conditions of Assumption 2.16. Let a_0 be given by (4.1) and assume that $\mathbb{U}(1, a_0) \subseteq g(\mathbb{U})$. Let $F^M_g : \mathbb{U}^2 \to \mathbb{C}^2$ be defined by

$$F^M_g(z) = \left(z_1 + a_0\left(1 - \frac{1}{M}\right)z_2^2, z_2\right), \quad z = (z_1, z_2) \in \mathbb{U}^2.$$

Is it true that $F^M_g \in \text{supp } S^0_g(\mathbb{U}^2, M) \setminus \left(\text{supp } S^0_g(\mathbb{U}^2) \cup \text{ex } S^0_g(\mathbb{U}^2)\right)$? In particular, if

$$F^M(z) = \left(z_1 + \left(1 - \frac{1}{M}\right)z_2^2, z_2\right), \quad z = (z_1, z_2) \in \mathbb{U}^2,$$

then is it true that $F^M \in \text{supp } S^0(\mathbb{U}^2, M) \setminus \left(\text{supp } S^0(\mathbb{U}^2) \cup \text{ex } S^0(\mathbb{U}^2)\right)$? Also, is it true that $F^M \in \text{ex } S^0(\mathbb{U}^2, M)$?

Acknowledgements Some of the research for this paper was carried out in April, 2016, when Gabriela Kohr visited the Department of Mathematics of the University of Toronto. She expresses her gratitude to the members of this department for their hospitality during that visit.

I. Graham was partially supported by the Natural Sciences and Engineering Research Council of Canada under Grant A9221.

H. Hamada was partially supported by JSPS KAKENHI Grant Number JP16K05217.

G. Kohr was supported by a grant of the Romanian National Authority for Scientific Research, CNCS-UEFISCDI, project number PN-II-ID-PCE-2011-3-0899.

References

1. M. Abate, F. Bracci, M. Contreras, S. Diaz-Madrigal, The evolution of Loewner's differential equations. Eur. Math. Soc. Newslett. **78**, 31–38 (2010)
2. L. Arosio, Resonances in Loewner equations. Adv. Math. **227**, 1413–1435 (2011)
3. L. Arosio, F. Bracci, F.E. Wold, Solving the Loewner PDE in complete hyperbolic starlike domains of $\mathbb{C}^n$. Adv. Math. **242**, 209–216 (2013)
4. F. Bracci, Shearing process and an example of a bounded support function in $S^0(\mathbb{B}^2)$. Comput. Methods Funct. Theory **15**, 151–157 (2015)
5. F. Bracci, O. Roth, Support points and the Bieberbach conjecture in higher dimension. Preprint (2016). arXiv: 1603.01532
6. F. Bracci, M.D. Contreras, S. Díaz-Madrigal, Evolution families and the Loewner equation II: complex hyperbolic manifolds. Math. Ann. **344**, 947–962 (2009)
7. F. Bracci, I. Graham, H. Hamada, G. Kohr, Variation of Loewner chains, extreme and support points in the class S^0 in higher dimensions. Constr. Approx. **43**, 231–251 (2016)
8. T. Chirilă, H. Hamada, G. Kohr, Extreme points and support points for mappings with g-parametric representation in $\mathbb{C}^n$. Mathematica (Cluj) **56**(79), 21–40 (2014)
9. P. Duren, I. Graham, H. Hamada, G. Kohr, Solutions for the generalized Loewner differential equation in several complex variables. Math. Ann. **347**, 411–435 (2010)
10. M. Elin, S. Reich, D. Shoikhet, Complex dynamical systems and the geometry of domains in Banach spaces. Diss. Math. **427**, 1–62 (2004)
11. S. Gong, *The Bieberbach Conjecture* (American Mathematical Society, Providence, RI, 1999)
12. I. Graham, G. Kohr, *Geometric Function Theory in One and Higher Dimensions* (Marcel Dekker Inc., New York, 2003)
13. I. Graham, H. Hamada, G. Kohr, Parametric representation of univalent mappings in several complex variables. Can. J. Math. **54**, 324–351 (2002)
14. I. Graham, H. Hamada, G. Kohr, M. Kohr, Asymptotically spirallike mappings in several complex variables. J. Anal. Math. **105**, 267–302 (2008)
15. I. Graham, H. Hamada, G. Kohr, M. Kohr, Extreme points, support points and the Loewner variation in several complex variables. Sci. China Math. **55**, 1353–1366 (2012)
16. I. Graham, H. Hamada, G. Kohr, A survey on extreme points, support points and Loewner chains in $\mathbb{C}^n$. Math. Rep. **15**(65), 411–423 (2013)
17. I. Graham, H. Hamada, T. Honda, G. Kohr, K.H. Shon, Growth, distortion and coefficient bounds for Carathéodory families in $\mathbb{C}^n$ and complex Banach spaces. J. Math. Anal. Appl. **416**, 449–469 (2014)
18. I. Graham, H. Hamada, G. Kohr, Extremal problems and g-Loewner chains in $\mathbb{C}^n$ and reflexive complex Banach spaces, in *Topics in Mathematical Analysis and Applications*, ed. by T.M. Rassias, L. Toth, vol. 94 (Springer, Cham, 2014), pp. 387–418
19. I. Graham, H. Hamada, G. Kohr, M. Kohr, Extremal properties associated with univalent subordination chains in $\mathbb{C}^n$. Math. Ann. **359**, 61–99 (2014)
20. I. Graham, H. Hamada, G. Kohr, M. Kohr, Support points and extreme points for mappings with A-parametric representation in $\mathbb{C}^n$. J. Geom. Anal. **26**, 1560–1595 (2016)

21. I. Graham, H. Hamada, G. Kohr, M. Kohr, Bounded support points for mappings with g-parametric representation in $\mathbb{C}^2$. J. Math. Anal. Appl. **454**, 1085–1105 (2017)
22. D.J. Hallenbeck, T.H. MacGregor, *Linear Problems and Convexity Techniques in Geometric Function Theory* (Pitman, Boston, 1984)
23. D.J. Hallenbeck, S. Perera, D.R. Wilken, Subordination, extreme points and support points. Complex Variables **11**, 111–124 (1989)
24. H. Hamada, Polynomially bounded solutions to the Loewner differential equation in several complex variables. J. Math. Anal. Appl. **381**, 179–186 (2011)
25. H. Hamada, Approximation properties on spirallike domains of $\mathbb{C}^n$. Adv. Math. **268**, 467–477 (2015)
26. H. Hamada, T. Honda, Sharp growth theorems and coefficient bounds for starlike mappings in several complex variables. Chin. Ann. Math. Ser. B **29**, 353–368 (2008)
27. H. Hamada, M. Iancu, G. Kohr, Extremal problems for mappings with generalized parametric representation in $\mathbb{C}^n$. Complex Anal. Oper. Theory **10**, 1045–1080 (2016)
28. T.H. MacGregor, D.R. Wilken, Extreme points and support points, in *Handbook of Complex Analysis: Geometric Function Theory*, ed. by R. Kühnau, vol. 1 (North-Holland, Amsterdam, 2002), pp. 371–392
29. J. Muir, T. Suffridge, Extreme points for convex mappings of $\mathbb{B}_n$. J. Anal. Math. **98**, 169–182 (2006)
30. J.A. Pfaltzgraff, Subordination chains and univalence of holomorphic mappings in $\mathbb{C}^n$. Math. Ann. **210**, 55–68 (1974)
31. Ch. Pommerenke, *Univalent Functions* (Vandenhoeck & Ruprecht, Göttingen, 1975)
32. T. Poreda, On the univalent holomorphic maps of the unit polydisc in $\mathbb{C}^n$ which have the parametric representation, I-the geometrical properties. Ann. Univ. Mariae Curie Skl. Sect. A **41**, 105–113 (1987)
33. T. Poreda, On the univalent holomorphic maps of the unit polydisc in $\mathbb{C}^n$ which have the parametric representation, II-the necessary conditions and the sufficient conditions. Ann. Univ. Mariae Curie Skl. Sect. A **41**, 115–121 (1987)
34. S. Reich, D. Shoikhet, *Nonlinear Semigroups, Fixed Points, and Geometry of Domains in Banach Spaces* (Imperial College Press, London, 2005)
35. K. Roper, T.J. Suffridge, Convexity properties of holomorphic mappings in $\mathbb{C}^n$. Trans. Am. Math. Soc. **351**, 1803–1833 (1999)
36. O. Roth, Pontryagin's maximum principle for the Loewner equation in higher dimensions. Can. J. Math. **67**, 942–960 (2015)
37. S. Schleissinger, On support points of the class $S^0(\mathbb{B}^n)$. Proc. Am. Math. Soc. **142**, 3881–3887 (2014)
38. S. Schleissinger, On the parametric representation of univalent functions on the polydisc. preprint (2017). arXiv: 1706.09784
39. T.J. Suffridge, *Starlikeness, Convexity and Other Geometric Properties of Holomorphic Maps in Higher Dimensions*. Lecture Notes in Mathematics, vol. 599 (Springer, Berlin, 1977)
40. M. Voda, Solution of a Loewner chain equation in several complex variables. J. Math. Anal. Appl. **375**, 58–74 (2011)
41. M. Voda, Loewner theory in several complex variables and related problems. Ph.D. thesis, University of Toronto, 2011
42. Q.H. Xu, T.S. Liu, Löwner chains and a subclass of biholomorphic mappings. J. Math. Anal. Appl. **334**, 1096–1105 (2007)

Evolution of States of a Continuum Jump Model with Attraction

Yuri Kozitsky

Abstract We study a model of an infinite system of point particles in $\mathbb{R}^d$ performing random jumps with attraction. The system's states are probability measures on the space of particle configurations, and their evolution is described by means of Kolmogorov and Fokker-Planck equations. Instead of solving these equations directly we deal with correlation functions evolving according to a hierarchical chain of differential equations, derived from the Kolmogorov equation. Under quite natural conditions imposed on the jump kernels—and analyzed in the paper— we prove that this chain has a unique classical sub-Poissonian solution on a bounded time interval. This gives a partial answer to the question whether the sub-Poissonicity is consistent with any kind of attraction. We also discuss possibilities to get a complete answer to this question.

Keywords Markov evolution • Configuration space • Stochastic semigroup • Sun-dual semigroup • Kawasaki model correlation function • Scale of Banach spaces

1991 Mathematics Subject Classification 34G10, 47D06, 60J80

1 Introduction

1.1 Setup

In this work, we deal with the model introduced and studied in [6]. It describes an infinite system of point particles placed in $\mathbb{R}^d$ which perform random jumps with attraction. To the best of our knowledge, [6] and the present research are the only works where the dynamics of an infinite particle system of this kind has been studied hitherto.

Y. Kozitsky (✉)
Instytut Matematyki, Uniwersytet Marii Curie-Skłodowskiej, 20-031 Lublin, Poland
e-mail: jkozi@hektor.umcs.lublin.pl

© Springer International Publishing AG 2018
M. Agranovsky et al. (eds.), *Complex Analysis and Dynamical Systems*,
Trends in Mathematics, https://doi.org/10.1007/978-3-319-70154-7_10

The phase space of the model is the set Γ of all subsets $\gamma \subset \mathbb{R}^d$ such that the set $\gamma \cap \Lambda$ is finite whenever $\Lambda \subset \mathbb{R}^d$ is compact. It is equipped with a topology, see below, and thus with a σ-field of measurable subsets. Thereby, one can consider probability measures on Γ as states of the system. Among them there are Poissonian states in which the particles are independently distributed over $\mathbb{R}^d$. Such states are completely characterized by the density of the particles. In *sub-Poissonian* states, the dependence between the positions of the particles is controlled in a certain way (see the next subsection), and the particles' density is still an important characteristic of the state. For an infinite particle system with repulsion, in [5] the evolution of the system's states $\mu_0 \mapsto \mu_t$ in the set of sub-Poissonian measures was shown to hold for $t < T$ with some $T < \infty$. Then in [3] this result was improved by constructing the global in time evolution of states. Thus, a paramount question regarding such models is whether the sub-Poissonicity is consistent with some sort of attraction, and—if yes—for which sort and on which time intervals. In this paper, we give a partial answer to this question. Namely, we present quite a reasonable condition on the attraction, see (2.15) below, under which—as we show—the correlation functions evolve $k_0 \mapsto k_t$ and remain sub-Poissonian on a bounded time interval. This result extends the corresponding result of [6] in the following directions: (1) the evolution $k_0 \mapsto k_t$ is constructed as a classical solution of the corresponding Cauchy problem, not in a weak sense; (2) our result is valid for much more general types of attraction (see Sects. 3.3 and 3.4 below). At the same time, the following problems remain open: (a) proving that each k_t is the correlation function of a unique sub-Poissonian state; (b) continuing the evolution $k_0 \mapsto k_t$ to all $t > 0$. In Sect. 3.4 below, we discuss possibilities to solve them.

1.2 Presenting the Result

States of an infinite particle system are usually characterized by means of their values $\mu(F)$ on *observables* $F : \Gamma \to \mathbb{R}$, defined as

$$\mu(F) = \int_\Gamma F d\mu.$$

The system's evolution is supposed to be Markovian and hence described by the Kolmogorov equation

$$\frac{d}{dt} F_t = LF_t, \qquad F_t|_{t=0} = F_0, \qquad (1.1)$$

where the operator L specifies the model. Alternatively, the evolution of states is derived from the Fokker-Planck equation

$$\frac{d}{dt} \mu_t = L^* \mu_t, \qquad \mu_t|_{t=0} = \mu_0, \qquad (1.2)$$

related to that in (1.1) by the duality $\mu_t(F_0) = \mu_0(F_t)$. For the model considered in this work, the operator L is

$$(LF)(\gamma) = \sum_{x \in \gamma} \int_{\mathbb{R}^d} a(x, y) \left[1 + \epsilon(x, y|\gamma)\right] \left[F(\gamma \setminus x \cup y) - F(\gamma)\right] dy, \quad (1.3)$$

with

$$\epsilon(x, y|\gamma) = \sum_{z \in \gamma \setminus x} b(x, y|z). \quad (1.4)$$

The quantity $b(x, y|z) \geq 0$ describes the increase of the jump rate from $x \in \gamma$ to $y \in \mathbb{R}^d$ caused by the particle located at $z \in \gamma \setminus x$. Then $\epsilon(x, y|\gamma)$ is the (multiplicative) increase of the corresponding jump rate caused by the whole configuration γ. For $\epsilon \equiv 0$, (1.3) turns into the generator of free jumps, see, e.g., [2].

As is usual for models of this kind, the direct meaning of (1.1) or (1.2) can only be given for states of finite systems, cf. [9]. In this case, the Banach space where the Cauchy problem in (1.2) is defined can be the space of signed measures with finite variation. For infinite systems, the evolution is described by means of correlation functions, see [5–7] and the references quoted in these works. In the present paper, we follow this approach the main idea of which can be outlined as follows. Let Θ be the set of all compactly supported continuous functions $\theta : \mathbb{R}^d \to (-1, 0]$. For a state μ, its *Bogoliubov* functional $B_\mu : \Theta \to \mathbb{R}$ is set to be

$$B_\mu(\theta) = \int_\Gamma \prod_{x \in \gamma} (1 + \theta(x)) \mu(d\gamma), \qquad \theta \in \Theta. \quad (1.5)$$

The function $\gamma \mapsto \prod_{x \in \gamma}(1 + \theta(x))$ is bounded and measurable for each $\theta \in \Theta$; hence, (1.5) makes sense for each measure. For the homogeneous Poisson measure $\pi_\varkappa$, $\varkappa > 0$, we have

$$B_{\pi_\varkappa}(\theta) = \exp\left(\varkappa \int_{\mathbb{R}^d} \theta(x) dx\right).$$

In state $\pi_\varkappa$, the particles are independently distributed over $\mathbb{R}^d$ with density $\varkappa$. The set of *sub-Poissonian* states $\mathcal{P}_{\exp}(\Gamma)$ is then defined as that containing all those states μ for which B_μ can be continued to an exponential type entire function of $\theta \in L^1(\mathbb{R}^d)$. This means that it can be written down in the form

$$B_\mu(\theta) = 1 + \sum_{n=1}^{\infty} \frac{1}{n!} \int_{(\mathbb{R}^d)^n} k_\mu^{(n)}(x_1, \ldots, x_n) \theta(x_1) \cdots \theta(x_n) dx_1 \cdots dx_n, \quad (1.6)$$

where $k_\mu^{(n)}$ is the n-th order correlation function of the state μ. It is a symmetric element of $L^\infty((\mathbb{R}^d)^n)$ for which

$$\|k_\mu^{(n)}\|_{L^\infty((\mathbb{R}^d)^n)} \leq C \exp(\vartheta n), \qquad n \in \mathbb{N}_0, \tag{1.7}$$

with some $C > 0$ and $\vartheta \in \mathbb{R}$. Sometimes, (1.7) is called *Ruelle bound*, cf. [11, Chapter 4]. Note that (1.6) can be viewed as an analog of the Taylor expansion of the characteristic function of a probability measure. That is why $k_\mu^{(n)}$ are also called *moment functions*. Their evolution is described by a chain of differential equations derived from that in (1.1). The central problem of this work is the existence of classical solutions of this chain satisfying (1.7) with possibly time-dependent C and ϑ. Its solution is given in Theorem 3.3, formulated in Sect. 3.2 and proved in Sect. 4. In Sect. 2, we give some necessary information on the methods used in the paper and specify the model. In Sect. 3.1, we place the mentioned chain of equations into suitable Banach spaces, that is mostly performed by defining the corresponding operators. Then we formulate Theorem 3.3 and analyze the assumptions regarding the jump kernels under which we then prove this statement. In Sect. 3.4, we give some comments on the result and the assumptions, including discussing open problems related to the model, and compare our result with the corresponding result of [6]. Section 4 is dedicated to the proof of Theorem 3.3.

2 Preliminaries and the Model

Here we briefly recall the main notions relevant to the subject—for further information we refer to [1, 5–7] and the literature quoted in these works.

2.1 Configuration Spaces

Let $\mathcal{B}(\mathbb{R}^d)$ and $\mathcal{B}_b(\mathbb{R}^d)$ denote the sets of all Borel and all bounded Borel subsets of $\mathbb{R}^d$, respectively. The configuration space Γ, equipped with the vague topology, is homeomorphic to a separable metric (Polish) space, cf. [1, 8]. Let $\mathcal{B}(\Gamma)$ be the corresponding Borel σ-field. For $\Lambda \in \mathcal{B}(\mathbb{R}^d)$, the set $\Gamma_\Lambda = \{\gamma \in \Gamma : \gamma \subset \Lambda\}$ is clearly in $\mathcal{B}(\Gamma)$, and hence

$$\mathcal{B}(\Gamma_\Lambda) := \{A \cap \Gamma_\Lambda : A \in \mathcal{B}(\Gamma)\}$$

is a sub-field of $\mathcal{B}(\Gamma)$. The projection $p_\Lambda : \Gamma \to \Gamma_\Lambda$ defined by $p_\Lambda(\gamma) = \gamma_\Lambda = \gamma \cap \Lambda$ is measurable. Then, for each Borel Λ and $A_\Lambda \in \mathcal{B}(\Gamma_\Lambda)$, we have that

$$p_\Lambda^{-1}(A_\Lambda) := \{\gamma \in \Gamma : p_\Lambda(\gamma) \in A_\Lambda\} \in \mathcal{B}(\Gamma).$$

Let $\mathcal{P}(\Gamma)$ denote the set of all probability measures on $(\Gamma, \mathcal{B}(\Gamma))$. For a given $\mu \in \mathcal{P}(\Gamma)$, its projection on $(\Gamma_\Lambda, \mathcal{B}(\Gamma_\Lambda))$ is defined as

$$\mu^\Lambda(A_\Lambda) = \mu\left(p_\Lambda^{-1}(A_\Lambda)\right), \qquad A_\Lambda \in \mathcal{B}(\Gamma_\Lambda). \tag{2.1}$$

Let Γ_0 be the set of all finite $\gamma \in \Gamma$. Then $\Gamma_0 \in \mathcal{B}(\Gamma)$ as each of $\gamma \in \Gamma_0$ lies in some $\Lambda \in \mathcal{B}_b(\mathbb{R}^d)$, and hence belongs to Γ_Λ. It can be proved that a function $G : \Gamma_0 \to \mathbb{R}$ is $\mathcal{B}(\Gamma)/\mathcal{B}(\mathbb{R})$-measurable if and only if, for each $n \in \mathbb{N}_0$, there exists a symmetric Borel function $G^{(n)} : (\mathbb{R}^d)^n \to \mathbb{R}$ such that

$$G(\eta) = G^{(n)}(x_1, \ldots, x_n), \tag{2.2}$$

for $\eta = \{x_1, \ldots, x_n\}$.

Definition 2.1 A measurable function $G : \Gamma_0 \to \mathbb{R}$ is said have bounded support if: (a) there exists $\Lambda \in \mathcal{B}_b(\mathbb{R}^d)$ such that $G(\eta) = 0$ whenever $\eta \cap \Lambda^c \neq \emptyset$; (b) there exists $N \in \mathbb{N}_0$ such that $G(\eta) = 0$ whenever $|\eta| > N$. Here $\Lambda^c := \mathbb{R}^d \setminus \Lambda$ and $|\cdot|$ stands for cardinality.

The Lebesgue-Poisson measure λ on $(\Gamma_0, \mathcal{B}(\Gamma_0))$ is defined by the following formula

$$\int_{\Gamma_0} G(\eta)\lambda(d\eta) = G(\emptyset) + \sum_{n=1}^\infty \frac{1}{n!} \int_{(\mathbb{R}^d)^n} G^{(n)}(x_1, \ldots, x_n) dx_1 \cdots dx_n, \tag{2.3}$$

which has to hold for all $G \in B_{bs}(\Gamma_0)$—the set of all bounded functions with bounded support.

In this work, we use the following (real) Banach spaces of functions $g : \Gamma_0 \to \mathbb{R}$. The first group consists of the spaces $\mathcal{G}_\vartheta = L^1(\Gamma_0, w_\vartheta d\lambda)$, indexed by $\vartheta \in \mathbb{R}$. Here we have set $w_\vartheta(\eta) = \exp(\vartheta|\eta|)$. Hence the norm of $\mathcal{G}_\vartheta$ is

$$|g|_\vartheta = \int_{\Gamma_0} |g(\eta)| w_\vartheta(\eta) \lambda(d\eta). \tag{2.4}$$

Along with this norm we also consider

$$\|g\|_\vartheta := \operatorname*{ess\,sup}_{\eta \in \Gamma_0} \left\{ |g(\eta)| \exp\left(-\vartheta|\eta|\right) \right\}, \tag{2.5}$$

and then set $\mathcal{K}_\vartheta = \{g : \Gamma_0 \to \mathbb{R} : \|g\|_\vartheta < \infty\}$. These spaces constitute the second group which we use in the sequel. From (2.4) and (2.5) we see that $\mathcal{K}_\vartheta$ is the dual space to $\mathcal{G}_\vartheta$ with the duality

$$(G, k) \mapsto \langle\!\langle G, k \rangle\!\rangle := \int_{\Gamma_0} G(\eta) k(\eta) \lambda(d\eta), \tag{2.6}$$

holding for $G \in \mathcal{G}_\vartheta$ and $k \in \mathcal{K}_\vartheta$. Note that $B_{bs}(\Gamma_0)$ is contained in each $\mathcal{G}_\vartheta$ and each $\mathcal{K}_\vartheta$, $\vartheta \in \mathbb{R}$.

For $G \in B_{bs}(\Gamma)$, we set

$$(KG)(\gamma) = \sum_{\eta \subset \gamma} G(\eta), \tag{2.7}$$

where the sum is taken over all finite η.

2.2 Correlation Functions

For a given $\mu \in \mathcal{P}_{exp}(\Gamma)$, similarly as in (2.2) we introduce $k_\mu : \Gamma_0 \to \mathbb{R}$ such that $k_\mu(\emptyset) = 1$ and $k_\mu(\eta) = k_\mu^{(n)}(x_1, \ldots, x_n)$ for $\eta = \{x_1, \ldots, x_n\}$, $n \in \mathbb{N}$, cf. (1.5) and (1.6). With the help of the measure introduced in (2.3), the formulas in (1.5) and (1.6) can be combined into the following

$$B_\mu(\theta) = \int_{\Gamma_0} k_\mu(\eta) \prod_{x \in \eta} \theta(x) \lambda(d\eta) =: \int_{\Gamma_0} k_\mu(\eta) e(\eta; \theta) \lambda(d\eta)$$

$$= \int_\Gamma \prod_{x \in \gamma} (1 + \theta(x)) \mu(d\gamma) =: \int_\Gamma F_\theta(\gamma) \mu(d\gamma).$$

Thereby, we can transform the action of L on F, as in (1.3), to the action of L^Δ on k_μ according to the rule

$$\int_\Gamma (LF_\theta)(\gamma) \mu(d\gamma) = \int_{\Gamma_0} (L^\Delta k_\mu)(\eta) e(\eta; \theta) \lambda(d\eta). \tag{2.8}$$

This will allow us to pass from (1.1) to the corresponding Cauchy problem for the correlation functions, cf. (3.5) below. The main advantage here is that k_μ is a function of *finite* configurations.

For $\mu \in \mathcal{P}_{exp}(\Gamma)$ and $\Lambda \in \mathcal{B}_b(\mathbb{R}^d)$, let μ^Λ be as in (2.1). Then μ^Λ is absolutely continuous with respect to the restriction λ^Λ to $\mathcal{B}(\Gamma_\Lambda)$ of the measure defined in (2.3), and hence we may write

$$\mu^\Lambda(d\eta) = R_\mu^\Lambda(\eta) \lambda^\Lambda(d\eta), \qquad \eta \in \Gamma_\Lambda. \tag{2.9}$$

Then the correlation function k_μ and the Radon-Nikodym derivative R_μ^Λ satisfy

$$k_\mu(\eta) = \int_{\Gamma_\Lambda} R_\mu^\Lambda(\eta \cup \xi) \lambda^\Lambda(d\xi).$$

By (2.7), (2.1), and (2.9) we get

$$\int_{\Gamma} (KG)(\gamma)\mu(d\gamma) = \langle\!\langle G, k_{\mu} \rangle\!\rangle,$$

holding for each $G \in B_{\mathrm{bs}}(\Gamma_0)$ and $\mu \in \mathcal{P}_{\mathrm{exp}}(\Gamma)$, see (2.6). Define

$$B_{\mathrm{bs}}^{\star}(\Gamma_0) = \{G \in B_{\mathrm{bs}}(\Gamma_0) : (KG)(\gamma) \geq 0 \text{ for all } \gamma \in \Gamma\}.$$

By Kondratiev and Kuna [8, Theorems 6.1 and 6.2 and Remark 6.3] one can prove the next statement.

Proposition 2.2 *Let a measurable function $k : \Gamma_0 \to \mathbb{R}$ have the following properties:*

$$(a) \quad \langle\!\langle G, k \rangle\!\rangle \geq 0, \qquad \text{for all } G \in B_{\mathrm{bs}}^{\star}(\Gamma_0);$$

$$(b) \quad k(\emptyset) = 1; \qquad (c) \quad k(\eta) \leq C^{|\eta|},$$

with (c) holding for some $C > 0$ and λ-almost all $\eta \in \Gamma_0$. Then there exists a unique $\mu \in \mathcal{P}_{\mathrm{exp}}(\Gamma)$ for which k is the correlation function.

2.3 The Model

The model which we study is specified by the operator given in (1.3). The jump kernel a is supposed to satisfy

$$a(x, y) = a(y, x) \geq 0, \qquad \sup_{y \in \mathbb{R}^d} \int_{\mathbb{R}^d} a(x, y)dx = 1. \tag{2.10}$$

Regarding the quantities in (1.4) we assume

$$\sup_{x,y \in \mathbb{R}^d} \int_{\mathbb{R}^d} b(x, y|z)dz =: \langle b \rangle < \infty, \qquad \sup_{x,y,z \in \mathbb{R}^d} b(x, y|z) =: \bar{b} < \infty, \tag{2.11}$$

Moreover, let us define

$$\phi_+(x, y) = \int_{\mathbb{R}^d} a(z, x)b(z, x| y)dz, \tag{2.12}$$

$$\phi_-(x, y) = \int_{\mathbb{R}^d} a(x, z)b(x, z| y)dz.$$

By (2.10) and (2.11) we have that

$$\phi_\pm(x, y) \leq \bar{b}, \qquad \text{for all } x, y \in \mathbb{R}^d. \tag{2.13}$$

Remark 2.3 The quantities defined in (2.12) can be given the following interpretation: $\phi_+(x, y)$ is the rate with which the particle located at y attracts other particles to jump (from somewhere) to x; $\phi_-(x, y)$ is the rate with which the particle located at y forces that located at x to jump (to anywhere). In the latter case, the particle at y 'pushes out' the one at x. Thus, $\phi_+(x, y)$ and $\phi_-(x, y)$ can be called attraction and repulsion rates, respectively.

Now we set

$$\Phi_\pm(\eta) = \sum_{x \in \eta} \sum_{y \in \eta \setminus x} \phi_\pm(x, y), \tag{2.14}$$

which can be interpreted as the total rates of attraction and repulsion of the configuration η, respectively. In addition to (2.10) and (2.11) we assume that the following holds

$$\exists \omega \geq 0 \ \forall \eta \in \Gamma_0 \qquad \Phi_+(\eta) \leq \Phi_-(\eta) + \omega|\eta|. \tag{2.15}$$

Note that, for some $c > 0$ and all $\eta \in \Gamma_0$, by (2.11) it follows that

$$\Phi_-(\eta) + \omega|\eta| \leq c|\eta|^2. \tag{2.16}$$

According to the condition in (2.15), the rate of the jumps from somewhere to points close to the configuration η (i.e., those which make η denser) is in a sense dominated by the rate of the jumps to anywhere, which thin it out.

3 The Result

3.1 The Operators

By means of (1.3) and (2.8) we calculate L^Δ and present it in the form

$$L^\Delta = A^\Delta + B^\Delta + C^\Delta + D^\Delta, \tag{3.1}$$

with the entries

$$(A^\Delta k)(\eta) = \sum_{y \in \eta} \int_{\mathbb{R}^d} a(x, y) \left(1 + \sum_{z \in \eta \setminus y} b(x, y|z) \right) k(\eta \setminus y \cup x) dx, \tag{3.2}$$

$$(B^\Delta k)(\eta) = -\Psi(\eta)k(\eta),$$

where

$$\Psi(\eta) = \sum_{x \in \eta} \int_{\mathbb{R}^d} a(x, y) dy + \Phi_-(\eta). \tag{3.3}$$

Furthermore,

$$(C^\Delta k)(\eta) = \sum_{y \in \eta} \int_{\mathbb{R}^d} \int_{\mathbb{R}^d} a(x, y) b(x, y|z) k(\eta \setminus y \cup \{x, z\}) dx dz, \tag{3.4}$$

$$(D^\Delta k)(\eta) = -\int_{\mathbb{R}^d} \left(\sum_{x \in \eta} \int_{\mathbb{R}^d} a(x, y) b(x, y|z) dy \right) k(\eta \cup z) dz.$$

As mentioned above, instead of directly dealing with the problem in (1.2) we pass from μ_0 to the corresponding correlation function k_{μ_0} and then consider the problem

$$\frac{d}{dt} k_t = L^\Delta k_t, \qquad k_t|_{t=0} = k_{\mu_0}, \tag{3.5}$$

with L^Δ given in (3.1)–(3.4). Our aim now is to place this problem into the corresponding Banach space. By (1.7) we conclude that $\mu \in \mathcal{P}_{\exp}(\Gamma)$ implies that $k_\mu \in \mathcal{K}_\vartheta$ for some $\vartheta \in \mathbb{R}$. Hence, we assume that k_{μ_0} lies in some $\mathcal{K}_{\vartheta_0}$. Then the formulas in (3.1)–(3.4) can be used to define an unbounded operator acting in some $\mathcal{K}_\vartheta$. Like in [5, 7] we take into account that, for each $\vartheta'' < \vartheta'$, the space $\mathcal{K}_{\vartheta''}$ is continuously embedded into $\mathcal{K}_{\vartheta'}$, see (2.5), and use the ascending scale of such spaces. This means that we are going to define (3.5) in a given $\mathcal{K}_\vartheta$ assuming that $k_{\mu_0} \in \mathcal{K}_{\vartheta_0} \hookrightarrow \mathcal{K}_\vartheta$.

For ω as in (2.15) and Ψ as in (3.3), we set

$$\Psi_\omega(\eta) = \omega|\eta| + \Psi(\eta). \tag{3.6}$$

In the sequel, along with those as in (3.2) and (3.3) we use the following operators

$$(B^{\Delta,\omega} k)(\eta) = -\Psi_\omega(\eta) k(\eta), \tag{3.7}$$

$$(C^{\Delta,\omega} k)(\eta) = (C^\Delta k)(\eta) + \omega|\eta| k(\eta).$$

Then the decomposition (3.1) can be rewritten

$$L^\Delta = A^\Delta + B^{\Delta,\omega} + C^{\Delta,\omega} + D^\Delta, \tag{3.8}$$

with A^Δ and D^Δ being as above.

For a given $\vartheta \in \mathbb{R}$, we define L^Δ in $\mathcal{K}_\vartheta$ by means of the following estimates. For $k \in \mathcal{K}_\vartheta$, by (2.5) we have that

$$|k(\eta)| \leq \|k\|_\vartheta e^{\vartheta|\eta|}, \qquad \text{for } \lambda - \text{a.a. } \eta \in \Gamma_0. \tag{3.9}$$

By means of the latter estimate and (2.10), (2.11) we obtain from (3.2), (3.3) and (3.7) that

$$\left|(C^{\Delta,\omega}k)(\eta)\right| \leq (\omega + \langle b\rangle)\,|\eta|\exp[\vartheta(|\eta|+1)]\cdot\|k\|_\vartheta, \tag{3.10}$$

$$\left|(C^\Delta k)(\eta)\right| \leq \langle b\rangle|\eta|\exp[\vartheta(|\eta|+1)]\cdot\|k\|_\vartheta,$$

$$\left|(D^\Delta k)(\eta)\right| \leq \langle b\rangle|\eta|\exp[\vartheta(|\eta|+1)]\cdot\|k\|_\vartheta,$$

Now we use (2.13) and (2.14) to obtain from (3.2), (3.3), (3.6) the following

$$\left|(A^\Delta k)(\eta)\right| \leq \left(|\eta| + \bar{b}|\eta|^2\right)e^{\vartheta|\eta|}\cdot\|k\|_\vartheta, \tag{3.11}$$

$$\left|(B^\Delta k)(\eta)\right| \leq \left(|\eta| + \bar{b}|\eta|^2\right)e^{\vartheta|\eta|}\cdot\|k\|_\vartheta.$$

$$\left|(B^{\Delta,\omega}k)(\eta)\right| \leq \left[(1+\omega)|\eta| + \bar{b}|\eta|^2\right]e^{\vartheta|\eta|}\cdot\|k\|_\vartheta.$$

The estimates in (3.10) and (3.11) allow us to define $(L_\vartheta^\Delta, \mathcal{D}_\vartheta^\Delta)$, where

$$\mathcal{D}_\vartheta^\Delta := \{k \in \mathcal{K}_\vartheta : |\cdot|^2 k \in \mathcal{K}_\vartheta\}, \tag{3.12}$$

Lemma 3.1 *For each $\vartheta'' < \vartheta$, it follows that $\mathcal{K}_{\vartheta''} \subset \mathcal{D}_\vartheta^\Delta$.*

Proof By means of (3.9) and the inequality $x\exp(-\sigma x) \leq 1/e\sigma$, $x, \sigma > 0$, we get from (3.10) and (3.11) the following estimate,

$$|\eta|^2\,|k(\eta)| \leq \frac{4}{e^2(\vartheta - \vartheta'')^2}\|k\|_{\vartheta''}e^{\vartheta|\eta|},$$

which yields the proof.   □

The same estimate and (3.10), (3.11) also yield

$$\|L^\Delta k\|_\vartheta \leq 2\left(\frac{1 + \langle b\rangle}{e(\vartheta - \vartheta'')} + \frac{4\bar{b}}{e^2(\vartheta - \vartheta'')^2}\right)\|k\|_{\vartheta''}, \tag{3.13}$$

which allows us to define a bounded linear operator $L_{\vartheta\vartheta''}^\Delta : \mathcal{K}_{\vartheta''} \to \mathcal{K}_\vartheta$ the norm of which can be estimated by means of (3.13). In what follows, we consider two types of operators defined by the expression in (3.1)–(3.4): (a) unbounded operators $(L_\vartheta^\Delta, \mathcal{D}_\vartheta^\Delta)$, $\vartheta \in \mathbb{R}$, with domains as in (3.12) and Lemma 3.1; (b) bounded operators

$L^\Delta_{\vartheta\vartheta''}$ as just described. These operators are related to each other in the following way:

$$\forall \vartheta'' < \vartheta \ \ \forall k \in \mathcal{K}_{\vartheta''} \qquad L^\Delta_{\vartheta\vartheta''}k = L^\Delta_\vartheta k. \tag{3.14}$$

3.2 The Statement

We assume that the initial state μ_0 is fixed, which determines $\vartheta_0 \in \mathbb{R}$ by the condition that k_{μ_0} lies in $\mathcal{K}_{\vartheta_0}$. Since $\mathcal{K}_{\alpha''} \hookrightarrow \mathcal{K}_{\alpha'}$ for $\vartheta'' < \vartheta'$, we take the least ϑ_0 satisfying this condition. Then for $\vartheta > \vartheta_0$, we consider in $\mathcal{K}_\vartheta$ the problem, cf. (3.5) and Lemma 3.1,

$$\frac{d}{dt}k_t = L^\Delta_\vartheta k_t, \qquad k_t|_{t=0} = k_{\mu_0} \in \mathcal{K}_{\vartheta_0}. \tag{3.15}$$

Definition 3.2 By a (classical) solution of (3.15) on a time interval, $[0, T)$, $T \le +\infty$, we mean a continuous map $[0, T) \ni t \mapsto k_t \in \mathcal{D}^\Delta_\vartheta$ such that the map $[0, T) \ni t \mapsto dk_t/dt \in \mathcal{K}_\vartheta$ is also continuous and both equalities in (3.15) are satisfied.

For $\omega \ge 0$ as in (3.6) and (3.8), we set, cf. (2.11),

$$T(\vartheta, \vartheta_0) = \frac{(\vartheta - \vartheta_0)e^{-\vartheta}}{\omega + 2\langle b \rangle}, \tag{3.16}$$

where ϑ and ϑ_0 are as in (3.15).

Theorem 3.3 *Let the conditions in (2.10)–(2.15) be satisfied. Then, for each $\vartheta > \vartheta_0$, the problem in (3.15) has a unique solution on the time interval $[0, T(\vartheta, \vartheta_0))$.*

The proof of this statement will be done in Sect. 4 below. Let us now analyze how to choose ϑ in an optimal way. Since the length $T(\vartheta, \vartheta_0)$ of the time interval in Theorem 3.3 depends on the choice of ϑ, we take $\vartheta = \vartheta_*$ defined by the condition

$$T(\vartheta_*, \vartheta_0) = \max_{\vartheta > \vartheta_0} T(\vartheta, \vartheta_0),$$

which by (3.16) yields $\vartheta_* = 1 + \vartheta_0$. Hence, the maximum length of the time interval is

$$\tau(\vartheta_0) = T(1 + \vartheta_0, \vartheta_0) = \frac{e^{-\vartheta_0}}{e(\omega + 2\langle b \rangle)}.$$

Note that $\tau(\vartheta_0) \to 0$ as $\vartheta_0 \to +\infty$.

3.3   Analyzing the Assumptions

Our main assumption in (2.15) looks like the stability condition (with stability constant $\omega \geq 0$) for the interaction potential $\phi = \phi_- - \phi_+$, see (2.12), used in the statistical mechanics of continuum systems of interacting particles, cf. [11, Chapter 3] and also [7, Section 3.3]. A particular case of the kernels is where they are translation invariant and b has the following form

$$b(x, y|z) = \kappa_1(x - z) + \kappa_2(y - z), \tag{3.17}$$

where $\kappa_i(x) = \kappa_i(-x) \geq 0$ belong to $L^1(\mathbb{R}^d)$. Then

$$\phi_+(x, y) = (\alpha * \kappa_1)(x - y) + \kappa_2(x - y), \tag{3.18}$$

$$\phi_-(x, y) = \kappa_1(x - y) + (\alpha * \kappa_2)(x - y),$$

where $\alpha(x) = a(0, x)$ and $*$ denotes the usual convolution. By (2.10) and (2.12) α and both κ_i are integrable. Thus, we can use their transforms

$$\hat{\alpha}(p) = \int_{\mathbb{R}^d} \alpha(x) \exp\left(i(p, x)\right) dx, \qquad p \in \mathbb{R}^d,$$

$$\hat{\phi}_\pm(p) = \int_{\mathbb{R}^d} \phi_\pm(0, x) \exp\left(i(p, x)\right) dx,$$

$$\hat{\kappa}_i(p) = \int_{\mathbb{R}^d} \kappa_i(x) \exp\left(i(p, x)\right) dx, \qquad i = 0, 1,$$

Note that the left-hand sides here are real. Moreover, $\hat{\alpha}(p) \leq \hat{\alpha}(0) = 1$ and $\hat{\kappa}_i(p) \leq \hat{\kappa}_i(0)$, $i = 1, 2$. Then a sufficient condition for (2.15) to be satisfied, see [11, Section 3.2, Proposition 3.2.7], is that the following holds: (a) both $\phi_\pm(0, x)$ are continuous; (b) $\hat{\phi}_-(p) \geq \hat{\phi}_+(p)$ for all $p \in \mathbb{R}^d$. The latter means that the potential $\phi = \phi_- - \phi_+$ is positive definite (in Bochner's sense). In view of (3.18), (b) turns into

$$\forall p \in \mathbb{R}^d \qquad (1 - \hat{\alpha}(p))\, (\hat{\kappa}_1(p) - \hat{\kappa}_2(p)) \geq 0. \tag{3.19}$$

Thus, a sufficient condition for the latter to hold is $\hat{\kappa}_1(p) \geq \hat{\kappa}_2(p)$ for all those p for which $\hat{\alpha}(p) < 1$. An example can be

$$\kappa_i(x) = \frac{1}{(2\pi)^{d/2}\sigma_i} \exp\left(-\frac{|x|^2}{2\sigma_i^2}\right), \qquad \sigma_1 < \sigma_2, \tag{3.20}$$

cf. [7, Proposition 3.8].

3.4 Comments

First we make some comments on the result of Theorem 3.3. For the model specified in (1.3) with a particular choice of b, which we discuss below, in [6, Theorem 2 and Proposition 1] there was constructed a weak solution of the problem in (3.5) on a bounded time interval. Our Theorem 3.3 yields a solution in the strongest sense—a classical one—see Definition 3.2, existing, however, also on a bounded time interval. At the same time, this solution k_t yet may not be the correlation function of a state. To prove this, one ought to develop a technique similar to that used in [7, Section 5] and based on the use of Proposition 2.2. Noteworthy, the fact, proved in [7], that k_t is a correlation function allowed there for continuing to all $t > 0$ the solution primarily obtained on a bounded time interval. For jump dynamics with repulsion, such continuation was realized in [3, 4], also by means of the corresponding property of k_t. However, for the model considered here for such a continuation to be done proving that the solution k_t is a correlation function—and hence is positive in a certain sense—might not be enough. If this is the case, then the attraction in the form as in (1.3) is not consistent with the sub-Poissonicity of the states and hence essentially changes the dynamics of the model. We plan to clarify this in our forthcoming work.

Now let us return to discussing the conditions imposed on the model. As mentioned above, in [6] there was studied the model specified in (1.3) with the choices of b (cf. [6, Eqs. (3)–(5)]) which in our notations can be presented as follows: (1) $b(x,y|z) = \kappa(x - z)$; (2) $b(x,y|z) = \kappa(y - z)$; (3) $b(x,y|z) = [\kappa(x-z)+\kappa(y-z)]/2$. Note that all the three are particular cases of (3.17). However, instead of our condition (2.15) there was imposed a stronger one, which in our context can be written down as

$$\forall x \in \mathbb{R}^d \qquad \phi_-(0,x) \geq \phi_+(0,x). \tag{3.21}$$

In case (1), (3.21) turns into $\kappa(x) \geq (\alpha * \kappa)(x)$, which is much stronger than $\hat{\kappa}(p) \geq 0$ that follows from (3.19) in this case. E.g., the latter clearly holds for the Gaussian kernel κ, see (3.20). In case (2), which corresponds to pure attraction, cf. Remark 2.3, (3.21) turns into $\kappa(x) \leq (\alpha * \kappa)(x)$, which, in fact, is equivalent to $\kappa = (\alpha * \kappa)$. The latter can be considered as the problem of the existence of strictly positive fixed points of the corresponding (positive) integral operator in $L^1(\mathbb{R}^d)$. In some cases, this problem has such points, e.g., if the operator is compact—by the Krein-Rutman theorem. In the symmetric case (3), we have $\phi_+ = \phi_-$, and hence (2.15) trivially holds.

4   The Proof

The main idea of proving Theorem 3.3 is to construct the family of bounded linear operators $Q_{\vartheta \vartheta_0}(t) : \mathcal{K}_{\vartheta_0} \to \mathcal{K}_{\vartheta}$ with $t \in [0, T(\vartheta, \vartheta_0))$ such that the solution of (3.15) is obtained in the form

$$k_t = Q_{\vartheta \vartheta_0}(t)k_0. \tag{4.1}$$

An important element of this construction is another family of bounded operators obtained by means of a substochastic semigroup constructed in the $\mathcal{G}_{\vartheta}$. We obtain this semigroup in the next subsection in Proposition 4.2.

4.1   An Auxiliary Semigroup

For a given $\vartheta \in \mathbb{R}$, the formulas in (3.11) allows one to define the corresponding unbounded operators in $\mathcal{K}_{\vartheta}$. The predual space of $\mathcal{K}_{\vartheta}$ is $\mathcal{G}_{\vartheta}$ equipped with the norm defined in (2.4). For A^{Δ} and $B^{\Delta,\omega}$, see (3.7), we introduce $\widehat{A}$ and $\widehat{B}^{\omega}$ by setting, cf. (2.6),

$$\langle\!\langle \widehat{A}G, k \rangle\!\rangle = \langle\!\langle G, A^{\Delta,\omega}k \rangle\!\rangle, \qquad \langle\!\langle \widehat{B}^{\omega}G, k \rangle\!\rangle = \langle\!\langle G, B^{\Delta,\omega}k \rangle\!\rangle.$$

This yields

$$(\widehat{A}G)(\eta) = \sum_{x \in \eta} \int_{\mathbb{R}^d} a(x,y) \left(1 + \sum_{z \in \eta \setminus x} b(x,y|z) \right) G(\eta \setminus x \cup y) dy, \tag{4.2}$$

$$(\widehat{B}^{\omega}G)(\eta) = -\Psi_{\omega}(\eta)G(\eta).$$

Now for a given ϑ, we set, cf. (3.12)

$$\widehat{\mathcal{D}}_{\vartheta}^{\omega} := \{G \in \mathcal{G}_{\vartheta} : \Psi_{\omega}G \in \mathcal{G}_{\vartheta}\}. \tag{4.3}$$

Clearly, the multiplication operator $\widehat{B}_{\vartheta}^{\omega} : \widehat{\mathcal{D}}_{\vartheta}^{\omega} \subset \mathcal{G}_{\vartheta} \to \mathcal{G}_{\vartheta}$ defined in the second line of (4.2) is closed. Moreover, it generates a C_0-semigroup $\{S_{\vartheta}^{(0)}(t)\}_{t \geq 0}$ of bounded multiplication operators $(S_{\vartheta}^{(0)}(t)G)(\eta) = \exp(-t\Psi_{\omega}(\eta))G(\eta)$. Note that each operator is a positive contraction, i.e., $S_{\vartheta}^{(0)}(t)$ maps

$$\mathcal{G}_{\vartheta}^+ := \{G \in \mathcal{G}_{\vartheta} : G(\eta) \geq 0, \ \lambda - \text{a.a. } \eta \in \Gamma_0\}$$

into itself and $|S_\vartheta^{(0)}(t)G|_\vartheta \leq |G|_\vartheta$, see (2.4). That is, $\{S_\vartheta^{(0)}(t)\}_{t\geq 0}$ is a *substochastic* semigroup.

For $G \in \widehat{\mathcal{D}}_\vartheta^{\omega,+} := \widehat{\mathcal{D}}_\vartheta^\omega \cap \mathcal{G}_\vartheta^+$, by (2.15) and (4.2) we have

$$|\widehat{A}G|_\vartheta = \int_{\Gamma_0} \left(\sum_{y \in \eta} \int_{\mathbb{R}^d} a(x,y)dx + \Phi_+(\eta) \right) G(\eta) e^{\vartheta |\eta|} \lambda(d\eta) \qquad (4.4)$$

$$\leq \int_{\Gamma_0} \left(\sum_{x \in \eta} \int_{\mathbb{R}^d} a(x,y)dy + \omega |\eta| + \Phi_-(\eta) \right) G(\eta) e^{\vartheta |\eta|} \lambda(d\eta)$$

$$= -\int_{\Gamma_0} \left(\widehat{B}^\omega G \right)(\eta) e^{\vartheta |\eta|} \lambda(d\eta).$$

Likewise, for $G \in \widehat{\mathcal{D}}_\vartheta^\omega$ we get

$$|\widehat{A}G|_\vartheta \leq \int_{\Gamma_0} \Phi_\omega(\eta) |G(\eta)| e^{\vartheta |\eta|} \lambda(d\eta), \qquad (4.5)$$

which means that $\widehat{A}$ can be defined on $\widehat{\mathcal{D}}_\vartheta^\omega$, see (4.3).

Lemma 4.1 *The closure $\widehat{T}_\vartheta$ of $(\widehat{A}+\widehat{B}^\omega, \widehat{\mathcal{D}}_\vartheta^\omega)$ in $\mathcal{G}_\vartheta$ is the generator of a substochastic semigroup.*

Proof We use the Thieme-Voigt perturbation technique [12], see also [9, Section 3]. For each $G \in \mathcal{G}_\vartheta^+$, we have that

$$|G|_\vartheta = \varphi_\vartheta(G) := \int_{\Gamma_0} G(\eta) e^{\vartheta |\eta|} \lambda(d\eta).$$

Clearly, φ_ϑ is a positive linear functional on $\mathcal{G}_\vartheta$, and thus the norm defined in (2.4) is additive on the cone $\mathcal{G}_\vartheta^+$. For $\vartheta' > \vartheta$, by (2.4) $\mathcal{G}_{\vartheta'}$ is densely and continuously embedded into $\mathcal{G}_\vartheta$. Moreover, the mentioned above semigroup $\{S_\vartheta^{(0)}(t)\}_{t\geq 0}$ has the property $S_\vartheta^{(0)}(t) : \mathcal{G}_{\vartheta'} \to \mathcal{G}_{\vartheta'}$, $t \geq 0$, and the restrictions $S_\vartheta^{(0)}(t)|_{\mathcal{G}_{\vartheta'}}$ constitute a C_0-semigroup, which is just $\{S_{\vartheta'}^{(0)}(t)\}_{t\geq 0}$ generated by $(\widehat{B}_{\vartheta'}^\omega, \widehat{\mathcal{D}}_{\vartheta'}^\omega)$. By (4.5) we have that $\widehat{A} : \widehat{\mathcal{D}}_{\vartheta'}^\omega \to \mathcal{G}_{\vartheta'}$. Then according to [12, Theorem 2.7], see also [9, Proposition 3.2], the proof will be done if we show that, for some $\vartheta' > \vartheta$, the following holds

$$\forall G \in \widehat{\mathcal{D}}_{\vartheta'}^{\omega,+} \qquad \varphi_{\vartheta'}((\widehat{A} + \widehat{B}^\omega)G) \leq \varphi_{\vartheta'}(G) - \varepsilon \varphi_\vartheta(\Psi_\omega G) \qquad (4.6)$$

with some $\varepsilon > 0$. Since (4.4) holds for each $\vartheta \in \mathbb{R}$, we have that

$$\varphi_{\vartheta'}((\widehat{A} + \widehat{B}^\omega)G) \leq 0.$$

Then (4.6) turns into $\varphi_\vartheta(\Psi_\omega G) \le (1/\varepsilon)\varphi_{\vartheta'}(G)$. By (2.16) the latter holds for each $\vartheta' > \vartheta$ and the correspondingly small ε. $\square$

Let $S_\vartheta := \{S_\vartheta(t)\}_{t\ge 0}$ be the semigroup as in Lemma 4.1. The semigroup which we need is the sun-dual to S_ϑ. It is introduced as follows. Let T_ϑ^* be the adjoint to the generator of S_ϑ with domain $\mathrm{Dom}(T_\vartheta^*) \subset \mathcal{K}_\vartheta$ defined in a standard way. That is,

$$\mathrm{Dom}(T_\vartheta^*) = \{k \in \mathcal{K}_\vartheta : \exists q \in \mathcal{K}_\vartheta \ \forall G \in \widehat{\mathcal{D}}_\vartheta^\omega \ \langle\langle \widehat{T}_\vartheta G, k \rangle\rangle = \langle\langle G, q \rangle\rangle\}.$$

For each $k \in \mathrm{Dom}(T_\vartheta^*)$, we have that

$$(T_\vartheta^* k)(\eta) = (A^\Delta k)(\eta) + (B^{\Delta,\omega} k)(\eta), \tag{4.7}$$

see (3.2) and (3.7). By (3.11) we then get that $\mathcal{K}_{\vartheta''} \subset \mathrm{Dom}(T_\vartheta^*)$ for each $\vartheta'' < \vartheta$. Let $\mathcal{Q}_\vartheta$ stand for the closure of $\mathrm{Dom}(T_\vartheta^*)$ in $\mathcal{K}_\vartheta$. Then

$$\mathcal{Q}_\vartheta := \overline{\mathrm{Dom}(T_\vartheta^*)} \supset \mathrm{Dom}(T_\vartheta^*) \supset \mathcal{K}_{\vartheta''}. \tag{4.8}$$

For each $t \ge 0$, the adjoint $(S_\vartheta(t))^*$ of $S_\vartheta(t)$ is a bounded operator in $\mathcal{K}_\vartheta$. However, the semigroup $\{(S_\vartheta(t))^*\}_{t\ge 0}$ is not strongly continuous. For $t > 0$, let $S_\vartheta^\odot(t)$ denote the restriction of $(S_\vartheta(t))^*$ to $\mathcal{Q}_\vartheta$. Since S_ϑ is the semigroup of contractions, for $k \in \mathcal{Q}_\vartheta$ and all $t \ge 0$, we have that

$$\|S_\vartheta^\odot(t)k\|_\vartheta = \|S^*(t)k\|_\vartheta \le \|k\|_\vartheta. \tag{4.9}$$

Proposition 4.2 *For every $\vartheta'' < \vartheta$ and any $k \in \mathcal{K}_{\vartheta''}$, the map*

$$[0, +\infty) \ni t \mapsto S_\vartheta^\odot(t)k \in \mathcal{K}_\vartheta$$

is continuous.

Proof By Pazy [10, Theorem 10.4, p. 39], the collection $S_\vartheta^\odot := \{S_\vartheta^\odot(t)\}_{t\ge 0}$ constitutes a C_0-semigroup on $\mathcal{Q}_\vartheta$ the generator of which, $T_\vartheta^\odot$, is the part of T_ϑ^* in $\mathcal{Q}_\vartheta$. That is, $T_\vartheta^\odot$ is the restriction of T_ϑ^* to the set

$$\mathrm{Dom}(T_\vartheta^\odot) := \{k \in \mathrm{Dom}(T_\vartheta^*) : T_\vartheta^* k \in \mathcal{Q}_\vartheta\},$$

cf. [10, Definition 10.3, p. 39]. The continuity in question follows by the C_0-property of the semigroup $\{S_\vartheta^\odot(t)\}_{t\ge 0}$ and (4.8). $\square$

By (3.11) it follows that $\mathrm{Dom}(T_{\vartheta'}^\odot) \supset \mathcal{K}_{\vartheta''}$, holding for each $\vartheta'' < \vartheta'$. Hence, see [10, Theorem 2.4, p. 4],

$$S_{\vartheta'}^\odot(t)k \in \mathrm{Dom}(T_{\vartheta'}^\odot),$$

and

$$\frac{d}{dt}S_{\vartheta'}^{\odot}(t)k = A_{\vartheta'}^{\odot}S_{\vartheta'}^{\odot}(t)k, \tag{4.10}$$

which holds for all $\vartheta' \in (\vartheta'', \vartheta]$ and $k \in \mathcal{K}_{\vartheta''}$.

4.2 Getting the Solutions

Here we construct the family of the operators $Q_{\vartheta\vartheta_0}(t)$ which appear in (4.1). To this end we use the semigroup as in Proposition 4.2.

For $\vartheta'' < \vartheta$, let $\mathcal{L}(\mathcal{K}_{\vartheta''}, \mathcal{K}_{\vartheta})$ denote the Banach space of bounded linear operators acting from $\mathcal{K}_{\vartheta''}$ to $\mathcal{K}_{\vartheta}$. By means of the estimates in (3.11) one can introduce $A_{\vartheta\vartheta''}^{\Delta}$ and $B_{\vartheta\vartheta''}^{\Delta,\omega}$, both in $\mathcal{L}(\mathcal{K}_{\vartheta''}, \mathcal{K}_{\vartheta})$. Then, cf. (3.14) and (4.7),

$$\forall k \in \mathcal{K}_{\vartheta''} \qquad T_{\vartheta}^{\odot}k = \left(A_{\vartheta\vartheta''}^{\Delta} + B_{\vartheta\vartheta''}^{\Delta,\omega}\right)k. \tag{4.11}$$

Let now $S_{\vartheta\vartheta''}(t)$, $t > 0$ be the restriction of $S_{\vartheta}^{\odot}(t)$ to $\mathcal{K}_{\vartheta''}$. Let also $S_{\vartheta\vartheta''}(0)$ be the embedding $\mathcal{K}_{\vartheta''} \hookrightarrow \mathcal{K}_{\vartheta}$. By (4.9) we have that the operator norm of such operators satisfy

$$\forall t \geq 0 \qquad \|S_{\vartheta\vartheta''}(t)\| \leq 1. \tag{4.12}$$

By Proposition 4.2 the map

$$[0, +\infty) \ni t \mapsto S_{\vartheta\vartheta''}(t) \in \mathcal{L}(\mathcal{K}_{\vartheta''}, \mathcal{K}_{\vartheta})$$

is continuous, and for each $\vartheta' \in (\vartheta'', \vartheta)$, the following holds, see (4.10) and (4.11),

$$\frac{d}{dt}S_{\vartheta\vartheta''}(t) = \left(A_{\vartheta\vartheta'}^{\Delta} + B_{\vartheta\vartheta'}^{\Delta,\omega}\right)S_{\vartheta'\vartheta''}(t). \tag{4.13}$$

Now by means of the estimates in (3.10) one concludes that the formulas in (3.4) and (3.7) can be used to introduce $C_{\vartheta\vartheta''}^{\Delta,\omega}$ and $D_{\vartheta\vartheta''}^{\Delta}$, both in $\mathcal{L}(\mathcal{K}_{\vartheta''}, \mathcal{K}_{\vartheta})$. Their operator norms satisfy, cf. (3.13),

$$\|C_{\vartheta\vartheta''}^{\Delta,\omega}\| \leq \frac{(\omega + \langle b \rangle)e^{\vartheta}}{e(\vartheta - \vartheta'')}, \qquad \|D_{\vartheta\vartheta''}^{\Delta}\| \leq \frac{\langle b \rangle e^{\vartheta}}{e(\vartheta - \vartheta'')}. \tag{4.14}$$

Let ϑ_0 be as in (3.15). Take $\vartheta > \vartheta_0$ and then define

$$\mathcal{A}(\vartheta, \vartheta_0) = \{(\vartheta_1, \vartheta_2, t) : \vartheta_0 \le \vartheta_1 < \vartheta_2 \le \vartheta, \ 0 \le t < T(\vartheta_2, \vartheta_1)\},$$

where $T(\vartheta_2, \vartheta_1)$ is as in (3.16).

Lemma 4.3 *For any* $(\vartheta_1, \vartheta_2, t) \in \mathcal{A}(\vartheta, \vartheta_0)$*, there exists* $Q_{\vartheta_2 \vartheta_1}(t) \in \mathcal{L}(\mathcal{K}_{\vartheta_1}, \mathcal{K}_{\vartheta_2})$ *such that the family* $\{Q_{\vartheta_2 \vartheta_1}(t) : (\vartheta_1, \vartheta_2, t) \in \mathcal{A}(\vartheta, \vartheta_0)\}$ *has the following properties:*

  (i) *the map* $[0, T(\vartheta_2, \vartheta_1)) \ni t \mapsto Q_{\vartheta_2 \vartheta_1}(t) \in \mathcal{L}(\mathcal{K}_{\vartheta_1}, \mathcal{K}_{\vartheta_2})$ *is continuous;*
  (ii) *the operator norm of* $Q_{\vartheta_2 \vartheta_1}(t)$ *satisfies*

$$\|Q_{\vartheta_2 \vartheta_1}(t)\| \le \frac{T(\vartheta_2, \vartheta_1)}{T(\vartheta_2, \vartheta_1) - t}; \tag{4.15}$$

(iii) *for each* $\vartheta_3 \in (\vartheta_1, \vartheta_2)$ *and* $t < T(\vartheta_3, \vartheta_1)$*, the following holds*

$$\frac{d}{dt} Q_{\vartheta_2 \vartheta_1}(t) = L^{\Delta}_{\vartheta_2 \vartheta_3} Q_{\vartheta_3 \vartheta_1}(t), \tag{4.16}$$

where $L^{\Delta}_{\vartheta_2 \vartheta_3}$ *is as in (3.14).*
The proof of this statement employs the following construction. For $l \in \mathbb{N}$ and $t > 0$, we set

$$\mathcal{T}_l := \{(t, t_1, \ldots, t_l) : 0 \le t_l \le \cdots \le t_1 \le t\},$$

fix some $\theta \in (\vartheta_1, \vartheta_2]$, and then take $\delta < \theta - \vartheta_1$. Next we divide the interval $[\vartheta_1, \theta]$ into subintervals with endpoints ϑ^s, $s = 0, \ldots, 2l + 1$, as follows. Set $\vartheta^0 = \vartheta_1$, $\vartheta^{2l+1} = \theta$, and

$$\vartheta^{2s} = \vartheta_1 + \frac{s}{l+1}\delta + s\epsilon, \qquad \epsilon = (\theta - \vartheta_1 - \delta)/l, \tag{4.17}$$

$$\vartheta^{2s+1} = \vartheta_1 + \frac{s+1}{l+1}\delta + s\epsilon, \qquad s = 0, 1, \ldots, l.$$

Then for $(t, t_1, \ldots, t_l) \in \mathcal{T}_l$, define

$$\Pi^{(l)}_{\theta \vartheta_1}(t, t_1, \ldots, t_l) = S_{\theta \vartheta^{2l}}(t - t_1) \left(C^{\Delta, \omega}_{\vartheta^{2l} \vartheta^{2l-1}} + D^{\Delta}_{\vartheta^{2l} \vartheta^{2l-1}} \right) \tag{4.18}$$

$$\times \cdots \times S_{\vartheta^{2s+1} \vartheta^{2s}}(t_{l-s} - t_{l-s+1}) \left(C^{\Delta, \omega}_{\vartheta^{2s} \vartheta^{2s-1}} + D^{\Delta}_{\vartheta^{2s} \vartheta^{2s-1}} \right)$$

$$\times \cdots \times S_{\vartheta^3 \vartheta^2}(t_{l-1} - t_l) \left(C^{\Delta, \omega}_{\vartheta^2 \vartheta^1} + D^{\Delta}_{\vartheta^2 \vartheta^{2l}} \right) S_{\vartheta^1 \vartheta_1}(t_l).$$

Proposition 4.4 *For each $l \in \mathbb{N}$, the operators defined in (4.18) have the following properties:*

(i) for each $(t, t_1, \ldots, t_l) \in \mathcal{T}_l$, $\Pi^{(l)}_{\theta \vartheta_1}(t, t_1, \ldots, t_l) \in \mathcal{L}(\mathcal{K}_{\vartheta_1}, \mathcal{K}_\theta)$, and the map

$$\mathcal{T}_l \ni (t, t_1, \ldots, t_l) \mapsto \Pi^{(l)}_{\theta \vartheta_1}(t, t_1, \ldots, t_l) \in \mathcal{L}(\mathcal{K}_{\vartheta_1}, \mathcal{K}_\theta)$$

is continuous;
(ii) for fixed $t_1, t_2, \ldots, t_l$, and each $\varepsilon > 0$, the map

$$(t_1, t_1 + \varepsilon) \ni t \mapsto \Pi^{(l)}_{\theta \vartheta_1}(t, t_1, \ldots, t_l) \in \mathcal{L}(\mathcal{K}_{\vartheta_1}, \mathcal{K}_{\vartheta_2})$$

is continuously differentiable and for each $\vartheta' \in (\vartheta_1, \theta)$ the following holds

$$\frac{d}{dt} \Pi^{(l)}_{\theta \vartheta_1}(t, t_1, \ldots, t_l) = \left(A^\Delta_{\theta \vartheta'} + B^{\Delta, \omega}_{\theta \vartheta'} \right) \Pi^{(l)}_{\vartheta' \vartheta_1}(t, t_1, \ldots, t_l). \tag{4.19}$$

Proof The first part of claim (i) follows by (4.18), (4.12), and (4.14). To prove the second part we apply Proposition 4.2 and (4.13), which yields (4.19). $\qquad \square$

Proof of Lemma 4.3 Take any $T < T(\vartheta_2, \vartheta_1)$ and then pick $\theta \in (\vartheta_1, \vartheta_2]$ and a positive $\delta < \theta - \vartheta_1$ such that

$$T < T_\delta := \frac{(\theta - \vartheta_1 - \delta)e^{-\vartheta_2}}{\omega + 2\langle b \rangle}.$$

For this δ, take $\Pi^{(l)}_{\theta \vartheta_1}$ as in (4.18), and then set

$$Q^{(n)}_{\theta \vartheta_1}(t) = S_{\theta \vartheta_1}(t) \tag{4.20}$$

$$+ \sum_{l=1}^{n} \int_0^t \int_0^{t_1} \cdots \int_0^{t_{l-1}} \Pi^{(l)}_{\theta \vartheta_1}(t, t_1, \ldots, t_l) dt_l \cdots dt_1, \quad n \in \mathbb{N}.$$

By (4.12), (4.14), and (4.17) the operator norm of (4.18) satisfies

$$\| \Pi^{(l)}_{\theta \vartheta_1}(t, t_1, \ldots, t_l; \mathbb{B}) \| \leq \left(\frac{l}{eT_\delta} \right)^l, \tag{4.21}$$

holding for all $l = 1, \ldots, n$. This yields in (4.20)

$$\| Q^{(n)}_{\theta \vartheta_1}(t) - Q^{(n-1)}_{\theta \vartheta_1}(t) \| \leq \frac{1}{n!} \left(\frac{n}{e} \right)^n \left(\frac{T}{T_\delta} \right)^n,$$

which implies

$$\forall t \in [0, T] \quad Q^{(n)}_{\theta \vartheta_1}(t) \to Q_{\theta \vartheta_1}(t) \in \mathcal{L}(\mathcal{K}_{\vartheta_1}, \mathcal{K}_\theta), \quad \text{as } n \to +\infty.$$

This proves claim (i). The estimate in (4.15) follows from that in (4.21). Now by (4.18), (4.13), and (4.19) we obtain

$$\frac{d}{dt} Q^{(n)}_{\vartheta_2 \vartheta_1}(t) = \left(A^\Delta_{\vartheta_2 \theta} + B^{\Delta,\omega}_{\vartheta_2 \theta} \right) Q^{(n)}_{\theta \vartheta_1}(t) + \left(C^{\Delta,\omega}_{\vartheta_2 \theta} + D^\Delta_{\vartheta_2 \theta} \right) Q^{(n-1)}_{\theta \vartheta_1}(t), \quad n \in \mathbb{N}.$$

Then the continuous differentiability of the limit and (4.16) follow by standard arguments.   □

Now let k_t be as in (4.1). Then by (3.14) and (4.16) we conclude that it has all the properties assumed in Definition 3.2 and hence solves (3.15). Then to complete the proof of Theorem 3.3 we have to show that this is a unique solution.

4.3   Proving the Uniqueness

Since the problem in (3.15) is linear, it is enough to show that its version with the zero initial condition has only the zero solution. Let $u_t \in \mathcal{K}_\vartheta$ be a solution of this version. Take some $\vartheta' > \vartheta$ and then $t > 0$ such that $t < T(\vartheta', \vartheta_0)$. Clearly, u_t solves (3.15) also in $\mathcal{K}_{\vartheta'}$. Thus, it can be written down in the following form

$$u_t = \int_0^t S_{\vartheta' \vartheta''}(t - s) \left(C^{\Delta,\omega}_{\vartheta'' \vartheta} + D^\Delta_{\vartheta'' \vartheta} \right) u_s ds, \tag{4.22}$$

where u_t on the left-hand side (resp. u_s on the right-hand side) is considered as an element of $\mathcal{K}_{\vartheta'}$ (resp. $\mathcal{K}_\vartheta$) and $\vartheta'' \in (\vartheta, \vartheta')$. Let us show that for all $t < T(\vartheta, \vartheta_0)$, $u_t = 0$ as an element of $\mathcal{K}_\vartheta$. In view of the embedding $\mathcal{K}_\vartheta \hookrightarrow \mathcal{K}_{\vartheta'}$, this will follow from the fact that $u_t = 0$ as an element of $\mathcal{K}_{\vartheta'}$. For a given $n \in \mathbb{N}$, we set $\epsilon = (\vartheta' - \vartheta)/2n$ and $\alpha^l = \vartheta + l\epsilon, l = 0, \ldots, 2n$. Then we repeatedly apply (4.22) and obtain

$$u_t = \int_0^t \int_0^{t_1} \cdots \int_0^{t_{n-1}} S_{\vartheta' \vartheta^{2n-1}}(t - t_1) \left(C^{\Delta,\omega}_{\vartheta^{2n-1} \vartheta^{2n-2}} + D^\Delta_{\vartheta^{2n-1} \vartheta^{2n-2}} \right)$$
$$\times \cdots \times S_{\vartheta^2 \vartheta^1}(t_{n-1} - t_n) \left(C^{\Delta,\omega}_{\vartheta^1 \vartheta} + D^\Delta_{\vartheta^1 \vartheta} \right) u_{t_n} dt_n \cdots dt_1.$$

Like in (4.21), we then get from the latter

$$\|u_t\|_{\vartheta'} \leq \frac{t^n}{n!} \prod_{l=1}^{n} \|C^{\Delta,\omega}_{\vartheta 2l-1,\vartheta 2l-2} + D^{\Delta}_{\vartheta 2l-1,\vartheta 2l-2}\| \sup_{s\in[0,t]} \|u_s\|_{\vartheta}$$

$$\leq \frac{1}{n!} \left(\frac{n}{e}\right)^n \left(\frac{2t(\omega + 2\langle b\rangle)e^{\vartheta'}}{\vartheta' - \vartheta}\right)^n \sup_{s\in[0,t]} \|v_s\|_{\vartheta}.$$

This implies that $u_t = 0$ for $t < (\vartheta' - \vartheta)/2(\omega + 2\langle b\rangle)e^{\vartheta'}$. To prove that $u_t = 0$ for all t of interest one has to repeat the above procedure appropriate number of times.

Acknowledgements The present research was supported by the European Commission under the project STREVCOMS PIRSES-2013-612669.

References

1. S. Albeverio, Y.G. Kondratiev, M. Röckner, Analysis and geometry on configuration spaces. J. Funct. Anal. **154**, 444–500 (1998)
2. J. Barańska, Y. Kozitsky, Free jump dynamics in continuum. Contemp. Math. **653**, 13–23 (2015)
3. J. Barańska, Y. Kozitsky, The global evolution of states of a continuum Kawasaki model with repulsion. arXiv:1509.02044 (Preprint, 2016)
4. J. Barańska, Y. Kozitsky, A Widom-Rowlinson jump dynamics in the continuum. arXiv:1604.07735 (Preprint, 2016)
5. C. Berns, Y. Kondratiev, Y. Kozitsky, O. Kutoviy, Kawasaki dynamics in continuum: micro- and mesoscopic descriptions. J. Dyn. Diff. Equat. **25**, 1027–1056 (2013)
6. C. Berns, Y. Kondratiev, O. Kutoviy, Markov jump dynamics with additive intensities in continuum: state evolution and mesoscopic scaling. J. Stat. Phys. **161**, 876–901 (2015)
7. Y. Kondratiev, Y. Kozitsky, The evolution of states in a spatial population model. J. Dyn. Diff. Equat. (2016). https://doi.org/10.1007/s10884-016-9526-6
8. Y. Kondratiev, T. Kuna, Harmonic analysis on configuration space. I. General theory. Infin. Dimens. Anal. Quantum Probab. Relat. Top. **5**, 201–233 (2002)
9. Y. Kozitsky, Dynamics of spatial logistic model: finite systems, in *Semigroups of Operators–Theory and Applications: Beedlewo, Poland, October 2013*, ed. by J. Banasiak, A. Bobrowski, M. Lachowicz. Springer Proceedings in Mathematics & Statistics, vol. 113 (Springer, Berlin, 2015), pp. 197–211
10. A. Pazy, *Semigroups of Linear Operators and Applications to Partial Differential Equations*. Applied Mathematical Sciences, vol. 44 (Springer, New York, 1983)
11. D. Ruelle, *Statistical Mechanics: Rigorous Results* (W.A. Benjamin, Inc., New York, 1969)
12. H.R. Thieme, J. Voigt, Stochastic semigroups: their construction by perturbation and approximation, in *Positivity IV – Theory and Applications*, ed. by M.R. Weber, J. Voigt (Tech. Univ. Dresden, Dresden, 2006), pp. 135–146

Problems on Weighted and Unweighted Composition Operators

Valentin Matache

Abstract This paper contains a collection of open problems related to composition operators and weighted composition operators acting on spaces of analytic functions, predominantly on the Hilbert Hardy space H^2 over the open unit disc. Some are related to the invariant subspaces of composition operators, some to the spectra and numerical ranges of such operators, others are related to the connection of certain weighted composition on H^2 and the unweighted composition operators on Hardy–Smirnov spaces, or the connection of composition operators with asymptotically Toeplitz operators. The problems raised are open to the knowledge of this author, and interesting, in his opinion.

Keywords Composition operators • Weighted composition operators • Asymptotic Toeplitz operators

2010 Mathematics Subject Classification Primary 47B33; Secondary 47B35

1 Introduction

This paper presents developments in the study of weighted or "unweighted" composition operators and a collection of open problems which resulted in the course of obtaining the results that make the object of those developments. While many other open problems related to that topic surely exist, we selected those that this author considers interesting, given his work in that field. This first section is dedicated to introducing the main concepts and basic notation.

Given a space S of scalar-valued functions, defined on a subset D of K, the field of scalars (usually $K = \mathbb{R}$ or $K = \mathbb{C}$), one of the basic operations on K, namely

V. Matache (✉)
Department of Mathematics, University of Nebraska, Omaha, NE 68182, USA
e-mail: vmatache@unomaha.edu

© Springer International Publishing AG 2018
M. Agranovsky et al. (eds.), *Complex Analysis and Dynamical Systems*,
Trends in Mathematics, https://doi.org/10.1007/978-3-319-70154-7_11

multiplication, naturally induces linear operators defined on S as follows:

$$M_\psi f := \psi f \qquad f \in S. \tag{1.1}$$

The map ψ in (1.1) is fixed. It has domain D and is called the *symbol* of M_ψ, the *multiplication operator* induced by ψ.

Composition of functions is another elementary operation that induces linear operators on function spaces, as follows:

$$C_\varphi f := f \circ \varphi \qquad f \in S. \tag{1.2}$$

The map φ in (1.2) is also fixed. It is a selfmap of D and is called the *symbol* of C_φ, the *composition operator* induced by φ. A composition operator followed by a multiplication operator is customarily called a *weighted composition operator*. More formally, the operator

$$T_{\psi,\varphi} := M_\psi C_\varphi \tag{1.3}$$

is called the *weighted composition operator* with *weight-symbol* ψ and *composition-symbol* φ. If λ is a constant, then clearly, $T_{\lambda,\varphi} = \lambda C_\varphi$. On the other hand, if φ is the identity, then $T_{\psi,\varphi} = M_\psi$. We will say a weighted composition operator is *nontrivial*, in all the other cases (that is when that operator is not a scalar multiple of a composition operator or a multiplication operator). Various notations for weighted composition operators are currently used. For instance, instead of $T_{\psi,\varphi}$ some authors use the symbol $W_{\psi,\varphi}$, with W staying probably for *weight*. Others use ψC_φ instead of $T_{\psi,\varphi}$. In our notation, T stays for Toeplitz, since multiplication operators are historically related to the name of Otto Toeplitz, and the first operator one reads left to right in definition (1.3) is a multiplication operator. To clarify why multiplication operators are related to O. Toeplitz recall that, if one considers $S = H^2$, where H^2 denotes the Hilbert Hardy space on the open unit disc $\mathbb{U} = \{z \in \mathbb{C} : |z| < 1\}$, one can easily introduce the operators originally studied by Toeplitz. What we mean here is that one takes $K = \mathbb{C}, D = \mathbb{U}$ and calls H^2 the space of all functions f which are analytic in $\mathbb{U}$ and satisfy the mean growth condition

$$\|f\|_2 := \sup \left\{ \sqrt{\int_\mathbb{T} |f(ru)|^2 \, dm(u)} : 0 \le r < 1 \right\} < \infty, \tag{1.4}$$

where m denotes the normalized arclength measure on the unit circle $\mathbb{T} = \partial\mathbb{U}$. The quantity $\| \quad \|_2$ is a Hilbert norm on H^2 which can be calculated with the alternate formula

$$\|f\|_2 = \sqrt{\sum_{n=0}^\infty |c_n|^2},$$

where $\{c_n\}$ is the sequence of Maclaurin coefficients of f. If one substitutes 2 by $1 \leq p < \infty$ in the above construction, one obtains the Hardy spaces H^p, which are only Banach spaces if $p \neq 2$. The notation H^∞ designates the Banach algebra of bounded analytic functions on $\mathbb{U}$ endowed with the supremum norm $\| \ \|_\infty$. A classical result of Hardy space theory says that, given a fixed $f \in H^p$, $1 \leq p \leq \infty$, then for almost all $u \in \mathbb{U}$, the radial limit $f(u) := \lim_{r \to 1^-} f(ru)$ exists. The radial limit function obtained that way is also denoted f and belongs to the Lebesgue space $L^p_{\mathbb{T}}(dm)$. Actually the map transforming $f \in H^p$ into its radial limit function is an isometric embedding of H^p into $L^p_{\mathbb{T}}(dm)$. That embedding allows one to denote by the same symbol a Hardy space function and its radial limit function, rely on the context in order to distinguish between them, and write $H^p \subseteq L^p_{\mathbb{T}}(dm)$. Under the circumstances, the projection P of H^2 onto $L^2_{\mathbb{T}}(dm)$ can be considered. Given a complex valued, essentially bounded, measurable function ψ on $\mathbb{T}$, one introduces the *Toeplitz operator* T_ψ with symbol ψ as follows:

$$T_\psi f = P\psi f \qquad f \in H^2. \tag{1.5}$$

If $\psi \in H^\infty$, then clearly, $T_\psi = M_\psi$, and this kind of operator is traditionally called an *analytic* Toeplitz operator. All composition operators on H^2 are bounded (a well known fact). So are the Toeplitz operators. In order that an operator of type $T_{\psi,\varphi}$ be bounded on H^2, one must require that $\psi \in H^2$, since $\psi = T_{\psi,\varphi}(1)$, but that necessary condition is not also sufficient, that is, if $\psi \in H^2$, it does not always follow that $T_{\psi,\varphi}$ is bounded for all analytic selfmaps φ of $\mathbb{U}$. We will introduce more terminology in posterior sections as necessary.

Composition operators appeared implicitly in works of B.O. Koopman dealing with mechanics published in the 1930s. Their systematic study began with papers [62] and [56] and remained an active field of research ever since. While there are several environments where all that took place, composition operators on spaces of analytic functions were by far, the most popular, their study exhibiting a nice interplay of functional analysis, measure theory, and complex analysis. The main directions of research were compactness, the study of the spectrum, and more recently, that of invariant subspaces. A good account of what was achieved is to be found in the two most popular monographs on composition operators [66] and [21]. Although weighted composition operators appeared as the isometries of non-Hilbert Hardy spaces [25], as early as 1964, they were not called weighted composition operators at that time, and their systematic study (on spaces of analytic functions), began rather recently, namely less than 10 years ago. It followed pretty much the same directions as the study of composition operators, but it is far from being a well developed body of knowledge.

We conclude the current introductory section by a brief outline of the other sections. Those sections contain problems on composition operators or weighted composition operators as follows.

Section 2 contains spectral problems. Section 3 contains problems related to numerical ranges. Section 4 deals with invariant subspaces of composition operators. In Sect. 5 we raise some problems related to weighted composition operators in general and composition operators on Hardy–Smirnov spaces in particular. In the last section, Sect. 6, we report on and raise questions about, composition operators which are asymptotically Toeplitz.

2 Spectral Problems

The study of spectra of composition operators started with [56], a paper where the spectra of automorphic composition operators acting on H^2 are determined. By *automorphic* composition operator, we mean of course, a composition operator whose symbol is a disc automorphism.

2.1 Maps Fixing a Point in $\mathbb{U}$

As one knows, the disc automorphisms with a unique fixed point in $\mathbb{U}$ are called *elliptic*. Denote by $\{\varphi^{[n]}\}$ the sequence of iterates of φ and recall the following [21]:

Theorem 1 (Denjoy–Wolff) *Let φ be an analytic selfmap of $\mathbb{U}$ other than the identity or an elliptic disc automorphism. Then the sequence of iterates $\{\varphi^{[n]}\}$ converges uniformly on compacts to a point $\omega \in \overline{\mathbb{U}}$ called the Denjoy–Wolff point of φ.*

An immediate consequence is the fact that an analytic selfmap φ of $\mathbb{U}$, other than the identity, can have at most one fixed point in $\mathbb{U}$.

For any composition operator, the constant function 1 is obviously an eigenfunction associated to the eigenvalue 1, by the obvious equality $1 \circ \varphi = 1$. If φ, not the identity, fixes $\omega \in \mathbb{U}$, then, by a simple computation, one establishes that the only other eigenvalues C_φ can possibly have are $(\varphi'(\omega))^n$, $n = 1, 2, \ldots$. We call those complex numbers the *Schröder eigenvalues* of C_φ, even when they are not eigenvalues of C_φ (which can happen). The reason for this terminology is E. Schröder's formulation of the eigenvalue functional equation for composition operators [63]. It is known that the Schröder eigenvalues always belong to the spectrum $\sigma(C_\varphi)$ of C_φ [21]. Denote by $\sigma_e(C_\varphi)$ the essential spectrum of C_φ. Combining the above considerations with G. Koenig's theorem on Schröder's equation [37] (see also Theorem 7 in this paper), one easily gets:

Remark 1 ([49, Remark 1]) Let φ be a non-automorphic analytic selfmap of $\mathbb{U}$ with a fixed point $\omega \in \mathbb{U}$. If $\sigma_e(C_\varphi)$ is simply connected then

$$\sigma(C_\varphi) = \sigma_e(C_\varphi) \cup \{(\varphi'(\omega))^n : n = 1, 2, \ldots\} \cup \{1\}. \qquad (2.1)$$

Equality (2.1) also holds if $\sigma_p(C_\varphi^*) \subseteq \{0, 1\}$ where $\sigma_p(C_\varphi^*)$ denotes the point spectrum of the adjoint of C_φ.

Thus, if one can find the essential spectrum of C_φ, the composition operator induced by some non-automorphic symbol φ fixing a point in $\mathbb{U}$, and $\sigma_e(C_\varphi)$ is simply connected, then the spectrum of C_φ can be easily obtained by adding 1 and the Schröder eigenvalues to the essential spectrum.

One trivial situation when $\sigma_e(C_\varphi)$ is simply connected is, of course, when C_φ is essentially quasinilpotent (that is C_φ has null essential spectral radius). Such composition operators are necessarily induced by non-automorphic symbols fixing a point [8].

There are two other well known cases when the spectrum of C_φ is the union of a closed disc centered at the origin, the set of Schröder eigenvalues, and 1. The first is when φ is non-automorphic, fixes a point, and has an analytic extension on an open neighborhood of the closed unit disc. The proof is due to Kamowitz [35]. See also [18]. The second case is when φ is univalent, non-automorphic, with a fixed point [20]. The circular disc involved in the aforementioned description of spectra is not known to be the essential spectrum of C_φ, though; not in all cases.

Relative to all that, let φ be a non-automorphic, analytic selfmap of $\mathbb{U}$ having a fixed point in $\mathbb{U}$. For such φ, one formulates the following:

Problem 1 Is $\sigma_e(C_\varphi)$ always simply connected? If so, is $\sigma_e(C_\varphi)$ always a disc centered at the origin?

The above problem is related to the following one, formulated in [5].

Problem 2 If the spectrum of C_φ equals the union of a circular disc centered at the origin, the Schröder eigenvalues, and 1, is that circular disc always equal to $\sigma_e(C_\varphi)$?

There are particular cases when $\sigma_e(C_\varphi)$ is known to be a disc (possibly degenerate) centered at the origin. Besides the obvious case of essentially quasinilpotent composition operators, this author proved that *symbols having orthogonal powers* (that is analytic selfmaps φ of $\mathbb{U}$ with the property that $\{1, \varphi, \varphi^2, \varphi^3, \ldots\}$ is an orthogonal subset of H^2) induce composition operators with circular essential spectrum and therefore their spectrum is given by (2.1) [45].

Also, P. Bourdon proves the answer to Problem 2 is affirmative in the particular case when φ is an essentially linear fractional symbol, fixing a point in $\mathbb{U}$ [5].

Let φ be a non-automorphic analytic selfmap of $\mathbb{U}$. Consider the *pull-back* measure $m\varphi^{-1}$ induced by φ, that is the Borel measure

$$m\varphi^{-1}(E) = m(\varphi^{-1}(E)),$$

and denote by ψ the Nikodym derivative $\psi := \frac{dm\varphi^{-1}}{dm}$. A recent result of this author is:

Proposition 1 ([49]) *Let φ be a non-automorphic analytic selfmap of $\mathbb{U}$ with a fixed point $\omega \in \mathbb{U}$. If $r_e(C_\varphi) \leq \sqrt{\text{ess inf } \psi}$, then*

$$\sigma(C_\varphi) = (\sqrt{\text{ess inf } \psi})\overline{\mathbb{U}} \cup \{(\varphi'(\omega))^n : n = 1, 2, \ldots\} \cup \{1\} \qquad (2.2)$$

and

$$\sigma_e(C_\varphi) = (\sqrt{\text{ess inf } \psi})\overline{\mathbb{U}}, \tag{2.3}$$

where $r_e(C_\varphi)$ denotes the essential spectral radius of C_φ.

The main known examples of symbols φ satisfying condition

$$r_e(C_\varphi) \leq \sqrt{\text{ess inf } \psi} \tag{2.4}$$

are the symbols with orthogonal powers and the symbols inducing essentially quasinilpotent operators. These ideas lead to formulating the following:

Problem 3 Exactly which analytic selfmaps φ of $\mathbb{U}$ satisfy condition (2.4)?

2.2 Maps Without Fixed Points in $\mathbb{U}$

The disc automorphisms with two distinct fixed points situated on $\mathbb{T}$ are called *hyperbolic* disc automorphisms. The Denjoy–Wolff point ω of such an automorphism φ is called the *attractive* fixed point of φ (as opposed to the other fixed point, which is called *repulsive*). One can show that $0 < \varphi'(\omega) < 1$, which means that $\varphi'(\omega)$ is a real number satisfying the previous inequalities. The disc automorphisms with exactly one fixed point situated on $\mathbb{T}$ are called *parabolic*. If φ is a parabolic disc automorphism and $\omega \in \mathbb{T}$ its fixed point, then $\varphi'(\omega) = 1$. The generalization of this circle of ideas leads to the introduction of angular derivatives.

Recall that the analytic selfmap φ of $\mathbb{U}$ has an angular derivative at $\omega \in \mathbb{T}$ if there is some $\eta \in \mathbb{T}$ so that the difference quotient $(\eta - \varphi(z))/(\omega - z)$ converges to the same value $\varphi'(\omega) \in \mathbb{C}$ as z tends to ω trough any angular approach region with vertex at ω. In that case, η is the angular limit of φ at ω. Whenever the radial limit at $\omega \in \mathbb{T}$ of some analytic selfmap φ of $\mathbb{U}$ exists and equals ω, one calls ω a *boundary fixed point* of φ. The following well known property leads to a trichotomic classification of analytic selfmaps of $\mathbb{U}$, other than the identity [21].

Theorem 2 *Let φ be an analytic selfmap of $\mathbb{U}$, without fixed points in $\mathbb{U}$. Then ω, the Denjoy–Wolff point of φ, is the unique boundary fixed point of φ where the angular derivative $\varphi'(\omega)$ exists and satisfies the condition $\varphi'(\omega) \leq 1$.*

For that reason, if φ is an analytic selfmap of $\mathbb{U}$, other than the identity, exactly one of the following is valid:

(A) The map φ has exactly one fixed point in $\mathbb{U}$.
(B) The map φ has no fixed points in $\mathbb{U}$ and satisfies condition $\varphi'(\omega) < 1$ at its Denjoy–Wolff point, $\omega \in \mathbb{T}$.
(C) The map φ has no fixed points in $\mathbb{U}$ and satisfies condition $\varphi'(\omega) = 1$ at its Denjoy–Wolff point, $\omega \in \mathbb{T}$.

If condition (B) holds, the map φ is called a map of *hyperbolic* type, whereas if φ satisfies (C) we call φ a map of *parabolic* type. The spectrum of composition operators induced by maps of parabolic or hyperbolic type, is known only in select cases. In all those cases, it coincides with the essential spectrum. This leads Paul Bourdon to raising the following [5]:

Problem 4 If φ is an analytic selfmap of parabolic or hyperbolic type, is it true that $\sigma(C_\varphi) = \sigma_e(C_\varphi)$?

2.3   Complex Dynamics and Spectra

The space H^2 is a space where the weak convergence of a sequence $\{f_n\}$ to some $f \in H^2$ is equivalent to the fact that the sequence $\{f_n\}$ is norm-bounded and convergent to f uniformly on compact subsets. Thus, the Denjoy–Wolff theorem says that the sequence $\{\varphi^{[n]}\}$ of the iterates of an analytic selfmap φ of $\mathbb{U}$ (other than the identity or an elliptic automorphism) tends weakly to some $\omega \in \overline{\mathbb{U}}$. The behavior of sequences of iterates of analytic selfmaps belongs to the field of *complex dynamics*. A normal question here, formulated from the point of view of that body of knowledge would be, does the Denjoy–Wolff point of φ attract the sequence $\{\varphi^{[n]}\}$ in a stronger sense, than just weakly? The answer is affirmative with an obvious exception:

Theorem 3 ([12]) *If φ is an analytic selfmap of $\mathbb{U}$ other than an inner function with a fixed point in $\mathbb{U}$, then $\|\varphi^{[n]} - \omega\|_2 \to 0$, where ω is the Denjoy–Wolff point of φ.*

It should be recalled that a map in H^∞ is called an *inner* function if it is unimodular a.e. on $\mathbb{T}$. Recently, this author used the above strong form of the Denjoy–Wolff theorem to prove that strong forms of attractiveness to the Denjoy–Wolff point influence the spectra of composition operators:

Proposition 2 ([49, Corollary 3]) *If φ is a non-inner analytic selfmap of $\mathbb{U}$, fixing a point $\omega \in \mathbb{U}$, and*

$$\limsup_{n\to+\infty} \sqrt[n]{\|\varphi^{[n]} - \omega\|_2} = 0, \qquad (2.5)$$

then $\sigma(C_\varphi)$ satisfies equality (2.1).

Analytic selfmaps satisfying condition (2.5) were shown to be maps for which ω must be a *hyperattracting* fixed point (a term used in complex dynamics which means that besides $\varphi(\omega) = \omega$, one also has that $\varphi'(\omega) = 0$). The major example known to this author of maps satisfying condition (2.5) is that of maps inducing essentially quasinilpotent composition operators. All these considerations raise the following:

Problem 5 If φ satisfies (2.5), must C_φ be essentially quasinilpotent? If not, characterize the selfmaps φ which satisfy condition (2.5) and find $\sigma_e(C_\varphi)$ in each case.

Finally, the class of maps φ inducing essentially quasinilpotent composition operators is repeatedly mentioned in this section. However, its exact characterization is not known and so, please note that Problem 5 hints to the following:

Problem 6 What is the description of the class of analytic selfmaps φ of $\mathbb{U}$ inducing essentially quasinilpotent composition operators C_φ?

Some comments are in order here. The answer to the above problem will describe a superset of the set of analytic selfmaps of $\mathbb{U}$ inducing compact composition operators. It is known that noncompact essentially quasinilpotent composition operators exist [8] and so, that superset does not coincide with the set of symbols which induce compact composition operators (which has known characterizations [65] respectively [17]).

3 Numerical Ranges

The numerical range $W(T)$ of a Hilbert space operator T is the (necessarily convex) set $W(T) := \{\langle Tx, x \rangle : \|x\| = 1\}$. The closure of $W(T)$ is known to contain the spectrum $\sigma(T)$ of T so, the description of the numerical range of an operator is related to the description of its spectrum, but is known as being significantly harder to get, in most cases. The latter statement is illustrated by the quest for describing numerical ranges of composition operators, where very few things are currently known. There is a small collection of papers addressing this issue (for composition operators acting on H^2): [1, 2, 9, 10, 41, 46]. What is known about the shape and size of the numerical range of a composition operator can be briefly described as follows. If φ is an analytic selfmap of $\mathbb{U}$ with a fixed point then the numerical range of C_φ can exhibit quite a variety of shapes [41]. If φ is inner, not the identity or a rotation, and the fixed point is the origin, then $W(C_\varphi) = \mathbb{U} \cup \{1\}$ [41]. The case when $\varphi(z) = \lambda z$, $|\lambda| = 1$, is very easy to handle, leading to a regular polygon inscribed in $\mathbb{T}$ if λ is a root of 1, respectively to $\mathbb{U} \cup \{\lambda^n : n \geq 0\}$, when λ is not a root of 1 [41]. There are only two kinds of quadratic composition operators: those with constant symbol and those having symbol of the form $\alpha_p(z) = (p-z)/(1-\bar{p}z)$, where $p \in \mathbb{U}$. They have elliptical numerical ranges (possibly degenerate, that is the elliptical disk can be reduced to the focal axis). The elliptical disk is closed in the case of constant symbols and open in the other case [42]. Composition operators of symbol α_p are automorphic composition operators whose symbol fixes a point (also known as elliptic automorphisms). For any elliptic automorphic symbol, that symbol is conformally conjugated to a rotation. If the fixed point is not the origin and the aforementioned rotation is not by a root of unity, then $\overline{W(C_\varphi)}$ is a disk about the origin, [9, Theorem 4.1], but the exact description of $W(C_\varphi)$ is currently unknown. Composition operators whose symbols are monomials fixing the origin can have polygonal or cone-like numerical ranges [41]. Composition operators whose symbol is an inner map of hyperbolic or parabolic automorphic type have circular numerical ranges [46]. It should be added that maps of parabolic type are classified as maps of *parabolic automorphic* type respectively *parabolic non-automorphic* type based on the separation of their orbits. More formally, let

$$\rho(z, w) = \left| \frac{w - z}{1 - \overline{w}z} \right| \qquad z, w \in \mathbb{U}$$

denote the pseudohyperbolic distance. Given φ an analytic selfmap of $\mathbb{U}$ of parabolic type, either

$$\lim_{n \to \infty} \rho(\varphi^{[n+1]}(z), \varphi^{[n]}(z)) > 0 \qquad z \in \mathbb{U} \tag{3.1}$$

or

$$\lim_{n \to \infty} \rho(\varphi^{[n+1]}(z), \varphi^{[n]}(z)) = 0 \qquad z \in \mathbb{U}. \tag{3.2}$$

Parabolic automorphisms have property (3.1), for which reason we call any analytic selfmap φ of $\mathbb{U}$ of parabolic type, a map of *parabolic automorphic type*, if φ has property (3.1), respectively, a map of *parabolic non-automorphic type*, if φ has property (3.2).

If T is a Hilbert space operator, the quantity $w(T) := \sup\{|\lambda| : \lambda \in W(T)\}$ is called the *numerical radius* of T. Very few numerical radius computations for composition operators exist. The simplest is $w(C_\varphi) = 1$ if $\varphi(0) = 0$. However, if φ fixes a point other than the origin, the numerical radius of C_φ is unknown in most cases. So, we formulate:

Problem 7 Find $w(C_\varphi)$ when φ is not the identity and there is $p \in \mathbb{U}$ so that $p \neq 0$ and $\varphi(p) = p$. Describe $W(C_\varphi)$.

Also, the authors of [9] obtained the only numerical radius computation known for hyperbolic automorphic composition operators: $w(C_\varphi) = \sqrt{\frac{1+|\varphi(0)|}{1-|\varphi(0)|}}$ if φ is a hyperbolic disc automorphism with antipodal fixed points.

Problem 8 What is the numerical radius of a hyperbolic automorphic composition operator when the fixed points of its symbol are not antipodal?

More generally, it was mentioned that composition operators with inner symbols of parabolic automorphic type, respectively hyperbolic type, have circular numerical ranges centered at the origin [46]. It is known that the numerical radius satisfies relation $w(C_\varphi) > 1$, but the exact value of $w(C_\varphi)$ is not known, except the case of hyperbolic automorphisms with antipodal fixed points. Nor is it known when the circular numerical range is an open disc or a closed one (again with the exception of hyperbolic automorphisms with antipodal fixed points, when the numerical range is open). It is natural to ask:

Problem 9 Given φ an inner function of hyperbolic or parabolic automorphic type (other than a hyperbolic automorphisms with antipodal fixed points), what is the value of $w(C_\varphi)$? When is $W(C_\varphi)$ open and when closed?

4 Invariant Subspaces

4.1 The Invariant Subspace Problem via Composition Operators

In this section, we call *operator* any bounded linear transformation of a Hilbert space into itself. By *invariant subspace* of an operator T, we mean a closed, linear manifold, left invariant by T. The collection $\mathrm{Lat}T$ of all invariant subspaces of T is a lattice (which is why we use notation $\mathrm{Lat}T$). If T is an operator, then the trivial elements of $\mathrm{Lat}T$ are the null subspace and the whole space. The following is a famous unsolved problem called the *invariant subspace problem*.

Problem 10 Does any Hilbert space operator acting on a complex, separable, infinite dimensional space, always have nontrivial invariant subspaces?

That problem has multiple equivalent reformulations [15, 58, 59]. One of them is in terms of automorphic, hyperbolic, composition operators. Let C_φ be such a composition operator. It was proved in [57] that

Theorem 4 ([57]) *The answer to* Problem 10 *is affirmative if and only if the only atoms contained by* $\mathrm{Lat}C_\varphi$ *are the* 1*-dimensional eigenspaces.*

By *atom* of $\mathrm{Lat}T$ or *minimal invariant subspace* of T, one means any nonzero subspace $\mathcal{L} \in \mathrm{Lat}T$ so that the restriction $T|\mathcal{L}$ has trivial invariant subspace lattice, that is $\mathrm{Lat}(T|\mathcal{L}) = \{0, \mathcal{L}\}$.

Given T a selfmap of a set and x an element of that set we call

$$O_T(x) := \{x, T(x), T(T(x)), T(T(T(x))), \dots\}$$

the *orbit* of x under T. If T is invertible, then

$$BO_T(x) := O_T(x) \cup O_{T^{-1}}(x)$$

is called the *bilateral orbit* of x under T (or under T^{-1}). This author made some easy remarks on the atoms in $\mathrm{Lat}T$ when T is invertible (as is the operator in Theorem 4) [38]. More exactly, it is easy to prove that an atom of $\mathrm{Lat}T$ is doubly invariant, that is, it is invariant under both T and T^{-1}. Then, let us use the notation

$$K_x^+ \quad K_x^-, \quad \text{and} \quad K_x \tag{4.1}$$

for the closed subspace spanned by $O_T(x)$, $O_{T^{-1}}(x)$, respectively $BO_T(x)$. If $\mathcal{L}$ is an atom of $\mathrm{Lat}T$ then $\mathcal{L} = K_x^+ = K_x^- = K_x$ $x \in \mathcal{L}$.

If $T = C_\varphi$ where φ is a hyperbolic automorphism, and $x = f \in H^2$, denote by $\mathcal{L}_f$ any of the spaces described in (4.1). This author raised the following [38]:

Problem 11 Given $f \in H^2$, can one tell, by the properties of f if $\mathcal{L}_f$ is an atom of $\mathrm{Lat}C_\varphi$ or not?

Since the only known atoms are, so far, the 1-dimensional eigenspaces, this led to the characterization of eigenfunctions of C_φ as follows: the inner eigenfunctions were characterized in [39] (see also [53] for earlier partial characterizations) and the outer eigenfunctions in [28] (see also an alternative point of view on inner eigenfunctions in [28]). The other direction followed in order to obtain partial answers to Problem 11, was identifying classes of functions $f \in H^2$ which should be ruled out, since they generate nonminimal invariant subspaces K_f. Most of those results can be found in [16, 26–28, 38, 39, 53, 54], and [69]; see [51], for a detailed description. In general, the non-minimality of an invariant subspace K_f is proved if $f \neq 0$ has nonzero nontangential limits at the fixed points of φ and is essentially bounded on open arcs containing those points or when f is extensible by continuity at least one of the fixed points of φ and either the extension is by a nonzero value or the extension is by 0 but then, some strong form of continuity is satisfied at that point. As noted in [51], two problems are nearly solved here, but not in full generality, namely:

Problem 12 If φ is a hyperbolic automorphism and u is inner, or more generally if $u \in H^\infty \setminus \{0\}$, then K_u is minimal invariant if and only if u is an eigenfunction of C_φ. True or false?

The other problem we want to raise here is:

Problem 13 If φ is a hyperbolic automorphism and $f \in H^2 \setminus \{0\}$, is continuously extendable at the fixed points of φ (or more generally, if f has a finite nontangential limit at each of the two fixed points of φ), then K_f is a minimal invariant subspace of C_φ if and only if f is a nonzero constant function. True or false?

As observed in [51], only functions continuously extendable by 0 at both fixed points could possibly generate a minimal invariant subspace K_f. However no such functions can be eigenfunctions of C_φ and if, the kind of continuity at the fixed points is "stronger" than mere continuity then, K_f is not minimal invariant because the restriction $C_\varphi | K_f$ has a large point spectrum.

4.2   Aleksandrov Operators

In [48], this author observed that an important theorem in geometric function theory, the Julia–Carathéodory theorem can be reformulated in terms of the action of a composition operator on the invariant subspace lattice of M_z, the multiplication operator with the coordinate function. The details are as follows. For any fixed unimodular constant λ and any nonnegative, finite, Borel measure μ, on the unit circle $\mathbb{T}$, which is singular with respect to the arclength measure denote

$$\lambda S_\mu(z) = \lambda e^{-\int_\mathbb{T}(u+z)/(u-z)\,d\mu(u)} \qquad z \in \mathbb{U}. \tag{4.2}$$

Any function of type (4.2) is called a *singular inner function*. If $\omega \in \mathbb{T}$, then δ_ω denotes the unit mass concentrated at ω. With this notation, the following is proved in [48].

Theorem 5 *The analytic selfmap φ of $\mathbb{U}$ has an angular derivative at $\omega \in \mathbb{T}$ if and only if there is some $\eta \in \mathbb{T}$ and $p > 0$ so that*

$$C_\varphi \left(S_{p\delta_\eta} H^2 \right) \subseteq S_{\delta_\omega} H^2. \tag{4.3}$$

If condition (4.3) holds, then η is the angular limit of φ at ω and

$$|\varphi'(\omega)| = \min\{p > 0 : p \text{ satisfies } (4.3)\}. \tag{4.4}$$

The theorem above raises the problem:

Problem 14 If φ is some analytic selfmap φ of $\mathbb{U}$ and μ, ν are nonzero singular measures on $\mathbb{T}$, then what characterization can be written for the condition

$$C_\varphi(S_\mu H^2) \subseteq S_\nu H^2? \tag{4.5}$$

Paper [48] where the problem above was initially raised contains only modest comments and partial answers. That paper also contains a formulation of the Julia–Carathéodory theorem in terms of what this author calls the Aleksandrov operator with symbol φ (where φ is an analytic selfmap of $\mathbb{U}$). Here are some details. For each $\omega \in \mathbb{T}$, we denote by τ_ω, the *Aleksandrov measure with index ω* of φ, that is the measure whose Poisson integral equals $P(\varphi(z), \omega)$ (where $P(z, u), z \in \mathbb{U}, u \in \mathbb{T}$ is the usual Poisson kernel). There exists a unique such measure, by the well-known Herglotz theorem [61, Theorem 11.19].

The *Aleksandrov operator A_φ with symbol φ* is the operator on the space $\mathcal{M}$ of complex Borel measures on $\mathbb{T}$ satisfying:

$$P_{A_\varphi(\mu)} = P_\mu \circ \varphi \qquad \mu \in \mathcal{M},$$

where for all measures ν, P_ν denotes the Poisson integral of ν. All Aleksandrov operators are bounded and the Julia–Crathéodory theorem can be formulated in terms of such operators as follows. Denote by σ_ω the singular part of τ_ω in its Lebesgue decomposition with respect to m. We call the *little Aleksandrov operator with symbol φ*, the operator $a_\varphi(\mu)$ equal to the singular part of $A_\varphi(\mu)$ with respect to m. With the notation above, $a_\varphi(\delta_\omega) = \sigma_\omega$, $\omega \in \mathbb{T}$. Paper [48] contains the following:

Theorem 6 *Condition (4.5) holds if and only if*

$$a_\varphi(\mu) \geq \nu. \tag{4.6}$$

As a consequence, the angular derivative of some analytic selfmap φ of $\mathbb{U}$ exists at some $\omega \in \mathbb{T}$ if and only if there is some $\eta \in \mathbb{U}$ so that

$$[a_\varphi(\delta_\eta)](\{\omega\}) > 0. \tag{4.7}$$

Clearly one has $[a_\varphi(\delta_\eta)](\{\omega\}) = [A_\varphi(\delta_\eta)](\{\omega\})$ and, if (4.7) holds, then η is the angular limit of φ at ω and $[a_\varphi(\delta_\eta)](\{\omega\}) = 1/|\varphi'(\omega)|$.

All these considerations raise the problem:

Problem 15 What conditions on φ are equivalent to condition (4.6)?

In [48], where the problem above was raised, a particular solution in the case of purely atomic singular, measures is presented. However, the problem remains open, in the general case.

4.3 Description of the Invariant Subspace Lattice

Describing up to an order isomorphism the invariant subspace lattice of a composition operator is most interesting when that operator is an automorphic, hyperbolic, composition operator, given Problem 10. While that is a long standing open problem, other composition operators might have invariant subspace lattices easier to understand. In paper [48], this author takes up the problem of describing those invariant subspace lattices and succeeds in some modest cases such as the case when the symbol is an inner map fixing a point and some very basic cases of composition operators induced by non-inner symbols. One of the latter cases raises some interest, namely composition operators induced by maps conformally equivalent to maps of type $\varphi(z) = \lambda z, 0 < |\lambda| < 1$. Such operators C_φ are compact and $\varphi'(p) = \lambda \neq 0$, where $p \in \mathbb{U}$ is their fixed point. For that reason, such operators possess a "Koenigs function" and, as shown in [48, Proposition 1.6], their invariant subspace lattice is spanned by the powers of that function. Here is what we mean here.

We already mentioned that E. Schröder considered the eigenvalue equation for composition operators, named the Schröder equation, because of his contribution. However it was Gabriel Koenigs who took the first important step towards solving it [37]:

Theorem 7 (Koenigs's Theorem) *Let φ be a non-automorphic, analytic selfmap of $\mathbb{U}$ fixing $p \in \mathbb{U}$. If $\varphi'(p) \neq 0$ then there is a nonzero function σ which satisfies equation*

$$C_\varphi \sigma = \lambda \sigma \tag{4.8}$$

for $\lambda = \varphi'(p)$. Consequently, for all $n = 1, 2, 3, \ldots$, the functions σ^n satisfy the same equation for $\lambda = (\varphi'(p))^n$, respectively. The eigenspaces of $\widetilde{C_\varphi}$ corresponding to the eigenvalues above are all 1-dimensional, where $\tilde{C_\varphi}$ is the composition operator induced by φ on the space of all holomorphic functions.

Under the assumptions in Theorem 7, the remarkable function σ, with property $\sigma'(p) = 1$ is called the *Koenigs* function of C_φ. Based on Theorem 7, one easily gets that [48, Remark 1.10]:

Remark 2 If φ, an analytic selfmap of $\mathbb{U}$, fixes $p \in \mathbb{U}$, induces a compact composition operator C_φ, and $\varphi'(p) \neq 0$, then the lattice of finite-dimensional invariant subspaces of C_φ consists of the subspaces

$$\text{Span}(\{\sigma^n : n \in E\}) \tag{4.9}$$

where E is any finite subset of $\mathbb{N}$, the set of positive integers. Hence

$$\overline{\{\text{Span}(\{\sigma^n : n \in E\})} : E \subseteq \mathbb{N}\} \subseteq \text{Lat}C_\varphi. \tag{4.10}$$

Above, we agree to say that the subspace spanned by $\emptyset \subseteq \mathbb{N}$ is the null subspace.

Thus, if the two sets involved in relation (4.10) are equal, one may say that "the invariant subspace lattice of C_φ is spanned by the powers of the Koenigs function of φ", but does that happen at all? There is a simple example when the answer is known to be affirmative [48, Proposition 1.6], namely if $p \in \mathbb{U}$ is the fixed point of φ, α_p is, as before, the selfinverse disc automorphism $\alpha_p(z) = (p - z)/(1 - \bar{p}z)$ and $\alpha_p \circ \varphi \circ \alpha_p(z) = \lambda z$, for some constant $0 < |\lambda| < 1$, then "Yes, the invariant subspace lattice of C_φ is spanned by the powers of the Koenigs function of φ". All that raises the question:

Problem 16 If a compact composition operator C_φ is induced by a selfmap φ of $\mathbb{U}$ fixing a point $p \in \mathbb{U}$, and $\varphi'(p) \neq 0$, then is it always true that $LatC_\varphi$ is spanned by the powers of the Koenigs function of φ? If not, what is the description of the invariant subspace lattice of C_φ?

It should be added here that only some selfmaps fixing a point in $\mathbb{U}$ can induce compact composition operators, or more generally:

Theorem 8 ([45, Theorem 3.3]) *An analytic selfmap φ of $\mathbb{U}$ satisfies condition*

$$r_e(C_\varphi) < 1$$

(where $r_e(C_\varphi)$ is the essential spectral radius of C_φ), if and only if φ is a noninner map with a fixed point in $\mathbb{U}$.

4.4 Strongly Cyclic Composition Operators

Given T an operator acting on a linear topological space, that operator is called *cyclic* if the closed subspace spanned by $O_T(x)$ is the whole space, for some vector x. If $O_T(x)$ is a dense subset of the space, then the term *hypercyclic* is used. Certain classes or composition operators on H^2 are known to be hypercyclic and, as observed in [49], hypercyclicity influences properties of the spectrum or numerical range. Only univalent symbols of parabolic or hyperbolic type can induce hypercyclic composition operators. According to [7], many of those of hyperbolic or parabolic automorphic type do. However, an exact description of the set of analytic selfmaps of $\mathbb{U}$ inducing hypercyclic composition operators on H^2 is not known, and so it is interesting to ask:

Problem 17 Which analytic selfmaps of the disc induce hypercyclic composition operators?

This problem is from [7], where several questions on the cyclicity of composition operators are raised, with comments and examples. We consider Problem 17 the most interesting of them.

5 Weighted Composition Operators

Recall that we call weighted composition operators $T_{\psi,\varphi}$ *nontrivial* when they do not coincide with analytic Toeplitz operators or to scalar multiples of composition operators. In this section, we will always assume weighted composition operators are nontrivial.

5.1 Boundedness and Compactness

Weighted composition operators of a special kind are the only isometric operators acting on non-Hilbert Hardy spaces. This result [25] due to Frank Forelli brought these operators to public attention. Weighted composition operators started to be seriously studied rather recently, though. Basic problems about such operators are still open or only partially solved. The first is of course boundedness. So:

Problem 18 Exactly when is $T_{\psi,\varphi}$ bounded and, if bounded, what is the formula for its norm?

The first comments here are the following. A weighted composition operator $T_{\psi,\varphi}$ cannot be bounded, unless $\psi \in H^2$. If ψ is in H^2 and φ is an inner function then one can calculate the norm of $T_{\psi,\varphi}$ and write a nice formula for $\|T_{\psi,\varphi}\|$ [44, Theorem 7]. Obviously isometric operators are bounded and have norm 1. In [47], it is shown that in order that $T_{\psi,\varphi}$ be isometric it is necessary that φ be inner and a necessary and sufficient condition that $T_{\psi,\varphi}$ be isometric is provided [47, Theorem 5]. That necessary and sufficient condition turns out to be exactly "Forelli's condition adapted to H^2". What we mean is, the isometries of H^p, $p \neq 2$ need to be weighted composition operators $T_{\psi,\varphi}$ where $\psi \in H^p$, φ is inner and

$$\int_X |\psi|^p \, dm = \int_X \frac{dm(u)}{P(\varphi(0), \varphi(u))} \qquad X = \varphi^{-1}(Y) \tag{5.1}$$

where Y is any Lebesgue measurable subset of the unit circle. Since H^2 is a Hilbert space, the isometric operators on H^2 need not be weighted composition operators. However, according to [47, Theorem 5], in order that $T_{\psi,\varphi}$ be isometric it is necessary and sufficient that $\psi \in H^2$, φ be inner, and condition (5.1) hold for $p = 2$.

Besides the case when φ is inner, and the case when ψ is bounded, the boundedness of $T_{\psi,\varphi}$ is currently understood only in some particular cases. Now, we are perfectly aware of the fact that no comprehensive collection of formulas for the norm of a composition operator ("unweighted"!) exists and so, Problem 18 is anything but easy, mostly when it comes to norm computations. More exactly, what is known is that the only situation when C_φ is a contraction is when $\varphi(0) = 0$, in which case $\|C_\varphi\| = 1$. Besides that case, the norm of C_φ is known when φ is inner, in which case $\|C_\varphi\| = \sqrt{\frac{1+|\varphi(0)|}{1-|\varphi(0)|}}$. There are some other (minor) situations when the norm of a composition operator is known, but it may be notoriously hard to compute, even in simple cases like that of some linear fractional symbols [11]. As for norm-computations for weighted composition operators, the little that is known is summarized, with small exceptions, in [44]. One notable small exception is the case of weighted composition operators which are unitarily equivalent to composition operators on Hardy–Smirnov spaces over half-plane (see Remark 3 in this paper).

Clearly, $T_{\psi,\varphi}$ is bounded when it is compact (since it transforms the closed unit ball of the space into a set with compact closure, hence into a bounded set!). Now, the case when $T_{\psi,\varphi}$ is Hilbert–Schmidt and hence compact is well understood [44, Theorem 9]. Theorems paralleling results known from the study of compactness for composition operators exist for weighted composition operators too (see [44, Theorems 10, 11]). Some emphasize the fact that the symbol ψ may make $T_{\psi,\varphi}$ compact even when C_φ is noncompact [44, Theorems 12, 13]. So, to make a long story short, the question is:

Problem 19 Exactly when is a weighted composition operator on H^2 compact?

In the case of composition operators, the problem above triggered a good decade of research, starting with Schwarz's dissertation [64], and ending with Shapiro's seminal paper [65], containing a full solution. Cima and Matheson came with a posterior interesting measure theoretical point of view in [17]. It would be interesting to see the weighted composition version of this body of knowledge. However it is very likely that it will be an intricate classification of the maps ψ and φ, since the case when ψ is constant but nonzero reduces to the case of compact composition operators. The main references where partial answers to Problem 19 can be found are [30] and [44]. Very little is known beyond that (in the interesting case when $T_{\psi,\varphi}$ is nontrivial!).

5.2   Invertibility and Spectra

There is a rather short collection of papers (known to this author), where the spectra of nontrivial weighted composition operators on H^2 are studied. We will list them all in this subsection. Historically, the first spectral computations for composition operators were in the case of invertible operators (and a composition operator is invertible if and only if its symbol is a disc automorphism) [56], followed by the case of compact operators [64] and [14]. Since the problem of the full description of the spectra of composition operators is still open, we will content ourselves to formulate minimal open questions about the spectra of weighted composition operators, following the little that is known about it. The author of [29], was able to sow that, if $T_{\psi,\varphi}$ is compact and φ has a fixed point $p \in \mathbb{U}$ then the spectrum of $T_{\psi,\varphi}$ is given by the formula

$$\sigma(T_{\psi,\varphi}) = \{0, \psi(p)\varphi'(p), \psi(p)(\varphi'(p))^2, \psi(p)(\varphi'(p))^3, \dots\}.$$

As was already mentioned, a compact composition operator C_φ is necessarily induced by a symbol φ fixing a point in $\mathbb{U}$. However, there are compact weighted composition operators $T_{\psi,\varphi}$, whose composition symbol φ fixes no point in $\mathbb{U}$ [29, 44]. Thus the first open question here is:

Problem 20 If $T_{\psi,\varphi}$ is compact, exactly when is it the case that φ must fix a point in $\mathbb{U}$?

Once the problem above is answered, the next problem we wish to raise is:

Problem 21 What is the description of the spectra of compact weighted composition operators whose composition symbols fix no points in $\mathbb{U}$?

The reason for focusing on compact operators is of course the well known structure of their spectra which consists of 0 and their eigenvalues (if any). So, this is most probably one of the easiest spectral computations one may expect to perform for weighted composition operators.

Although the first spectral computation for composition operators was performed for invertible composition operators, the case of nontrivial, invertible weighted composition operators was different, the first descriptions of the spectra for such operators [31, 33], being posterior to the partial computation of the spectra of nontrivial, compact, weighted composition operators. There is a rather simple reason for that. Indeed, a weighted composition operator $T_{\psi,\varphi}$ is invertible if and only if φ is a disc automorphism and ψ is a zero-free, bounded analytic function, whose reciprocal is also bounded [31] (hence ψ must be a particular kind of outer function). Now, the weight symbol ψ having the freedom of being that particular kind of outer function, and of being paired with any disc automorphism φ, makes things complicated here. Since there are three kinds of disc automorphisms: elliptic, parabolic, and hyperbolic, the results known on the description of the spectra of invertible weighted composition operators fall in those three categories. The papers known to this author which contain relevant results along those lines are just

[6, 31, 33]. If $T_{\psi,\varphi}$ is unitary, $\sigma(T_{\psi,\varphi})$ is completely described in [6]. Interestingly, φ may be any of the three kinds of disc automorphism, in this particular case. It should be added that, as is well known, the spectrum of a nonunitary Hilbert space isometry coincides with $\overline{\mathbb{U}}$ (and the nonunitary, isometric, weighted composition operators on H^2 are determined in [47]).

In general, if φ is an elliptic disc automorphism, then φ is conformally conjugated to a rotation. If that rotation is by a root of 1 and $T_{\psi,\varphi}$ is invertible, then $\sigma(T_{\psi,\varphi})$ is known [31, Theorem 3.1.1]. However, if φ is not conformally conjugated to a rotation by a root of unity, then $\sigma(T_{\psi,\varphi})$ is known only if ψ belongs to the disc algebra [31, Theorem 3.2.1] (that is ψ has a continuous extension to $\overline{\mathbb{U}}$).

The description of $\sigma(T_{\psi,\varphi})$ when $T_{\psi,\varphi}$ is invertible, and φ is a parabolic or hyperbolic disc automorphism is also known only if ψ belongs to the disc algebra [33]. Thus, the normal problem is to complete the description of the spectra of invertible, weighted composition operators or more generally to answer the following:

Problem 22 What is the description of $\sigma(T_{\psi,\varphi})$, when φ is a disc automorphism and ψ is a nonconstant bounded analytic function?

As the reader can easily observe, if φ is a disc automorphism, then $T_{\psi,\varphi}$ is bounded if and only if ψ is a bounded analytic function.

Some particular computations of spectra and essential spectra of weighted composition operators can also be found in [5]. There the composition symbol φ is assumed to be an essentially linear fractional map, that is a special type of horocyclic map, i.e. a map with range contained in a proper horocycle (a proper disc internally tangent to the unit circle).

Relative to the essential spectrum, besides the essential spectral computations found in [5], very little work exists in the literature on the essential spectrum of a weighted composition operator on H^2. So:

Problem 23 What is the essential spectrum of $T_{\psi,\varphi}$ when φ is not an essentially linear fractional map? In particular, what is the essential spectrum of an invertible weighted composition operator?

As we know, the closure of the numerical range of an operator is a superset of the convex hull of its spectrum. It goes without saying that knowing the description of the numerical ranges of nontrivial, weighted composition operators, is interesting, when talking about their spectra. However, research on the numerical ranges of weighted composition operators is only at the beginning, since the numerical ranges of "unweighted" composition operators are so poorly known. There is a short list of papers studying the numerical ranges of nontrivial weighted composition operators (on H^2), namely [32, 47, 50]. In the case of trivial weighted composition operators, one should add that, if the composition symbol is the identity and so, the weighted composition operator induced is an analytic Toeplitz operator, one can find the description of the numerical range in [36].

5.3   Normality Concepts

Composition operators on H^2 are normal only in the rather trivial case when the symbol φ is of form $\varphi(z) = \lambda z$, for some constant $|\lambda| \leq 1$, hermitian if $\lambda \in \mathbb{R}$, and unitary only when $|\lambda| = 1$. It is interesting to observe that weighted composition operators can be normal in significantly more situations. There exist complete characterizations of the cases when a weighted composition operator on H^2 is unitary [6, Theorem 6]. According to that result φ must be a disc automorphism and ψ a scalar multiple of the reproducing kernel function at $\varphi(0)$ (that is a scalar multiple of $1/(1 - \overline{\varphi(0)}z)$). The same paper, contains a complete characterization of normal weighted composition operators $T_{\psi,\varphi}$ induced by composition symbols φ which have a fixed point in $\mathbb{U}$ [6, Theorem 10]. In the case of the latter theorem, the symbol φ must be linear-fractional and the weight symbol ψ must be well paired with φ via reproducing kernel functions. Actually, it is known when a weighted composition operator with linear fractional composition symbol φ is normal. There are examples of such operators so that φ does not fix a point in $\mathbb{U}$. Nevertheless, nearly nothing is known about the case when φ, the composition symbol of $T_{\psi,\varphi}$ has no fixed points in $\mathbb{U}$ and is not a linear fractional selfmap of $\mathbb{U}$. Finally, it should be added that the characterization of hermitian weighted composition operators is known in the particular case when ψ, the weight symbol, is bounded [19, Theorem 2.1]. Both the composition symbol and the weight symbol must be linear fractional in that case. Besides the minimal bibliography outlined in these comments, very little is known about:

Problem 24   Exactly when is a weighted composition operator on H^2 normal?

Weaker normality concepts (co-hyponormality to be exact) have been considered too [22].

5.4   Hardy–Smirnov Spaces

Weighted and unweighted composition operators have been studied on a vast collection of spaces. So far we chose to consider them only on H^2, though, since a lot of work has been done in that space and a lot of hard open problems were left behind. We turn now to Hardy–Smirnov spaces a class of spaces of analytic functions where weighted and unweighted composition operators meet in an interesting way. Here is what we mean.

Let $G \subseteq \mathbb{C}$ be a simply connected domain, $G \neq \mathbb{C}$. Let γ be a conformal isomorphism of $\mathbb{U}$ onto G. For each $0 < p < \infty$, the Hardy–Smirnov space $H^p(G)$ is by definition, the collection of all functions f analytic in G that satisfy the condition

$$\sup_{0<r<1} \left(\int_{\Gamma_r} |f(z)|^p \, |dz| \right)^{1/p} < \infty, \tag{5.2}$$

where, for each r, Γ_r is the image under γ of the circle of radius r about the origin. Although condition (5.2) seems to produce spaces that depend on the conformal isomorphism γ, it is well known that, for each $0 < p < \infty$, $H^p(G)$ depends only on G.

We refer to the spaces $H^p(G)$ as Hardy–Smirnov spaces on G. If G is a half-plane, then a result of this author states that $H^p(G)$ does not support compact composition operators [40]. This result raised the question; "Exactly which Hardy–Smirnov spaces can support compact composition operators and which can't?" The answer was obtained by Shapiro and Smith [70]: "The space $H^p(G)$ supports compact composition operators if and only if the one-dimensional Hausdorf measure of ∂G, the boundary of G, is finite, a fact that obviously does not happen when G is a half-plane." Let $G = \Pi^+$ be the right open half-plane. Unlike the case of Hardy spaces of a disc (where all analytic selfmaps of that disc induce bounded composition operators), $H^2(\Pi^+)$ supports fewer bounded composition operators. More exactly, only analytic selfmaps ϕ of Π^+ which fix the point at infinity and have a finite angular derivative at that point induce bounded composition operators C_ϕ on $H^2(\Pi^+)$ [44] and the norm of such an operator is computable by the formula

$$\|C_\phi\| = \sqrt{\phi'(\infty)}, \tag{5.3}$$

[24]. It is a common place to say that the first natural question about a class of operators is "When are they compact?" and the next "When are they normal?" Paper [50] shows that only some Möbius selfmaps (or linear fractional selfmaps) of Π^+ can induce, invertible, hermitian, or Fredholm composition operators. Exact characterizations of the symbols inducing such operators are provided. That paper also contains a characterization of the Möbius selfmaps of Π^+ which induce normal composition operators. However, the following question is left unanswered:

Problem 25 Does an analytic selfmap of Π^+ inducing a normal composition operator on $H^2(\Pi^+)$ have to be linear fractional? If not, what is the characterization of normal composition operators acting on $H^2(\Pi^+)$?

Relative to our opening statement in this subsection, saying that Hardy–Smirnov spaces are a class of spaces of analytic functions where weighted and unweighted composition operators meet in an interesting way, we would like to observe that composition operators on such spaces are unitarily equivalent to certain weighted composition operators on H^2 [44], and the results obtained by this author on composition operators acting on Hardy–Smirnov spaces were established by studying the aforementioned weighted composition operators on H^2. Thus, Problem 25 is a sub-problem of Problem 24. Furthermore, the study of boundedness of composition operators on Hardy–Smirnov spaces and formula (5.3) solve Problem 18 in the following particular case:

Remark 3 If $\alpha \in \mathbb{T}$, φ is an analytic selfmap of $\mathbb{U}$ and $\psi(z) = (\alpha - z)/(\alpha - \varphi(z))$, then $T_{\psi,\varphi}$ is bounded if and only if φ has a finite angular derivative $\varphi'(\alpha)$ at α and the angular limit of φ at α is equal to α. In that case, the norm of $T_{\psi,\varphi}$ is computable with formula $\|T_{\psi,\varphi}\| = \sqrt{\varphi'(\alpha)}$.

It is worth noting that:

Remark 4 While $H^2(\Pi^+)$ cannot support compact composition operators, that space does support compact weighted composition operators.

Indeed, for any $\psi \in H^2(\Pi^+)$ and any constant selfmap ϕ of Π^+, one can see that $T_{\psi,\phi}$ is compact, by recalling that a sequence of functions in $H^2(\Pi^+)$ is weakly convergent if and only if that sequence is norm-bounded and pointwise convergent. We conclude this subsection by stating that a serious study of weighted composition operators on $H^2(\Pi^+)$ treating as comprehensively as possible such issues as boundedness, compactness, normality, the computation of spectra and numerical range does not exist, but is interesting to produce, encouraged by Remark 4.

5.5 Brennan's Conjecture and a Bergman Version of Hardy–Smirnov Spaces

Let τ denote an analytic, univalent transform of the open unit disk $\mathbb{U}$ onto some simply connected domain $G \subseteq \mathbb{C}$, $G \neq \mathbb{C}$. We denote $g = \tau^{-1}$ and refer to g as a *Riemann transform of G onto $\mathbb{U}$*, given Riemann's well known conformal equivalence theorem. Using this terminology, we recall the following important conjecture in univalent function theory:

Brennan's Conjecture *If g is a Riemann transform of a simply connected domain $G \subsetneq \mathbb{C}$ onto $\mathbb{U}$ and $4/3 < p < 4$, then*

$$\int_G |g'|^p \, dA < \infty. \tag{5.4}$$

Above, dA denotes the area measure. The fact that (5.4) holds when $4/3 < p < 3$ is an easy consequence of the Koebe distortion theorem. Brennan [4] extended this to $4/3 < p < 3 + \delta$ for some small $\delta > 0$, and conjectured it to hold for $4/3 < p < 4$. This range of p cannot be extended, as shown by the example $G = \mathbb{C} \setminus (-\infty, -1]$. The upper bound of those p for which (5.4) is known to hold has been increased by several authors, in particular to approximately 3.78 by S. Shimorin in [71].

Let us consider the Hilbert Bergman space $L_a^2(\mathbb{U})$, that is the space of all analytic functions on $\mathbb{U}$ that are square integrable dA. Brennan's conjecture can be easily reformulated in terms of $\tau = g^{-1}$. Indeed, elementary computations lead to the following equivalent formulation of Brennan's conjecture:

If τ is a Riemann transform of $\mathbb{U}$ onto a simply connected domain $G \subsetneq \mathbb{C}$ and $-1/3 < p < 1$, then $1/(\tau')^p \in L_a^2(\mathbb{U})$.

Brennan's conjecture can also be formulated in terms of the compactness of some special weighted composition operators. Denote by $A_{\varphi,p}$ the weighted composition operator

$$A_{\varphi,p} = T_{(\tau'\circ\varphi/\tau')^p,\varphi}.$$

The main result in [72] is:

Theorem 9 ([72, Theorem 1.1]) $1/(\tau')^p \in L_a^2(\mathbb{U})$ *if and only if there is some analytic selfmap φ of $\mathbb{U}$ so that $A_{\varphi,p}$ is a compact operator on $L_a^2(\mathbb{U})$.*

One can formulate Brennan's conjecture in terms of ("unweighted") composition operators as follows. Denote by $\mathcal{H}(G)$ the space of holomorphic functions on G. One introduces the function spaces

$$L_a^2(\mu_p) := \left\{ F \in \mathcal{H}(G) : \int_G |F|^2 \, d\mu_p < +\infty \right\}$$

where p is any fixed real number and $d\mu_p = |g'|^{2p+2}dA$. It is known that the space $L_a^2(\mu_p)$ does not depend on the Riemann map chosen from G onto $\mathbb{U}$.

An analytic selfmap of G is denoted by ϕ and $\varphi = g \circ \phi \circ g^{-1}$ denotes the analytic selfmap of $\mathbb{U}$ that is conformally conjugate to ϕ. With this notation we recall the following:

Proposition 3 ([52]) *The composition operator C_ϕ on $L_a^2(\mu_p)$ is unitarily equivalent to $A_{\varphi,p}$ on $L_a^2(\mathbb{D})$ and hence, Brennan's conjecture is equivalent to the statement that the spaces $L_a^2(\mu_p)$ endowed with the norm*

$$\|F\| = \sqrt{\frac{1}{\pi} \int_G |F|^2 |g'|^{2p+2} dA}$$

support compact composition operators if $-1/3 < p < 1$.

The problem raised here is;

Problem 26 Is there a smart functional analysis argument allowing one to solve Brennan's conjecture via composition operators or weighted composition operators, as suggested above?

6   Asymptotic Toeplitz Concepts

An operator T on H^2 is a Toeplitz operator if and only if $M_z^* T M_z = T$, [13]. Composition operators rarely have that property (the identity operator is the only Toeplitz composition operator [68]). However, an interesting connection of those operators and composition operators was exhibited in [55]: the asymptotic Toeplitz concepts due to Barria and Halmos [3] do have an impact on composition operators (see also [68]). More formally, if T is a Toeplitz operator, then by our previous considerations, the sequence $\{M_z^{*n}TM_z^n\}$ must converge to T in all usual topologies on the space of bounded operators. Thus, given an arbitrary Hilbert space operator T, it makes sense to call T a *uniformly asymptotically Toeplitz* operator (UAT) if $\{M_z^{*n}TM_z^n\}$ is convergent in the uniform operator topology, respectively to use the terms *strongly asymptotically Toeplitz* (SAT) and *weakly asymptotically Toeplitz* (WAT), if convergence takes place with respect to those topologies. No matter which topology is considered, the limit A of $\{M_z^{*n}TM_z^n\}$ necessarily satisfies relation $M_z^*AM_z = A$ and is therefore a Toeplitz operator induced by some φ, an essentially bounded measurable function called the asymptotic symbol of T. Nazarov and Shapiro initiated the study of the asymptotic Toeplitzness of composition operators [55] (see also[67] and [68]). This author followed in that line of research observing that M_z is just a forward shift [43], that is an isometry S with the property $S^{*n} \to 0$ strongly, and so, one can use the ideas of [60], where an operator T is called S-Toeplitz (where S is any forward shift) if equation $S^*TS = T$ is satisfied, and introduce S-Toeplitzness asymptotic concepts by using as above, the sequence $\{S^{*n}TS^n\}$. In particular, one can take S of the form $S = M_u$ where u is any nonconstant inner function, since, as proved in [43], such operators are forward shifts. One of the results in [43] is that, while a composition operator C_φ may or may not be asymptotically Toeplitz, the operator $C_\varphi^*C_\varphi$ is always WAT. Recently, the authors of [23] strengthened that result by proving that actually, $C_\varphi^*C_\varphi$ is always SAT. All the ideas above were recently applied to the study of the asymptotic Toeplitz properties of weighted composition operators [34].

Replacing M_z by an arbitrary forward shift is motivated by the matricial description of a Toeplitz operator: the matrix of such an operator has constant main diagonal (that is the same entry is repeated on its main diagonal), and the diagonals parallel to the main diagonal are constant as well. In case one uses any shift S instead of M_z and $S^*TS = T$ then, the S-Toeplitz operator T must have a block matrix with constant main diagonal and constant parallel diagonals [60]. If one studies M_u-Toeplitzness, where u is a nonconstant inner function, the collection of M_u-Toeplitz composition operators becomes much wider than just the identity. It is proved that only composition operators with inner symbol can be M_u-Toeplitz, and u must be an invariant inner function of C_φ, in that case [43, Theorem 1]. It is shown in [46], that inner functions of hyperbolic type and those of parabolic automorphic type have a rich assortment of inner invariant functions.

The composition operators which are UAT are also few, only the identity and the compact composition operators are UAT [55]; see also [43, Corollary 2]. However, weaker asymptotic Toeplitz concepts are more generous. For instance:

Proposition 4 ([55, Proposition 2.1]) *Let φ be an analytic selfmap of $\mathbb{U}$ with property $\varphi(0) = 0$, other than the identity or a rotation, then C_φ is WAT with null asymptotic symbol.*

It is conjectured in paper [55], and reiterated in [68], that the answer to the following is affirmative:

Problem 27 Is it true that any analytic selfmap φ of $\mathbb{U}$ other than a rotation or the identity induces a *WAT* composition operator C_φ with null asymptotic symbol?

That is, can the restriction $\varphi(0) = 0$ be removed? The fact that the answer to Problem 27 might be affirmative is supported by the particular examples of maps φ with property C_φ is WAT although $\varphi(0) \neq 0$. For instance, one can consider inner maps (other than rotations or the identity) with negligible singular set (that is the subset of $\mathbb{T}$ where the inner map does not extend by continuity must be negligible), or maps satisfying the condition $|\varphi| < 1$ a.e. on $\mathbb{T}$ (hence all maps inducing compact composition operators) [55]. While it is known that nontrivial composition operators which are not SAT do exist (e.g. those induced by inner symbols other than the identity [55, Theorem 3.3]), the answer to the following problem is currently unknown [68].

Problem 28 Characterize the analytic selfmaps of $\mathbb{U}$ which induce *SAT* composition operators.

One can find examples of such maps in [55] (e.g. if $|\varphi| < 1$ a.e. on $\mathbb{T}$, C_φ is actually SAT, [55, Proposition 1.1], but condition $|\varphi| < 1$ a.e. on $\mathbb{T}$ is not necessary, meaning that SAT composition operators which do not satisfy it, do exist [55, Theorem 3.4]). Paper [43] extends some of the results in [55], from the case of "M_z-asymptotic concepts" to that of "M_u-asymptotic concepts", were u is a nonconstant inner function other than the identity. Obviously problems similar to those formulated in this subsection for the shift M_z can be formulated for shifts of type M_u.

For the time being, we contend ourselves to bringing up operator $C_\varphi^* C_\varphi$ which, as we stated above, is always a SAT operator [23], whose asymptotic symbol is the Nykodym derivative ψ of the pull back measure $m\varphi^{-1}$ induced by φ [49, Theorem 6]. In the aforementioned theorem, it is also shown that $\sqrt{\text{esssup } \psi}$ equals the essential norm of C_φ when $C_\varphi^* C_\varphi$ is UAT. Motivated by all that, we raise the following:

Problem 29 Characterize the analytic selfmaps φ of $\mathbb{U}$ which induce composition operators C_φ such that $C_\varphi^* C_\varphi$ is *UAT*.

While a sufficient condition for $C_\varphi^* C_\varphi$ to be UAT exists [43, Theorem 5], a complete answer to Problem 29 is currently not known.

References

1. A. Abdollahi, The numerical range of a composition operator with conformal automorphism symbol. Linear Algebra Appl. **408**, 177–188 (2005)
2. A. Abdollahi, M.T. Heydari, The numerical range of finite order elliptic automorphism composition operators. Linear Algebra Appl. **483**, 128–138 (2015)

3. J. Barria, P.R. Halmos, Asymptotic Toeplitz operators. Trans. Am. Math. Soc. **273**(2), 621–630 (1982)
4. J.E. Brennan, The integrability of the derivative in conformal mapping. J. Lond. Math. Soc. (2) **18**, 261–272 (1978)
5. P.S. Bourdon, Spectra of some composition operators and some associated weighted composition operators. J. Oper. Theory **67**(2), 537–560 (2012)
6. P.S. Bourdon, S.K. Narayan, Normal weighted composition operators on the Hardy space. J. Math. Anal. Appl. **367**(1), 278–286 (2010)
7. P.S. Bourdon, J.H. Shapiro, Cyclic composition operators on H^2, in *Operator Theory: Operator Algebras and Applications, Part 2 (Durham, NH, 1988)*. Proceedings of Symposia in Pure Mathematics, vol. 51, Part 2 (American Mathematical Society, Providence, RI, 1990), pp. 43–53
8. P.S. Bourdon, J.H. Shapiro, Riesz composition operators. Pac. J. Math. **181**(2), 231–246 (1997)
9. P.S. Bourdon, J.H. Shapiro, The numerical ranges of automorphic composition operators. J. Math. Anal. Appl. **251**(2), 839–854 (2000)
10. P.S. Bourdon, J.H. Shapiro, When is zero in the numerical range of a composition operator? Integr. Equ. Oper. Theory **44**(4), 410–441 (2002)
11. P.S. Bourdon, E.E. Fry, C. Hammond, C.H. Spofford, Norms of linear fractional composition operators. Trans. Am. Math. Soc. **356**(6), 2459–2480 (2003)
12. P.S. Bourdon, V. Matache, J.H. Shapiro, On convergence to the Denjoy-Wolff point. Ill. J. Math. **49**(2), 405–430 (2005)
13. A. Brown, P.R. Halmos, Algebraic properties of Toeplitz operators. J. Reine Angew. Math. **213**, 89–102 (1963/1964)
14. J.G. Caughran, H.J. Schwartz, Spectra of compact composition operators. Proc. Am. Math. Soc. **51**, 127–130 (1975)
15. I. Chalendar, J.R. Partington, *Modern Approaches to the Invariant-Subspace Problem*. Cambridge Tracts in Mathematics, vol. 188 (Cambridge University Press, Cambridge, 2011)
16. V. Chkliar, Eigenfunctions of the hyperbolic composition operator. Integr. Equ. Oper. Theory **29**(3), 364–367 (1997)
17. J.A. Cima, A.L. Matheson, Essential norms of composition operators and Aleksandrov measures. Pac. J. Math. **179**, 59–64 (1997)
18. C.C. Cowen, Composition operators on H^2. J. Oper. Theory **9**, 77–106 (1983)
19. C.C. Cowen, E. Ko, Hermitian weighted composition operators on H^2. Trans. Am. Math. Soc. **362**(11), 5771–5801 (2010)
20. C.C. Cowen, B.D. MacCluer, Spectra of some composition operators. J. Funct. Anal. **125**(1), 223–251 (1994)
21. C.C. Cowen, B.D. MacCluer, *Composition Operators on Spaces of Analytic Functions* (CRC Press, Boca Raton, 1995)
22. C.C. Cowen, S. Jung, E. Ko, Normal and cohyponormal weighted composition operators on H^2, in *Operator Theory in Harmonic and Non-commutative Analysis*. Operator Theory: Advances and Applications, vol. 240 (Birkhäuser/Springer, Basel/Cham, 2014), pp. 69–85
23. C. Duna, M. Gagne, C. Gu, J. Shapiro, Toeplitzness of products of composition operators and their adjoints. J. Math. Anal. Appl. **410**(2), 577–584 (2014)
24. S. Elliott, M.T. Jury, Composition operators on Hardy spaces of a half-plane. Bull. Lond. Math. Soc. **44**(3), 489–495 (2012)
25. F. Forelli, The isometries of H^p. Can. J. Math. **16**, 721–728 (1964)
26. E.A. Gallardo-Gutiérrez, P. Gorkin, Cyclic Blaschke products for composition operators. Rev. Mat. Iberoam. **25**(2), 447–470 (2009)
27. E.A. Gallardo-Gutiérrez, P. Gorkin, Minimal invariant subspaces for composition operators. J. Math. Pures Appl. (9) **95**(3), 245–259 (2011)
28. E.A. Gallardo-Gutiérrez, P. Gorkin, D. Suárez, Orbits of non-elliptic disc automorphisms on H^p. J. Math. Anal. Appl. **388**(2), 1013–1026 (2012)
29. G. Gunatillake, Spectrum of a compact weighted composition operator. Proc. Am. Math. Soc. **135**(2), 461–467 (2007)

30. G. Gunatillake, Compact weighted composition operators on the Hardy space. Proc. Am. Math. Soc. **136**(8), 2895–2899 (2008)
31. G. Gunatillake, Invertible weighted composition operators. J. Funct. Anal. **261**(3), 831–860 (2011)
32. G. Gunatillake, M. Jovovic, W. Smith, Numerical ranges of weighted composition operators. J. Math. Anal. Appl. **413**(1), 485–475 (2014)
33. O. Hyvarinen, M. Lindstrom, I. Nieminen, E. Saukko, Spectra of weighted composition operators with automorphic symbols. J. Funct. Anal. **265**, 1749–1777 (2013)
34. S. Jung, E. Ko, On T_u-Toeplitzness of weighted composition operators on H^2. Complex Var. Elliptic Equ. **60**(11), 1522–1538 (2015)
35. H. Kamowitz, The spectra of composition operators on H^p. J. Funct. Anal. **18**, 32–50 (1975)
36. E.M. Klein, The numerical range of a Toeplitz operator. Proc. Am. Math. Soc. **35**, 101–103 (1972)
37. G. Koenigs, Recherches sur les intégrales de certaines équations fonctionnelles. Ann. Sci. Éc. Norm. Supér. (3) **1**(Suppl.), 3–41 (1884)
38. V. Matache, On the minimal invariant subspaces of the hyperbolic composition operator. Proc. Am. Math. Soc. **119**(3), 837–841 (1993)
39. V. Matache, The eigenfunctions of a certain composition operator. Contemp. Math. **213**, 121–136 (1998)
40. V. Matache, Composition operators on Hardy spaces of a half-plane. Proc. Am. Math. Soc. **127**, 1483–1491 (1999)
41. V. Matache, Numerical ranges of composition operators. Linear Algebra Appl. **331**(1–3), 61–74 (2001)
42. V. Matache, Distances between composition operators. Extracta Math. **22**(1), 19–33 (2007)
43. V. Matache, S-Toeplitz composition operators, in *Complex and Harmonic Analysis* (DEStech Publications Inc., Lancaster, PA, 2007), pp. 189–204
44. V. Matache, Weighted composition operators on H^2 and applications. Compl. Anal. Oper. Theory **2**(1), 169–197 (2008)
45. V. Matache, Composition operators whose symbols have orthogonal powers. Houst. J. Math. **37**(3), 845–857 (2011)
46. V. Matache, Numerical ranges of composition operators with inner symbols. Rocky Mountain J. Math. **42**(1), 235–249 (2012)
47. V. Matache, Isometric weighted composition operators. New York J. Math. **20**, 711–726 (2014)
48. V. Matache, Invariant subspaces of composition operators. J. Oper. Theory **73**(1), 243–264 (2015)
49. V. Matache, On spectra of composition operators. Oper. Matrices **9**(2), 277–303 (2015)
50. V. Matache, Invertible and normal composition operators on Hardy spaces of a half-plane. Concr. Oper. **3**, 77–84 (2016)
51. V. Matache, Nonminimal cyclic invariant subspaces of hyperbolic composition operators, in *Complex Analysis and Dynamical Systems VII*. Contemporary Mathematics, vol. 699 (American Mathematical Society, Providence, 2017), pp. 247–262
52. V. Matache, W. Smith, Composition operators on a class of analytic function spaces related to Brennan's conjecture. Compl. Anal. Oper. Theory **6**(1), 139–162 (2012)
53. R. Mortini, Cyclic subspaces and eigenvectors of the hyperbolic composition operator, in *Travaux mathématiques, Fasc. VII*. Sém. Math. Luxembourg (Centre Univ. Luxembourg, Luxembourg, 1995), pp. 69–79
54. R. Mortini, Superposition operators, B-universal functions, and the hyperbolic composition operator. Acta Math. Hungar. **138**(3), 267–280 (2013)
55. F. Nazarov, J.H. Shapiro, On the Toeplitzness of composition operators. Complex Var. Elliptic Equ. **52**(2–3), 193–210 (2007)
56. E.A. Nordgren, Composition operators. Can. J. Math. **20**, 442–449 (1968)
57. E.A. Nordgren, P. Rosenthal, F.S. Wintrobe, Invertible composition operators on H^p. J. Funct. Anal. **73**(2), 324–344 (1987)

58. C.M. Pearcy, *Some Recent Developments in Operator Theory*. Regional Conference Series in Mathematics, No. 36 (American Mathematical Society, Providence, RI, 1978)

59. H. Radjavi, P. Rosenthal, *Invariant Subspaces* (Springer, New York, 1973)

60. M. Rosenblum, J. Rovnyak, *Hardy Classes and Operator Theory*. Oxford Mathematical Monographs (Oxford Science Publications/The Clarendon Press/Oxford University Press, New York, 1985)

61. W. Rudin, *Real and Complex Analysis* (McGraw-Hill, New York, 1966)

62. J.V. Ryff, Subordinate H^p functions. Duke Math. J. **33**, 347–354 (1966)

63. E. Schröder, Über unendlich viele Algorithmen zur Auflösung der Gleichungen. Math. Ann. **2**, 317–365 (1870)

64. H.G. Schwartz, Composition operators on H^p, Dissertation, University of Toledo, 1969

65. J.H. Shapiro, The essential norm of a composition operator. Ann. Math. (2) **125**(2), 375–404 (1987)

66. J.H. Shapiro, *Composition Operators and Classical Function Theory* (Springer, Berlin, 1993)

67. J.H. Shapiro, Every composition operator is (mean) asymptotically Toeplitz. J. Math. Anal. Appl. **333**(1), 523–529 (2007)

68. J.H. Shapiro, Composition operators ♡ Toeplitz operators, in *Five Lectures in Complex Analysis*. Contemporary Mathematics, vol. 525 (American Mathematical Society, Providence, RI, 2010), pp. 117–139

69. J.H. Shapiro, The invariant subspace problem via composition operators-redux, in *Topics in Operator Theory. Volume 1. Operators, Matrices and Analytic Functions*. Operator Theory: Advances and Applications, vol. 202 (Birkhäuser Verlag, Basel, 2010), pp. 519–534

70. J.H. Shapiro, W. Smith, Hardy spaces that support no compact composition operators. J. Funct. Anal. **205**, 62–89 (2003)

71. S. Shimorin, Weighted composition operators associated with conformal mappings. Quadrature domains and their applications. Oper. Theory Adv. Appl. **156**, 217–237 (2005)

72. W. Smith, Brennan's conjecture for weighted composition operators. Contemp. Math. **393**, 209–214 (2006)

Harmonic Measures of Slit Sides, Conformal Welding and Extremum Problems

Dmitri Prokhorov

Abstract We present a list of 3 conjectures and 12 questions in geometric function theory. All of them have a connection with the Loewner theory for subordinate chains of conformal maps.

Keywords Harmonic measure • Analytic cusp • Conformal welding • Bombieri numbers • Loewner equation

2010 Mathematics Subject Classification Primary 30C55; Secondary 30C75, 30C80

1 Introduction

We present a list of 3 conjectures and 12 questions in geometric function theory. This list contains 4 different areas: harmonic measures of slit sides, boundary behavior of a conformal map in the neighborhood of an analytic cusp, asymptotic conformal welding and Bombieri's conjectures for coefficients of univalent functions close to the Koebe function.

The only idea that joins these items concerns their relation with the Loewner evolution of simply connected domains in the complex plane. Paper references can serve a guideline for background definitions and initial concepts as well as a proposition to use either the known or original approaches to solve the problems proposed.

D. Prokhorov (✉)
Department of Mechanics and Mathematics, Saratov State University, 83 Astrakhanskaya Str., Saratov 410012, Russia

Petrozavodsk State University, 33 Lenin Str., Petrozavodsk, Republic of Karelia, Russia
e-mail: ProkhorovDV@info.sgu.ru

© Springer International Publishing AG 2018   219
M. Agranovsky et al. (eds.), *Complex Analysis and Dynamical Systems*,
Trends in Mathematics, https://doi.org/10.1007/978-3-319-70154-7_12

2 Harmonic Measures of Slit Sides and Singular Solutions to the Loewner Equation

The concept of harmonic measure relates to the Dirichlet problem for harmonic functions. For a simply connected domain $D \subset \mathbb{C}$ with at least three boundary points and $w_0 \in D$, there exists a conformal map $w = f(z)$ from $\mathbb{D} = \{z : |z| < 1\}$ onto $D, f(0) = w_0$. Suppose that the boundary ∂D of D consists only of accessible points. In this case f is extended continuously onto the closure of $\mathbb{D}$. Let $E \subset D$ be such that E is the image of a measurable set $A \subset \mathbb{T} := \{z : |z| = 1\}$ under f. Then the harmonic measure $\omega(w_0; E, D)$ of E at w_0 with respect to D equals (meas $A)/2\pi$.

Recall Problem 6.32 from the collection [1] (posed by Lucas) of estimating the ratio of harmonic measures for the sides of a slit E at $0, 0 \notin E$, with respect to $\mathbb{C}$ when the slit goes to infinity, and at each of its points, the slit and the radial direction form an angle which does not exceed $\alpha\pi, 0 \le \alpha < \frac{1}{2}$. This problem was solved in [19]. If hmeas(E_k) denotes the harmonic measures of the two sides $E_k, k = 1, 2$, of E, then

$$\frac{1 - 2\alpha}{1 + 2\alpha} \le \frac{\text{hmeas}(E_1)}{\text{hmeas}(E_2)} \le \frac{1 + 2\alpha}{1 - 2\alpha}.$$

We will focus on slits in the upper half-plane $\mathbb{H} := \{z : \text{Im } z > 0\}$. Let $\gamma(t)$, $0 \le t \le T$, be a simple curve in $\mathbb{H} \cup \{\gamma(0)\}$. Without loss of generality, assume $\gamma(0) = 0$. There is a unique conformal map $f(z, t) : \mathbb{H} \setminus \gamma[0, t] \to \mathbb{H}$ with the hydrodynamic normalization

$$f(z, t) = z + \frac{2t}{z} + O\left(\frac{1}{|z|^2}\right), \quad z \to \infty.$$

Here $2t$ is the half-plane capacity of $\gamma[0, t]$; see, e.g., [11, p. 69]. The mapping $f(z, t)$ can be extended continuously onto $\mathbb{H} \cup \mathbb{R}$. The extended function $f(z, t)$ maps $\gamma[0, t]$ onto a segment $I(t) \subset \mathbb{R}$, while $\mathbb{R}$ is mapped onto $\mathbb{R} \setminus I(t)$. The slit $\gamma[0, t]$ has two sides $\gamma_1[0, t]$ and $\gamma_2[0, t]$ which consist of the same points but determine different prime ends, except for its tip, $f(\gamma_k[0, t]) = I_k(t), k = 1, 2, I_1(t) \cup I_2(t) = I(t)$, $I_1(t) \cap I_2(t)$ is the image of the tip.

The harmonic measures $\omega(f^{-1}(i, t); \gamma_k[0, t], \mathbb{H} \setminus \gamma[0, t])$ of γ_k at $f^{-1}(i, t)$ with respect to $\mathbb{H} \setminus \gamma[0, t]$ are defined by the functions ω_k which are harmonic on $\mathbb{H} \setminus \gamma[0, t]$ and continuously extended on $(\mathbb{H} \setminus \gamma[0, t]) \cup \gamma_1[0, t]$ and on $(\mathbb{H} \setminus \gamma[0, t]) \cup \gamma_2[0, t]$, $\omega_k|_{\gamma_k[0, t]} = 1, \omega_k|_{\mathbb{R} \cup (\gamma[0, t] \setminus \gamma_k[0, t])} = 0, k = 1, 2$. Denote

$$m_k(t) = \omega(f^{-1}(i, t); \gamma_k[0, t], \mathbb{H} \setminus \gamma[0, t]), \quad k = 1, 2.$$

It was proved in [26] that if $\gamma(t)$ is continuously differentiable on $[0, T]$ and tangential at 0 to the straight line under the angle $\frac{\pi}{2}(1 - c)$ to $\mathbb{R}, -1 < c < 1$,

then

$$\lim_{t \to 0+} \frac{m_1(t)}{m_2(t)} = \frac{1 \mp c}{1 \pm c}$$

provided the asymptotic relation

$$s = s(t) = C\sqrt{t} + o(\sqrt{t}), \quad s \to 0+, \quad C \neq 0, \tag{1}$$

for the arclength parameter $s = s(t)$ is valid. Wu and Dong showed in [30] that (1) holds if $\gamma(t)$ is continuously differentiable on $(0, T]$, $\gamma'(t) \neq 0$ and

$$\lim_{t \to 0+} \arg \gamma'(t) = \frac{\pi}{2}(1 - c).$$

Note that, under assumptions in [26] and [30], the slit $\gamma(t)$ is generated by a driving function $\lambda(t)$ which belongs to the class $\mathrm{Lip}(\frac{1}{2})$ on $[0, T]$, that is, $f(z, t)$ solves the Loewner ordinary differential equation

$$\frac{df(z, t)}{dt} = \frac{2}{f(z, t) - \lambda(t)}, \quad f(z, 0) \equiv z, \quad z \in \mathbb{H}, \quad 0 \leq t \leq T_z; \tag{2}$$

see, e.g., [15].

The case of the curve $\gamma(t)$ tangential to $\mathbb{R}$ at 0 corresponds to $c = \pm 1$ and is more complicated. For the partial case of a circular arc $\gamma(t)$ tangential at 0 to $\mathbb{R}$, it was shown in [25] that $\gamma(t)$ is generated by a driving function $\lambda(t) \in \mathrm{Lip}(\frac{1}{3})$ in the Loewner equation (2). More precisely, $\lambda(t)$ is expanded in powers of $\sqrt[3]{t}$. Besides, if the circular arc $\gamma(t)$ of radius 1 is situated in the upper half-plane, and $m_1(t)$ and $m_2(t)$ denote the harmonic measures of the "right" and the "left" sides of $\gamma[0, t]$, respectively, then

$$\lim_{t \to 0+} \frac{m_1(t)}{m_2^2(t)} = \frac{1}{2\pi}.$$

Observe that the circular arc $\gamma(t)$ has first order tangency to $\mathbb{R}$, i.e.,

$$\gamma(0) = \gamma'(0) = 0, \quad \gamma''(0) \neq 0,$$

or, in other terminology, $\gamma(t) = x(t) + iy(t)$, where

$$y(t) = ax^2(t) + o(x^2), \quad x \to 0, \quad a \neq 0.$$

We say that $\gamma(t)$ has pth order tangency at 0 to $\mathbb{R}$, $p > 0$, if

$$y(t) = ax^{p+1}(t) + o(x^{p+1}), \quad x \to 0, \quad a \neq 0. \tag{3}$$

The circular slit example and reasonings in [22] make it reasonable to propose the following conjectures.

Conjecture 1 If a curve $\gamma(t) \subset \mathbb{H} \cup \{0\}$ is (right side) tangential of order $p > 0$ at 0 to $\mathbb{R}$, and $m_1(t)$ is the harmonic measure of the "right" side of $\gamma[0, t]$ and $m_2(t)$ is the harmonic measure of the "left" side of $\gamma[0, t]$, then

$$0 < \lim_{t \to 0+} \frac{m_1(t)}{m_2^{p+1}(t)} < \infty.$$

Conjecture 2 If a curve $\gamma(t) \subset \mathbb{H} \cup \{0\}$ is tangential of order $p > 0$ at 0 to $\mathbb{R}$, then the function $f(z, t) : \mathbb{H} \setminus \gamma[0, t] \to \mathbb{H}$ with the hydrodynamical normalization at infinity is a solution to the Loewner differential equation (2) with the driving function $\lambda(t)$ satisfying the asymptotic expansion

$$\lambda(t) = b^{p+2}\!\sqrt{t} + o(^{p+2}\!\sqrt{t}), \quad t \to 0+, \quad b \neq 0.$$

Conjecture 3 The driving function

$$\lambda(t) = {}^{p+2}\!\sqrt{t} + o({}^{p+2}\!\sqrt{t^{1+\epsilon}}), \quad t \to 0+, \quad p > 0, \quad \epsilon > 0,$$

in the Loewner differential equation (2), for $t > 0$ small enough, generates a slit solution $f(z, t) : \mathbb{H} \setminus \gamma[0, t] \to \mathbb{H}$, where the slit $\gamma(t)$, $\gamma(0) = 0$, is tangential of order p at 0 to $\mathbb{R}$.

The harmonic measures of slit sides for solutions of the Loewner equation are described by singular solutions to (2) which do not satisfy the local uniqueness conditions. Singular solutions appear at the moment T_z when the denominator in the right-hand side of (2) vanishes, $f(z, T_z) = \lambda(T_z)$. The initial singular solutions correspond to $z = 0$ and $t = T_0 = 0$. We distinguish the maximal singular solution $f_1(0, t)$ and the minimal singular solution $f_2(0, t)$, $f_2(0, t) < \lambda(t) < f_1(0, t)$ for $t > 0$. The intervals $I_1(t) = [\lambda(t), f_1(0, t)]$ and $I_2(t) = [f_2(0, t), \lambda(t)]$ are mapped by $f(z, t)$ onto the "right" and the "left" sides of $\gamma[0, t]$, respectively.

There are not so many sources to evaluate singular solutions. The method of power series is one of them provided $\lambda(t)$ is expanded in powers of t^α for rational α. For example, the well known case $\lambda(t) = c\sqrt{t}$ immediately gives $f_1(0, t) = a_1\sqrt{t}$ and $f_2(0, t) = a_2\sqrt{t}$ with a_1 and a_2 satisfying the quadratic equation $a^2 - ac - 4 = 0$. From the other side, the case $\lambda(t) = \sqrt[3]{t}$ considered in [22] leads to two series

$$f_k(0, t) = \sum_{n=1}^{\infty} a_n^{(k)}\sqrt[3]{t^n}, \quad k = 1, 2,$$

but the series for $f_1(0, t)$ diverges for any $t \neq 0$. However, it admits summation by the Borel regular method, see [3]. This implies that $g_1(\tau) := f_1(0, \tau^3)$, $\tau = \sqrt[3]{t}$, is infinitely differentiable at $\tau = 0$. Similarly, $g_2(\tau) := f_2(0, \tau^3)$ is also infinitely differentiable at $t = 0$.

Return to the circular arc $\gamma(t)$. It follows from [25] that in this case

$$\lambda(t) = f_1(0, t) + 2f_2(0, t), \quad f_1(0, t) = f_2(0, t) + 2\sqrt{-\pi f_2(0, t)}$$

and $\sqrt{-f_2(0, t)}$ satisfies the 4th order algebraic equation

$$3w^2 + 4w\sqrt{-\pi w} + 6t = 0.$$

The point $t = 0$ is critical for the algebraic function defined by this algebraic equation. There is exactly one function among four functional elements which has the fixed point $t = 0$ and is real-valued for $t > 0$. This function denoted by $\sqrt{-f_2(0, t)}$ is represented by the convergent Puiseux series in powers of $\sqrt[3]{t}$. Hence $f_2(0, t), f_1(0, t)$ and $\lambda(t)$ also are represented by convergent series in powers of $\sqrt[3]{t}$.

Comparing the two examples we ask the following questions.

Question 1 Suppose that $\lambda(t)$ is represented by a convergent series in powers of $\sqrt[N]{t}, N \in \mathbb{N}, N \geq 3$. Is it true that the singular solutions $f_k(0, t), k = 1, 2$, to the Loewner differential equation (2) with the given driving function $\lambda(t)$ are infinitely differentiable as functions of $\sqrt[N]{t}$ on $[0, T]$ for $T > 0$ small enough.

Question 2 What are the conditions on the driving function $\lambda(t)$ in the Loewner differential equation (2) that provide analyticity of both its singular solutions as functions of $\sqrt[N]{t}, N \in \mathbb{N}, N \geq 3$.

Question 3 Suppose that $h(t), h(0) = 0$, is infinitely differentiable (or analytic) as a function of $\sqrt[N]{t}, N \in \mathbb{N}, N \geq 3$, on $[0, T]$ for $T > 0$ small enough. Is it possible to restore a driving function $\lambda(t)$ for the Loewner differential equation (2) so that $h(t)$ is one of the singular solutions $f_k(0, t)$ to (2) in a neighborhood of $t = 0$.

It is worth attention that an analytic slit emanating from the origin which is tangential to the real axis forms the analytic cusp, i.e., the domain with analytic boundary curves intersecting under the zero inner angle. We will discuss the asymptotic behavior of a conformal mapping from the half-plane onto a cusp.

3   Conformal Mapping onto an Analytic Cusp

A conformal mapping from $\mathbb{D}$ or $\mathbb{H}$ onto a simply connected domain $D \subset \mathbb{C}$ with an analytic arc Γ on the boundary ∂D of D can be extended analytically through the preimage of Γ. The situation is more complicated when Γ consists of two arcs Γ_1 and Γ_2 meeting at a boundary point even if both Γ_1 and Γ_2 are analytic curves. Without restrictions assume that Γ_1 and Γ_2 intersect at the origin. We refer to the book [18] by Pommerenke, Sections 3.3–3.4, for an overview. In particular, Pommerenke gives in [18, p. 57], the asymptotic behavior of $f : \mathbb{D} \to f(\mathbb{D})$ in the case when $\partial f(\mathbb{D})$ has a Dini-smooth corner of opening $\pi\alpha, 0 < \alpha < 2$.

Warschawski [28] proved a geometric criterion under which a mapping function $f(z)$ behaves like z^α for a domain with opening angle $\pi\alpha$, $0 < \alpha \le 2$; see also the results of Lehman [14] about a mapping function developed in a certain generalized power series.

There are not so many works on determining the behavior of the mapping function at a cusp of a domain with a piecewise analytic boundary. We say that a domain $D \subset \mathbb{C}$ with $0 \in \partial D$ has an *analytic cusp* at 0 if ∂D at 0 consists of two regular analytic curves such that the interior angle of D at 0 vanishes. For instance, in [29], Warschawski considered conformal mappings of infinite strips, which correspond to domains with two zero interior angles, i.e., cusps, at infinity.

Kaiser investigated in [9] analytic cusps possessing a perturbation property. For a domain D, denote the boundary curves of an analytic cusp at $0 \in \partial D$ by Γ_1 and Γ_2, $\Gamma_k = \Gamma_k[0, \epsilon] = te^{i\sphericalangle_k(t)}$ where $\sphericalangle_k(t)$ are real power series convergent in a neighborhood of $t = 0$, $\sphericalangle_k(0) = 0$, $k = 1, 2$, $\epsilon > 0$, $\sphericalangle_D(t) := \sphericalangle_2(t) - \sphericalangle_1(t)$. For $h(t) = \sum_{n=1}^{\infty} a_n t^n$, $\mathrm{ord}(h) := \min\{n \in \mathbb{N} : a_n \neq 0\}$. Let $d := \mathrm{ord}(\sphericalangle_D)$ and $a > 0$ be such that $\lim_{t \to 0} \sphericalangle_D(t)/at^d = 1$. We say that D has small perturbations of angles if $\min\{\mathrm{ord}(\sphericalangle_1), \mathrm{ord}(\sphericalangle_2)\} = d$ and $\mathrm{ord}(\sphericalangle_D(t) - at^d) > 2d$. Kaiser proved in [9] that, for $D \subset \mathbb{C}$ with an analytic cusp $0 \in \partial D$ and having small perturbations of angles, the conformal map $f : \mathbb{H} \to D, f(0) = 0$, satisfies the asymptotic relation

$$\lim_{z \to 0} f(z) \left(-\frac{\pi}{da \log |z|} \right)^{-\frac{1}{d}} = 1, \quad \mathrm{Im}\, z > 0. \tag{4}$$

Kaiser's parametrization of the cusp boundary curves means that Γ_1 and Γ_2 have dth order tangency at 0, $d \in \mathbb{N}$. The evident idea is to generalize the asymptotic relation (4) for all $d > 0$.

Question 4 Suppose that a domain $D \subset \mathbb{C}$ has an analytic cusp at the origin, and the cusp boundary curves Γ_1 and Γ_2 have dth order tangency at 0. Under what conditions, if any, does relation (4) hold for a conformal map $f : \mathbb{H} \to D, f(0) = 0$, with a certain $a > 0$ and $d > 0$?

Assume that $\Gamma_2 = (0, \epsilon) \subset \mathbb{R}$, $\epsilon > 0$. A parametrization of Γ_1 can be motivated by the explicit formula $f : \mathbb{H} \to \mathbb{H} \setminus \gamma$ in [25] with the hydrodynamic normalization at infinity where γ is the circular arc tangential to $\mathbb{R}$. Set $f(0) = 0$ and rewrite $f(z)$ from [25] in the form

$$\frac{1}{f(z)} = \frac{1}{2\pi} \left[\log \frac{z + \alpha}{z} + \frac{2\pi - \alpha}{z + \alpha} \right], \tag{5}$$

$\alpha > 0$ depends on the second endpoint of γ.

On the boundary of $\mathbb{H} \setminus \gamma$ we see the corner of opening π at $w = 0$ and the cusp with the mutual boundary arc γ. Representation (5) allows us to expand $f(z)$ near

$z = 0, z \in \mathbb{H}$,

$$f(z) = \frac{2\pi}{-\log z + \log(z + \alpha) + (2\pi - \alpha)/(z + \alpha)} =$$

$$\frac{-2\pi}{\log z}\left[1 + \sum_{n=1}^{\infty}\left(\frac{\log(z + \alpha) + (2\pi - \alpha)/(z + \alpha)}{\log z}\right)^n\right] = \frac{-2\pi}{\log z}\left(1 + \sum_{n=1}^{\infty}\frac{c_n(\Phi_n(z))^n}{\log^n z}\right)$$

with functions Φ_n which are analytic in a neighborhood of $z = 0$ and real-valued for real z, and real coefficients c_n, $n \geq 1$. All $\Phi_n(z)$ and c_n can be computed explicitly. We ask what are the generalizations of this representation for an arbitrary cusp.

Question 5 Suppose that f, satisfying $\lim_{z \to 0} f(z) = 0$, $z \in \mathbb{H}$, is a conformal map from $\mathbb{H}$ onto a cusp with a boundary segment $[0, \delta] \subset \mathbb{R}$ and a boundary curve $\gamma \subset \mathbb{H} \cup \{0\}$ having dth order tangency at 0 to $\mathbb{R}$,

$$\gamma(z) = \left[\frac{-2\pi}{a \log z}\left(1 + \sum_{n=1}^{\infty}\frac{c_n(\Phi_n(z))^n}{\log^n z}\right)\right]^{\frac{1}{d}}, \quad a > 0, \quad z \in (-\epsilon, 0), \quad \epsilon > 0, \quad (6)$$

$c_n \in \mathbb{R}$, $n \geq 1$, and the series in the right-hand side of (6) converges in a neighborhood of the origin. What are the conditions, if any, for functions Φ_n, $n \geq 1$, so that (4) holds?

Question 6 Suppose that f satisfies the conditions of Question 5 for a certain boundary curve γ of an analytic cusp. What representations of γ different from (6), if any, are possible for γ so that (4) holds?

4 Asymptotic Conformal Welding via Loewner-Kufarev Evolution

For the unit disk $\mathbb{D}$ and the complement $\mathbb{D}^* = \{z : |z| > 1\}$ to the closure of $\mathbb{D}$, let $f : \mathbb{D} \to \Omega$ and $F : \mathbb{D}^* \to \Omega^*$ be conformal maps where the domain Ω is bounded by a closed Jordan curve Γ, and Ω^* is the unbounded complementary component of Γ. The composition $F^{-1} \circ f$ determines a homeomorphism of the unit circle $\mathbb{T} = \partial\mathbb{D} = \partial\mathbb{D}^*$ which is called a conformal welding. Suppose that $0 \in \Omega$, $f(0) = 0, f'(0) > 0$, and $F(\infty) = \infty, F'(\infty) > 0$.

An asymptotic conformal welding for domains close to $\mathbb{D}$ was proposed in [21]. It is based on asymptotic formulas for conformal mappings onto these domains. The bounded version of $f : \mathbb{D} \to \Omega$ was obtained by Siryk in [27], and the unbounded version of $F : \mathbb{D}^* \to \Omega^*$ is given in [21].

Another approach proposed in [23] uses the Loewner-Kufarev evolution which produces asymptotics for mappings onto domains close to $\mathbb{D}$. The Loewner equation [16] is a differential equation obeyed by a family of continuously varying univalent functions $f(z,t)$, $f(0,t) = 0$, from $\mathbb{D}$ onto a domain with a slit formed by a continuously increasing arc. The real parameter t characterizes the length of the arc and can be chosen so that $f(z,t) = e^{-t}z + \dots$, $t \geq 0$. Kufarev [10] and Pommerenke [17] generalized this idea to a wider class of domains. We present here the "decreasing" version of the Loewner-Kufarev evolution, see [5] for details of the connection between "decreasing" and "increasing" cases in the Loewner-Kufarev theory. Given a chain of domains $\Omega(t)$, $0 \in \Omega(t_2) \subset \Omega(t_1)$, $0 \leq t_1 < t_2 < T$, and functions $w = f(z,t) : \mathbb{D} \to \Omega(t)$ normalized as above, there exist functions $p(z,t)$, $p(\cdot,t)$ are analytic in $\mathbb{D}$, $p(z,\cdot)$ are measurable for $0 \leq t < T$, and p are from the Carathéodory class which means that

$$p(z,t) = 1 + p_1(t)z + p_2(t)z^2 + \dots, \quad \operatorname{Re} p(z,t) > 0, \quad z \in \mathbb{D}, \quad 0 \leq t < T,$$

such that

$$\frac{\partial f(z,t)}{\partial t} = -z \frac{\partial f(z,t)}{\partial z} p(z,t) \tag{7}$$

for $z \in \mathbb{D}$ and for almost all $t \in [0,T)$, T may be ∞.

In case when $\Omega(t)$ are bounded and $\Omega^*(t)$ is the exterior of $\Omega(t)$, let $w = F(z,t)$ be the unique conformal map from $\mathbb{D}^*$ onto $\Omega^*(t)$ such that $F(\infty,t) = \infty$, $F'(\infty,t) > 0$. Normalize the maps so that

$$F(z,t) = e^{-\tau(t)}z + b_0(t) + b_1(t)z^{-1} + \dots, \quad z \to \infty,$$

with a differentiable real function $\tau = \tau(t)$, $\tau(0) = 0$, $\tau'(t) > 0$, $t \geq 0$. Then $F(z,t)$ satisfies the Loewner-Kufarev equation

$$\frac{\partial F(z,t)}{\partial t} = -z \frac{\partial F(z,t)}{\partial z} q(z,t) \frac{d\tau(t)}{dt}, \quad z \in \mathbb{D}^*, \quad 0 \leq t < T, \tag{8}$$

where $q(\cdot,t)$ are analytic in $\mathbb{D}^*$, $q(z,\cdot)$ are measurable for $0 \leq t < T$, and

$$q(z,t) = 1 + \frac{q_1(t)}{z} + \frac{q_2(t)}{z^2} + \dots, \quad \operatorname{Re} q(z,t) > 0, \quad z \in \mathbb{D}^*, \quad 0 \leq t < T.$$

The following theorem was proved in [23].

Theorem 1 *Let the driving function $p(\cdot,t)$ from the Carathéodory class in (7) be C^2 in $\overline{\mathbb{D}}$ for $0 \leq t < T$, $p(z,\cdot)$ be continuous in $[0,T)$ for $z \in \overline{\mathbb{D}}$, $p(z,t)$, $p'(z,t)$ and $p''(z,t)$ be bounded in $\overline{\mathbb{D}} \times [0,T)$. Then, for solutions $f(z,t)$ to (7) with $\Omega(0) = \mathbb{D}$, $\Omega(t) = f(\mathbb{D},t)$, $\partial\Omega(t) = \Gamma(t)$, and the corresponding functions $F(\cdot,\tau(t))$, $F(\mathbb{T},\tau(t)) = f(\mathbb{T},t)$, the conformal welding $\varphi : \mathbb{T} \to \mathbb{T}$ of the curve $\Gamma(t)$,*

$\varphi = \varphi(\tilde\varphi)$, *satisfies the relation*

$$\varphi = \tilde\varphi + 2\, Im\, p(e^{i\tilde\varphi}, 0)t + o(t), \quad t \to 0+ . \tag{9}$$

It is quite interesting what characteristics are contained in the next asymptotic term.

Question 7 What is the second term in the asymptotic representation (9)?

Note that the pair (f, F) maps $\mathbb{D}$ and $\mathbb{D}^*$ onto nonoverlapping domains $\Omega(t)$ and $\Omega^*(t)$, respectively. Compare expansions for f and F and apply Lebedev's theorem [12], see also [13, p. 223], which states, that $\tau(t) \le t$ with equality only in the case when $f(\mathbb{D}, t)$ is a disk centered at the origin. It was proved in [23] that

$$\tau(t) = t + o(t), \quad t \to 0+ . \tag{10}$$

Question 8 What is the second term in the asymptotic representation (10)?

To solve the inverse problem to Theorem 1 we have to find conditions on the driving function $p(\cdot, t)$ in the Loewner-Kufarev equation (7) which are implied by the asymptotic relations (9) and (10). In other words, Theorem 1 provides the sufficient conditions for the asymptotic conformal welding while in the inverse problem we are looking for necessary conditions.

Question 9 What are the necessary conditions on the driving function in the Loewner-Kufarev equation for the asymptotic relations (9) and (10)?

5 The Bombieri Conjecture for Univalent Functions

Let S stand for the class of all holomorphic and univalent functions

$$f(z) = z + \sum_{n=2}^{\infty} a_n z^n, \quad z \in \mathbb{D}.$$

During the long history of univalent functions, the famous Koebe function

$$K(z) = \frac{z}{(1-z)^2} = \sum_{n=1}^{\infty} n z^n \in S$$

was known to be extremal in many problems.

Bombieri [2] posed the problem to find *Bombieri's numbers*

$$\sigma_{mn} := \liminf_{a_m \to m} \frac{n - Re\, a_n}{m - Re\, a_m} = \liminf_{S \ni f \to K} \frac{n - Re\, a_n}{m - Re\, a_m}, \quad m, n \ge 2,$$

$f \to K$ locally uniformly in $\mathbb{D}$. Bombieri conjectured that $\sigma_{mn} = B_{mn}$, where

$$B_{mn} := \min_{\theta \in [0,2\pi]} \frac{n \sin \theta - \sin(n\theta)}{m \sin \theta - \sin(m\theta)}.$$

Bshouty and Hengartner [4] proved Bombieri's conjecture for the class S_R of functions $f \in S$ with real coefficients. Since $\sigma_{mn} \geq 0$ and $B_{mn} \geq 0$ by de Branges' proof [6] of the Bieberbach conjecture $|a_n| \leq n, n \geq 2$, it follows from [4] that $\sigma_{mn} = 0$ provided $B_{mn} = 0$. Indeed, in this case,

$$\sigma_{mn} = \liminf_{S \ni f \to K} \frac{n - \operatorname{Re} a_n}{m - \operatorname{Re} a_m} \leq \liminf_{S_R \ni f \to K} \frac{n - \operatorname{Re} a_n}{m - \operatorname{Re} a_m} = B_{mn} = 0.$$

In particular, $\sigma_{43} = B_{43} = \sigma_{23} = B_{23} = 0$, which can be verified directly. In general, $B_{mn} = 0$ for even m and odd n. Besides, Bombieri proved in [2] that

$$\liminf_{a_2 \to 2} \frac{3 - \operatorname{Re} a_3}{\sqrt{(2 - \operatorname{Re} a_2)^3}} = \frac{8}{3}.$$

For (m, n) such that $\sigma_{mn} = 0$, define

$$\beta_{mn} := \sup \left\{ \alpha : \liminf_{f \to K} \frac{n - \operatorname{Re} a_n}{(m - \operatorname{Re} a_m)^\alpha} = 0, f \to K \right\}.$$

The above result of Bombieri shows that $\beta_{23} = \frac{3}{2}$ and makes it sensible to ask the following question.

Question 10 For (m, n) such that $\sigma_{mn} = 0$, is it true that

$$0 < \liminf_{f \to K} \frac{n - \operatorname{Re} a_n}{(m - \operatorname{Re} a_m)^{\beta_{mn}}} < \infty?$$

We believe that the answer to Question 10 is affirmative, and the problem to evaluate β_{mn} becomes actual.

Bombieri's conjecture in the whole class S was disproved by Greiner and Roth [7] for $n = 2$ and $m = 3$. They showed that

$$\sigma_{32} = \frac{e - 1}{4e} < \frac{1}{4} = B_{32}.$$

In continuation of this contribution, it was shown in [24] that

$$\sigma_{42} = 0.050057 \ldots < B_{42} = 0.1, \quad \sigma_{24} = 0.969556 \ldots < B_{24} = 1,$$

$$\sigma_{34} = 0.791557 \ldots < B_{34} = 0.828427 \ldots .$$

Such examples encourage us to ask whether this is a general rule.

Question 11 For (m, n) such that $\sigma_{mn} > 0$, is it true that

$$\sigma_{mn} < B_{mn}?$$

The search of Bombieri's numbers $\sigma_{mn} > 0$ is reduced to estimates of linear functionals over S; see [7, 24]. Namely,

$$\sigma_{mn} = \sup\{\lambda \in \mathbb{R} : \text{Re}(a_n - \lambda a_m) \text{ is locally maximized on } S \text{ by } K(z)\}.$$

In case Question 10 is answered affirmatively, the problem is reduced to estimating

$$\sup\{\lambda \in \mathbb{R} : \text{Re}(n - a_n) - \lambda (\text{Re}(m - a_m))^{\beta_{mn}} \text{ is locally minimized on } S \text{ by } K(z)\}.$$

The methods developed in [20] and applied in [24] can be useful also when $\sigma_{mn} = 0$.

Another problem posed earlier by Bombieri (cf. [8, problem 6.3]) relates to estimates

$$|n - |a_n|| \le d_n(2 - |a_2|), \quad f \in S, \quad n \ge 3,$$

if such constants $d_n > 0$ exist. Though these problems are close to each other, they require their own proofs which appear sometimes to be rather different.

Question 12 What is common and what is different between the evaluation problems for

$$\liminf_{f \to K} \frac{n - |a_n|}{m - |a_m|} \quad \text{and} \quad \liminf_{f \to K} \frac{n - \text{Re } a_n}{m - \text{Re } a_m}, \quad m, n \ge 2?$$

Acknowledgements This work has been supported by the Russian Science Foundation under grant 17-11-01229.

References

1. I.M. Anderson, K.F. Barth, D.A. Brannan, Research problems in complex analysis. Bull. Lond. Math. Soc. **9**, 129–162 (1977)
2. E. Bombieri, On the local maximum property of the Koebe function. Invent. Math. **4**, 26–67 (1967)
3. E. Borel, Mémoire sur les séries divergents. Ann. Sc. Éc. Norm. Super. (3) **16**, 9–131 (1899); See also: Borel E. Leçons sur les Séries Divergents, Paris, 1928
4. D. Bshouty, W. Hengartner, A variation of the Koebe mapping in a dense subset of S. Can. J. Math. **39**, 54–73 (1987)
5. M.D. Contreras, S. Díaz-Madrigal, P. Gumenyuk, Local duality in Loewner equations. J. Nonlinear Convex Anal. **15**, 269–297 (2014)
6. L. de Branges, A proof of the Bieberbach conjecture. Acta Math. **154**(1–2), 137–152 (1985)
7. R. Greiner, O. Roth, On support points of univalent functions and a disproof of a conjecture of Bombieri. Proc. Am. Math. Soc. **129**, 3657–3664 (2001)

8. W.K. Hayman, *Research Problems in Function Theory* (The Athlone Press, University of London, London, 1967)
9. T. Kaiser, Asymptotic behaviour of the mapping function at an analytic cusp with small perturbation of angles. Comput. Methods Funct. Theory **10**, 35–47 (2010)
10. P.P. Kufarev, On one-parameter families of analytic functions. Rec. Math. [Mat. Sbornik] N.S. **13**(55), 87–118 (1943)
11. G.F. Lawler, *Conformally Invariant Processes in the Plane* (American Mathematical Society, Providence, 2005)
12. N.A. Lebedev, Application of the area principle to the problems for nonoverlapping domains. Tr. Math. Inst. Acad. Sci. USSR **60**, 211–231 (1961)
13. N.A. Lebedev, *The Area Principle in the Theory of Univalent Functions* (Nauka, Moscow, 1975)
14. R.S. Lehman, Development of the mapping function at an analytic corner. Pac. J. Math. **7**, 1437–1449 (1957)
15. J. Lind, D.E. Marshall, S. Rohde, Collisions and spirals of Loewner traces. Duke Math. J. **154**, 527–573 (2010)
16. K. Löwner, Untersuchungen über schlichte konforme Abbildungen des Einheitskreises. I. Math. Ann. **89**, 103–121 (1923)
17. Ch. Pommerenke, Über die Subordination analytischer Funktionen. J. Reine Angew. Math. **218**, 159–173 (1965)
18. Ch. Pommerenke, *Boundary Behaviour of Conformal Maps* (Springer, Berlin, 1992)
19. D.V. Prokhorov, Relation of measures of boundary sets under a mapping of a disk onto a plane with cuts. Izv. Vyssh. Uchebn. Zaved. Mat. (4), 53–55 (1985) (Russian); English translation: Soviet Math. (Iz. VUZ) **29**(4), 66–69 (1985)
20. D.V. Prokhorov, Sets of values of systems of functionals in classes of univalent functions. Mat. Sbornik **181**(12), 1659–1677 (1990); English translation: Math USSR Sbornik **71**(2), 499–516 (1992)
21. D. Prokhorov, Conformal welding for domains close to a disk. Anal. Math. Phys. **1**, 101–114 (2011)
22. D. Prokhorov, Exponential driving function for the Löwner equation, in *60 Years of Analytic Functions in Lublin*, ed. by J. Szynal (Innovatio Press Scientific, Lublin, 2012), pp. 69–83
23. D. Prokhorov, Asymptotic conformal welding via Löwner-Kufarev evolution. Comput. Methods Funct. Theory **13**, 37–46 (2013)
24. D. Prokhorov, A. Vasil'ev, Optimal control in Bombieri's and Tammi's conjectures. Georgian Math. J. **12**, 743–761 (2005)
25. D. Prokhorov, A. Vasil'ev, Singular and tangent slit solutions to the Löwner equation, in *Analysis and Mathematical Physics*, ed. by B. Gustafsson, A. Vasil'ev (Birkhäuser, Basel, 2009), pp. 455–463
26. D. Prokhorov, A. Zakharov, Harmonic measures of sides of a slit perpendicular to the domain boundary. J. Math. Anal. Appl. **394**, 738–743 (2012)
27. G.V. Siryk, On a conformal mapping of near domains. Uspekhi Matem. Nauk **9**, 57–60 (1956)
28. S.E. Warschawski, Über das Randverhalten der Ableitung der Abbildungsfunktion bei konformer Abbildung. Math. Z. **35**, 321–456 (1932)
29. S.E. Warschawski, On conformal mapping of infinite strips. Trans. Am. Math. Soc. **51**, 280–335 (1942)
30. H.H. Wu, X.H. Dong, Driving functions and traces of the Loewner equation. Sci. China Math. **57**, 1615–1624 (2014)

Comparison Moduli Spaces of Riemann Surfaces

Eric Schippers and Wolfgang Staubach

Dedicated to the memory of Alexander Vasil'ev

Abstract We define a kind of moduli space of nested surfaces and mappings, which we call a comparison moduli space. We review examples of such spaces in geometric function theory and modern Teichmüller theory, and illustrate how a wide range of phenomena in complex analysis are captured by this notion of moduli space. The paper includes a list of open problems in classical and modern function theory and Teichmüller theory ranging from general theoretical questions to specific technical problems.

Keywords Conformal invariants • Conformal mappings • Extremal problems • Geometric function theory • Kernel functions • Moduli spaces • Riemann surfaces • Quasiconformal maps • Teichmuller theory • Weil-Petersson metric

1991 Mathematics Subject Classification 30C35, 30C40, 30C55, 30C75, 30F30, 30F60, 32G15, 32Q15, 37F30, 46G20, 53B35, 53C55

1 Introduction

Theorems in complex analysis should be conformally invariant. In any branch of mathematics, one does not distinguish objects with respect to some equivalence; in complex analysis, this is conformal equivalence. Thus, if one alters all objects in the

E. Schippers (✉)
Department of Mathematics, University of Manitoba, Winnipeg, MB, Canada R3T 2N2
e-mail: eric_schippers@umanitoba.ca

W. Staubach
Department of Mathematics, Uppsala University, Box 480, Uppsala 751 06, Sweden
e-mail: wulf@math.uu.se

© Springer International Publishing AG 2018
M. Agranovsky et al. (eds.), *Complex Analysis and Dynamical Systems*,
Trends in Mathematics, https://doi.org/10.1007/978-3-319-70154-7_13

hypothesis of a theorem by applying a biholomorphism, then the theorem should apply to the new situation.

That is not to say that theorems which are not manifestly conformally invariant are not interesting. Indeed any pair of points on two distinct Riemann surfaces are contained in locally biholomorphic neighbourhoods by the Riemann mapping theorem, so local phenomena need not ever refer to conformal invariance. Furthermore we have powerful uniformization theorems at our disposal which reduce problems involving Riemann surfaces or mappings between them to canonical cases. This accounts for the fact that many deep results in complex analysis need not mention conformal invariance. However, it is still profitable to attempt to formulate—or reformulate—complex analytic results invariantly.

It is not always obvious how to do this. In the case of results about compact Riemann surfaces, it is straightforward. These Riemann surfaces are characterized by a finite set of conformal invariants, or by a finite-dimensional function space which is easily defined conformally invariantly, such as the vector space of holomorphic abelian differentials. However, things change radically when the Riemann surfaces have boundary curves. For example, the function spaces become infinite-dimensional, boundary values might enter their definition, and furthermore one must become concerned with boundary regularity of the biholomorphism. Thus many important complex analytic phenomena do not easily fit into the framework which is so successful for compact surfaces.

An example will illustrate the point. The universal Teichmüller space can be identified with the set of suitably normalized domains in the sphere bounded by quasicircles, and can be thought of as the set of deformations of the complex structure of the unit disk. Consider the following reasonable objection to this definition, encountered occasionally by the authors: by the Riemann mapping theorem, these domains (or deformations) are all conformally equivalent. So why would one care about such a space? In other words, by conformal invariance, the universal Teichmüller space is of little interest. However, if one takes this objection too seriously, one will never find out that the universal Teichmüller space contains all Teichmüller spaces, or learn the beautiful geometric and analytic theory of quasidisks and quasiconformal mappings. Thus this naive approach to conformal invariance is not appropriate for quasiconformal Teichmüller theory.

How then do we introduce conformal invariance into this picture? This is done by taking into account the fact that the universal Teichmüller space involves two Riemann surfaces: the quasidisk and the sphere itself. Thus it is a moduli space of embeddings of disks into the sphere.

In this paper, we outline a general notion of moduli spaces of pairs of surfaces, or the closely related moduli spaces of mappings. We call them comparison moduli spaces. This notion of moduli spaces and the corresponding notion of conformal invariance capture a wide range of complex analytic phenomena. We show how comparison moduli spaces are present in classical geometric function theory, and along the way draw attention to some deep ideas of Nehari and Schiffer which in our opinion have received too little attention. We also review some modern manifestations of the idea in quasiconformal Teichmüller theory and conformal field theory, with special attention to the so-called Weil-Petersson class Teichmüller space. The paper also contains a list of open problems.

In attempting to convey a unifying viewpoint, we have chosen to paint a broad rather than a specific picture. Thus, our choice of topics is selective; especially so when we are dealing with older topics. We have not attempted to make a survey of current or past research on each topic, and have confined ourselves to providing a few key references which should be viewed as entry points to the literature. However, since the Weil-Petersson class Teichmüller theory of non-compact surfaces is rather new, we made an exception for this case and gave a more comprehensive literature review.

The organization of this paper is as follows. In Sect. 2, we define and give examples of comparison moduli spaces and the corresponding notion of conformal invariance. In Sect. 3 we review examples of such moduli spaces in geometric function theory, in the work of Nehari and Schiffer. In Sect. 4 we give a brief overview of quasiconformal Teichmüller theory, and the rigged moduli space arising in conformal field theory. We describe the correspondence between these two, which also demonstrates that Teichmüller space is, up to a discontinuous group action, a comparison moduli space. Finally, in Sect. 5, we give an overview of new developments in Weil-Petersson Teichmüller theory. We also briefly list some of the applications of this new field.

2 Comparison Moduli Spaces

By a comparison moduli space, we mean a moduli space of one of two types, which we now describe. Our purpose is to illustrate the similarity of a wide range of ideas, and how the notion of comparison moduli space is useful in capturing complex analytic phenomena. Our purpose is *not* to give a Bourbaki styled axiomatization. Although the definitions of the comparison moduli spaces can easily be made precise, this introduces unnecessary abstractions. Thus, rather than taking this approach, we give them informal but general definitions. Precise definitions are reserved for the many examples throughout the paper.

2.1 Moduli Space of Nested Surfaces

Let R_1 and R_2 be Riemann surfaces such that $R_2 \subseteq R_1$. We will also usually equip the nested surfaces with extra data $a, b, c, \ldots$. We assume that for any biholomorphism $g : S_1 \to R_1$, there is an operation g^* on the data denoted $a \mapsto g^*(a)$ which takes data on (R_1, R_2) to data on (S_1, S_2). For example, a might be a specified point in S_1 and $g^*(a) = g^{-1}(a)$, while b might be a harmonic function on $S_1 \backslash \{a\}$ and $g^*b = b \circ g$. This operation should be invertible and respect composition: that is,

$(g_1 \circ g_2)^*(a) = g_2^*(g_1^*(a))$ and $(g^{-1})^* = (g^*)^{-1}$. We call this operation the *pull-back* under g, and call the set $\{(R_1, R_2, a, b, c, \ldots)\}$ the *configuration space* $\mathfrak{C}$.

Two elements of the configuration space are equivalent

$$(R_1, R_2, a, b, c, \ldots) \sim (S_1, S_2, A, B, C, \ldots)$$

if and only if there is a biholomorphism $g : S_1 \to R_1$ such that

$$g(S_2) = R_2, \quad g^*(a) = A, \quad g^*(b) = B, \quad g^*(c) = C, \ldots.$$

The *moduli space of nested surfaces* $\mathfrak{M}$ is the set of equivalence classes

$$\mathfrak{M} = \mathfrak{C}/\sim.$$

A *conformal invariant* is a function on the configuration space $I : \mathfrak{C} \to \mathbb{C}$ (or into $\mathbb{R}$) such that for any biholomorphic map $g : S_1 \to R_1$ such that $g(S_2) = R_2$ one has

$$I(R_1, R_2, a, b, c, \ldots) = I(S_1, S_2, g^*(a), g^*(b), g^*(c), \ldots). \tag{2.1}$$

Equivalently, it is a function on $\mathfrak{M}$.

Remark 2.1 The definition (2.1) can be generalized easily to allow for the possibility that I is a differential of order n on R_2 (that is, given in local coordinates by $h(z)dz^n$).

Finally, one might place restrictions on the Riemann surfaces (e.g. bounded by quasicircles, fixed genus) or the data (e.g. harmonic functions of finite Dirichlet energy, singularity of specified form at a point). It is required that these conditions are conformally invariant.

Example 2.1 Let $\mathfrak{C} = \{(R_1, R_2, z)\}$ such that R_1 and R_2 are hyperbolic Riemann surfaces, $R_2 \subseteq R_1$ and $z \in R_2$. Let $\lambda_i(z)^2 |dz|^2$ denote the hyperbolic metric on R_1 in any fixed local coordinate (with curvature normalized to -1 say). For a biholomorphism $g : S_1 \to R_1$ define $g^*(z) = g^{-1}(z)$. Define

$$I(R_1, R_2, z) = \lambda_1(z)/\lambda_2(z).$$

This is independent of the choice of coordinate and conformally invariant.

Another example of a conformal invariant is the following [62]. Defining $\Gamma_i = 2\frac{\partial}{\partial z} \log \lambda_i$ in some local coordinate, the quantity

$$I_2(R_1, R_2, z) = \lambda_2^{-1}(z)|\Gamma_2(z) - \Gamma_1(z)|$$

is independent of choice of coordinate and conformally invariant.

Example 2.2 Let $\mathfrak{C} = \{(R_1, R_2)\}$ where R_1 is a simply connected Riemann surface which is biholomorphic to the disk and $R_2 \subseteq R_1$ is bounded by a quasicircle in R_1. The condition that R_2 is bounded by a quasicircle in R_1 is conformally invariant.

Example 2.3 Let $\mathfrak{C} = \{(R_1, R_2, h)\}$ where R_1 and R_2 are doubly connected Riemann surfaces (i.e. biholomorphic to $r < |z| < R$ for $0 < r < R < \infty$), $R_2 \subseteq R_1$ and h is a harmonic function on R_1 of finite Dirichlet energy:

$$\iint_{R_1} dh \wedge \overline{dh} < \infty.$$

These conditions are conformally invariant. For a biholomorphism $g : S_1 \to R_1$ define $g^*h = h \circ g$. The quantity

$$I(R_1, R_2, h) = \iint_{R_1 \setminus R_2} dh \wedge \overline{dh}$$

is a conformal invariant.

Example 2.4 Let $\mathfrak{C} = \{(R_1, R_2, a, b, c)\}$ where R_1 is a Riemann surface biholomorphic to the Riemann sphere, $R_2 \subseteq R_1$ is bordered by a quasicircle in R_1, and a, b, c are points on ∂R_2. For a biholomorphism $g : S_1 \to R_1$ we define $g^*(a) = g^{-1}(a)$, and similarly for b and c. The moduli space $\mathfrak{M}$ is then the universal Teichmüller space. This will be discussed in Sect. 4.2 ahead.

2.2 Moduli Spaces of Mappings

We consider also moduli spaces of mappings between Riemann surfaces. That is, we define a configuration space $\mathfrak{C} = \{(R_0, f, R, a, b, c, \ldots)\}$ where R_0 is a fixed Riemann surface, R is a variable Riemann surface, $f : S \to R$ is an injective holomorphic map, and $a, b, c, \ldots$ are data as above. Pull-backs under biholomorphisms are defined as above. We define an equivalence relation $(R_0, f, R, a, b, c, \ldots) \sim (R_0, F, S, A, B, C, \ldots)$ if and only if there is a biholomorphism $g : S \to R$ such that $f = g \circ F$, $g^*(a) = A$, $g^*(b) = B$, $g^*(c) = C, \ldots$. The moduli space of mappings is

$$\mathfrak{M} = \mathfrak{C}/\sim.$$

A conformal invariant is a quantity I on $\mathfrak{C}$ such that for any biholomorphism $g : S \to R$

$$I(R_0, g \circ F, R, a, b, c, \ldots) = I(R_0, F, S, g^*(a), g^*(b), g^*(c), \ldots). \tag{2.2}$$

Equivalently, it is a function on $\mathfrak{M}$.

We give further examples. Denote the unit disk by $\mathbb{D} = \{z : |z| < 1\}$, the complex plane by $\mathbb{C}$, and the Riemann sphere by $\bar{\mathbb{C}}$.

Example 2.5 Let $\mathfrak{C} = \{(\mathbb{D}, f, R, z)\}$ where R is a simply connected Riemann surface conformally equivalent to $\mathbb{D}$, f is an injective holomorphic mapping such that $f(0) = z$. Then $\mathfrak{M}$ is in one-to-one correspondence with the set of bounded univalent functions $f : \mathbb{D} \to \mathbb{D}$ such that $f(0) = 0$.

Note that if we set $R_1 = R$ and $R_2 = f(\mathbb{D})$ then $\mathfrak{C}$ and $\mathfrak{M}$ corresponds to the comparison moduli spaces of Example 2.1, modulo an action by $\mathbb{S}^1$ (since the domain $f(\mathbb{D})$ does not uniquely determine f). Setting $z = 0$ and $R = \mathbb{D}$ we can compute the conformal invariant I_2 explicitly in terms of f, namely $I_2(\mathbb{D}, f, R, z) = |f''(0)/f'(0)^2|$. Since this is invariant under the $\mathbb{S}^1$-action $f(z) \mapsto e^{-i\theta}f(e^{i\theta}z)$, the invariant I_2 is well-defined on the configuration $\mathfrak{C}$ considered here. The general expression in terms of z is more complicated.

Remark 2.2 The notion of comparison moduli spaces is required in order to formulate statements regarding higher-order derivatives of maps conformally invariantly [66].

Example 2.6 Let $\mathfrak{C} = \{(\mathbb{D}, f, R)\}$ where R is conformally equivalent to the sphere $\bar{\mathbb{C}}$, a, b, c are points in R, and $f : \mathbb{D} \to R$ is an injective holomorphic map with a quasiconformal extension to R. Then $\mathfrak{M}$ is easily seen to be in one-to-one correspondence with Example 2.4, and thus is the universal Teichmüller space. Other choices of normalization on f can be imposed.

Example 2.7 Let $\mathfrak{C} = \{(\mathbb{D}, f_1, \ldots, f_n, R, p_1, \ldots, p_n)\}$ where R is a compact Riemann surface of genus g, $p_1, \ldots, p_n \in R$, $f_i : \mathbb{D} \to R$ such that f_i are one-to-one holomorphic maps on the closure of the disk $\mathrm{cl}\mathbb{D}$, the closures of the images of $\mathbb{D}$ do not overlap, and $f_i(0) = p_i$. Then $\mathfrak{M}$ is a moduli space in conformal field theory due to Friedan, Shenker, Segal, and Vafa, see Huang [25] for references. The case where analyticity on $\mathrm{cl}\mathbb{D}$ is weakened to quasiconformal extendibility will be considered at length in this paper.

As one can also see from these examples, given a moduli space of nested surfaces, there is a corresponding moduli space of mappings, if one adds the correct data.

3   Some Examples in Geometric Function Theory

3.1   Nehari Monotonicity Theorems and Generalizations

Classically, Riemann surfaces are characterized by their spaces of functions or differentials with specified singularities. The existence of functions or differentials with specified singularities is closely related to the Dirichlet problem on Riemann surfaces. In the case of compact surfaces, algebraic geometric techniques nearly (but not quite) eliminate the need to consider the Dirichlet problem. On the other hand, for nested surfaces with boundaries, there is not sufficient rigidity for these algebraic geometric techniques to determine the families of differentials, and the Dirichlet problem again takes centre stage.

Nehari [38] defined a class of functionals on subdomains of Riemann surfaces, obtained from the Dirichlet integrals of harmonic functions of specified singularities. It is closely related to the so-called contour integral and area techniques in function theory. We will not review these connections since they can be found in other sources.

Nehari considered the comparison moduli space $\mathfrak{C}$ (he did not use the term), which in our terminology is given by $\mathfrak{C} = \{(R_1, R_2, z_1, z_2, \ldots, z_n, h)\}$ where R_1 and R_2 are Riemann surfaces, $R_2 \subseteq R_1$ is bounded by analytic curves in the interior of R_1, $z_1, \ldots, z_n \in R_2$ and h is a harmonic function on R_1 with singularities at z_i. He did not precisely specify the nature of the singularities of h, but in all examples they were such that h was locally the real part of a meromorphic function with poles at z_i. If we define pull-back on points and the harmonic function in the obvious way, these conditions are conformally invariant and thus we obtain the moduli space $\mathfrak{M}$.

Below, we shall give a collection of monotonic functionals on $\mathfrak{M}$ defined by Nehari. We will also shorten the notation of the configuration space to (R_1, R_2, h) (since the information of the location of the singularities are encoded in the domain of the harmonic function h).

Definition 3.1 (Nehari Functional) Let R be a Riemann surface, and let D be a subdomain of R bounded by finitely many analytic curves. Let h be a harmonic function with a finite number of isolated singularities all of which are in D. Let q be the unique function on D which is constant on each component of ∂D, such that $q + h$ is harmonic on D, and such that for any closed contour Γ in D,

$$\int_\Gamma \frac{\partial(q + h)}{\partial n} ds = 0.$$

Here n denotes the unit outward normal and ds denotes the unit arc length. We define the "Nehari functional" as

$$N(R, D, h) = \int_{\partial D} h \frac{\partial q}{\partial n} ds.$$

Remark 3.2 Although the outward normal is coordinate dependent, the expression $h \frac{\partial q}{\partial n} ds$ is not.

It is immediately evident that the Nehari functional is conformally invariant in the sense that if $g : R \to R'$ is a biholomorphism of Riemann surfaces then $N(R, D, h) = N(R', g(D), h \circ g^{-1})$.

Theorem 3.3 (Nehari [38]) *N is a monotonic functional in the sense that whenever $D_1 \subseteq D_2$ it holds that $N(R, D_2, h) \leq N(R, D_1, h)$.*

This follows from the Dirichlet principle. It can be generalized to collections of non-overlapping domains.

Nehari showed that this concept unifies a large number of results in function theory, including the Grunsky inequalities, many coefficient estimates for univalent functions, and estimates on capacity. Diverse applications from the point of view of capacitance can also be found in the monograph of Dubinin [11].

We restrict now to the case that both domains are simply connected and hyperbolic. In [62, 63], Schippers obtained many estimates on conformal invariants involving bounded univalent functions from Nehari's method. The order of the estimates are determined by the order of the singularity. By Teichmüller's principle [27], extremal problems are in general associated with quadratic differentials (this principle was enunciated by Jenkins who attributes it to O. Teichmüller). If the estimate involves $f^{(n)}(z)$ at a fixed point z, then the associated quadratic differential has a pole of order $n + 1$ at z.

Thus, one is led to

1. formulate Nehari's functional in terms of quadratic differentials.
2. consider quadratic differentials with poles of arbitrary order.

In constructing this reformulation in terms of quadratic differentials, we restrict to the case that the outside domain R is conformally equivalent to a disk. Since we are dealing with domains with a global coordinate, we will use the notation $Q(z)dz^2$ for quadratic differentials. In [67] Schippers showed that, by altering h, every Nehari functional is up to a constant equal to a Nehari functional such that the boundary of R is a trajectory of the quadratic differential $Q(z)dz^2 = h'(z)^2 dz^2$. Furthermore, this particular choice of h is precisely the one so that $N(R, R, h) = 0$.

Of course, this strongly suggests that the Nehari functional can be generalized to arbitrary quadratic differentials, not just those of the form $h'(z)^2 dz^2$. This can indeed be done by passing to a double cover $\pi : \tilde{R} \to R$ branched at odd-order poles and zeros [66, 67]. For any quadratic differential $Q(z)dz^2$ such that ∂R is a trajectory, define

$$m(R, D, Q(z)dz^2) = \int_{\partial \tilde{D}} h \frac{\partial q}{\partial n} ds \tag{3.1}$$

where $\tilde{D} = \pi^{-1}(D)$, $h = \mathrm{Re} \int \sqrt{Q(\pi(z))\pi'(z)^2} \, dz$, and q is the unique harmonic function on $\tilde{D}$ such that $q = h$ on $\partial \tilde{D}$. We also assume that the boundary of D is sufficiently regular, but the functional extends to the case that D is bounded by a Jordan curve. Some effort is required to make this definition rigorous; one must deal with branch points, show that this integral is single-valued and independent of choice of branch of the square root, etc. Once all these issues are settled, one obtains the following result:

Theorem 3.4 ([67]) *Let R and D be conformally equivalent to the unit disk, such that $D \subseteq R$, let $Q(z)dz^2$ be a quadratic differential such that ∂R is a trajectory, all of whose poles are contained in D. Then the following statements hold:*

(1) $m(R, D, Q(z)dz^2)$ is conformally invariant in the sense that if $g : R \rightarrow R'$ is a conformal bijection and $g(D) = D'$ then $m(R', D', Q(z)dz^2) = m(R, D, Q(g(z))g'(z)^2 dz^2)$;

(2) if $D_1 \subseteq D_2$ then $m(R, D_1, Q(z)dz^2) \leq m(R, D_2, Q(z)dz^2)$;

(3) for any D, $m(R, D, Q(z)dz^2) \leq 0$;

(4) if $m(R, D, Q(z)dz^2) = 0$ then D is R minus arcs of trajectories of $Q(z)dz^2$;

(5) if $Q(z)dz^2 = h'(z)^2 dz^2$ for some harmonic function h with singularities, then $m(R, D, Q(z)dz^2) = (-1/2)N(R, D, h)$.

Remark 3.5 These invariants can be extended to domains slit by analytic arcs.

By choosing $R = \mathbb{D}$ and $D = f(\mathbb{D})$ for a univalent function f, many inequalities for bounded univalent functions follow [66, 67]. The functional associated with $Q(z)dz^2$ is in accordance with Teichmüller's principle. That is, if the quadratic differential has a pole of order $n + 1$ at z, the nth derivative of the mapping function $f^{(n)}(z)$ arises in the functional.

We now pose the following problem:

Problem 3.1 Generalize this to the case that the inner domain is bordered by quasicircles.

The geometric motivation for this problem will be discussed ahead; see Problems 4.1, 4.3, and 4.4.

The next problems are much harder, and are at the root of many problems in geometric function theory. In the remainder of the section, we restrict to the case of two nested simply connected hyperbolic Riemann surfaces (although it is clear that the problems have natural generalizations). By conformal invariance we may choose $R = \mathbb{D}$. We have that any conformal map from $\mathbb{D}$ into $\mathbb{D}$ which is admissible for a quadratic differential $Q(z)dz^2$ attains the upper bound of the functional $m(\mathbb{D}, f(\mathbb{D}), Q(z)dz^2)$.

Now suppose that we are given a functional in advance, which is not obviously of the form $m(\mathbb{D}, f(\mathbb{D}), Q(z)dz^2)$—for example, a coefficient functional for bounded univalent functions. How does one choose a quadratic differential so that $m(\mathbb{D}, f(\mathbb{D}), Q(z)dz^2)$ is the desired functional? Thus we have two problems.

Problem 3.2 Let $\Phi_{Q(z)dz^2}(f) = m(\mathbb{D}, f(\mathbb{D}), Q(z)dz^2)$. Describe the map $Q(z)dz^2 \mapsto \Phi_{Q(z)dz^2}$ algebraically.

Conversely we ask, in the same spirit,

Problem 3.3 Describe the set of functionals arising from quadratic differentials (say with a pole at the origin, and no other poles). Describe the inverse map $\Phi_{Q(z)dz^2} \mapsto Q(z)dz^2$.

There are many interpretations of "algebraically", of course. A satisfactory answer will almost certainly involve the Lie-theoretic properties of the classes of (sufficiently regular) bounded and unbounded univalent functions; see e.g. Kirillov and Yuri'ev [29], Markina, Prokhorov, and Vasil'ev [33], and Schippers [64].

The heart of the problem is that the relation between quadratic differentials and functionals has been imprecise since its introduction by Schiffer. It is a hard problem to construct a method for producing inequalities. It is still harder to obtain an inequality which one decides on at the outset, no matter which of the existing methods is used. This problem has never been satisfactorily resolved, in spite of the many successes of the theory of extremal problems.

It has long been known that there is a correspondence between quadratic differentials, boundary points of function spaces, and extremals of functions. The following references could serve as a starting point: Duren [12], Pommerenke [41], Schaeffer and Spencer [55], and Tammi [73, 74]. Although there are many heuristic principles and concrete theorems, the precise relation has never been established. This may be partly because the proof of the Bieberbach conjecture by deBranges has had the unfortunate effect of diverting attention from the problem (unjustly, since the Bieberbach conjecture involves only a single point on the boundary).

A full exposition of these problems would take more space than we have here, so we will content ourselves by quoting an elegant result of Pfluger. Since this result strongly suggests that much more remains to be discovered, we will follow it with a few related problems.

Theorem 3.6 ([40]) *Consider the class of mappings $f(z) = z + a_2 z^2 + a_3 z^3 + \cdots$ where $f : \mathbb{D} \to \mathbb{C}$ is holomorphic and one-to-one. For every $\lambda \in \mathbb{C} \backslash \{1\}$, there are precisely two mappings $f_\lambda(z)$ and $-f_\lambda(-z)$ maximizing the functional $f \mapsto \mathrm{Re}(a_3 - \lambda a_2^2)$. For every $A \in \mathbb{C} \backslash \{0\}$, with the exception of the Koebe function, every function f mapping onto the plane minus trajectories of the quadratic differential $(1 + Aw)w^{-4} dw^2$ is extremal for precisely one such functional.*

The class of functions is closely related to the universal Teichmüller space. In fact, the set of such functions which are additionally supposed to be quasiconformally extendible can be identified with the Bers fibre space over the Teichmüller space [72], which is also a comparison moduli space. In some sense Pfluger's result gives information about the boundary of this moduli space. Some natural questions follow.

Problem 3.4 Does Pfluger's result hold for higher-order functionals? For example, is there an analogous result for functionals of the form $\mathrm{Re}(a_4 + \lambda a_3 a_2 + \mu a_2^3)$ and quadratic differentials of the form $((1 + Aw + Bw^2)/w^5) dw^2$?

The form of the functional is obtained by demanding that the expression be homogeneous with respect to the transformation $f(z) \mapsto e^{-i\theta} f(e^{i\theta} z)$, while the form of the quadratic differential can be obtained using Schiffer's variational theorem to differentiate the functional [41, Chapter 7].

Similarly, we can ask whether Pfluger's results hold in some sense for the comparison moduli space considered throughout this section:

Problem 3.5 Is there a version of Pfluger's result for bounded one-to-one holomorphic mappings $f : \mathbb{D} \to \mathbb{D}$ such that $f(0) = 0$?

We observe that Pfluger's result was obtained by combining Schiffer variation with Jenkins' quadratic differential/extremal metric methods. Possibly the generalization of Nehari's results to quadratic differentials might shed further light on these questions.

Finally, we state some general problems on the relation to Schiffer variation. The Schiffer variational method (in its various forms) says that for various classes of conformal maps, the extremal map of a (reasonably regular) fixed functional maps onto the target domain minus trajectories of a quadratic differential [12, 41]. However, the quadratic differential is not precisely determined and depends to some extent on the unknown extremal function. In some sense, the Schiffer variational method produces the functional derivative of a functional at an extremal and relates it to the quadratic differential (see [41, Theorem 7.4]).

In the case of the invariants (3.1), the extremal condition suggests that the quadratic differential obtained by Schiffer variation equals the original quadratic differential. We are thus led to ask:

Problem 3.6 What is the relation between the functional derivative of $m(\mathbb{D}, f(\mathbb{D}), Q(z)dz^2)$ and the quadratic differential $Q(z)dz^2$?

By using the Loewner method to obtain functional derivatives, it was shown in [65] that in the case of Nehari functionals (that is, for quadratic differentials of the form $h'(z)^2 dz^2$) with poles only at the origin, the functional derivative at an extremal is the pull-back of the original quadratic differential under the extremal map. We conjecture that this holds in much greater generality, and also ask whether a stronger statement can be made.

3.2 A Few Remarks on the Extremal Metric Method

Jenkins' extremal metrics method associates functionals to quadratic differentials [27]. This method is based on length-area inequalities of Grötzsch [21] and Teich-müller [76], which grew into the theory of extremal length. A modern exposition can be found in the monograph of Vasiliev [77]. According to Jenkins [27], one can interpret these estimates as involving reduced modules; see Schmidt [68] for an enlightening exposition in special cases. However, in practice, this statement is difficult to make precise except for quadratic differentials of order two or lower (see e.g. Dubinin [11, Chapter 6]). Also, Jenkins' method does not manifestly produce conformally invariant functionals for quadratic differentials of order strictly greater than two. Thus one is led to the following problems:

Problem 3.7 Find a systematic definition of conformally invariant reduced modules in terms of extremal length of curve families, which produces estimates on arbitrary-order derivatives of conformal maps.

This is deliberately imprecise, since it has not even been done in the case of bounded univalent maps. Here is a more precise formulation in that special case.

Problem 3.8 Let $D_2 \subseteq D_1$ be simply connected domains and let z be a point in D_2. Define conformally invariant reduced modules of curve families which for $D_1 = \mathbb{D}$ and $z = 0$, explicitly involving the nth derivative of the conformal map $f : \mathbb{D} \to D_1$

taking 0 to 0. Prove estimates on these modules using extremal length methods of Grötzsch/Teichmüller/Jenkins.

The boundaries of the domains could be chosen as regular as necessary, although we note in passing that the quasiconformal invariance of the notion of prime ends should allow a quite general formulation.

These reduced modules should be associated with quadratic differentials through their curve families.

Problem 3.9 Systematically associate a conformally invariant reduced module to each quadratic differential (with poles of arbitrary order).

Again, this can be formulated for various classes of maps or pairs of domains, but it would be of great interest even for bounded univalent maps/simply connected domains.

In some sense Theorem 3.4 of the first author is an answer to Problem 3.9, but with Nehari's Dirichlet energy technique replacing extremal length. Now by a theorem of A. Beurling, the Dirichlet energy of a mixed Dirichlet/Neumann boundary value problem is the reciprocal of the extremal length of a curve family [4]. This immediately leads to the following problem.

Problem 3.10 Formulate and prove a generalization of Beurling's theorem for harmonic functions with singularities.

This problem was posed earlier in [63], before the discovery of Theorem 3.4. The answer to this problem will certainly involve the generalized Nehari invariants (3.1) and Jenkins' method of quadratic differentials.

3.3 Schiffer Comparison Theory of Domains

Another form of the comparison moduli space idea is Schiffer's comparison theory of nested domains, where one compares the kernel functions of two domains E and $\mathcal{E}$ where $E \subset \mathcal{E}$. The ideas go back to his foundational paper with Bergman [6] and his earlier paper [56]. Those papers deal with the special case $\mathbb{C}$; in this case the importance of the role of the outer domain is not obvious. The comparison theory of kernels on nested domains was principally investigated by Schiffer [59, 60], so we will attach his name to it. Schiffer's theory reached its final published form in his appendix to the book of Courant [8]. This section draws mainly on that source. We add some simple but important observations on conformal invariance.

Let E be a planar domain with Green's function g. Define two kernel functions as follows.

Definition 3.7 The Bergman kernel of a domain E with Green's function g is

$$K(w, z) = -\frac{2}{\pi} \frac{\partial^2 g}{\partial \bar{w} \partial z}.$$

The Schiffer kernel of E is

$$L(w, z) = -\frac{2}{\pi} \frac{\partial^2 g}{\partial w \partial z}.$$

The first appearance of the L-kernel is to our knowledge [56], hence we call it the Schiffer kernel. Clearly

$$L(z, w) = L(w, z)$$

and

$$K(z, w) = \overline{K(w, z)}.$$

These kernels are conformally invariant in the sense that if $g : E \to E'$ is a conformal bijection then

$$L_{E'}(g(w), g(z))g'(w)g'(z) = L_E(w, z) \quad \text{and} \quad K_{E'}(g(w), g(z))\overline{g'(w)}g'(z) = K_E(w, z).$$

Strictly speaking it is best to view the kernel functions as differentials $L(w, z)dwdz$ and $K(w, z)d\bar{w}dz$.

We will define the Bergman and Schiffer kernels of the plane as follows.

Definition 3.8 The Bergman kernel of $\mathbb{C}$ is defined to be $K(z, w) = 0$. The Schiffer kernel of $\mathbb{C}$ is

$$L_{\mathbb{C}}(z, w) = \frac{1}{\pi} \frac{1}{(z - w)^2}.$$

These definitions can be motivated by computing the Schiffer kernel of the disk of radius r and letting $r \to \infty$. Another motivation is that with that definition, the identities and inequalities continue to hold in the case that $\mathcal{E} = \mathbb{C}$.

Next we give two important properties of the kernel functions. Denote the Bergman space of a domain E by

$$A_1^2(E) = \left\{ h : E \to \mathbb{C} \text{ holomorphic} \; : \; \iint_E |h|^2 dA \right\} < \infty.$$

Remark 3.9 The Bergman space is best viewed as a space of one-differentials, but for now we view it as a function space to be consistent with the majority of the function theory literature.

Proposition 3.10 *Let $f \in A_1^2(E)$. Then*

$$\iint_E K(\zeta, \eta) f(\eta) \, dA_\eta = f(\zeta)$$

and

$$\iint_E L(\zeta, \eta)\overline{f(\eta)} \, dA_\eta = 0.$$

Let E, E^c and $\mathcal{E}$ be domains bounded by finitely many smooth curves, satisfying $E \subset \mathcal{E}$ and $E^c \subset \mathcal{E}$, in such a way that $\mathcal{E}$ is the union of E and E^c together with their shared boundary curves. Note that E^c is not the literal complement in $\mathcal{E}$ but rather its interior. Denote by $E \bigsqcup E^c$ the disjoint union of E and E^c. The inner products on each domain will be denoted $(,)_E$ etc.

Schiffer considered integral operators naturally associated with a configuration of nested domains. Let K, K^c and $\mathcal{K}$ be the Bergman kernels of E, E^c and $\mathcal{E}$ respectively. Similarly, let L, L^c and $\mathcal{L}$ be the Schiffer L-kernels associated with E, E^c and $\mathcal{E}$.

Definition 3.11

$$T^1_{E,\mathcal{E}} : A^2_1(E) \to A^2_1(\mathcal{E})$$

$$f \mapsto \iint_E \mathcal{K}(\cdot, w)f(w) \, dA_w$$

Next we consider the operator associated with the Schiffer L-kernel.

Definition 3.12

$$T^2_{E,\mathcal{E}} : \overline{A^2_1(E)} \to A^2_1(E \sqcup E^c)$$

$$\overline{f} \mapsto \iint_E \mathcal{L}(\cdot, w)\overline{f(w)} \, dA_w.$$

Here the inner product on $A^2_1(E \sqcup E^c)$ is of course

$$(h_1, h_2)_{E \sqcup E^c} = \iint_{E \sqcup E^c} h_1 \overline{h_2}.$$

Remark 3.13 Schiffer defines the operator $T^2_{E,\mathcal{E}}$ as a map from $A^2_1(E)$ into $A^2_1(E \sqcup E^c)$. This is not convenient, since with that convention the operator is complex anti-linear.

There is a relation between $T^1_{E,\mathcal{E}}$ and $T^2_{E,\mathcal{E}}$.

Theorem 3.14 *For all* $f, g \in \overline{A^2_1(E)}$,

$$(T^2_{E,\mathcal{E}}g, T^2_{E,\mathcal{E}}f)_{E \sqcup E^c} = (g, f)_E - (T^1_{E,\mathcal{E}}g, T^1_{E,\mathcal{E}}f)_{E \sqcup E^c}.$$

Thus

$$T_{E,\mathcal{E}}^2{}^\dagger T_{E,\mathcal{E}}^2 = I - T_{E,\mathcal{E}}^1{}^\dagger T_{E,\mathcal{E}}^1.$$

Theorem 3.14 immediately implies the *bounded Grunsky inequalities* (in Schiffer's form) for the T operator.

Corollary 3.15 *For $f \in A_1^2(E)$ one has that*

$$\|T_{E,\mathcal{E}}^2 f\|_E^2 = \|f\|_E^2 - \|T_{E,\mathcal{E}}^1 f\|_{E \cup E^c}^2 - \|T_{E,\mathcal{E}}^2 f\|_{E^c}^2.$$

In particular,

$$\|T_{E,\mathcal{E}}^2 f\|_{E^c} \leq \|f\|_E.$$

Remark 3.16 In the case that E is simply connected, the classical Grunsky inequalities can be obtained by setting $E = g(\mathbb{D})$ for some conformal map g, $\mathcal{E} = \mathbb{C}$ and f to be a polynomial in $1/z$ in the above corollary. This can be found in Bergman-Schiffer [6, Section 9].

In Schiffer's comparison theory discussed above, one can identify a comparison moduli space as follows. We set $\mathcal{C} = \{(\mathcal{E}, E, h) : h \in A_1^2(E)\}$, where as before $E \subseteq \mathcal{E}$ are planar domains. Let $g : \mathcal{E} \to \mathcal{E}'$ be a bijective holomorphic map. Define the pull-back by

$$g^* : A_1^2(E') \to A_1^2(E)$$
$$h \mapsto h \circ g \cdot g'$$

(cf Remark 3.9) and similarly $g^* \overline{h} = \overline{h \circ g \cdot g'}$. Then we have

Proposition 3.17 *The operators T_1 and T_2 are conformally invariant in the following sense: If g is a one-to-one holomorphic map from E onto E', then for all $h \in A_1^2(E')$*

$$T_{E,\mathcal{E}}^1 (g^* h) = g^* T_{E',\mathcal{E}'}^1 (h) \quad and \quad T_{E,\mathcal{E}}^2 (g^* \overline{h}) = g^* T_{E',\mathcal{E}'}^2 (\overline{h}).$$

Thus if we define

$$I^1(\mathcal{E}, E, h) = \|T_{E,\mathcal{E}}^1 h\|_{A_1^2(E)} \quad and \quad I^2(\mathcal{E}, E, h) = \|T_{E,\mathcal{E}}^2 \overline{h}\|_{A_1^2(E)}$$

we obtain conformal invariants on $\mathcal{C}$. Similarly the norms of $T_{E,\mathcal{E}}^i$ over $A_1^2(E^c)$ are conformally invariant.

Remark 3.16 shows that it is possible to derive coefficient inequalities from Corollary 3.15. We summarize the familiar procedure [12, 41] in this notation. Setting $\mathcal{E} = \mathbb{C}$ and $E = g(\mathbb{D})$ for a univalent function $g : \mathbb{D} \to \mathbb{C}$, it is possible to write the operators $T_{g(\mathbb{D}),\mathbb{C}}^2$ and $T_{g(\mathbb{D}),\mathbb{C}}^1$ in terms of the coefficients of g. Assuming that $g(0) = 0$, for every choice of test function h which is polynomial in $1/z$, one obtains a distinct inequality from Corollary 3.15. The same can be done for bounded univalent functions by replacing $\mathbb{C}$ with $\mathbb{D}$. The inequalities derived in this manner are conformally invariant.

Remark 3.18 The similarity to the Nehari-type invariants is of course not a coincidence.

Some generalizations of Schiffer's comparison theory to Riemann surfaces can be found in Schiffer and Spencer [61].

3.4 Fredholm Determinant and Fredholm Eigenvalues

The connection between the so-called Fredholm eigenvalues and Grunsky matrices was first observed by Schiffer, see e.g. [58]. One of Schiffer's accomplishments was to place Fredholm's real integral equations of potential theory in the complex setting. We draw on his insights freely here without justification, and refer the interested reader to [57, 58].

Given a domain D in the plane, bounded by finitely many C^3 Jordan curves, consider the operator defined for $f \in A_1^2(D)$ by

$$T_D f := (T_{D,\mathbb{C}}^2 f)|_D, \tag{3.2}$$

where T^2 is the operator defined in Definition 3.12. The operator T_D is a bounded operator from $\overline{A_1^2(D)}$ to $A_1^2(D)$. We note that this can even be extended to disconnected domains [57].

Moreover T_D is a Hilbert-Schmidt operator. This yields that $T_D \overline{T_D}$ is a positive trace class operator, to which the spectral theorem is applicable ($T_D \overline{T_D}$ is compact and self-adjoint). According to this theorem, there is an orthogonal basis f_n of eigenfunctions of $T_D \overline{T_D}$ with non-negative eigenvalues λ_n^2. The non-zero λ_n are called the *Fredholm eigenvalues* of D. The reader should however observe that this is not standard in the literature, as some authors refer to $1/\lambda_n$ as the Fredholm eigenvalues. Now since $T_D \overline{T_D}$ is a trace class operator, one can define the Fredholm determinant of the domain D as

$$\Delta_D := \det(I - T_D \overline{T_D}) = \prod_n (1 - \lambda_n^2). \tag{3.3}$$

Using variational techniques, Schiffer showed that the Fredholm determinant is conformally invariant in the sense of (2.1). More precisely, let $D_1, \ldots, D_n$ be a collection of bounded simply connected domains in D, each bounded by a C^3 Jordan curve. Let E be the interior of the complement of $D_1 \cup \cdots \cup D_n$ in $\mathbb{C}$. Fix an $l \in \{1, \ldots, n\}$ and let $\mathcal{E} = \mathbb{C} \backslash D_l$. Let $\mathfrak{C} = \{(\mathcal{E}, E)\}$ where E and $\mathcal{E}$ are of the above form. In that case, the quantity

$$I(\mathcal{E}, E) = \frac{\Delta_{\mathcal{E}}}{\Delta_E}$$

is a conformal invariant in the sense of (2.1). Observe that this striking invariance property must involve pairs of domains, and therefore is a result regarding comparison moduli spaces.

Remark 3.19 Although we do not doubt the validity of this result, Schiffer's proof is rather heuristic, since it infers invariance under all conformal maps from invariance under a specific set of variations.

In the case that D is simply connected, T_D is a version of the Grunsky operator, as we have observed above. By work of Takhajan and Teo [72], the Fredholm determinant relates to the Weil-Petersson metric. We will re-visit this in Sect. 5.4 ahead.

4 Teichmüller Space as a Comparison Moduli Space

4.1 Spaces of Non-overlapping Maps and the Rigged Moduli Space

Sets of maps with non-overlapping images are often considered in geometric function theory. In the general case of the sphere with n non-overlapping maps, these are sometimes called the "Goluzin-Lebedev class" [19]. As we saw in Example 2.7, sets of non-overlapping conformal maps appear in conformal field theory. We will be concerned with the case that the mappings are quasiconformally extendible and the closures do not intersect.

Let

$$A_1^{\infty}(\mathbb{D}) = \{f : \mathbb{D} \to \mathbb{C} \text{ holomorphic} : \sup_{z \in \mathbb{D}}(1 - |z|^2)|f(z)| < \infty\}$$

and recall that $A_1^2(\mathbb{D})$ is the Bergman space on the disk.

Definition 4.1 Let $\mathcal{O}^{qc}$ denote the set of maps $f : \mathbb{D} \to \mathbb{C}$ such that f is an injective holomorphic map, $f(0) = 0$ and f is quasiconformally extendible to a neighbourhood of cl$\mathbb{D}$. Let $\mathcal{O}_{WP}^{qc} = \{f \in \mathcal{O}^{qc} : f''/f' \in A_1^2(\mathbb{D})\}$.

Here "cl" denotes closure, and "WP" stands for Weil-Petersson. The terminology and motivation for this definition, as well as a review of the literature, will be given in Sect. 5 ahead.

Definition 4.2 Let Σ be a Riemann surface of finite genus with n ordered punctures $(p_1, \ldots, p_n)$. $\mathcal{O}^{qc}(\Sigma)$ is the set of n-tuples of one-to-one holomorphic mappings $(f_1, \ldots, f_n)$ such that

(1) $f_i : \mathbb{D} \to \Sigma$ are one-to-one, holomorphic for $i = 1, \ldots, n$
(2) $f_i(0) = p_i$
(3) f_i has a quasiconformal extension to a neighbourhood of the closure of $\mathbb{D}$ for all i
(4) $f_i(\mathrm{cl}\mathbb{D}) \cap f_j(\mathrm{cl}\mathbb{D})$ is empty whenever $i \neq j$.

This defines a configuration space of mappings in the sense of Sect. 2.2. We will call such an n-tuple a "rigging" of Σ.

Observe also that $(f_1, \ldots, f_n) \in \mathcal{O}^{qc}(\Sigma)$ if and only if there is at least one n-tuple of local coordinates $(\zeta_1, \ldots, \zeta_n)$ on open neighbourhoods U_i of $\mathrm{cl}f(\mathbb{D})$ such that $\zeta_i \circ f_i \in \mathcal{O}^{qc}$ for $i = 1, \ldots, n$. We call such a chart an n-chart. If the condition holds for one n chart, then it holds for every n chart so that each U_i contains $\mathrm{cl}f_i(\mathbb{D}) = f_i(\mathrm{cl}\mathbb{D})$.

Radnell and Schippers showed that the set of riggings is a complex Banach manifold.

Theorem 4.3 ([44]) *Let Σ be a Riemann surface with n punctures. $\mathcal{O}^{qc}(\Sigma)$ is a complex Banach manifold locally modelled on $\bigoplus^n \left(\mathbb{C} \oplus A_1^\infty(\mathbb{D}) \right)$.*

The charts of this manifold are

$$(f_1, \ldots, f_n) \mapsto \left((\zeta_1 \circ f_1)'(0), \frac{(\zeta_1 \circ f_1)''}{(\zeta_1 \circ f_1)'}, \ldots, (\zeta_n \circ f_n)'(0), \frac{(\zeta_n \circ f_n)''}{(\zeta_n \circ f_n)'} \right)$$

for some n-chart $(\zeta_1, \ldots, \zeta_n)$.

Similarly we have the Weil-Petersson class riggings.

Definition 4.4 Let Σ be a Riemann surface with n punctures. $\mathcal{O}^{qc}_{WP}(\Sigma)$ is the set of $(f_1, \ldots, f_n) \in \mathcal{O}^{qc}(\Sigma)$ such that there is an n-chart such that $\zeta_i \circ f_i \in \mathcal{O}^{qc}_{WP}$.

Again, this defines a configuration space of mappings in the sense of (2.2).

Analogously to Theorem 4.3, Radnell, Schippers and Staubach showed that $\mathcal{O}^{qc}_{WP}(\Sigma)$ is a complex Hilbert manifold.

Theorem 4.5 ([50]) *Let Σ be a Riemann surface of finite genus with n ordered punctures. $\mathcal{O}^{qc}_{WP}(\Sigma)$ is a complex Hilbert manifold locally modelled on $\bigoplus^n \left(\mathbb{C} \oplus A_1^2(\mathbb{D}) \right)$.*

The charts are obtained in the same way as for $\mathcal{O}^{qc}(\Sigma)$.

We now define the rigged moduli space, which appears in conformal field theory and is closely related to the theory of vertex operator algebras [25].

Definition 4.6 The *quasiconformally rigged moduli space* is the set of equivalence classes

$$\widetilde{\mathcal{M}}(g, n) = \{(R, \mathbf{f})\}/ \sim$$

where R is a Riemann surface of genus g with n punctures, $\mathbf{f} \in \mathcal{O}^{qc}(R)$ and the equivalence relation is defined by $(R, \mathbf{f}) \sim (S, \mathbf{g})$ if and only if there is a biholomorphism $\sigma : R \to S$ such that $\sigma \circ \mathbf{f} = \mathbf{g}$. The Weil-Petersson class rigged moduli space is the moduli space $\widetilde{\mathcal{M}}_{WP}(g, n)$ defined as above with $\mathcal{O}^{qc}(R)$ replaced with $\mathcal{O}^{qc}_{WP}(R)$.

By $\sigma \circ \mathbf{f} = \mathbf{g}$ we mean that $\sigma \circ f_i = g_i$ for $i = 1, \ldots, n$ where $\mathbf{f} = (f_1, \ldots, f_n)$ and $\mathbf{g} = (g_1, \ldots, g_n)$.

In two-dimensional conformal field theory, various analytic choices for the riggings exist. We will see in the next section that the choices above lead to deep connections with Teichmüller theory.

A problem of immediate interest is the following:

Problem 4.1 Extend the invariants (3.1) to functions on $\widetilde{\mathcal{M}}(g, n)$; that is, to configuration spaces of maps into Riemann surfaces whose images are bounded by quasicircles.

This problem requires some clarification. The extension to Riemann surfaces was already accomplished by Nehari [38] for harmonic functions, and the methods of Schippers [66, 67] for quadratic differentials extend without difficulty to fairly arbitrary Riemann surfaces. The main issue is analytic, and amounts to extending the functionals to Riemann surfaces bordered by quasicircles. However, although the functionals can be extended continuously from maps with analytic boundaries to roughly bounded ones by exhaustion, it is highly desirable to have a natural definition which does not require special pleading. This could perhaps be done by replacing the contour integrals with appropriate integrals with respect to harmonic measure on the boundary of each quasicircle.

A closely related problem is the following. Since the extremals of the functionals map onto R minus trajectories of the quadratic differential, we ask

Problem 4.2 Can the rigged moduli space be endowed with a boundary which includes riggings that are maps onto R minus trajectories of the quadratic differential? This must be done in such a way that the functionals extend continuously to this boundary.

4.2   A Brief Primer on Quasiconformal Teichmüller Theory

We give an overview of quasiconformal Teichmüller theory; for details see Lehto [31], Nag [34], or Hubbard [26]. First we define the Teichmüller space of a Riemann surface. Let Σ be a Riemann surface whose universal cover is the disk. Given another Riemann surface Σ_1, a *marking* of this surface by Σ is a quasiconformal map $f : \Sigma \to \Sigma_1$. Note that in particular this implies that Σ_1 is quasiconformally equivalent to Σ (and thus in particular they are homeomorphic). We will denote marked Riemann surfaces by triples (Σ, f, Σ_1).

We say that two quasiconformally equivalent marked Riemann surfaces are Teichmüller equivalent

$$(\Sigma, f_1, \Sigma_1) \sim (\Sigma, f_2, \Sigma_2)$$

if there is a biholomorphism $\sigma : \Sigma_1 \to \Sigma_2$ such that $f_2^{-1} \circ \sigma \circ f_1$ is homotopic to the identity rel boundary. The term "rel boundary" means that the homotopy fixes the boundary pointwise. Making the meaning of boundary and homotopy rel boundary precise requires some effort in terms of the lift to the universal cover; we refer the readers to [26, 31] for a thorough treatment. The Teichmüller space can now be defined as follows.

Definition 4.7 Let Σ be a fixed Riemann surface. The *Teichmüller space* of a Riemann surface Σ is the set of equivalence classes

$$T(\Sigma) = \{(\Sigma, f, \Sigma_1)\}/ \sim$$

where $f : \Sigma \to \Sigma_1$ is a quasiconformal marking of a Riemann surface Σ_1 and $\sim$ denotes Teichmüller equivalence. Denote equivalence classes by $[\Sigma, f, \Sigma_1]$.

An equivalent definition is given in terms of Beltrami differentials. A $(-1, 1)$-differential is one which is given in a local coordinate z by

$$\mu = h(z)\frac{d\bar{z}}{dz}$$

for a Lebesgue-measurable complex-valued function h, and which transforms under a change of coordinate $z = g(w)$ by

$$\tilde{h}(w) = h(g(w))\frac{\overline{g'(w)}}{g'(w)}; \quad \text{that is} \quad h(z)\frac{d\bar{z}}{dz} = \tilde{h}(w)\frac{d\bar{w}}{dw}.$$

It is evident from the transformation rule that the essential supremum of a Beltrami differential is well-defined. For a Riemann surface Σ define

$$L^{\infty}_{-1,1}(\Sigma)_1 = \{ \text{measurable } (-1,1) \text{ differentials } \mu \text{ on } \Sigma : \|\mu\|_{\infty} < 1\}. \quad (4.1)$$

Given any $\mu \in L^{\infty}_{-1,1}(\Sigma)_1$ there is a quasiconformal solution f to the Beltrami differential equation

$$\mu = \frac{\overline{\partial}f}{\partial f} \quad (4.2)$$

which is unique up to post-composition by a conformal map. We call μ in (4.2) the Beltrami differential of f. We say two Beltrami differentials μ, ν are Teichmüller equivalent $\mu \sim \nu$ iff the corresponding solutions $[\Sigma, f_\mu, \Sigma_\mu]$ and $[\Sigma, f_\nu, \Sigma_\nu]$ are

Teichmüller equivalent in the sense of Definition 4.7. The Teichmüller space can thus be identified with

$$T(\Sigma) = \{\mu \in L^{\infty}_{-1,1}(\Sigma)_1\}/ \sim \tag{4.3}$$

via the map

$$[\Sigma, f, \Sigma_1] \mapsto [\mu].$$

In this paper, we restrict our attention to two possible kinds of Riemann surfaces: those of genus g with n punctures, and those of genus g and n boundary curves homeomorphic to the circle. In the border case we assume that the boundary curves are borders in the sense of Ahlfors and Sario [5]. We refer to these surfaces as punctured Riemann surfaces of type (g, n) or bordered surfaces of type (g, n).

We now describe the universal Teichmüller space, and in doing so fill out the details of Examples 2.4 and 2.6. Let $\mathbb{D}^* = \{z : |z| > 1\} \cup \{\infty\}$. The universal Teichmüller space $T(\mathbb{D}^*)$ can be represented as follows. For a given quasiconformal marking $(\mathbb{D}^*, f, \Omega)$ we let $\hat{f} : \bar{\mathbb{C}} \to \bar{\mathbb{C}}$ be the quasiconformal map of the Riemann sphere which is conformal on $\mathbb{D}$ and whose Beltrami differential equals that of f on $\mathbb{D}^*$. That is,

$$\frac{f_{\bar{z}}}{f_z} = \frac{\hat{f}_{\bar{z}}}{\hat{f}_z} \quad \text{a.e. on } \mathbb{D}^*.$$

The resulting map $\hat{f}$ is unique up to post-composition with a Möbius transformation (and is traditionally specified through three normalizations). Stated in terms of the rigged moduli space, we have that the map

$$\Phi : T(\mathbb{D}^*) \to \mathcal{M}(0, 1)$$

$$[\mathbb{D}^*, f, \Omega] \mapsto [\bar{\mathbb{C}}, \hat{f}]$$

is a bijection.

Given $\mu \in L^{\infty}_{-1,1}(\mathbb{D}^*)$ as above, let $\hat{f}_{\mu} : \bar{\mathbb{C}} \to \bar{\mathbb{C}}$ be a quasiconformal map with Beltrami differential equal to μ in $\mathbb{D}^*$ and to 0 in $\mathbb{D}$. However, now we uniquely specify $\hat{f}_{\mu}$ with the normalizations $\hat{f}_{\mu}(0) = 0, \hat{f}_{\mu}'(0) = 1$ and $\hat{f}_{\mu}(\infty) = \infty$. Denote

$$f_{\mu} = \hat{f}_{\mu}\Big|_{\mathbb{D}}$$

which is a conformal map; this map f_{μ} is independent of the choice of representative in the Teichmüller equivalence class. It can be shown that two Beltrami differentials μ and ν on $\mathbb{D}^*$ represent the same point of $T(\mathbb{D}^*)$ if and only if $f_{\mu} = f_{\nu}$ [31]. Thus, $T(\mathbb{D}^*)$ can be identified with the moduli space of Example 2.6 up to a change of normalization, by applying the uniformization theorem to identify R with $\bar{\mathbb{C}}$.

Define the spaces of abelian differentials

$$A_1^\infty(\mathbb{D}) = \{\alpha(z)dz \text{ holomorphic on } \mathbb{D} : \sup_{z \in \mathbb{D}}(1 - |z|^2)|\alpha(z)| < \infty\}$$

and quadratic differentials

$$A_2^\infty(\mathbb{D}) = \{Q(z)dz^2 \text{ holomorphic on } \mathbb{D} : \sup_{z \in \mathbb{D}}(1 - |z|^2)^2|Q(z)| < \infty\}.$$

Denote the Schwarzian derivative of a conformal map f by

$$S(f) = \frac{f'''}{f'} - \frac{3}{2}\left(\frac{f''}{f'}\right)^2.$$

The Bers embedding is defined by

$$\beta : T(\mathbb{D}^*) \to A_2^\infty(\mathbb{D})$$

$$[\mu] \mapsto S(f_\mu)dz^2.$$

where we use the Beltrami differential model of $T(\mathbb{D}^*)$. That is, $\beta = S \circ \Phi$. It is a classical theorem of Bers that $\beta(T(\mathbb{D}^*))$ is an open subset of $A_2^\infty(\mathbb{D})$; in fact this is a homeomorphism with respect to the so-called Teichmüller metric. Thus $T(\mathbb{D}^*)$ can be given a complex structure from $A_2^\infty(\mathbb{D})$.

Next, we recall another model of the universal Teichmüller space $T(\mathbb{D}^*)$ in terms of quasisymmetries.

Definition 4.8 An orientation-preserving homeomorphism h of $\mathbb{S}^1$ is called a *quasisymmetric mapping*, iff there is a constant $k > 0$, such that for every α and every β not equal to a multiple of 2π, the inequality

$$\frac{1}{k} \leq \left|\frac{h(e^{i(\alpha+\beta)}) - h(e^{i\alpha})}{h(e^{i\alpha}) - h(e^{i(\alpha-\beta)})}\right| \leq k$$

holds. Let $QS(\mathbb{S}^1)$ denote the set of quasisymmetric maps from $\mathbb{S}^1$ to $\mathbb{S}^1$.

The boundary values of a quasiconformal map are in general quasisymmetries [31]. By a classical result due to Beurling and Ahlfors [31], any quasisymmetry h has a quasiconformal extension to $\mathbb{D}^*$. Another quasiconformal extension $E(h)$ was given by Douady and Earle [34] with the property that $E(T \circ h \circ S) = T \circ E(h) \circ S$ for any disk automorphisms T and S.

Two Beltrami differentials μ and ν are Teichmüller equivalent if and only if the quasiconformal solutions $f^\mu : \mathbb{D}^* \to \mathbb{D}^*$ and $f^\nu : \mathbb{D}^* \to \mathbb{D}^*$ to the Beltrami equation on $\mathbb{D}^*$ are equal on $\mathbb{S}^1$ up to post-composition by a Möbius transformation of $\mathbb{S}^1$. That is, if and only if there is a Möbius transformation $T : \mathbb{D}^* \to \mathbb{D}^*$ such that $T \circ f^\mu|_{\mathbb{S}^1} = f^\nu|_{\mathbb{S}^1}$. Thus we may identify

$$T(\mathbb{D}^*) \cong QS(\mathbb{S}^1)/\text{Möb}(\mathbb{S}^1). \tag{4.4}$$

The identification is given by $[\mu] \mapsto f^\mu|_{\mathbb{S}^1}$. The inverse is obtained by applying the Ahlfors-Beurling extension theorem.

It is an important result that $QS(\mathbb{S}^1)$ and $QS(\mathbb{S}^1)/\text{Möb}(\mathbb{S}^1)$ are groups under composition. Although the universal Teichmüller space has a topological structure (determined for example by the Teichmüller metric), it is not a topological group, since while right composition is continuous, left composition is not.

A remarkable property of the universal Teichmüller space is that it contains the Teichmüller spaces of all surfaces whose universal cover is the disk. This fact is obtained through the representation of surfaces by quotients of the disk by Fuchsian groups. Teichmüller theory can also be viewed as the deformation theory of Fuchsian groups [26, 31, 34]. In this paper we take instead the equivalent view of Teichmüller theory as the space of deformations of complex structures.

4.3 The Teichmüller/Rigged Moduli Space Correspondence

We saw in Examples 2.4 and 2.6 that the universal Teichmüller space is a comparison moduli space. In fact, by work of Radnell and Schippers [43], this holds (up to a $\mathbb{Z}^n$ action) for Teichmüller spaces of Riemann surfaces with more boundary curves and higher genus. More precisely, the quotient of the Teichmüller space of genus g surfaces bordered by n closed curves by a $\mathbb{Z}^n$ action is the rigged moduli space $\mathcal{M}(g, n)$. We outline these results here.

The case of $T(\mathbb{D}^*)$ as the moduli space of maps into $\bar{\mathbb{C}}$ is special, since all Riemann surfaces which are homeomorphic to the sphere are biholomorphic to the sphere. We now deal with the case that g is non-zero. Given a bordered Riemann surface Σ^B of type (g, n) for $g \geq 1$ and $n \geq 1$, represent it as a subset of a punctured Riemann surface Σ^P of type (g, n), in such a way that the boundary $\partial\Sigma^B$ consists of n quasicircles each encircling one puncture. This can always be done using a sewing procedure [43]. Given an element $[\Sigma^B, f, \Sigma_1^B] \in T(\Sigma^B)$, let $\hat{f} : \Sigma^P \to \Sigma_1^P$ be a quasiconformal map such that $\hat{f}$ is conformal on $\Sigma^P \backslash \text{cl}\Sigma^B$ and equal to f on Σ^B. Now fix a collection of conformal maps $\tau_i : \mathbb{D} \to \Sigma^P$ onto each of the connected components of the complement $\Sigma^P \backslash \text{cl}\Sigma^B$. We then have the map

$$\Phi : T(\Sigma^B) \to \widetilde{\mathcal{M}}(g, n)$$

$$[\Sigma^B, f, \Sigma_1^B] \mapsto \left[\Sigma_1^P, \left(\hat{f} \circ \tau_1, \ldots, \hat{f} \circ \tau_n\right)\right].$$

The pure Teichmüller modular group $\text{PMod}(\Sigma^B)$ is the set of equivalence classes of quasiconformal maps $\rho : \Sigma^B \to \Sigma^B$ such that $\rho(\partial_i \Sigma^B) = \partial_i \Sigma^B$ as a set. Two such maps ρ_1, ρ_2 are equivalent $\rho_1 \sim \rho_2$ if and only if they are homotopic rel boundary. The modular group acts discontinuously via $[\rho][\Sigma^B, f, \Sigma_1^B] = [\Sigma^B, f \circ \rho^{-1}, \Sigma_1^B]$, and for any $[\rho]$ the induced map of $T(\Sigma^B)$ is a biholomorphism. Let DB denote the subset of $\mathrm{PMod}(\Sigma^B)$ generated by quasiconformal maps which are the identity on every boundary curves and are homotopic to the identity map. This group can be pictured as the set of elements twisting each boundary curve an integer number of times, and is isomorphic to $\mathbb{Z}^n$ except when $g = 0$ and $n = 1$ or $n = 2$, in which case it is trivial or isomorphic to $\mathbb{Z}$ respectively.

We then have the following.

Theorem 4.9 ([43]) *Let* Σ^B *be a bordered Riemann surface. The map* $\Phi :$ $T(\Sigma^B) \rightarrow \widetilde{\mathcal{M}}(g, n)$ *is a bijection up to a discrete group action. That is, if* $\Phi([\Sigma^B, f_1, \Sigma_1^B]) = \Phi([\Sigma^B, f_2, \Sigma_2^B])$ *then there is a* $[\rho] \in$ DB *such that* $[\rho][\Sigma^B, f_1, \Sigma_1^B] = [\Sigma^B, f_2, \Sigma_2^B]$.

Here is a specific example.

Example 4.1 ([46]) Let A be an annulus of finite modulus. The Teichmüller space $T(A)/\mathbb{Z}$ can be identified with the rigged moduli space $\widetilde{\mathcal{M}}(0, 2)$.

Interestingly this has a semigroup structure (it is known as the *Neretin-Segal semigroup*) and every element can be decomposed as the product of a quasisymmetry of $\mathbb{S}^1$ and a bounded univalent function.

This example suggests the following problems, closely related to Problem 4.1:

Problem 4.3 Lift the conformally invariant functionals (3.1) to $T(A)$.

Of course this problem extends to Teichmüller space.

Problem 4.4 Extend the functionals (3.1) to Teichmüller spaces of bordered Riemann surfaces of type (g, n).

And again we can ask, similarly to Problem 4.2:

Problem 4.5 Does the Teichmüller space of bordered surfaces of type (g, n) have a boundary, which corresponds under Φ to those riggings which map onto analytic arcs of quadratic differentials?

Problem 4.6 Can Theorem 4.9 be extended to Teichmüller spaces of more general surfaces, say those with infinite genus and/or infinitely many riggings? What conditions are necessary to make this work?

It is likely that a bound on the lengths of closed geodesics is a necessary condition; see Sect. 5.3 and Problem 5.3 ahead.

4.4 Some Applications of the Teichmüller Space/Rigged Moduli Space Correspondence

This identification of the moduli spaces appearing in conformal field theory (CFT) and in Teichmüller theory has far-reaching consequences. In particular, one can endow the rigged moduli space with a complex structure and show that the operation of sewing bordered Riemann surfaces together with quasisymmetries is holomorphic [43]. These are required for the construction of two-dimensional conformal field theory from vertex operator algebras. A comprehensive review can be found in [52]. The sewing technique is also of independent interest in Teichmüller theory.

On the other hand, the correspondence can be applied to transfer structures in CFT to Teichmüller theory. For example, one can obtain a fibre structure on Teichmüller space as follows. Let Σ^B be a fixed Riemann surface. Define the map

$$\mathcal{C} : T(\Sigma^B) \to T(\Sigma^P)$$

$$[\Sigma^B, f, \Sigma_1^B] \mapsto [\Sigma^P, \hat{f}, \Sigma_1^P]$$

where Σ^P, $\hat{f}$ and Σ_1^P are determined from $[\Sigma^B, f, \Sigma_1^B]$ as above. Radnell and Schippers showed the following.

Theorem 4.10 ([45]) *Let Σ^B be a bordered surface of type (g, n), and Σ^P be a punctured surface of type (g, n). Assume that $\Sigma^B \subset \Sigma^P$ and $\partial\Sigma^B$ consists of n quasicircles each enclosing a distinct puncture in Σ^P.*

(1) *$\mathcal{C}$ is a holomorphic map with local holomorphic sections. The fibres $\mathcal{C}^{-1}(p)$ for $p \in T(\Sigma^P)$ are complex submanifolds of $T(\Sigma^B)$.*
(2) *Let $p \in T(\Sigma^P)$. If $(\Sigma^P, g, \Sigma_1^P)$ is a representative of p, then the quotient $\mathcal{C}^{-1}(p)/DB$ is canonically bijective with $\mathcal{O}^{qc}(\Sigma_1^P)$.*
(3) *The bijection in (2) is a biholomorphism.*

In particular, this can be used to give holomorphic coordinates on $T(\Sigma^B)$. The proof uses a variational technique of Gardiner [15], which was based on an idea of Schiffer. The use of Gardiner-Schiffer variation to address holomorphicity issues in the rigged moduli space originates with Radnell [42].

Problem 4.7 Are there generalizations of Theorem 4.10 for Teichmüller spaces to more general Riemann surfaces, such as those with infinite genus and/or infinite number of boundary curves?

5 Weil-Petersson Class Teichmüller Space

5.1 The Weil-Petersson Class Teichmüller Space

There exist many refinements of Teichmüller space, including for example the asymptotically conformal Teichmüller space [13], BMO-Teichmüller space [10], and L^p Teichmüller spaces (references below). The L^2 case is usually referred to as the Weil-Petersson class Teichmüller space, which we motivate in this section. From now on we abbreviate this as WP-class.

The Weil-Petersson metric is a metric in the sense of Riemannian/Hermitian geometry, that is, an inner product at every tangent space. It is based on a representation of the tangent space to Teichmüller space by a set of quadratic differentials, which we must now describe. Fix a Riemann surface Σ. Given a holomorphic curve $t \mapsto \mu_t$ in the set of Beltrami differentials $L^\infty_{-1,1}(\Sigma)_1$, the derivative $\frac{d}{dt}\mu$ is in $L^\infty_{-1,1}(\Sigma)$. However, there is an enormous amount of redundancy in this model, since the difference between any pair of these might be tangent to the Teichmüller equivalence relation, and hence represent the same direction in Teichmüller space.

The following decomposition remedies the problem. Let λ^2 denote the hyperbolic metric on Σ. Define the harmonic Beltrami differentials by

$$\Omega_{-1,1}(\Sigma) = \{\mu \in L^\infty_{-1,1}(\Sigma) : \mu = \alpha/\lambda^2, \quad \alpha \text{ a quadratic differential}\}. \tag{5.1}$$

It can easily be verified that a quotient of a quadratic differential by the hyperbolic metric is a Beltrami differential, by writing them in local coordinates and verifying that the quotient satisfies the correct transformation rule. It is a classical result that we have the Banach space decomposition

$$L^\infty_{-1,1}(\Sigma)_1 = \Omega_{-1,1}(\Sigma) \oplus \mathcal{N}$$

where $\mathcal{N}$ is the set of Beltrami differentials which are tangent to the Teichmüller equivalence relation. These are the so-called "infinitesimally trivial Beltrami differentials", which can be characterized precisely [31, 34].

Thus, the tangent space to $T(\Sigma)$ at $[\Sigma, \mathrm{Id}, \Sigma]$ can be identified with $\Omega_{-1,1}(\Sigma)$. Furthermore, there is an open subset $U \subset \Omega_{-1,1}(\Sigma)$ containing 0 such that

$$\Psi : U \to T(\Sigma)$$

$$\mu \mapsto [\Sigma, f, \Sigma_1], \quad \text{where } \bar{\partial}f/\partial f = \mu \tag{5.2}$$

is a biholomorphism from an open neighbourhood of 0 to an open neighbourhood of 0. Applying right compositions by quasiconformal maps, one obtains a system of coordinates on $T(\Sigma)$ which is compatible with that obtained from the Bers embedding β. This is closely related to the so-called Ahlfors-Weill reflection. The tangent spaces at other points can be obtained by right composing by quasiconformal maps. These constructions can also all be lifted to the universal cover and expressed in terms of differentials invariant under Fuchsian groups.

For compact surfaces or compact surfaces with punctures, the Weil-Petersson metric is defined on the tangent space at the identity by the L^2 pairing of Beltrami differentials

$$\langle \mu, \nu \rangle = \iint_\Sigma \bar{\mu}\nu \, dA_{\mathrm{hyp}}, \quad \mu, \nu \in \Omega_{-1,1}(\Sigma) \tag{5.3}$$

where dA_{hyp} is the hyperbolic area measure [2]. The integral (5.3) is finite, because on compact surfaces L^∞ Beltrami differentials are also L^2. This Hermitian inner product can be transferred to tangent spaces at other points in the Teichmüller space by right composition. In the compact/compact with punctures case, the Weil-Petersson metric has been much studied [78]. Ahlfors showed that the WP-metric is Kähler [2] and later computed its Ricci and scalar curvatures of holomorphic sections, and showed they are negative [3].

The Weil-Petersson inner product had not been defined on any other Teichmüller space until the turn of the millennium. The obstacle was that if the Riemann surface is not compact or compact with punctures, then the L^2 pairing need not be finite. Thus in order to define a Weil-Petersson metric one must restrict to a smaller Teichmüller space, so that tangent directions generate only L^2 Beltrami differentials. Nag and Verjovsky [37] showed that if one restricts to the subset of the universal Teichmüller space corresponding to diffeomorphisms of $\mathbb{S}^1$, then the WP metric converges on tangent directions. Equivalently, the corresponding representative quasisymmetries are smooth. However, this is only a heuristic principle, and the smoothness assumption is rather artificial from the point of view of Teichmüller theory. The correct analytic condition giving the largest universal Teichmüller space on which the WP metric converges is the subject of the next section.

5.2 Weil-Petersson Class Universal Teichmüller Space

The investigation of L^2 Teichmüller spaces is due independently to Cui [9] and Takhtajan and Teo [72]. Guo [22] and Tang [75] extended some of the results to L^p Teichmüller spaces.

The definitions of the WP-class Teichmüller space given by Takhtajan/Teo and Cui are rather different, but equivalent. Takhtajan/Teo give the definitions and complex structure in terms of L^2 harmonic Beltrami differentials, while Cui's definition is in terms of an L^2 condition on the quadratic differentials in the image of the Bers embedding.

Remark 5.1 In fact Takhtajan and Teo's approach defines a foliation of $T(\mathbb{D}^*)$ by right translates of the WP-class Teichmüller space.

Recall the space of differentials

$$A_1^2(\mathbb{D}) = \left\{ \alpha(z)dz \text{ holomorphic on } \mathbb{D} : \iint_{\mathbb{D}} |\alpha(z)|^2 < \infty \right\}$$

and quadratic differentials

$$A_2^2(\mathbb{D}) = \left\{ Q(z)dz^2 \text{ holomorphic on } \mathbb{D} : \iint_{\mathbb{D}} (1 - |z|^2)^2 |Q(z)|^2 < \infty \right\}.$$

The coefficients of the differentials in $A_1^2(\mathbb{D})$ are in the Bergman space, and thus we use the same notation as in Sect. 3.3 (see Remark 3.9).

We begin with Cui's definition.

Definition 5.2 The WP-*class universal Teichmüller space* is defined to be

$$T_{\mathrm{WP}}(\mathbb{D}^*) = \{[\mu] \in T(\mathbb{D}^*) \; : \; \beta([\mu]) \in A_2^2(\mathbb{D})\}.$$

Guo, Cui and Takhtajan/Teo also showed that $[\mu] \in T_{\mathrm{WP}}(\mathbb{D}^*)$ if and only if $f_\mu''/f_\mu' \in A_1^2(\mathbb{D})$. Thus we have that

$$[\mu] \in T_{\mathrm{WP}}(\mathbb{D}^*) \Leftrightarrow f_\mu \in \mathcal{O}_{\mathrm{WP}}^{\mathrm{qc}}(\mathbb{D}).$$

Although Takhtajan and Teo used a different definition, they also showed the equivalence of their definition with Definition 5.2, so we attribute the theorems below jointly.

Takhtajan/Teo and Cui independently showed that

Theorem 5.3 ([9, 72]) $A_2^2(\mathbb{D}) \subset A_2^\infty(\mathbb{D})$ *and the inclusion is holomorphic.*

Combined with the following, we get an analogue of the Bers embedding for $T_{\mathrm{WP}}(\mathbb{D}^*)$.

Theorem 5.4 ([9, 72]) $\beta(T_{\mathrm{WP}}(\mathbb{D}^*)) = \beta(T(\mathbb{D}^*)) \cap A_2^2(\mathbb{D})$. *In particular,* $\beta(T_{\mathrm{WP}}(\mathbb{D}^*))$ *is open.*

Thus we have that $T_{\mathrm{WP}}(\mathbb{D}^*)$ has a complex structure inherited from $A_2^2(\mathbb{D})$.

Guo and Cui showed that the fact that $\beta([\mu]) \in A_2^2(\mathbb{D})$ is equivalent to the existence of a representative Beltrami differential which is L^2 with respect to the hyperbolic metric. Define for any Riemann surface Σ with a hyperbolic metric λ^2

$$L_{\mathrm{hyp}}^2(\Sigma) = \left\{ \mu \in L_{-1,1}^\infty(\Sigma) \; : \; \iint_\Sigma |\mu|^2 dA_{\mathrm{hyp}} < \infty \right\}$$

where dA_{hyp} denotes hyperbolic area measure. In particular

$$L_{\mathrm{hyp}}^2(\mathbb{D}^*) = \left\{ \mu \in L_{-1,1}^\infty(\mathbb{D}^*) \; : \; \iint_{\mathbb{D}^*} \frac{|\mu|^2}{(1 - |z|^2)^2} dA < \infty \right\}.$$

We then have the following.

Theorem 5.5 ([9, 72]) $[\mu] \in T_{\mathrm{WP}}(\mathbb{D}^*)$ *if and only if* $[\mu]$ *has a representative* $\mu \in L_{\mathrm{hyp}}^2(\mathbb{D}^*)$.

In fact, Guo and Cui gave stronger results in two directions. Guo showed such a result for L^p differentials with respect to the hyperbolic metric, $p \geq 1$. Cui showed that the Douady-Earle extension of the boundary values of any representative conformal map f_μ has this property, and in fact satisfies a stronger integral estimate.

Tang showed that the Douady-Earle extension is in L^p and that the Bers embedding is holomorphic with respect to the intersection norm on $L^p \cap L^\infty$.

If $f_\mu \in \mathcal{O}^{qc}$, then the image of $\mathbb{S}^1$ under the unique homeomorphic extension of f_μ to $\mathrm{cl}\mathbb{D}$ is a quasicircle. Although there are an astonishing number of non-trivially equivalent characterizations of quasicircles [18], there is no known geometric characterization of Weil-Petersson class quasicircles.

Problem 5.1 Characterize quasicircles of the form $f_\mu(\mathbb{S}^1)$ for $[\mu] \in T_{\mathrm{WP}}(\mathbb{D})$ (equivalently, for $f_\mu \in \mathcal{O}^{qc}_{\mathrm{WP}}$) via analytic or geometric conditions on the set itself.

Radnell, Schippers, and Staubach [49] showed that a Weil-Petersson class quasicircle is a rectifiable chord-arc curve, but this is unlikely to be sufficient.

We also define the WP-class quasisymmetries as follows.

Definition 5.6 We say that $\phi \in QS(\mathbb{S}^1)$ is *Weil-Petersson class* if its corresponding Teichmüller space representative is in $T_{\mathrm{WP}}(\mathbb{D}^*)$. Denote the set of WP-class quasisymmetries by $QS_{\mathrm{WP}}(\mathbb{S}^1)$.

Cui and Takhtajan/Teo showed that, like quasisymmetries, these are closed under composition and inverse.

Theorem 5.7 ([9, 72]) $QS_{\mathrm{WP}}(\mathbb{S}^1)$ *is a group.*

A stronger result was obtained by Takhtajan and Teo:

Theorem 5.8 ([72]) $QS_{\mathrm{WP}}(\mathbb{S}^1)/M\ddot{o}b(\mathbb{S}^1)$ *is a topological group.*

This is in contrast to $QS(\mathbb{S}^1)/M\ddot{o}b(\mathbb{S}^1)$, which is not a topological group.

Remark 5.9 Takhtajan and Teo also showed that $QS_{\mathrm{WP}}(\mathbb{S}^1)/\mathbb{S}^1$ is a topological group, and it can be identified naturally with the WP-class universal Teichmüller curve.

The WP-class quasisymmetries were characterized by Shen as follows, answering a problem posed by Takhtajan and Teo.

Theorem 5.10 ([71]) *Let* $\phi : \mathbb{S}^1 \to \mathbb{S}^1$ *be a homeomorphism.* $\phi \in QS_{\mathrm{WP}}(\mathbb{S}^1)$ *if and only if* ϕ *is absolutely continuous and* $\log \phi'$ *is in the Sobolev space* $H^{1/2}(\mathbb{S}^1)$.

Further characterizations (e.g. in terms of the composition operator associated with ϕ) were given by Hu and Shen [24].

It was also shown that right composition is biholomorphic.

Theorem 5.11 ([9, 72]) *Right composition (mod $M\ddot{o}b(\mathbb{S}^1)$) in $T(\mathbb{D}^*)$ by a fixed element $h \in QS_{\mathrm{WP}}(\mathbb{D}^*)$ is a biholomorphism.*

This theorem combined with Theorem 5.4 was used by Cui to define the Weil-Petersson pairing on any tangent space, by using the pairing in $A_2^2(\mathbb{D})$ at the identity, and then applying the above theorem to define a right-invariant metric.

Takhtajan and Teo's approach, on the other hand, defined the complex structure in terms of local charts into the space of harmonic Beltrami differentials, which are defined by

$$H_{-1,1}(\mathbb{D}^*) = \{\mu = (1 - |z|^2)^2 \overline{Q(z)} : \mu \in L^2_{\mathrm{hyp}}(\mathbb{D}^*)\}.$$

This is non-trivial as it must be shown that the transition functions of the charts are biholomorphisms. As in the classical case, their proof relies on the use of the Ahlfors-Weill reflection.

Their construction also gives a description of the tangent space in terms of harmonic Beltrami differentials. By Theorem 5.3 we immediately obtain

Theorem 5.12 ([72])

$$L^{\infty}_{-1,1}(\mathbb{D}^*) \cap L^2_{\mathrm{hyp}}(\mathbb{D}^*) = H_{-1,1}(\mathbb{D}^*) \oplus (\mathcal{N} \cap L^2_{\mathrm{hyp}}(\mathbb{D}^*)).$$

This also shows that the tangent space at [0] to $T_{\mathrm{WP}}(\mathbb{D}^*)$ can be identified with $H_{-1,1}(\mathbb{D}^*)$. This has finite Weil-Petersson pairing (5.3) by definition. By applying the holomorphic right translation, Takhtajan and Teo obtain a right invariant inner product at all points in $T_{\mathrm{WP}}(\mathbb{D}^*)$.

In fact, they extended this complex structure and Hermitian metric to all of $T(\mathbb{D}^*)$ in the following way. A neighbourhood of [0] in $T(\mathbb{D}^*)$ can be obtained under the map Ψ (5.2), with $\Omega_{-1,1}(\mathbb{D}^*)$ replaced by $H_{-1,1}(\mathbb{D}^*)$. Using right composition, the charts patch together to give a complex structure compatible with the Bers embedding. However, the topology and complex structure are not equivalent to the standard one on $T(\mathbb{D}^*)$. Indeed, $T(\mathbb{D}^*)$ consists of uncountably many disjoint translates of $T_{\mathrm{WP}}(\mathbb{D}^*)$, each of which is a connected component of $T(\mathbb{D}^*)$ with this new topology. In their definition, $T_{\mathrm{WP}}(\mathbb{D}^*)$ appears as the connected component of the identity. This accounts for the convergence of the WP pairing, since at any given point there are far fewer tangent vectors than there are when $T(\mathbb{D}^*)$ is given the standard complex structure. That is, the directions on which the WP pairing diverges are excluded.

It must be emphasized that these constructions require a great deal of analysis; one cannot simply make small adjustments to classical theorems of Teichmüller theory.

Finally, Takhtajan and Teo showed (improving on Nag and Verjovsky's result, which held only in the diffeomorphism case)

Theorem 5.13 ([37, 72]) *The Weil-Petersson metric on $T(\mathbb{D}^*)$ (with the new complex structure and topology) is Kähler. In particular the Weil-Petersson metric on $T_{\mathrm{WP}}(\mathbb{D}^*)$ is Kähler.*

In fact they gave different explicit formulas for the Kähler potential, and computed the Ricci curvatures. We will return to this in Sect. 5.4.

5.3 Higher Genus Weil-Petersson Class Teichmüller Spaces

It is possible to extend the Weil-Petersson metric to a much wider class of surfaces, again by obtaining an L^2 theory. Radnell, Schippers and Staubach [47, 48, 50, 51] did this for bordered surfaces of type (g, n). Yanagishita [79] extended the L^p theory of Guo [22] and Tang [75] to surfaces satisfying "Lehner's condition", which includes bordered surfaces of type (g, n). In the L^2-setting the two Teichmüller spaces are the same as sets, but the constructions of the complex structures are rather different. Yanagishita constructs the complex structure from L^p quadratic differentials under the image of the Bers embedding, following the approach of Cui, Guo and Tang. Radnell, Schippers, and Staubach constructed the complex structure in two equivalent ways: using harmonic Beltrami differentials (along the lines of Takhtajan/Teo), and by refining the fiber structure of Theorem 4.10. As in the classical L^∞ case, the complex structures arising from the Bers embedding into quadratic differentials and from harmonic Beltrami differentials should be equivalent, but this has not yet been established (see footnote 1, page 35).

The Weil-Petersson class Teichmüller space is specified by boundary behaviour. To describe this we need a local coordinate near the boundary.

Definition 5.14 Let Σ be a bordered surface of type (g, n). A collar neighbourhood of a boundary curve $\partial_i \Sigma$ is a doubly connected open set in Σ, one of whose boundaries is $\partial_i \Sigma$ and the other is an analytic curve in the interior of Σ. A collar chart of a bordered surface Σ is a conformal map $\zeta : U \to \{1 < |z| < r\}$ for some $r > 1$ which extends continuously to $\partial_i \Sigma$.

It is possible to show that the chart extends to a conformal map of an open neighbourhood of $\mathbb{S}^1$ into the double of Σ.

We define the WP-class Teichmüller space in two steps.

Definition 5.15 Let Σ and Σ_1 be bordered surfaces of type (g, n). A quasiconformal map $f : \Sigma \to \Sigma_1$ is called refined if for each pair of boundary curves $\partial_i \Sigma$, $\partial_j \Sigma_1$ such that $f(\partial_i \Sigma) = \partial_j \Sigma_1$, there are collar charts ζ_i, η_j of $\partial_i \Sigma$ and $\partial_j \Sigma_1$ respectively such that $\eta_j \circ f \circ \zeta_i^{-1}|_{\mathbb{S}^1} \in \mathrm{QS_{WP}}(\mathbb{S}^1)$. Denote the set of such quasiconformal maps by $\mathrm{QC}_r(\Sigma)$.

If the condition holds for one collar chart at a boundary $\partial_i \Sigma$, then it holds for all of them.

Definition 5.16 Let Σ be a bordered surface of type (g, n). The WP-*class Teichmüller space* of Σ is

$$T(\Sigma) = \{(\Sigma, f, \Sigma_1) : f \in \mathrm{QC}_r(\Sigma)\} / \sim$$

where $\sim$ is the usual Teichmüller equivalence.

A different definition was given by Yanagishita [79], for L^p spaces for $p \geq 1$. It was phrased in terms of Fuchsian groups satisfying a condition he terms "Lehner's condition". We will restate Yanagishita's approach in its equivalent form on the Riemann surface, for consistency of presentation. Let Σ^* denote the double of the Riemann surface Σ (if Γ is the Fuchsian group such that $\Sigma = \mathbb{D}^* / \Gamma$, then $\Sigma^* = \mathbb{D}/\Gamma$).

Definition 5.17 (Lehner's Condition) A Riemann surface Σ covered by the disk $\mathbb{D}^*$ satisfies *Lehner's condition* if the infimum of the hyperbolic lengths of the simple closed geodesics is strictly greater than 0.

Definition 5.18 Let Σ be a Riemann surface covered by the disk satisfying Lehner's condition. The *p-integrable Teichmüller space* $T^p(\Sigma)$ is the subset of $T(\Sigma)$ consisting of elements $[\Sigma, f, \Sigma_1]$ such that there is a representative (Σ, f, Σ_1) such that the Beltrami differential of f is in L^p with respect to the hyperbolic metric.

Yanagishita also showed that the L^p representative is given by the Douady-Earle extension of the boundary values of the lift to $\mathbb{D}^*$. We will see shortly that it agrees with the definition above in the special case of $p = 2$ and bordered surfaces of type (g, n).

Radnell, Schippers and Staubach obtained the following analogue of Theorem 5.5, using sewing techniques and the lambda lemma.

Theorem 5.19 ([47]) *Let Σ be a bordered Riemann surface of type (g, n). Then $f \in QC_r(\Sigma)$ if and only if it is homotopic rel boundary to a quasiconformal map whose Beltrami differential is in $L^2_{\text{hyp}}(\Sigma)$.*

Since bordered surfaces of type (g, n) satisfy Lehner's condition, by Theorem 5.19, the Definitions 5.16 and 5.18 are equivalent for $p = 2$ and bordered surfaces of this type.

The two approaches to the complex structure are rather different. Yanagishita's approach involves the following theorem. Let Σ^* denote the double of Σ. Define $L^p_{\text{hyp}}(\Sigma)$ to be the set of Beltrami differentials on Σ which are L^p with respect to the hyperbolic area measure. Following [79] denote

$$Ael^p(\Sigma) = L^\infty_{-1,1}(\Sigma)_1 \cap L^p_{\text{hyp}}(\Sigma).$$

The intersection norm $\|\cdot\|_p + \|\cdot\|_\infty$ induces a topology on $T^p(\Sigma)$. Furthermore let

$$A^p_2(\Sigma^*) = \left\{\alpha \text{ a quadratic differential on } \Sigma^* : \iint_{\Sigma^*} \lambda^{-2}|\alpha|^p < \infty\right\}.$$

(Recall that λ^2 is the hyperbolic metric.)

Theorem 5.20 ([79]) *For $p \geq 2$, the restriction of the Bers embedding β to $T^p(\Sigma)$ is a homeomorphism onto its image in $A^p_2(\Sigma^*)$ with respect to the Ael^p norm.*

This induces a complex structure on $T^p(\Sigma)$, and in particular on $T_{\text{WP}}(\Sigma)$ for bordered Riemann surfaces of type (g, n). Furthermore, Yanagishita showed that right composition is a biholomorphism with respect to this structure, for $p \geq 2$. Thus, although the tangent space structure is not treated in the paper [79], it is possible to define a WP-pairing on the tangent space at $[0]$ using the Γ-invariant subspace of $A^2_2(\mathbb{D})$ as a model, and then using right composition to obtain a right-invariant metric at every point.

Radnell, Schippers, and Staubach gave two other complex structures for bordered surfaces of type (g, n), which are equivalent to each other. In [48, 50], it was shown that the fiber structure of Theorem 4.10 passes down to $T_{\text{WP}}(\Sigma)$ for bordered Riemann surfaces of type (g, n). That is, one may view $T_{\text{WP}}(\Sigma)$ as fibred over $T(\Sigma^P)$ for a compact surface with punctures Σ^P, such that the fibres $\mathcal{C}^{-1}(p)$

modulo a discrete group action are biholomorphic to $\mathcal{O}_{WP}^{qc}(\Sigma_1^P)$. This can be used to construct a Hausdorff, second countable topology on $T_{WP}(\Sigma)$, and a complex Hilbert manifold structure [48]. The advantage of this approach is that it is very flexible and constructive, and explicit coordinates can be given in terms of Gardiner-Schiffer variation.

In [51], Radnell, Schippers, and Staubach showed that this fiber structure is compatible with that obtained by harmonic Beltrami differentials and right translation, analogous to both the classical case and Takhtajan and Teo's approach on $T_{WP}(\mathbb{D}^*)$. We briefly describe this below, as well as the description of the tangent spaces.

In all of the following theorems, Σ is a bordered surface of type (g, n).

Theorem 5.21 ([51]) $H_{-1,1}(\Sigma) \subseteq \Omega_{-1,1}(\Sigma)$ and inclusion is holomorphic. Furthermore

$$L_{-1,1}^\infty(\Sigma) \cap L_{\text{hyp}}^2(\Sigma) = H_{-1,1}(\Sigma) \oplus (\mathcal{N} \cap L_{\text{hyp}}^2(\Sigma)).$$

Theorem 5.22 ([51]) Let α_t be any holomorphic curve in $T_{WP}(\Sigma)$ for t in a disk centered at 0, such that $\alpha_0 = [0]$. There is an open disk D centered at zero so that for each $t \in D$, α_t has a representative which is in $L_{\text{hyp}}^2(\Sigma) \cap L_{-1,1}^\infty(\Sigma)$, which is a holomorphic curve in the Hilbert space $L_{\text{hyp}}^2(\Sigma)$ and the Banach space $L_{-1,1}^\infty(\Sigma)$.

Thus the tangent space at the identity is described by $H_{-1,1}(\Sigma)$. This can be right translated, and also leads to a complex structure as in the the classical case.

The harmonic Beltrami differentials also induce local coordinates. For $\mu \in H_{-1,1}(\Sigma)$ let $f_\mu : \Sigma \to \Sigma_1$ denote a quasiconformal solution to the Beltrami equation.

Theorem 5.23 ([51]) There is an open neighbourhood of U of 0 in $H_{-1,1}(\Sigma)$ such that the map $\mu \mapsto [\Sigma, f_\mu, \Sigma_1] \in T_{WP}(\Sigma)$ obtained by solving the Beltrami equation is a biholomorphism onto its image. Furthermore, change of base point is a biholomorphism. This describes a system of complex coordinates which endows $T_{WP}(\Sigma)$ with a complex Hilbert manifold structure. This complex structure is compatible with the complex structure obtained from the fiber structure.

Corollary 5.24 ([51]) $T_{WP}(\Sigma)$ has a finite Weil-Petersson pairing on each tangent space.

Remark 5.25 In particular, this shows that if one had defined the complex structure using harmonic Beltrami differentials in the first place, then $T_{WP}(\Sigma)$ would have a holomorphic fiber structure with fibers (mod DB) biholomorphic to $\mathcal{O}_{WP}^{qc}(\Sigma_1^P)$ (and hence locally biholomorphic to $(\mathcal{O}_{WP}^{qc})^n$). The analogous result for the classical L^∞ case is Theorem 4.10 above; even in the L^∞ case it is non-trivial.

The above discussion leads naturally to the following problems.

Problem 5.2 Show that the complex structure on $T_{WP}(\Sigma)$ induced by harmonic Beltrami differentials $H_{-1,1}(\Sigma)$ and right composition is equivalent to that induced by the Bers embedding into hyperbolically L^2 quadratic differentials (see footnote 1, page 35).

Problem 5.3 Can the holomorphic fiber structure of Theorem 4.10 be extended to more general surfaces satisfying Lehner's condition, in the WP-class/L^p case? For example, to surfaces of infinite genus and/or infinitely many boundary curves?

Since Takhtajan and Teo obtain a foliation of $T(\mathbb{D}^*)$ by translates of $T_{\mathrm{WP}}(\mathbb{D}^*)$ we are led to ask:

Problem 5.4 Can $T(\Sigma)$ be endowed with a complex Hilbert manifold structure, such that $T_{\mathrm{WP}}(\Sigma)$ is the connected component of the identity, and the other connected components are right translates of $T_{\mathrm{WP}}(\Sigma)$?

5.4 Kähler Potential of Weil-Petersson Metric

In this section we give a brief overview of some geometric problems associated with the Weil-Petersson metric.

Ahlfors [2] showed that for compact Riemann surfaces the Weil-Petersson metric is Kähler. In the commentary to his collected works he stated that André Weil also had a proof but had not published it. Later Ahlfors computed the curvatures of holomorphic sections and the Ricci curvature [3].

Kirillov and Yuriev [29] sketched a generalization of the period mapping in Teichmüller theory to $\mathrm{Diff}(\mathbb{S}^1)/\mathrm{M\ddot{o}b}(\mathbb{S}^1)$. Some aspects were filled out and extended to the full universal Teichmüller space by Nag [35] and Nag and Sullivan in [36]. In this formulation, the period map is a map from Teichmüller space into an infinite-dimensional Siegel disk, which is a set of bounded, symmetric operators Z such that $I - Z\overline{Z}$ is positive-definite, analogous to the period mapping for compact Riemann surfaces. Nag and Sullivan [36] indicated that this period map is holomorphic by proving Gâteaux holomorphicity. The first complete proof of holomorphicity of the Kirillov-Yuriev-Nag-Sullivan period mapping was given by Takhtajan and Teo [72], in both the Weil-Petersson and classical setting.

In [23] Hong and Rajeev showed that the Siegel disk possesses a natural Kähler metric, whose Kähler potential is given, up to a multiplicative constant, by $\log \det(I - Z\overline{Z})$. Kirillov and Yuriev [29] and Nag [35] showed that the pull-back of a natural Kähler metric on the infinite Siegel disk is the Weil-Petersson metric. Thus $\log \det(I - Z\overline{Z})$ is also a Kähler potential for the Weil-Petersson metric. Hong and Rajeev noted that $\mathrm{Diff}(\mathbb{S}^1)/\mathrm{M\ddot{o}b}(\mathbb{S}^1)$ was not complete with respect to the Kähler metric, which indicated that it was not the correct analytic setting for the Weil-Petersson metric.

The completion of $\mathrm{Diff}(\mathbb{S}^1)/\mathrm{M\ddot{o}b}(\mathbb{S}^1)$ with respect to the Weil-Petersson metric is $T_{\mathrm{WP}}(\mathbb{D}^*)$, as was demonstrated in [72]. In [72] Takhtajan and Teo also proved the striking result that the period mapping Z is in fact the Grunsky operator. In particular, this implies that a constant multiple of the Fredholm determinant (3.3) is a Kähler potential for the Weil-Petersson metric. Finally, Takhtajan and Teo [72] and Shen [70] independently showed that the Grunsky operator Z is Hilbert-Schmidt if and only if the corresponding Teichmüller representative is WP-class. This is a very satisfying result because this is exactly the condition required in order for $Z\bar{Z}$ to be trace-class and hence that $\det(I - Z\bar{Z})$ exists.

This leads us to the following natural problems:

Problem 5.5 Is the Weil-Petersson metric on the L^2 Teichmüller space of bordered surfaces of type (g, n) Kähler? More generally, is this true for surfaces satisfying Lehner's condition?

Problem 5.6 Compute the sectional or Ricci curvatures of the Weil-Petersson metric for bordered surfaces of type (g, n), or more generally those satisfying Lehner's condition.[1]

These problems are closely related to the following.

Problem 5.7 Is there a generalization of the Kirillov-Yuriev-Nag-Sullivan period mapping to Teichmüller spaces of bordered surfaces of type (g, n), or more general surfaces? Can one obtain an analogous Kähler potential from this period mapping?

Note that generalizations of the Grunsky matrix to higher genus surfaces have been obtained by Reimer and Schippers [54], and for genus zero surfaces with n boundary curves by Radnell, Schippers, and Staubach [53].

5.5 Applications of Weil-Petersson Class Teichmüller Theory

The Weil-Petersson Teichmüller space has recently attracted a great deal of attention, in part because of its intrinsic importance to Teichmüller theory, and in part because of its many applications. We sketch some of these now.

In potential theory, the Weil-Petersson class domains are precisely those on which the Fredholm determinant of function theory exists, and therefore on which it is a viable tool in potential theory. In Teichmüller theory, the Weil-Petersson metric has been an important tool in the investigation of the geometry of Teichmüller space, see e.g. Wolpert [78]. It is now available in vastly greater generality.

There are also various physical applications. It has been suggested by several authors that the universal Teichmüller space could serve as a basis for a nonperturbative formulation of bosonic string theory. See Hong and Rajeev [23],

---

[1] After this chapter was submitted in May 2016, the paper [80] of Yanagishita appeared, in which it was shown that for surfaces satisfying Lehner's condition, the Weil-Petersson metric is indeed Kähler and the sectional and Ricci curvatures are negative. The convergent Weil-Petersson metric was obtained independently of Radnell, Schippers and Staubach [47, 51]. Yanagishita [80] also showed that the complex structure from harmonic Beltrami differentials is compatible with the complex structure from the Bers embedding. When combined with the results of [51], this apparently shows that these two complex structures are equivalent to that obtained from fibrations over the compact surfaces for surfaces of type (g, n).

Bowick and Rajeev [7], and the (somewhat dated) review of Pekonen [39]; for a review of connections to conformal field theory see Markina and Vasil'ev [32]. The g-loop scattering amplitude in string theory can be expressed as an integral over Teichmüller space [23], and thus a non-perturbative formulation might be given on the universal Teichmüller space, since it contains all other Teichmüller spaces. Hong and Rajeev also propose that the computation of the scattering amplitudes should involve the exponential of the Kähler potential discussed in the previous section. As observed above, it was known already to Hong and Rajeev that $\mathrm{Diff}(\mathbb{S}^1)/\mathrm{M\ddot{o}b}(\mathbb{S}^1)$ is not complete with respect to the Kähler metric (which we now know to be the Weil-Petersson metric). The correct analytic setting for Hong and Rajeev's proposal thus appears to be the Weil-Petersson class Teichmüller space.

The Weil-Petersson Teichmüller space also has deep connections with two-dimensional conformal field theory as formulated by Segal, Kontsevich, Vafa and others (see [25] for a review of the literature in the formulation of CFT). Work of Radnell, Schippers and Staubach has established that the Weil-Petersson class rigged moduli space is the completion of the analytically rigged moduli space of Friedan/Shenker/Segal/Vafa, and is also the largest space on which constructions in conformal field theory can be carried out. These include for example sewing properties of the determinant line bundle over the rigged moduli space and the existence of local holomorphic sections. A review of this work can be found in [52].

Finally, there are applications to fluid mechanics and infinite dimensional groups of diffeomorphisms. The setting for this is a deep insight of Arnol'd, namely that the geodesic equations on infinite-dimensional diffeomorphism groups are analogous to the Euler equations of fluid mechanics [28]. Different choices of groups and metrics lead to different geodesic equations, which in turn are different systems of partial differential equations [28, Table 4.1]. See Grong, Markina, and Vasil'ev [20] for a survey of choices on $\mathrm{Diff}(\mathbb{S}^1)/\mathrm{M\ddot{o}b}(\mathbb{S}^1)$ and their relation to sub-Riemannian geometry. In the case of the Weil-Petersson metric on $\mathrm{Diff}(\mathbb{S}^1)/\mathbb{S}^1$, the geodesic equations are related to the KdV equation. Schonbek, Todorov, and Zubelli [69] were able to obtain long-term solutions to the KdV equation using this connection. Gay-Balmaz [16] was able to obtain global existence and uniqueness of the geodesics, and applied the Euler-Poincaré reduction process to obtain the spatial representation of the geodesics. Further important and interesting applications in this direction were given by Figalli [14], Gay-Balmaz and Ratiu [17], and Kushnarev [30].

We note that an important technical problem in this direction was solved by Shen [71]. Much of the analysis in the fluid mechanical models above has involved the assumption that the corresponding quasisymmetries were in $H^{3/2-\epsilon}(\mathbb{S}^1)$ for $\epsilon > 0$, and it was an open question whether the quasisymmetries in the Weil-Petersson class Teichmüller space would be precisely those in $H^{3/2}(\mathbb{S}^1)$. Theorem 5.10 of Shen above gave the correct characterization, and in the same paper [71] he also showed that there are WP-class quasisymmetries which are not in $H^{3/2}(\mathbb{S}^1)$.

6 Conclusion

6.1 Concluding Remarks

In this paper, we have given a number of examples of the general phenomenon of comparison moduli spaces in geometric function theory and moduli spaces of Riemann surfaces. We have seen that this concept spontaneously arises in both modern and classical complex analysis. We have also attempted to illustrate how this notion of moduli space captures many complex analytic phenomena in a simple way.

We would like to conclude with another observation. One is struck by the pervasive relevance of classical function theory. We have seen, for example, the unwitting re-invention in two-dimensional conformal field theory of the Teichmüller space of bordered surfaces and conformal welding, and the use of Schiffer's variational technique to construct a complex structure on Teichmüller space and the rigged moduli space of conformal field theory. We have also seen the Fredholm determinant of classical potential theory—as reformulated by Schiffer—emerge as a fundamental geometric object on moduli spaces of Riemann surfaces. These geometric problems in turn require the formulation and solution of analytic problems which can only be approached with function theory. Other examples spring readily to the mind of any mathematician with their ear to the ground.

We cannot express this in any better way than Ahlfors did [1]: "We start out from a purely classical problem, and we place it in a much more general modern setting, sometimes in a form that would not have been available to a classical mathematician. When the generalized problem is analyzed, it turns out to lead forcefully to a new and evidently significant problem in the original purely classical framework. In other words, we are faced with new evidence of the scope and fertility of classical analysis."

Acknowledgements The authors are grateful to David Radnell for a fruitful collaboration and valuable discussions through the years, and for his comments and suggestions concerning the initial draft of the manuscript.

Eric Schippers and Wolfgang Staubach are grateful for the financial support from the Wenner-Gren Foundations. Eric Schippers is also partially supported by the National Sciences and Engineering Research Council of Canada.

References

1. L.V. Ahlfors, Classical and contemporary analysis. SIAM Rev. **3**, 1–9 (1961)
2. L.V. Ahlfors, Some remarks on Teichmller's space of Riemann surfaces. Ann. Math. (2) **74**, 171–191 (1961)
3. L.V. Ahlfors, Curvature properties of Teichmüller's space. J. Anal. Math. **9**, 161–176 (1961/1962)

4. L.V. Ahlfors, *Conformal Invariants: Topics in Geometric Function Theory.* McGraw-Hill Series in Higher Mathematics (McGraw-Hill Book, New York-Düsseldorf-Johannesburg, 1973)
5. L.V. Ahlfors, L. Sario, *Riemann Surfaces* (Princeton University Press, Princeton, NJ, 1960)
6. S. Bergman, M. Schiffer, Kernel functions and conformal mapping. Comp. Math. **8**, 205–249 (1951)
7. M.J. Bowick, S.G. Rajeev, The holomorphic geometry of closed bosonic string theory and Diff($\mathbb{S}^1$)/$\mathbb{S}^1$. Nucl. Phys. B **293**(2), 348–384 (1987)
8. R. Courant, *Dirichlet's Principle, Conformal Mapping, and Minimal Surfaces.* With an appendix by M. Schiffer. Reprint of the 1950 original (Springer, New York, Heidelberg, 1977)
9. G. Cui, Integrably asymptotic affine homeomorphisms of the circle and Teichmüller spaces, Sci. China Ser. A **43**(3), 267–279 (2000)
10. G. Cui, M. Zinsmeister, BMO-Teichmüller spaces. Ill. J. Math. **48**(4), 1223–1233 (2004)
11. V.N. Dubinin, *Condenser Capacities and Symmetrization in Geometric Function Theory.* Translated from the Russian by Nikolai G. Kruzhilin (Springer, Basel, 2014)
12. P. Duren, *Univalent Functions.* Grundlehren der Mathematischen Wissenschaften, vol. 259 (Springer, New York, 1983)
13. C.J. Earle, F.P. Gardiner, N. Lakic, *Asymptotic Teichmller Space. I. The Complex Structure.* In the tradition of Ahlfors and Bers (Stony Brook, NY, 1998). Contemporary Mathematics, vol. 256 (American Mathematical Society, Providence, RI, 2000), pp. 17–38
14. A. Figalli, On flows of $H^{\frac{3}{2}}$-vector fields on the circle. Math. Ann. **347**, 43–57 (2010)
15. F.P. Gardiner, Schiffer's interior variation and quasiconformal mapping. Duke Math. J. **42**, 371–380 (1975)
16. F. Gay-Balmaz, Infinite dimensional geodesic flows and the universal Teichmller space. Thèse École polytechnique fédérale de Lausanne EPFL, vol. 4254 (2009). https://doi.org/10.5075/epfl-thesis-4254
17. F. Gay-Balmaz, T.S. Ratiu, The geometry of the universal Teichmüller space and the Euler-Weil-Petersson equation. Adv. Math. **279**, 717–778 (2015)
18. F.W. Gehring, K. Hag, *The Ubiquitous Quasidisk.* With contributions by Ole Jacob Broch. Mathematical Surveys and Monographs, vol. 184 (American Mathematical Society, Providence, RI, 2012)
19. A.Z. Grinshpan, Logarithmic geometry, exponentiation, and coefficient bounds in the theory of univalent functions and non-overlapping domains, in *Handbook of Complex Analysis: Geometric Function Theory*, ed. by R. Kühnau, vol. I (North-Holland, Amsterdam, 2002)
20. E. Grong, I. Markina, A. Vasil'ev, Sub-Riemannian structures corresponding to Kählerian metrics on the universal Teichmüller space and curve, in *"60 years of Analytic Functions in Lublin" in Memory of Our Professors and Friends Jan G. Krzyż, Zdzisław Lewandowski and Wojciech Szapiel*, vol. 97116. Monographs University of Economics and Innovation in Lublin (Innovatio Press Scientific Publishing House, Lublin, 2012)
21. H. Grötzsch, Über einige Extremalprobleme der konformen Abbildung. II. Ber. Verh. sächs. Akad. Wiss. Leipzig, Math.-Phys. Kl. **80**, 497–502 (1928)
22. H. Guo, Integrable Teichmüller spaces. Sci. China Ser. A **43**(1), 47–58 (2000)
23. D.K. Hong, S.G. Rajeev, Universal Teichmller space and Diff($\mathbb{S}^1$)/$\mathbb{S}^1$. Commun. Math. Phys. **135**(2), 401–411 (1991)
24. Y. Hu, Y. Shen, On quasisymmetric homeomorphisms. Isr. J. Math. **191**, 209–226 (2012)
25. Y.-Z. Huang, *Two-Dimensional Conformal Geometry and Vertex Operator Algebras.* Progress in Mathematics, vol. 148 (Birkhäuser, Boston, MA, 1997)
26. J.H. Hubbard, *Teichmüller Theory and Applications to Geometry, Topology, and Dynamics, Vol. 1. Teichmüller Theory.* With contributions by Adrien Douady, William Dunbar, Roland Roeder, Sylvain Bonnot, David Brown, Allen Hatcher, Chris Hruska and Sudeb Mitra. With forewords by William Thurston and Clifford Earle (Matrix Editions, Ithaca, NY, 2006)
27. J.A. Jenkins, *Univalent Functions and Conformal Mapping.* Ergebnisse der Mathematik und ihrer Grenzgebiete. Neue Folge, Heft 18. (Springer, Berlin, 1958)

28. B. Khesin, R. Wendt, *The Geometry of Infinite-Dimensional Groups*. Ergebnisse der Mathematik und ihrer Grenzgebiete. 3. Folge. A Series of Modern Surveys in Mathematics, vol. 51 (Springer, Berlin, 2009)
29. A.A. Kirillov, D.V. Yuriev, Representations of the Virasoro algebra by the orbit method. J. Geom. Phys. **5**(3), 351–363 (1988)
30. S. Kushnarev, Teichons: soliton-like geodesics on universal Teichmüller space. Exp. Math. **18**, 325–336 (2009)
31. O. Lehto, *Univalent Functions and Teichmüller Spaces*. Graduate Texts in Mathematics, vol. 109 (Springer, New York, 1987)
32. I. Markina, A. Vasil'ev, Virasoro algebra and dynamics in the space of univalent functions, in *Five lectures in complex analysis*. Contemporary Mathematics, vol. 525, pp. 85–116 (American Mathematical Society, Providence, RI, 2010)
33. I. Markina, D. Prokhorov, A. Vasil'ev, Sub-Riemannian geometry of the coefficients of univalent functions. J. Funct. Anal. **245**(2), 475–492 (2007)
34. S. Nag, *The Complex Analytic Theory of Teichmüller Spaces*. Canadian Mathematical Society Series of Monographs and Advanced Texts (Wiley, New York, 1988)
35. S. Nag, A period mapping in universal Teichmller space. Bull. Amer. Math. Soc. (N.S.) **26**(2), 280–287 (1992)
36. S. Nag, D. Sullivan, Teichmüller theory and the universal period mapping via quantum calculus and the $H^{1/2}$ space on the circle. Osaka J. Math. **32**(1), 1–34 (1995)
37. S. Nag, A. Verjovsky, Diff($\mathbb{S}^1$) and the Teichmüller spaces. Commun. Math. Phys. **130**(1), 123–138 (1990)
38. Z. Nehari, Some inequalities in the theory of functions. Trans. Am. Math. Soc. **75**, 256–286 (1953)
39. O. Pekonen, Universal Teichmüller space in geometry and physics, J. Geom. Phys. **15**(3), 227–251 (1995)
40. A. Pfluger, On the Functional $a_3 - \lambda a_2^2$ in the class $\mathcal{S}$. Complex Var. **10**, 83–95 (1988)
41. C. Pommerenke, *Univalent functions*, With a chapter on quadratic differentials by Gerd Jensen. Studia Mathematica/Mathematische Lehrbücher, Band XXV (Vandenhoeck & Ruprecht, Göttingen, 1975)
42. D. Radnell, Schiffer variation in Teichmüller space, determinant line bundles and modular functors. Ph.D. thesis, Rutgers University, New Brunswick, NJ (2003)
43. D. Radnell, E. Schippers, Quasisymmetric sewing in rigged Teichmller space. Commun. Contemp. Math. **8**(4), 481–534 (2006)
44. D. Radnell, E. Schippers, A complex structure on the set of quasiconformally extendible non-overlapping mappings into a Riemann surface. J. Anal. Math. **108**, 277–291 (2009)
45. D. Radnell, E. Schippers, Fiber structure and local coordinates for the Teichmüller space of a bordered Riemann surface. Conform. Geom. Dyn. **14**, 14–34 (2010)
46. D. Radnell, E. Schippers, The semigroup of rigged annuli and the Teichmüller space of the annulus. J. Lond. Math. Soc. (2) **86**(2), 321–342 (2012)
47. D. Radnell, E. Schippers, W. Staubach, Quasiconformal maps of bordered Riemann surfaces with L^2 Beltrami differentials. J. Anal. Math. (2014, to appear). Part of arXiv:1403.0868
48. D. Radnell, E. Schippers, W. Staubach, A Hilbert manifold structure on the Weil-Petersson class Teichmüller space of bordered surfaces. Commun. Contemp. Math. **17**(4) (2015). https://doi.org/10.1142/S0219199715500169
49. D. Radnell, E. Schippers, W. Staubach, Dirichlet problem and Sokhotski-Plemelj jump formula on Weil-Petersson class quasidisks. Ann. Acad. Sci. Fenn. Math. **41**, 119–127 (2016)
50. D. Radnell, E. Schippers, W. Staubach, Weil-Petersson class non-overlapping mappings into a Riemann surface. Commun. Contemp. Math. **18**(4) (2016). Part of arXiv:1207.0973
51. D. Radnell, E. Schippers, W. Staubach, Convergence of the Weil-Petersson metric on the Teichmüller space of bordered surfaces. Commun. Contemp. Math. **19** (2017). https://doi.org/10.1142/S0219199716500255. Part of arXiv:1403.0868 (2014)

52. D. Radnell, E. Schippers, W. Staubach, Quasiconformal Teichmüller theory as an analytical foundation for two dimensional conformal field theory, in *Lie Algebras, Vertex Operator Algebras and Related Topics*, ed. by K. Barron, E. Jurisich, A. Milas, K. Misra. Contemporary Mathematics, vol. 695 (American Mathematical Society, Providence, 2017)

53. D. Radnell, E. Schippers, W. Staubach, Dirichlet spaces of domains bounded by quasicircles. Preprint (2017). arXiv:1705.01279v1

54. K. Reimer, E. Schippers, Grunsky inequalities for mappings into a compact Riemann surface. Compl. Anal. Oper. Theory **9**(8), 1663–1679 (2015)

55. A. Schaeffer, D. Spencer, *Coefficient Regions for Schlicht Functions*. American Mathematical Society Colloquium Publications, vol. 35 (American Mathematical Society, New York, 1950)

56. M.M. Schiffer, The kernel function of an orthonormal system. Duke Math. J. **13**, 529–540 (1946)

57. M.M. Schiffer, Fredholm eigenvalues and conformal mappings. Rend. Mat. Appl. (5) **22**, 447–468 (1963)

58. M.M. Schiffer, Fredholm eigenvalues and Grunsky matrices. Ann. Polon. Math. **39**, 149–164 (1981)

59. M.M. Schiffer, *Menahem Max Schiffer: Selected Papers*, ed. by P. Duren, L. Zalcman, vol. 1. Contemporary Mathematicians (Birkhäuser/Springer, New York, 2013)

60. M.M. Schiffer, *Menahem Max Schiffer: Selected Papers*. Contemporary Mathematicians, vol. 2, ed. by P. Duren, L. Zalcman (Birkhuser/Springer, New York, 2014)

61. M.M. Schiffer, D. Spencer, *Functionals of Finite Riemann Surfaces* (Princeton University Press, Princeton, NJ, 1954)

62. E. Schippers, Conformal invariants and higher-order Schwarz lemmas. J. Anal. Math. **90**, 217–241 (2003)

63. E. Schippers, Conformal invariants corresponding to pairs of domains, in *Future Trends in Geometric Function Theory*, Reports, University of Jyväskylä. Department of Mathematics and Statistics, vol. 92 (University of Jyväskylä, Jyväskylä, 2003), pp. 207–219

64. E. Schippers, The power matrix, coadjoint action and quadratic differentials. J. Anal. Math. **98**, 249–277 (2006)

65. E. Schippers, The derivative of the Nehari functional. Ann. Acad. Sci. Fenn. **35**(1), 291–307 (2010)

66. E. Schippers, Quadratic differentials and conformal invariants. J. Anal. **24**, 209–228 (2016)

67. E. Schippers, Conformal invariants corresponding to quadratic differentials. Isr. J. Math. (to appear)

68. H.G. Schmidt, Some examples of the method of quadratic differentials in the theory of univalent functions. Matematisk Institut, Aarhus Universitet, Preprint No. 35 (1970)

69. M.E. Schonbek, A.N. Todorov, J.P. Zubelli, Geodesic flows on diffeomorphisms of the circle, Grassmannians, and the geometry of the periodic KdV equation. (English summary) Adv. Theor. Math. Phys. **3**(4), 1027–1092 (1999)

70. Y. Shen, On Grunsky operator, Sci. China Ser. A **50**(12), 1805–1817 (2007)

71. Y. Shen, Weil-Petersson Teichmüller space. Preprint (2013). arXiv:1304.3197

72. L. Takhtajan, L.-P. Teo, Weil-Petersson metric on the universal Teichmüller space. Mem. Am. Math. Soc. **183**(861), i–vii, 1–119 + front and back matter (2006)

73. O. Tammi, *Extremum Problems for Bounded Univalent Functions. I*. Lecture Notes in Mathematics, vol. 646 (Springer, Berlin-New York, 1978)

74. O. Tammi, *Extremum Problems for Bounded Univalent Functions. II*. Lecture Notes in Mathematics, vol. 913 (Springer, Berlin-New York, 1982)

75. S. Tang, Some characterizations of the integrable Teichmüller space. Sci. China Math. **56**(3), 541–551 (2013)

76. O. Teichmüller, Ungleichungen zwischen den Koeffizienten schlichter Funktionen, in *Sitzungsberichte der Preussischen Akademie der Wissenschaften, Physikalisch-Mathematische Klasse* (1938), pp. 363–375 Stzgsber. preuss. Akad. Wiss. Math.-Naturwiss. Kl. (1938), pp. 363–375

77. A. Vasil'ev, *Moduli of Families of Curves for Conformal and Quasiconformal Mappings*. Lecture Notes in Mathematics, vol. 1788 (Springer, Berlin, 2002)

78. S. Wolpert, Chern forms and the Riemann tensor for the moduli space of curves. Invent. Math. **85**(1), 119–145 (1986)
79. M. Yanagishita, Introduction of a complex structure on the p-integrable Teichmüller space. Ann. Acad. Sci. Fenn. Math. **39**(2), 947–971 (2014)
80. M. Yanagishita, Kählerity and negativity of Weil-Petersson metric on square integrable Teichmüller space. J. Geom. Anal. **27**(3), 1995–2017 (2017)

Asymptotic Ratio of Harmonic Measures of Sides of a Boundary Slit

Alexander Solynin

In the memory of Sasha Vasiliev,
an excellent mathematician, a wonderful person and a friend
from my youth

Abstract Let l be a Jordan arc in the upper half-plane $\mathbb{H}$ with the initial point at $z = 0$. For $\zeta \in l$, let $\omega^+(a, \zeta)$ and $\omega^-(a, \zeta)$ denote the harmonic measures of the left and right shores of the slit $l_\zeta \subset \mathbb{H}$ along the portion of the arc l travelled from 0 to ζ. In one of the seminars within the "*Complex Analysis and Integrable Systems*" semester held at the Mittag-Leffler Institute in 2011, D. Prokhorov suggested a study of the limit behavior of the quotient $\omega^-(a, \zeta)/\omega^+(a, \zeta)$ as $\zeta \to 0$ along l. In recent publications, D. Prokhorov and his coauthors discussed this problem and proved several results for smooth slits.

In this paper, we study the limit behavior of $\omega^+(a, \zeta)/\omega^-(a, \zeta)$ for a broader class of continuous slits which includes radially and angularly oscillating slits. Some related questions concerning behavior of the driving term of the corresponding chordal Löwner equation are also discussed.

Keywords Harmonic measure • Boundary slits • Löwner equation

2010 Mathematics Subject Classification 30C20, 30C75

A. Solynin (✉)
Department of Mathematics and Statistics, Texas Tech University, Box 41042, Lubbock, TX 79409, USA
e-mail: alex.solynin@ttu.edu

© Springer International Publishing AG 2018
M. Agranovsky et al. (eds.), *Complex Analysis and Dynamical Systems*,
Trends in Mathematics, https://doi.org/10.1007/978-3-319-70154-7_14

1 Main Results

Let D be a simply connected domain on $\mathbb{C}$, $D \neq \mathbb{C}$. Suppose that ∂D contains an open smooth Jordan arc γ_0 which has a neighborhood free of points of $\partial D \setminus \gamma_0$ and let $\zeta_0 \in \gamma_0$. Suppose further that l is a closed Jordan arc in $D \cup \{\zeta_0\}$ with the initial point at ζ_0 and terminal point at $\zeta_1 \in D$. For $\zeta \in l$, let l_ζ denote the Jordan Subaru of l with the initial point at ζ_0 and terminal point at ζ. We call l_ζ a ζ-*tail* of l obtained by *truncation* of l at the point ζ. Let $D_\zeta = D \setminus l_\zeta$. Then D_ζ is a simply connected domain, and l_ζ is a *boundary slit*. Each point ζ' on l_ζ represents two distinct boundary points of D_ζ except for the case when ζ' is the tip of the slit. Therefore, each ζ-tail l_ζ, considered as a boundary set of D_ζ, has two *shores*. By l_ζ^+ and l_ζ^- we denote the boundary arcs of D_ζ corresponding, respectively, to the left and right shores of the slit along l_ζ when a point traverses l_ζ from ζ_0 to ζ. See Fig. 1, which illustrates some of our notation.

In this paper, we denote by $\omega(z, E, D)$ the harmonic measure of a Borel set $E \subset \partial D$ with respect to the domain D evaluated at $z \in D$. Let D and l be as above. For $\zeta \in l$ and $a \in D_\zeta$, let

$$\omega^+(a, \zeta) = \omega(a, l_\zeta^+, D_\zeta), \quad \omega^-(a, \zeta) = \omega(a, l_\zeta^-, D_\zeta).$$

Thus, $\omega^+(a, \zeta)$ and $\omega^-(a, \zeta)$ are the harmonic measures of the left and right sides of the slit l_ζ, respectively.

Our goal here is to study the limit behavior of the quotient $\omega^-(a, \zeta)/\omega^+(a, \zeta)$ as $\zeta \to \zeta_0$ along l. Of course, this limit behavior depends on geometric properties of l near ζ_0. In particular, we consider smooth and monotonic slits as well as not necessarily smooth oscillating slits. By a smooth arc we mean an open or closed arc γ having a unit tangent vector $\vec{v}(\zeta)$ at every point $\zeta \in \gamma$ such that $\vec{v}(\zeta)$ is continuous on γ, including its end points if the arc is closed. A simple arc l is called monotonic (with respect to its initial point ζ_0) if it admits a parametrization $\zeta = \zeta(\tau)$, $0 \leq \tau \leq \tau_0$, such that $\zeta(0) = \zeta_0$ and $|\zeta(\tau) - \zeta_0|$ is a non-decreasing function of τ. Of course, if l is smooth, then l_ζ is monotonic for all ζ sufficiently close to ζ_0.

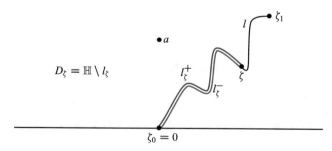

Fig. 1 Domain D_ζ and its boundary arcs l_ζ^- and l_ζ^+

Our first result is the following theorem about monotonic slits in the upper half-plane, i.e., in the case when $D = \mathbb{H} := \{z : \Im z > 0\}$.

Theorem 1 *Let l be a monotonic slit in $\mathbb{H} \cup \{0\}$ with its initial point at $\zeta_0 = 0$. Suppose that l forms an angle $\alpha\pi$, $0 < \alpha < 1$, with the positive real axis. Then for every $a \in \mathbb{H}$,*

$$\frac{\omega^-(a, \zeta)}{\omega^+(a, \zeta)} \to \frac{\alpha}{1 - \alpha} \quad \text{as } \zeta \to \zeta_0 \text{ along } l. \tag{1.1}$$

Convergence in (1.1) is uniform on compact subsets of $\mathbb{H}$.

Theorem 1 answers a question raised by Dmitri Prokhorov during informal discussions of participants of the scientific program *"Complex Analysis and Integrable Systems"* held at the Institut Mittag-Leffler, Djursholm, Sweden, during the Fall Semester, 2011. More precisely, he asked this question for the case when l is a smooth slit. Recently, D. Prokhorov and his collaborators returned to this problem and made essential progress toward its solution. In 2012, D. Prokhorov and A. Zakharov [5] proved Theorem 1 for perpendicular slits in $\mathbb{H}$, i.e., for the case $\alpha = 1/2$. Then D. Prokhorov and D. Ukrainskii [4] studied this problem for any $0 < \alpha < 1$ and proved a weaker version of Theorem 1 under the additional assumption that $l \in C^4$ (i.e., l has smoothness of order four and has at least fourth order tangency at ζ_0 to the straight line segment forming an angle of opening $\alpha\pi$ with ∂D). The proof presented in [4] is based on the theory of the Löwner differential equation and therefore requires sufficient smoothness of the driving term of this equation. Our proof does not require any smoothness except the existence of a tangent direction at the initial point of the slit. Moreover, our proof is conceptually simpler; its main ingredients are the Carathéodory convergence theorem and a simple trick with the Schwarz reflection principle for harmonic functions.

D. Prokhorov raised this question in relation to his study of the chordal Löwner equation in the upper half-plane $\mathbb{H}$, that is,

$$\frac{\partial f(z, t)}{\partial t} = \frac{2}{f(z, t) - \lambda(t)}, \quad f(z, 0) \equiv z, \quad t \ge 0. \tag{1.2}$$

Here $f(z, t)$ denotes the mapping from the domain $D_\zeta = \mathbb{H} \setminus l_\zeta$ onto $\mathbb{H}$ normalized at infinity by the condition

$$f(z, t) = z + \frac{2t}{z} + O\left(\frac{1}{z^2}\right). \tag{1.3}$$

In Eq. (1.3), $t = t(\zeta) > 0$ is a real-valued function of the point $\zeta \in l$ that is known as the *half-plane capacity* of the ζ-tail l_ζ; see [2, Chap. 3.4]. It is also well-known that $t(\zeta)$ strictly increases as ζ runs along the slit l starting at its initial point ζ_0 on the real axis $\mathbb{R}$. Therefore, the slit l can be parameterized as $l = \{\zeta = \zeta(t), \, 0 \le t \le T\}$ with some $T > 0$.

The real-valued function $\lambda(t)$ present in (1.2) is commonly known as a driving term of the chordal Löwner equation. To clarify its relation with l_ζ, we note that $f(z, t)$ maps l_ζ in the sense of boundary correspondence onto some closed interval $[a(t), b(t)]$ of $\mathbb{R}$. Then $\lambda(t)$ is the image of the tip $z = \zeta$ of the slit l_ζ under this mapping. Hence, $a(t) < \lambda(t) < b(t)$.

Since $f(z, t) \to z$ uniformly on compact subsets of $\mathbb{H}$ as $t \to 0$ it follows that each of the functions $a(t)$ and $b(t)$ converges to the initial point of the slit l at ζ_0. Thus, in this setting, the limit behavior of $\lambda(t)$ is clear. Therefore, we consider a renormalized family of mappings given by

$$g(z, \zeta) = \frac{2}{b(t) - a(t)} \left(f(z, t) - \frac{b(t) + a(t)}{2} \right), \qquad t = t(\zeta). \tag{1.4}$$

Now, for each $\zeta \in l$, $g(z, \zeta)$ maps l_ζ onto the interval $I_0 = [-1, 1]$; therefore, the image of the tip of l_ζ is the point

$$\Lambda(\zeta) = \frac{2}{b(t) - a(t)} \left(\lambda(t) - \frac{b(t) + a(t)}{2} \right), \tag{1.5}$$

which varies in the interior of I_0. Below, we refer to $\Lambda(\zeta)$ as the *scaled driving term*. The set $T(l)$ consisting of all limit points of $\Lambda(\zeta)$ as $\zeta \to \zeta_0$ will be called the *trace of the driving term* in I_0. As the cluster set of a continuous function at some point, the trace $T(l)$ is a closed subinterval of I_0. In this new setting, the question of the behavior of $\Lambda(\zeta)$ as ζ approaches ζ_0 along l and the question of the exact position of $T(l)$ within the interval I_0 become non-trivial and interesting. Theorem 2 below answers this question for simple monotonic slits in $\mathbb{H}$.

Theorem 2 *Suppose that l is a simple monotonic slit in $\mathbb{H} \cup \{0\}$. Suppose further that l forms an angle $\alpha\pi$, $0 < \alpha < 1$, with the positive real axis at $z = 0$. Then*

$$\Lambda(\zeta) \to 1 - 2\alpha \quad as \ \zeta \to 0 \ along \ l. \tag{1.6}$$

In other words, Eq. (1.6) says that under conditions of Theorem 2 the scaled driving term $\Lambda(\zeta)$ stabilizes at the point $x_\alpha = 1 - 2\alpha$ as $\zeta \to 0$ and that x_α is the only point in the trace $T(l)$.

The proofs of Theorems 1 and 2 are given in Sects. 5 and 6, respectively. In Sect. 2, we discuss an application of the Carathéodory convergence theorem to the problems on the harmonic measures of boundary slits. In Sect. 3, we present a form of the Schwarz reflection principle convenient for our study of the harmonic measure of boundary arcs. Section 4 contains technical calculations of some quantities related to harmonic measures of radial slits.

Since our proofs of Theorems 1 and 2 do not require any smoothness of the slit l, except for existence of the tangent direction at $z = 0$, it is tempting to relax the assumptions of these theorems further. One possibility is to consider slits l such that the unit tangent vector $\overrightarrow{v}(\zeta)$ exists for all $\zeta \in l$ but is not necessarily continuous at ζ_0. Or one may wonder if the monotonicity assumption is actually needed for the limit relation in (1.6). In Sect. 7, we present some examples, which show that under these relaxed conditions, Eqs. (1.1) and (1.6) are not valid in general. However, Theorems 4 and 3 given in this section show that certain bounds for the quotient of the harmonic measures $\omega^-(a, \zeta)$ and $\omega^+(a, \zeta)$ can be obtained for slits with restricted radial or angular oscillations.

In the last section, we discuss some questions concerning the limit behavior of the quotient of the harmonic measures of slits, their scaled driving terms $\Lambda(\zeta)$, and some other quantities.

2 Harmonic Measure and Convergence of Domains to the Kernel

In this section, we discuss a form of the Carathéodory theorem on the convergence of sequences of domains to the kernel, which is convenient for the purposes of this paper. For necessary definitions and results related to this convergence theorem we refer to [1, Chap. 5] and [3, Chap. 2]. For $k = 1, 2, \ldots$ and $k = \infty$, let D_k be a Jordan domain on $\mathbb{C}$ and let $z_0 \in D_k$ for all k. Suppose further that there are two points a and b such that $a, b \in \partial D_k$ for all k. It follows from the Riemann mapping theorem that for each domain D_k, there is a unique function f_k which maps D_k conformally and one-to-one onto the horizontal strip $\Pi = \{w : 0 < \Im w < 1\}$ such that $\Re f_k(z_0) = 0$ and

$$\lim_{\overline{D_k} \ni z \to a} \Re f_k(z) = -\infty, \qquad \lim_{\overline{D_k} \ni z \to b} \Re f_k(z) = +\infty. \tag{2.1}$$

Proposition 1 *Suppose that the sequence of domains D_k, $k = 1, 2, \ldots$, converges to the kernel D_∞. Suppose further that for every $\varepsilon > 0$, there exists an integer $N > 0$ such that for every $k \geq N$, there is a continuous one-to-one mapping $s_k : \partial D_k \to \partial D_\infty$ such that $|s_k(\zeta) - \zeta| < \varepsilon$ for all $\zeta \in \partial D_k$. Then*

$$f_k(z) \to f_\infty(z) \quad \text{and} \quad f_k^{(n)}(z) \to f_\infty^{(n)}(z) \quad \text{for } n = 1, 2, \ldots \tag{2.2}$$

uniformly on compact subsets of D_∞.

Proof Let $g_\infty(z)$ be a conformal mapping from D_∞ onto the unit disk $\mathbb{D} = \{w : |w| < 1\}$ such that $g_\infty(a) = -1$, $g_\infty(b) = 1$, and $\Re g_\infty(z_0) = 0$. The existence of such a mapping is an easy consequence of the Riemann mapping theorem. Let $h_\infty = g_\infty^{-1}$, $\hat{z} = h_\infty(0)$, and $\alpha = \arg g_\infty'(\hat{z})$. By the Riemann mapping theorem, for every sufficiently large k (such that $\hat{z} \in D_k$), there exists a unique function $g_k(z)$, which maps D_k conformally and one-to-one onto $\mathbb{D}$ such that $g_k(\hat{z}) = 0$ and $\arg g_k'(\hat{z}) = \alpha$. Let $h_k = g_k^{-1}$. Since D_k is a Jordan domain, the function h_k can be extended to a function continuous on the closed unit disk $\overline{\mathbb{D}}$, which we still denote by

h_k. Let $\tau_k^a = g_k(a)$, $\tau_k^b = g_k(b)$, and $w_k = g_k(z_0)$. Then of course, $|\tau_k^a| = |\tau_k^b| = 1$. By the Rado convergence theorem (see [1, Chap. 2, §5]), $h_k(w) \to h_\infty(w)$ uniformly on the closed unit disk $\overline{\mathbb{D}}$. The latter implies that

$$\tau_k^a \to -1, \quad \tau_k^b \to 1, \quad \text{and} \quad w_k \to g_\infty(z_0) \text{ as } k \to \infty. \qquad (2.3)$$

Let φ_k denote the Möbius automorphism of $\mathbb{D}$ such that $\varphi_k(\tau_k^a) = -1$, $\varphi_k(\tau_k^b) = 1$, and $\Re\varphi_k(w_k) = 0$, and let

$$\psi_0(w) = \frac{1}{\pi} \log \frac{1+w}{1-w} + \frac{i}{2}.$$

The function f_k can be represented as the composition $f_k = \psi_0 \circ \varphi_k \circ g_k$.

We note that $g_k(z) \to g_\infty(z)$ uniformly on compact subsets of D_∞ by the Carathéodory convergence theorem. Furthermore, Eq. (2.3) implies that $\varphi_k(w) \to w$ uniformly on $\overline{\mathbb{D}}$. Since $f_k = \psi_0 \circ \varphi_k \circ g_k$, we conclude that $f_k(z) \to f_\infty(z)$ uniformly on compact subsets of D_∞. This proves the first relation in (2.2), and the second relation follows from the classical Weierstrass theorem on the convergence of sequences of analytic functions. $\qquad \Box$

Proposition 1 is stated for the case in which D_k are Jordan domains. It can be easily extended to the case in which each domain D_k is a complement to a Jordan arc l_k with end points at $z = c$ and $z = d$, i.e., when $D_k = \overline{\mathbb{C}} \setminus l_k$, $k = 1, 2, \ldots$ or $k = \infty$. Suppose that the points $z = a$ and $z = b$ are boundary points of D_k for all k. For our purposes, we have to consider the case in which one of the points $z = a$ or $z = b$ lies at infinity. In this case, we use the spherical distance $d_s(z_1, z_2)$ rather than Euclidean distance. We recall that the spherical distance is given by $d_s(z_1, z_2) = \frac{|z_1-z_2|}{\sqrt{(1+|z_1|^2)(1+|z_2|^2)}}$ if $z_1 \neq \infty$, $z_2 \neq \infty$ and by $d_s(z_1, \infty) = \frac{1}{\sqrt{1+|z_1|^2}}$ if $z_2 = \infty$.

We assume, as above, that $z_0 \in D_k$ for all k. Let $f_k(z)$ denote the Riemann mapping function from D_k onto the strip Π satisfying conditions (2.1) and such that $\Re f_k(z_0) = 0$ for all k.

Corollary 1 *Suppose that the sequence of domains $D_k = \overline{\mathbb{C}} \setminus l_k$, $k = 1, 2, \ldots$, converges to the kernel $D_\infty = \overline{\mathbb{C}} \setminus l_\infty$. Suppose further that for every $\varepsilon > 0$ there exists an integer $N > 0$ such that for every $k \geq N$, there is a continuous one-to-one mapping $s_k : \partial D_k \to \partial D_\infty$ such that $d_s(s_k(\zeta), \zeta) < \varepsilon$ for all $\zeta \in \partial D_k$. Then the sequence of functions $f_k(z)$ satisfies the convergence relations given by formulas (2.2).*

Proof We use an auxiliary conformal mapping $g(z)$ to reduce our problem to the previous case, which was discussed in Proposition 1. Here $g(z)$ is essentially the same function, which appears in the proof of the Riemann mapping theorem. Namely, let $g(z) = \psi_2(\psi_1(z))$, where $\psi_1(z) = \sqrt{(z-c)/(z-d)}$ with some fixed branch of the radical and $\psi_2(z) = \varepsilon/(z + \psi_1(z_0))$ with some $\varepsilon >$. If ε is sufficiently small, then $g(z)$ maps D_k onto a Jordan domain $\widetilde{D}_k$ in the unit disk. It follows from the assumptions of Corollary 1 that the sequence of domains $\widetilde{D}_k$ and points $A = g(a)$

and $B = g(b)$ satisfy the assumptions of Proposition 1. Therefore, the sequence of functions $g^{-1}(f_k(z))$ converges to $g^{-1}(f_\infty(z))$ uniformly on compact subsets of $\widetilde{D}_\infty$. The latter implies that the functions $f_k(z)$ satisfy relations (2.2), as required. □

Remark 1 Proposition 1 and its Corollary 1 can be extended further to include sequences of domains D_k such that each complement $\overline{\mathbb{C}} \setminus D_k$ consists of the same number of closed Jordan arcs such that the end points of corresponding arcs are the same for all k. This form could be useful when studying harmonic measures on some trees or fractals.

The points a and b divide the boundary of D_k into two parts. By γ_{ab}^k we denote the part which corresponds (in the sense of boundary correspondence) to the line $\{w : \Im w = 1\}$ under the mapping f_k. The harmonic measure of γ_{ab}^k admits a very simple expression in terms of the function $f_k(z)$. Namely, $\omega(z, \gamma_{ab}^k, D_k) = \Im f_k(z)$; and therefore $\frac{\partial}{\partial z}\omega(z, \gamma_{ab}^k, D_k) = -\frac{i}{2}f_k'(z)$, where $\frac{\partial}{\partial z} = \frac{1}{2}\left(\frac{\partial}{\partial x} - i\frac{\partial}{\partial y}\right)$ is the formal complex partial derivative with respect to z. Now, Proposition 1 and Corollary 1 imply the following corollary on the convergence of harmonic measures.

Corollary 2 *Under the assumptions of Proposition 1 or Corollary 1,*

$$\omega(z, \gamma_{ab}^k, D_k) = \Im f_k(z) \to \Im f_\infty(z) = \omega(z, \gamma_{ab}^\infty, D_\infty) \quad as\ k \to \infty \tag{2.4}$$

and

$$\frac{\partial}{\partial z}\omega(z, \gamma_{ab}^k, D_k) = -\frac{i}{2}f_k'(z) \to -\frac{i}{2}f_\infty'(z) = \frac{\partial}{\partial z}\omega(z, \gamma_{ab}^\infty, D_\infty) \quad as\ k \to \infty. \tag{2.5}$$

Convergence in (2.4) and (2.5) is uniform on compact subsets of D_∞.

3 Reflection Principle for Harmonic Measures

Let Ω_+ be a finitely connected domain in the upper half-plane $\mathbb{H} := \{z : \Im z > 0\}$ with non-degenerate boundary components. Suppose that the boundary $\partial\Omega_+$ contains the interval (a, b), $a < b$, of the real axis and that there are no other boundary points of Ω_+ in some vicinity of this interval. Let E_+ be a closed Borel set on $\partial\Omega_+ \setminus (a, b)$ such that $0 < \omega(z, E_+, \Omega_+) < 1$ for $z \in \Omega_+$.

Let Ω_- and E_- denote, respectively, the reflections of Ω_+ and E_+ with respect to the real axis and let $\Omega = \Omega_+ \cup \Omega_- \cup (a, b)$. Figure 2 illustrates the notation used in this section. The following formulas are immediate consequences of the Schwarz reflection principle for harmonic functions:

$$\omega(z, E_+, \Omega_+) = \omega(z, E_+, \Omega) - \omega(z, E_-, \Omega) \quad for\ z \in \Omega_+, \tag{3.1}$$

$$\omega(z, E_+, \Omega_+) = \omega(z, E_+, \Omega) - \omega(\bar{z}, E_+, \Omega) \quad for\ z \in \Omega_+, \tag{3.2}$$

$$\frac{\partial}{\partial y}\omega(x, E_+, \Omega_+) = 2\frac{\partial}{\partial y}\omega(x, E_+, \Omega) > 0 \quad for\ x \in (a, b). \tag{3.3}$$

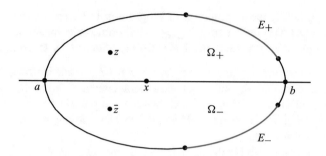

Fig. 2 Reflection principle for harmonic measures

The strict inequality in (3.3) follows from Hopf's lemma, also known as *Hopf's boundary maximum principle*. A form of Hopf's lemma, which is convenient when working with reflections and symmetrization, was given in [7].

An important advantage provided by formulas (3.1)–(3.3) is that they allow us to reduce some *problems on the boundary values of harmonic measures to problems on the behavior of harmonic measures at interior points of a domain*. The interior problems are usually much easier. The idea of such reduction is not new and goes back to the classical period of Complex Analysis. This author [9] used a similar reduction to an interior problem to prove a conjecture on the compactness of the set of critical points of Green's functions of a multiply connected domain, which was suggested by Ahmed Sebbar and Thérèse Falliero [6].

4   Radial Slit

The simplest configuration consisting of a domain D and a slit l as in Theorem 1 occurs when D is the upper half-plane $\mathbb{H} := \{z : \Im z > 0\}$ and l is the radial segment $I_\alpha := \{te^{i\alpha\pi} : 0 \le t \le 1\}$ with some α, $0 < \alpha < 1$. Let $H_\alpha = \mathbb{H} \setminus I_\alpha$. By I_α^- and I_α^+ we denote the boundary arcs of H_α, which correspond to the two shores of the slit along I_α such that $I_\alpha^- = \{te^{i(\alpha\pi-0)} : 0 \le t \le 1\}, I_\alpha^+ = \{te^{i(\alpha\pi+0)} : 0 \le t \le 1\}$.

For $\varepsilon > 0$ and $0 < \alpha < 1$, let

$$u^-(\varepsilon, \alpha) = \omega(i\varepsilon^{-1}, I_\alpha^-, H_\alpha), \quad u^+(\varepsilon, \alpha) = \omega(i\varepsilon^{-1}, I_\alpha^+, H_\alpha).$$

Finally, let

$$\Omega(\varepsilon, \alpha) = \frac{u^-(\varepsilon, \alpha)}{u^+(\varepsilon, \alpha)}. \tag{4.1}$$

The goal of this section is to find an explicit expression for $\Omega(\varepsilon, \alpha)$ for small $\varepsilon > 0$ and especially to find its limit as $\varepsilon \to 0^+$. Our derivation involves one particular conformal mapping, expressed in terms of elementary functions, the properties of which were discussed in several textbooks on a complex variable. To have a self-contained exposition we present all necessary details. It is more convenient to work with unbounded intervals $I^\alpha = \{te^{i\alpha\pi} : 1 \le t \le \infty\}$, with corresponding domains $H^\alpha = \mathbb{H} \setminus I^\alpha$, and with corresponding boundary segments $I_-^\alpha = \{te^{i(\alpha\pi-0)} : 1 \le t \le \infty\}$ and $I_+^\alpha = \{te^{i(\alpha\pi+0)} : 1 \le t \le \infty\}$. Since the harmonic measure is invariant under conformal mappings and reflections with respect to straight lines, $u^-(\varepsilon, \alpha) = \omega(i\varepsilon, I_-^\alpha, H^\alpha)$ and $u^+(\varepsilon, \alpha) = \omega(i\varepsilon, I_+^\alpha, H^\alpha)$. Thus, we may work with the domains H^α and segments I^α instead of H_α and I_α.

For negative α, $-1 < \alpha < 0$, we put $I^\alpha = \{z : \bar{z} \in I^{-\alpha}\}$, $I_-^\alpha = \{z : \bar{z} \in I_-^{-\alpha}\}$, $I_+^\alpha = \{z : \bar{z} \in I_+^{-\alpha}\}$, and $H^\alpha = \{z : \bar{z} \in D^{-\alpha}\}$.

For $0 < \alpha < 1$, we use the symmetric domains $G^\alpha = \mathbb{C} \setminus (I^\alpha \cup I^{-\alpha})$ and the following related notation:

$$v_1^- = v_1^-(\varepsilon, \alpha) = \omega(i\varepsilon, I_-^\alpha, G^\alpha), \quad v_2^- = v_2^-(\varepsilon, \alpha) = \omega(i\varepsilon, I_-^{-\alpha}, G^\alpha),$$

$$v_1^+ = v_1^+(\varepsilon, \alpha) = \omega(i\varepsilon, I_+^\alpha, G^\alpha), \quad v_2^+ = v_2^+(\varepsilon, \alpha) = \omega(i\varepsilon, I_+^{-\alpha}, G^\alpha).$$

Using the reflection principle for harmonic measure expressed by formula (3.1), we obtain

$$u^-(\varepsilon, \alpha) = v_1^-(\varepsilon, \alpha) - v_2^-(\varepsilon, \alpha) \quad \text{and} \quad u^+(\varepsilon, \alpha) = v_1^+(\varepsilon, \alpha) - v_2^+(\varepsilon, \alpha). \tag{4.2}$$

Therefore,

$$\Omega(\varepsilon, \alpha) = \frac{v_1^-(\varepsilon, \alpha) - v_2^-(\varepsilon, \alpha)}{v_1^+(\varepsilon, \alpha) - v_2^+(\varepsilon, \alpha)}. \tag{4.3}$$

To find $v_k^+(\varepsilon, \alpha)$ and $v_k^-(\varepsilon, \alpha)$, $k = 1, 2$, we consider the function

$$z = f_\alpha(w) = C \frac{w}{(1-w)^{2\alpha}(1+w)^{2(1-\alpha)}} \quad \text{with some } C > 0, \tag{4.4}$$

which is the integrated form of the Schwarz-Christoffel integral. It is a standard exercise to verify that f_α maps the unit disk $\mathbb{D} = \{w : |w| < 1\}$ conformally onto the domain G^α.

The preimage $w_\alpha = e^{i\theta_\alpha}$ of the tip at $z = e^{i\alpha\pi}$ can be found from the equation $f_\alpha'(w) = 0$, which after calculation gives the quadratic equation

$$w^2 - 2(1 - 2\alpha)w + 1 = 0. \tag{4.5}$$

Solving the latter equation, we find that $w_\alpha = 1 - 2\alpha + 2i\sqrt{\alpha(1-\alpha)}$ and therefore for $0 < \alpha < 1/2$,

$$\theta_\alpha = \arctan \frac{2\sqrt{\alpha(1-\alpha)}}{1 - 2\alpha}. \tag{4.6}$$

The value of C in (4.4) can be found from the equation $|f_\alpha(w_\alpha)| = 1$, which yields

$$C = |1 - w_\alpha|^{2\alpha}|1 + w_\alpha|^{2(1-\alpha)} = 4\alpha^\alpha(1 - \alpha)^{1-\alpha}. \tag{4.7}$$

For $\varphi_1 < \varphi_2 < \varphi_1 + 2\pi$ and $w \in \mathbb{D}$ the harmonic measure $\omega(w, \gamma(\varphi_1, \varphi_1), \mathbb{D})$ of the arc $\gamma(\varphi_1, \varphi_2) := \{e^{i\theta} : \varphi_1 < \theta < \varphi_2\}$ is given by

$$\omega(w, \gamma(\varphi_1, \varphi_2), \mathbb{D}) = \frac{1}{2\pi} \arg\left(\frac{e^{i\varphi_2} - w}{1 - \bar{w}e^{i\varphi_2}} \frac{1 - \bar{w}e^{i\varphi_1}}{e^{i\varphi_1} - w}\right). \tag{4.8}$$

Let $w^\varepsilon = f_\alpha^{-1}(i\varepsilon)$. Then it follows from (4.4) that

$$w^\varepsilon = iC^{-1}\varepsilon + O(\varepsilon^2) \quad \text{as } \varepsilon \to 0^+. \tag{4.9}$$

Using (4.8), we find

$$v_1^- - v_2^- = \frac{1}{2\pi}\left(\arg\frac{e^{i\theta_\alpha} - w^\varepsilon}{1 - \bar{w}^\varepsilon e^{i\theta_\alpha}} - \arg\frac{e^{i\theta_\alpha} - \bar{w}^\varepsilon}{1 - w^\varepsilon e^{i\theta_\alpha}} - 2\arg\frac{1 - w^\varepsilon}{1 - \bar{w}^\varepsilon}\right), \tag{4.10}$$

$$v_1^+ - v_2^+ = \frac{1}{2\pi}\left(\arg\frac{e^{i(\pi-\theta_\alpha)} + \bar{w}^\varepsilon}{1 + w^\varepsilon e^{i(\pi-\theta_\alpha)}} - \arg\frac{e^{i(\pi-\theta_\alpha)} + w^\varepsilon}{1 + \bar{w}^\varepsilon e^{i(\pi-\theta_\alpha)}} - 2\arg\frac{1 + \bar{w}^\varepsilon}{1 + w^\varepsilon}\right). \tag{4.11}$$

Substituting (4.9) into (4.10) and (4.11), we obtain

$$v_1^- - v_2^- = \frac{2\varepsilon}{\pi C}(1 - \cos\theta_\alpha) + O(\varepsilon^2) \quad \text{as } \varepsilon \to 0^+, \tag{4.12}$$

$$v_1^+ - v_2^+ = \frac{2\varepsilon}{\pi C}(1 + \cos\theta_\alpha) + O(\varepsilon^2) \quad \text{as } \varepsilon \to 0^+. \tag{4.13}$$

Using formulas (3.1) and (3.2) and applying the intermediate value theorem, we can represent the differences $v_1^- - v_2^-$ and $v_1^+ - v_2^+$ in the form

$$v_1^- - v_2^- = \omega(i\varepsilon, I_-^\alpha, G^\alpha) - \omega(-i\varepsilon, I_-^\alpha, G^\alpha) = 2\varepsilon\frac{\partial}{\partial y}\omega(iv_-(\varepsilon), I_-^\alpha, G^\alpha), \tag{4.14}$$

$$v_1^+ - v_2^+ = \omega(i\varepsilon, I_+^\alpha, G^\alpha) - \omega(-i\varepsilon, I_+^\alpha, G^\alpha) = 2\varepsilon\frac{\partial}{\partial y}\omega(iv_+(\varepsilon), I_+^\alpha, G^\alpha) \tag{4.15}$$

with some real-valued functions $v_-(\varepsilon)$ and $v_+(\varepsilon)$ such that $|v_-(\varepsilon)| < \varepsilon, |v_+(\varepsilon)| < \varepsilon$.

Now, Eq. (4.3) combined with relations (4.12)–(4.15), yields

$$\Omega(\varepsilon,\alpha) = \frac{\frac{\partial}{\partial y}\omega(iv_-(\varepsilon), I_-^\alpha, G^\alpha)}{\frac{\partial}{\partial y}\omega(iv_+(\varepsilon), I_+^\alpha, G^\alpha)} = \frac{1-\cos\theta_\alpha}{1+\cos\theta_\alpha} + O(\varepsilon) = \tan^2(\theta_\alpha/2) + O(\varepsilon) \quad \text{as } \varepsilon \to 0^+.$$

(4.16)

Substituting the expression for θ_α given by (4.6) in Eq. (4.16) and taking the limit, we obtain the required limit relation:

$$\lim_{\varepsilon\to0^+} \Omega(\varepsilon,\alpha) = \frac{\frac{\partial}{\partial y}\omega(0, I_-^\alpha, G^\alpha)}{\frac{\partial}{\partial y}\omega(0, I_+^\alpha, G^\alpha)} = \frac{\alpha}{1-\alpha}.$$

(4.17)

We emphasize here that the first of these relations follows from the fact that the partial derivatives in Eq. (4.16) are continuous functions and that the denominator $\frac{\partial}{\partial y}\omega(0, I_+^\alpha, G^\alpha)$ is not zero by formula (3.3).

To study the harmonic measures of oscillating slits in Sect. 7, we need a generalization of the limit relation (4.17). Below we describe the necessary changes.

Let $\lambda > 1$. Then the point $z = \lambda e^{i\alpha}$ represents two boundary points, $z^+ = z^+(\alpha, \lambda) \in I_+^\alpha$ and $z^- = z^-(\alpha, \lambda) \in I_-^\alpha$, of the domain H^α. Each of these points divides I^α into two boundary arcs of H^α. Accordingly, we introduce boundary sets

$$I_{\alpha,\lambda}^{+-} = \{z \in I_+^\alpha : |z| > \lambda\}, \quad I_{\alpha,\lambda}^{++} = I_+^\alpha \cup \{z \in I_-^\alpha : |z| < \lambda\}, \tag{4.18}$$

$$I_{\alpha,\lambda}^{--} = \{z \in I_-^\alpha : |z| > \lambda\}, \quad I_{\alpha,\lambda}^{-+} = I_-^\alpha \cup \{z \in I_+^\alpha : |z| < \lambda\}. \tag{4.19}$$

By $u^{+-}(\varepsilon,\alpha,\lambda)$, $u^{++}(\varepsilon,\alpha,\lambda)$, $u^{--}(\varepsilon,\alpha,\lambda)$, and $u^{-+}(\varepsilon,\alpha,\lambda)$, we denote the harmonic measures of the boundary sets $I_{\alpha,\lambda}^{+-}, I_{\alpha,\lambda}^{++}, I_{\alpha,\lambda}^{--}$, and $I_{\alpha,\lambda}^{-+}$, respectively, with respect to the domain H^α evaluated at $z = i\varepsilon$. With this notation we introduce the functions

$$\Omega^+(\varepsilon,\alpha,\lambda) = \frac{u^{-+}(\varepsilon,\alpha,\lambda)}{u^{+-}(\varepsilon,\alpha,\lambda)}, \quad \Omega^-(\varepsilon,\alpha,\lambda) = \frac{u^{--}(\varepsilon,\alpha,\lambda)}{u^{++}(\varepsilon,\alpha,\lambda)}.$$

(4.20)

Let $w_{\alpha,\lambda}^+ = e^{i\theta_{\alpha,\lambda}^+}$ and $w_{\alpha,\lambda}^- = e^{i\theta_{\alpha,\lambda}^-}$ denote the preimages of the boundary points $z^+(\alpha,\lambda)$ and $z^-(\alpha,\lambda)$ under the mapping (4.4). Then the angles $\theta = \theta_{\alpha,\lambda}^+$ and $\theta = \theta_{\alpha,\lambda}^-$ can be found from the equation $|f_\alpha(e^{i\theta})| = \lambda$, which is equivalent to the equation

$$\frac{\tan^{2\alpha}\frac{\theta}{2}}{1+\tan^2\frac{\theta}{2}} = \frac{C}{4\lambda},$$

(4.21)

where C is defined by (4.7). It is an elementary calculus exercise to show that (4.21) has two solutions: $0 < \theta_{\alpha,\lambda}^- < \theta_\alpha$ and $\theta_\alpha < \theta_{\alpha,\lambda}^+ < \pi$.

Now, the same argument used to derive (4.17) leads us to equations

$$\lim_{\varepsilon \to 0^+} \Omega^+(\varepsilon, \alpha, \lambda) = \tan^2 \frac{\theta^+_{\alpha,\lambda}}{2}, \quad \lim_{\varepsilon \to 0^+} \Omega^-(\varepsilon, \alpha, \lambda) = \tan^2 \frac{\theta^-_{\alpha,\lambda}}{2}. \quad (4.22)$$

5 Proof of Theorem 1

Although the notation needed for this proof is rather lengthy, the idea of the proof is simple: for each ζ-tail l_ζ, we construct two geometrically simple *comparison slits*, shown in Fig. 3, which provide us with lower and upper bounds for the quotient of harmonic measures in Eq. (1.1). Then we apply the Carathéodory convergence theorem in the form discussed in Sect. 2 to find the limit values of these bounds. We note here that other appropriate *comparison slits* (or *comparison arcs*) are also used in our proofs of Theorems 3 and 4 in Sect. 7.

Since the slit l is monotonic and forms an angle $\alpha\pi$ with the positive real axis, it follows that for every $\zeta = \varepsilon e^{i(\alpha\pi + \nu)} \in l$, there is $\delta = \delta(\varepsilon) > 0$ such that $l_\zeta \subset S_\alpha(\varepsilon, \delta(\varepsilon))$. Here and elsewhere below by $S_\alpha(\varepsilon, \delta)$ we denote the circular sector $\{z : 0 \le |z| \le \varepsilon, |\arg z - \alpha\pi| \le \delta\}$; see Fig. 3.

Furthermore, since l is tangent to the segment I_α at $\zeta_0 = 0$, the following limit relation holds:

$$\delta(\varepsilon)/\varepsilon \to 0 \qquad \text{as } \varepsilon \to 0. \quad (5.1)$$

For $\zeta = \varepsilon e^{i(\alpha\pi + \nu)}$ and $\delta = \delta(\varepsilon)$, consider the following boundary arcs of the sector $S_\alpha(\varepsilon, \delta)$:

$$l_1(\alpha, \varepsilon, \delta) = \{te^{i(\alpha\pi - \delta)} : 0 \le t \le \varepsilon\} \cup \{\varepsilon e^{i\theta} : \alpha\pi - \delta \le \theta \le \arg \zeta\}, \quad (5.2)$$

$$l_2(\alpha, \varepsilon, \delta) = \{te^{i(\alpha\pi + \delta)} : 0 \le t \le \varepsilon\} \cup \{\varepsilon e^{i\theta} : \arg \zeta \le \theta \le \alpha\pi + \delta\}. \quad (5.3)$$

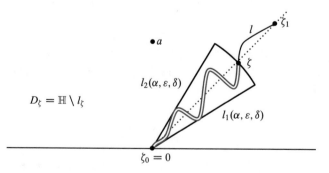

Fig. 3 Domain D_ζ and comparison arcs $l_1(\alpha, \varepsilon, \delta)$ and $l_2(\alpha, \varepsilon, \delta)$

For $a \in \mathbb{H} \setminus S_\alpha(\varepsilon, \delta)$, we have the following inequalities, which easily follow from Carleman's extension principle for harmonic measure [1, Chap. 8, §4]:

$$\omega(a, l_2^-(\alpha, \varepsilon, \delta), \mathbb{H} \setminus l_2(\alpha, \varepsilon, \delta)) \leq \omega(a, l_\zeta^-, \mathbb{H}_\zeta) \leq \omega(a, l_1^-(\alpha, \varepsilon, \delta), \mathbb{H} \setminus l_1(\alpha, \varepsilon, \delta)),$$
(5.4)

$$\omega(a, l_2^+(\alpha, \varepsilon, \delta), \mathbb{H} \setminus l_2(\alpha, \varepsilon, \delta)) \geq \omega(a, l_\zeta^+, \mathbb{H}_\zeta) \geq \omega(a, l_1^+(\alpha, \varepsilon, \delta), \mathbb{H} \setminus l_1(\alpha, \varepsilon, \delta)),$$
(5.5)

Inequalities (5.4) and (5.5) allow us to *compare* harmonic measures of the arcs l_ζ^+ and l_ζ^- with appropriate harmonic measures of the slits $l_1(\alpha, \varepsilon, \delta)$ and $l_2(\alpha, \varepsilon, \delta)$. The latter justifies our choice of the term "*comparison slit*".

Inequalities (5.4) and (5.5) lead to the following bounds for the quotient of harmonic measures in Eq. (1.1):

$$\frac{\omega(a, l_2^-(\alpha, \varepsilon, \delta), \mathbb{H} \setminus l_2(\alpha, \varepsilon, \delta))}{\omega(a, l_2^+(\alpha, \varepsilon, \delta), \mathbb{H} \setminus l_2(\alpha, \varepsilon, \delta))} \leq \frac{\omega^-(a, \zeta)}{\omega^+(a, \zeta)} \leq \frac{\omega(a, l_1^-(\alpha, \varepsilon, \delta), \mathbb{H} \setminus l_1(\alpha, \varepsilon, \delta))}{\omega(a, l_1^+(\alpha, \varepsilon, \delta), \mathbb{H} \setminus l_1(\alpha, \varepsilon, \delta))}.$$
(5.6)

To find the limit values of the bounds in (5.6) as $\varepsilon \to 0$, we change variables via $z \mapsto \varphi(z)$, where

$$\varphi(z) = A \frac{\varepsilon}{z} + B$$
(5.7)

with

$$A = \frac{\sin \alpha \pi}{\sin(\alpha \pi + \nu)} = 1 - \nu \cot \alpha \pi + O(\delta^2),$$
(5.8)

$$B = \cos \alpha \pi - \sin \alpha \pi \, \cot(\alpha \pi + \nu) = \nu \csc \alpha \pi + O(\delta^2).$$
(5.9)

Recall here that $|\nu| \leq \delta = \delta(\varepsilon)$ and that $\delta(\varepsilon)$ satisfies (5.1).

First we work with the upper bound in (5.6). Let $\Omega_+(\varepsilon) = \varphi(\mathbb{H} \setminus l_1(\alpha, \varepsilon, \delta))$, $L_+(\varepsilon) = \varphi(l_1(\alpha, \varepsilon, \delta))$, $L_+^-(\varepsilon) = \varphi(l_1^-(\alpha, \varepsilon, \delta))$ and $L_+^+(\varepsilon) = \varphi(l_1^+(\alpha, \varepsilon, \delta))$. It follows from the definition of $l_1(\alpha, \varepsilon, \delta)$ and formula (5.7) that for every $\zeta = \varepsilon e^{i(\alpha \pi + \nu)} \in l$, $L_+(\varepsilon)$ is a slit in $\mathbb{H}$ from the point $a = e^{i\alpha}$ to the point $b = \infty$ and that

$$\varphi(a) = A \varepsilon \frac{a}{|a|^2} + B = \varepsilon \frac{a}{|a|^2}(1 - \nu \cot \alpha \pi) + \nu \csc \alpha \pi + O(\delta^2),$$
(5.10)

where $O(\delta^2)/\delta^2$, considered as a function of a, is uniformly bounded on compact subsets of $\mathbb{H} \setminus l$.

The mapping φ defined by (5.7) preserves harmonic measures. Thus, the limit of the quotient in the right hand-side of (5.6) exists if and only if the following limit exists:

$$L = \lim_{\varepsilon \to 0+} \frac{\omega(\varphi(a), L_+^-(\varepsilon), \Omega_+(\varepsilon))}{\omega(\varphi(a), L_+^+(\varepsilon), \Omega_+(\varepsilon))}. \tag{5.11}$$

Let $\Omega_-(\varepsilon)$ be the reflection of $\Omega_+(\varepsilon)$ with respect to the real axis and let $\Omega(\varepsilon) = \Omega_+(\varepsilon) \cup \Omega_-(\varepsilon) \cup \mathbb{R}$. Using Eq. (3.2) and applying the mean value theorem for corresponding differences, we can rewrite Eq. (5.11) as

$$L = \lim_{\varepsilon \to =0+} \frac{\omega(\varphi(a), L_+^-(\varepsilon), \Omega(\varepsilon)) - \omega(\overline{\varphi(a)}, L_+^-(\varepsilon), \Omega(\varepsilon))}{\omega(\varphi(a), L_+^+(\varepsilon), \Omega(\varepsilon)) - \omega(\overline{\varphi(a)}, L_+^+(\varepsilon), \Omega(\varepsilon))} \tag{5.12}$$

$$= \lim_{\varepsilon \to =0+} \frac{\frac{\partial}{\partial y}\omega(w_1(\varepsilon), L_+^-(\varepsilon), \Omega(\varepsilon))}{\frac{\partial}{\partial y}\omega(w_2(\varepsilon), L_+^+(\varepsilon), \Omega(\varepsilon))},$$

where $w_1(\varepsilon)$ and $w_2(\varepsilon)$ are points on the vertical interval with end points at $\varphi(a)$ and $\overline{\varphi(a)}$.

Since $\delta(\varepsilon)/\varepsilon \to 0$ as $\varepsilon \to 0$, it follows from the definitions of the sets $l_1(\alpha, \varepsilon, \delta)$ and $L_+(\varepsilon)$ that for every sequence of positive numbers $\varepsilon_n \to 0$, the corresponding sequence of simply connected domains $\Omega(\varepsilon_n)$ converges to the kernel $G^\alpha = \mathbb{C} \setminus (I^\alpha \cup I^{-\alpha})$ considered in Sect. 4. Also, the corresponding sequences of boundary arcs $L_+^-(\varepsilon_n)$ and $L_+^+(\varepsilon_n)$ converge in the sense of Corollary 1 to the sets I_-^α and I_+^α, which are boundary arcs of the domain G^α. Furthermore, the set $L_+(\varepsilon)$ consists of a ray and a short circular arc. This fact and an elementary geometric argument show that there is a one-to-one continuous mapping $h(z)$ from $L_+(\varepsilon)$ onto I_+^α such that the spherical distance between points $z \in L_+(\varepsilon)$ and their images $h(z) \in I_+^\alpha$ is uniformly small if ε is sufficiently small. From these observations we conclude that all assumptions of Corollary 2 are satisfied for the sequence of domains $\Omega(\varepsilon_n)$ and corresponding sequences of boundary arcs. Therefore, we can apply Eq. (2.5) to conclude that

$$\lim_{\varepsilon \to 0+} \frac{\partial}{\partial y}\omega(w_1(\varepsilon), L_+^-(\varepsilon), \Omega(\varepsilon)) = \frac{\partial}{\partial y}\omega(0, I_-^\alpha, G^\alpha) \tag{5.13}$$

and

$$\lim_{\varepsilon \to 0+} \frac{\partial}{\partial y}\omega(w_2(\varepsilon), L_+^+(\varepsilon), \Omega(\varepsilon)) = \frac{\partial}{\partial y}\omega(0, I_+^\alpha, G^\alpha). \tag{5.14}$$

Equations (5.6), (5.12)–(5.14), combined with Eq. (4.17), imply that

$$\lim_{\varepsilon \to 0+} \frac{\omega^-(a, \zeta)}{\omega^+(a, \zeta)} \leq \frac{\lim_{\varepsilon \to 0+} \frac{\partial}{\partial y} \omega(w_1(\varepsilon), L_+^-(\varepsilon), \Omega(\varepsilon))}{\lim_{\varepsilon \to 0+} \frac{\partial}{\partial y} \omega(w_2(\varepsilon), L_+^+(\varepsilon), \Omega(\varepsilon))} \tag{5.15}$$

$$= \frac{\frac{\partial}{\partial y} \omega(0, I_-^\alpha, G^\alpha)}{\frac{\partial}{\partial y} \omega(0, I_+^\alpha, G^\alpha)} = \frac{\alpha}{1 - \alpha}.$$

The same argument as above can be used to find the limit value as $\varepsilon \to 0^+$ of the lower bound in (5.6), which gives

$$\lim_{\varepsilon \to 0+} \frac{\omega^-(a, \zeta)}{\omega^+(a, \zeta)} \geq \lim_{\varepsilon \to 0+} \frac{\omega(a, l_2^-(\alpha, \varepsilon, \delta), \mathbb{H} \setminus l_2(\alpha, \varepsilon, \delta))}{\omega(a, l_2^+(\alpha, \varepsilon, \delta), \mathbb{H} \setminus l_2(\alpha, \varepsilon, \delta))} = \frac{\alpha}{1 - \alpha}. \tag{5.16}$$

Furthermore, convergence in our formulas (5.13) and (5.14) and in similar formulas needed for the proof of formula (5.16) is uniform on compact subsets of $\mathbb{H} \setminus l$. Combining inequalities (5.15) and (5.16) and taking into account the latter remark on the uniform convergence, we complete our proof of Theorem 1.   □

6   Proof of Theorem 2

Let $\tau = g(z, \zeta)$ be a mapping defined by (1.4). Consider the function $h(z, \zeta) = \varphi(g(z, \zeta))$, where

$$\varphi(\tau) = \sqrt{\frac{1 + \tau}{1 - \tau}} \quad \text{with } \varphi(0) = 1. \tag{6.1}$$

Then $h(z, \zeta)$ maps $\mathbb{H}_\zeta$ conformally and one-to-one onto the first quadrant $Q = \{w : \Re w > 0, \Im w > 0\}$ in such a way that the slit l_ζ is mapped onto the positive real axis. Let $p_\zeta = h(\zeta, \zeta) = \varphi(\Lambda(\zeta))$ denote the image of the tip $z = \zeta$ of the slit l_ζ and let $w_\zeta = h(i, \zeta)$. It follows from (1.4) and (6.1) that $w_\zeta \to i$ as $\zeta \to 0$ along l.

Since harmonic measure is conformally invariant, it follows from Theorem 1 that

$$\frac{\omega(w_\zeta, [p_\zeta, \infty), Q)}{\omega(w_\zeta, [0, p_\zeta], Q)} \to \frac{\alpha}{1 - \alpha} \quad \text{as } \zeta \to 0. \tag{6.2}$$

Applying the reflection formula (3.2) (this time we reflect with respect to the imaginary axis), we can rewrite (6.2) in the following equivalent form:

$$\frac{\omega(w_\zeta, [p_\zeta, \infty), \mathbb{H}) - \omega(-\overline{w_\zeta}, [p_\zeta, \infty), \mathbb{H})}{\omega(w_\zeta, [0, p_\zeta], \mathbb{H}) - \omega(-\overline{w_\zeta}, [0, p_\zeta], \mathbb{H})} \to \frac{\alpha}{1 - \alpha} \quad \text{as } \zeta \to 0. \tag{6.3}$$

Using the mean-value theorem as in the proof of Theorem 1, we conclude that (6.3) is equivalent to the equation

$$\frac{\frac{\partial}{\partial x}\omega(w_1(\zeta), [p_\zeta, \infty), \mathbb{H})}{\frac{\partial}{\partial x}\omega(w_2(\zeta), [0, p_\zeta], \mathbb{H})} \to \frac{\alpha}{1-\alpha} \qquad \text{as } \zeta \to 0. \qquad (6.4)$$

Here, $w_1(\zeta)$ and $w_2(\zeta)$ are points on the horizontal interval joining the points w_ζ and $-\overline{w_\zeta}$.

Using the well-known formula

$$\omega(z, [a, b], \mathbb{H}) = \frac{1}{\pi} \arg \frac{z-b}{z-a} = \frac{1}{\pi} \Im \log \frac{z-b}{z-a},$$

which evaluates the harmonic measure of the interval $[a, b] \subset \mathbb{R}$ with respect to the upper half-plane $\mathbb{H}$, we find that (6.4) is equivalent to the equation

$$\frac{-\Im \frac{1}{w_1(\zeta)-p_\zeta}}{\Im\left(\frac{1}{w_2(\zeta)-p_\zeta} - \frac{1}{w_2(\zeta)}\right)} \to \frac{\alpha}{1-\alpha} \qquad \text{as } \zeta \to 0. \qquad (6.5)$$

Since $w_1(\zeta) \to i$ and $w_2(\zeta) \to i$ as $\zeta \to 0$, we conclude from (6.5) that

$$p_\zeta^2 \to \frac{1-\alpha}{\alpha} \qquad \text{as } \zeta \to 0 \text{ along } l.$$

Since $p_\zeta^2 = \varphi^2(\Lambda(\zeta)) = \frac{1+\Lambda(\zeta)}{1-\Lambda(\zeta)}$, the latter equation implies (1.6). $\square$

7   Oscillating Slits

Our proof of Theorem 1 is essentially geometrical and therefore can be easily modified to find bounds for the quotient of harmonic measures $\omega^-(a, \zeta)/\omega^+(a, \zeta)$ and for the trace $T(l)$ under less restrictive assumptions on the geometry of the slit. In this section, we discuss possible extensions of our results for the case of oscillating slits.

(a) **Radial oscillations.**   Suppose that l is a simple slit in $\mathbb{H} \cup \{0\}$ tangent to I_α at its initial point at $\zeta_0 = 0$. We say that l has radial oscillation of magnitude λ, $1 \le \lambda \le \infty$, at $\zeta_0 = 0$ if

$$1 \le \limsup_{l \ni \zeta \to 0} \max_{z \in l_\zeta} \left|\frac{z}{\zeta}\right| = \lambda. \qquad (7.1)$$

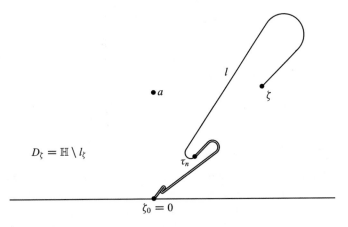

$D_\zeta = \mathbb{H} \setminus l_\zeta$

$\zeta_0 = 0$

Fig. 4 Slit with radial oscillation of magnitude $\lambda = 4/3$

Example 1 As an example, we construct a slit $l(\lambda)$ with radial oscillation of magnitude $\lambda > 1$, which will be used to demonstrate the sharpness of the upper and lower bounds in the inequalities (7.2) of Theorem 3; see Fig. 4.
For real α and λ such that $0 < \alpha \leq \frac{1}{2}, \lambda > 1$, let $I_0'' = \{z = te^{i\frac{3}{2}\alpha} : \lambda^{-1} \leq t \leq 1\}$ and for any integer $n \geq 1$, let I_n' and I_n'' be the line segments defined by

$$I_n' = \left\{z = te^{i\alpha} : 2^{-\frac{n(n-1)}{2}}\lambda^{-n} \leq t \leq 2^{-\frac{n(n-1)}{2}}\lambda^{-(n-1)}\right\},$$

$$I_n'' = \left\{z = te^{i\alpha\left(1+\frac{(-1)^n}{n+1}\right)} : 2^{-\frac{(n+1)n}{2}}\lambda^{-(n+1)} \leq t \leq 2^{-\frac{n(n-1)}{2}}\lambda^{-(n-1)}\right\}.$$

Furthermore, let C_n', $n \geq 1$, be a circular arc, which is tangent to I_n' and I_n'' at its endpoints $z = 2^{-\frac{n(n-1)}{2}}\lambda^{-(n-1)}e^{i\alpha}$ and $z = 2^{-\frac{n(n-1)}{2}}\lambda^{-(n-1)}e^{i\alpha\left(1+\frac{(-1)^n}{n+1}\right)}$. Similarly, let C_n'', $n \geq 0$, be a circular arc, which is tangent to I_n'' and I_{n+1}' at its endpoints $z = 2^{-\frac{(n+1)n}{2}}\lambda^{-(n+1)}e^{i\alpha\left(1+\frac{(-1)^n}{n+1}\right)}$ and $z = 2^{-\frac{(n+1)n}{2}}\lambda^{-(n+1)}e^{i\alpha}$. We assume here that the arcs C_n' and C_n'' form angles of opening π with corresponding segments at their endpoints.
Finally, we define $l(\lambda)$ as

$$l(\lambda) = \left(\cup_{n=1}^{\infty}\left(I_n' \cup C_n' \cup I''n \cup C_n''\right)\right) \cup \{0\}.$$

Then $l(\lambda)$ is a Jordan arc in $\mathbb{H} \cup \{0\}$, which has a tangent vector $\overrightarrow{v}(\zeta)$ at every point $\zeta \in l(\lambda)$, including the point at $\zeta = 0$. Moreover, this tangent vector is continuous at every point $\zeta \in l(\lambda)$ except at the point $\zeta = 0$. An elementary calculation shows that $l(\lambda)$ has radial oscillation of magnitude λ.
When a slit l tangent to the segment I_α at $\zeta_0 = 0$ has radial oscillation of magnitude $\lambda < \infty$, the comparison sets for the problem on the quotient of the harmonic measures are boundary arcs defined by formulas (4.18) and (4.19).

In this case, we have the following result.

Theorem 3 *Let l be a simple slit in $\mathbb{H} \cup \{0\}$, which is tangent to the segment I_α at $\zeta_0 = 0$, and let this slit have a radial oscillation of magnitude $\lambda < \infty$. Then for every $a \in \mathbb{H}$, the following sharp inequalities hold true:*

$$\tan^2 \theta_{\alpha,\lambda}^- \le \liminf_{\zeta \to 0} \frac{\omega^-(a,\zeta)}{\omega^+(a,\zeta)} \le \limsup_{\zeta \to 0} \frac{\omega^-(a,\zeta)}{\omega^+(a,\zeta)} \le \tan^2 \theta_{\alpha,\lambda}^+, \tag{7.2}$$

where $\theta_{\alpha,\lambda}^-$ and $\theta_{\alpha,\lambda}^+$ are solutions of Eq. (4.21).

Proof We present all the main steps of this proof, which is similar to the proof of Theorem 1, while some technical details are left to the interested reader.

For $\zeta \in l$ close enough to ζ_0, let $\varepsilon = \varepsilon(\zeta) = \max_{z \in l_\zeta} |z|$. Then the ζ-tail l_ζ lies in the sector $S_\alpha(\varepsilon, \delta) = \{z : |z| \le \varepsilon, |\arg z - \alpha\pi| < \delta\}$, where $\delta = \delta(\zeta) > 0$ is such that $\delta(\zeta)/\varepsilon(\zeta) \to 0$ as $\zeta \to \zeta_0$ along l. Let $\zeta^* \in l_\zeta$ be a point on l_ζ such that $|\zeta^*| = \varepsilon$ and $|z| < \varepsilon$ for all $z \in l_{\zeta^*}$. By ζ_+^* we denote the boundary point of l_ζ defined by the limit $\zeta_+^* = \lim_{r \to 1+}(r\zeta^*)$.

(a) First, we consider the case when $\zeta_+^* \in l_\zeta^+$. Then we may use the arc $l_1(\alpha, \varepsilon, \delta)$ defined by (5.2) (with ζ replaced by ζ^*) as a comparison slit for the corresponding quotient of the harmonic measures; see Fig. 5. In this case, we obtain the same upper bound as in (5.6):

$$\frac{\omega^-(a,\zeta)}{\omega^+(a,\zeta)} \le \frac{\omega(a, l_1^-(\alpha, \varepsilon, \delta), \mathbb{H} \setminus l_1(\alpha, \varepsilon, \delta))}{\omega(a, l_1^+(\alpha, \varepsilon, \delta), \mathbb{H} \setminus l_1(\alpha, \varepsilon, \delta))}. \tag{7.3}$$

Now, we can use the same argument as in the proof of Theorem 1 to conclude that

$$\limsup_{\zeta \to 0} \frac{\omega^-(a,\zeta)}{\omega^+(a,\zeta)} \le \tan^2 \frac{\theta_\alpha}{2} \le \tan^2 \theta_{\alpha,\lambda}^+,$$

where limsup is taken over all points $\zeta \in l$ such that the condition $\zeta_+^* \in l_\zeta^+$ is satisfied.

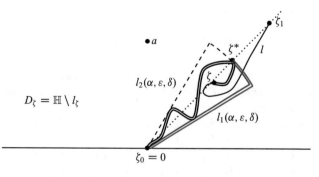

Fig. 5 Comparison arcs for radially oscillating slit. Case **(a)**

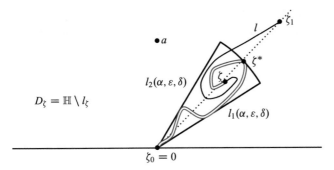

Fig. 6 Radially oscillating slit. Case **(b)**

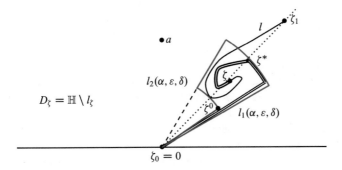

Fig. 7 Comparison arcs for radially oscillating slit. Case **(b)**

(b) Next, we consider the case when the boundary point ζ_+^* belongs to the set l_ζ^-; see Fig. 6. In this case, an appropriate comparison curve can be constructed as follows. Let $l_{\zeta,1}$ denote the arc consisting of the following three pieces: the radial segment $\{z = te^{i(\alpha-\delta)} : 0 \le t \le \varepsilon\}$, the circular arc $\{z = \varepsilon e^{i\theta} : \alpha-\delta \le \theta \le \arg(\zeta^*)\}$, and the portion of the ζ-tail l_ζ with endpoints ζ^* and ζ. The arc $l_{\zeta,1}$ will be our first comparison curve in the case under consideration; see Fig. 7. Applying Carleman's extension principle for the harmonic measures for the arcs l_ζ and $l_{\zeta,1}$ evaluated at $a \in \mathbb{H} \setminus S_\alpha(\varepsilon, \delta)$, we obtain the inequalities

$$\omega^-(a, l_\zeta) \le \omega^-(a, l_{\zeta,1}) \quad \text{and} \quad \omega^+(a, l_\zeta) \ge \omega^+(a, l_{\zeta,1}). \tag{7.4}$$

Now for α, ε, and δ as above and for λ_1 such that $\lambda_1 > \lambda$ and λ_1 is close enough to λ, we construct our second comparison curve $l_{\zeta,2}$ as a union of the arcs $\gamma_1 = \{z = te^{i(\alpha-2\delta)} : 0 \le t \le \varepsilon\}$, $\gamma_2 = \{z = \varepsilon e^{i\theta} : \alpha-2\delta \le \theta \le \alpha+\delta\}$, $\gamma_3 = \{z = te^{i(\alpha+\delta)} : (\varepsilon/\lambda_1) \le t \le \varepsilon\}$, $\gamma_4 = \{z = (\varepsilon/\varepsilon)e^{i\theta} : \alpha-\delta \le \theta \le \alpha+\delta\}$, and $l_5 = \{z = te^{i(\alpha-\delta)} : 0 \le t \le (\varepsilon/\lambda_1)\}$. Then $l_{\zeta,2}$ is a closed Jordan curve. Since l is a slit of bounded radial oscillation of magnitude λ and $\lambda_1 > \lambda$, it follows that $l_{\zeta,1}$ is a subset of the closure of the bounded component of $\mathbb{H} \setminus l_{\zeta,2}$ if $\zeta \in l$ is close enough to $\zeta_0 = 0$.

Let $\zeta^0 = (\varepsilon/\lambda_1)e^{i(\alpha-\delta)}$. Then the points $\zeta_0 = 0$ and ζ^0 divide $l_{\zeta,2}$ into two boundary arcs, which we denote by $l_{\zeta,2}^-$ and $l_{\zeta,2}^+$ assuming that the point $z = \varepsilon e^{i(\alpha-2\delta)}$ belongs to $l_{\zeta,2}^-$. Let $H_{\zeta,2}$ denote the unbounded component of $\mathbb{H} \setminus l_{\zeta,2}$. Applying Carleman's extension principle once more, we conclude that

$$\omega^-(a, l_{\zeta,1}) \leq \omega(a, l_{\zeta,2}^-, H_{\zeta,2}) \quad \text{and} \quad \omega^+(a, l_{\zeta,1}) \geq \omega(a, l_{\zeta,2}^+, H_{\zeta,2}) \tag{7.5}$$

for all $a \in H_{\zeta,2}$.

Combining inequalities (7.4) and (7.5), we obtain the upper bound

$$\frac{\omega^-(a, \zeta)}{\omega^+(a, \zeta)} \leq \frac{\omega(a, l_{\zeta,2}^-, H_{\zeta,2})}{\omega(a, l_{\zeta,2}^+, H_{\zeta,2})}. \tag{7.6}$$

Now we can complete the proof as in the proof of Theorem 1. Namely, first we change variables in the right-hand side of (7.5) via $z \to \varepsilon/\bar{z}$ to transform bounded slits $l_{\zeta,2}$ into unbounded slits starting at ∞. Then we take the limit as $\zeta \to 0$ and as $\lambda_1 \to \lambda$. Then the desired limit inequality in the right-hand side of (7.2) will follow from a version of the Carathéodory convergence theorem, which was discussed in Sect. 2; see Corollary 2.

The proof of the lower bound in (7.2) follows exactly the same lines and is therefore omitted.

To demonstrate the sharpness of the bounds in (7.2), we consider the slit $l = l(\lambda)$ defined in Example 1. Consider the sequence of points

$$\zeta_n = e^{i\alpha} 2^{-(2n-1)(n-1)} \lambda^{-(2n-1)}, \qquad n = 1, 2, \ldots$$

Then ζ_n is the lower endpoint of the segment l'_{2n-1}, which is a part of the slit $l = l(\lambda)$.

Consider the sequence of truncated slits l_{ζ_n}, $n \geq 1$, and the corresponding sequence of scaled slits $L_n = \{z : 2^{(2n-1)(n-1)}\lambda^{2n-1} \cdot z \in l_{\zeta_n}\}$. Let $H_n = \mathbb{H} \setminus L_n$. Then it is not difficult to see that the sequence of domains H_n converges to the domain $H_\alpha = H \setminus I_\alpha$ in the sense of the Carathéodory convergence theorem. Furthermore, the sequences of boundary arcs L_n^+, $n \geq 1$, and L_n^-, $n \geq 1$ converge (in an appropriate sense) to the boundary arcs $\{z = te^{i(\alpha+0)} : 0 \leq t \leq \lambda^{-1}\}$ and $\{z = te^{i(\alpha-0)} : 0 \leq t \leq 1\} \cup \{z = te^{i(\alpha+0)} : \lambda^{-1} \leq t \leq 1\}$, respectively. Taking this consideration into account and applying the version of the Carathéodory convergence theorem stated in Corollary 2 as applied in the proof of Theorem 1, we find that

$$\frac{\omega(a, l_{\zeta_n}^-, H_{\zeta_n})}{\omega(a, l_{\zeta_n}^+, H_{\zeta_n})} \to \tan^2 \theta_{\alpha,\lambda}^+.$$

This establishes the sharpness of the upper bound in (7.2). To prove that the lower bound in (7.2) is also sharp, we truncate the slit $l = l(\lambda)$ at the points

$\tau_n = e^{i\alpha} 2^{-n(2n-1)} \lambda^{-2n}$, which are the lower endpoints of the segments I'_{2n}. Then we proceed with the same arguments used to establish the sharpness of the upper bound. $\square$

(b) Angular oscillations.   Now we turn to monotonic slits with angular oscillations. We say that a slit l with initial point at $\zeta_0 = 0$ oscillates in the interval $[\alpha\pi, \beta\pi]$, $0 \leq \alpha < \beta \leq 1$, if

$$\alpha\pi = \liminf_{l \ni z \to 0} \arg z \leq \limsup_{l \ni z \to 0} \arg z = \beta\pi. \tag{7.7}$$

If l is monotonic and satisfies assumptions (7.7), then for all $\zeta \in l$ close enough to $\zeta_0 = 0$, the slit l_ζ lies in the sector $S_{\alpha_0}(\varepsilon, \delta_0) = \{z = te^{i\theta} : 0 \leq t \leq \varepsilon_0, |\theta - \alpha_0| \leq \delta_0\}$ with $\varepsilon = |\zeta|$, $\alpha_0 = \frac{\alpha+\beta}{2}$, and with $\delta_0 > \frac{\beta-\alpha}{2\pi}$ sufficiently close to $\frac{\beta-\alpha}{2\pi}$.
To estimate the harmonic measures of l_ζ^+ and l_ζ^- in the case of monotonic slits with angular oscillations, we have to consider comparison slits consisting of a radial segment and a circular arc. Thus, we introduce the following notation. For τ_1 and τ_2 such that $0 < \tau_1 < \tau_2 < 1$, let $J_1 = J_1(\tau_1, \tau_2)$ and $J_2 = J_2(\tau_1, \tau_2)$ be the slits starting at ∞ defined by

$$\begin{aligned} J_1 &= J_1(\tau_1, \tau_2) = \{te^{i\tau_1\pi} : t \geq 1\} \cup \{e^{i\theta} : \tau_1\pi \leq \theta \leq \tau_2\pi\}, \\ J_2 &= J_2(\tau_1, \tau_2) = \{te^{i\tau_2\pi} : t \geq 1\} \cup \{e^{i\theta} : \tau_1\pi \leq \theta \leq \tau_2\pi\}. \end{aligned} \tag{7.8}$$

The sets J_1 and J_2 play the same role in our study of slits with angular oscillations as the radial slits I^α played in our discussion in part **(a)** of radially oscillating slits. As before, J_l^+ and J_l^- denote the right shore and the left shore of the slit J_l when a point ζ travels along J_l starting at ∞.
Let $H_l(\tau_1, \tau_2) = \mathbb{H} \setminus J_l(\tau_1, \tau_2)$, $l = 1, 2$. By $u_l^+(a, \tau_1, \tau_2)$ and $u_l^-(a, \tau_1, \tau_2)$ we denote the harmonic measures at the point $a \in H_l(\tau_1, \tau_2)$ of the boundary sets $J_l^+(\tau_1, \tau_2)$ and $J_l^-(\tau_1, \tau_2)$, respectively, with respect to the domain $H_l(\tau_1, \tau_2)$, $l = 1, 2$.

Example 2 In this example, we construct a monotonic slit which oscillates in the interval $[\alpha\pi, \beta\pi]$; see Fig. 8. This example is used to demonstrate the sharpness of the inequalities in Eq. (7.9) below. For real α and β such that $0 < \alpha < \beta < 1$ and an integer $n \geq 1$, let I'_n and I''_n be the line segments defined by

$$\begin{aligned} I'_n &= \{z = te^{i\alpha} : 2^{-(2n+1)n} \leq t \leq 2^{-n(2n-1)}\}, \\ I''_n &= \{z = te^{i\beta} : 2^{-n(2n-1)} \leq t \leq 2^{-(2n-1)(n-1)}\}. \end{aligned}$$

Let C_n be the circular arc

$$C_n = \{z = 2^{-\frac{n(n-1)}{2}} e^{i\theta} : \alpha \leq \theta \leq \beta\}.$$

Define the slit

$$l(\alpha, \beta) = \left(\bigcup_{n=1}^{\infty} \left(C_n \cup I''_n \cup C_{n+1} \cup I'_n \right) \right) \cup \{0\}.$$

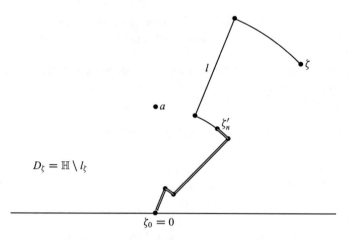

Fig. 8 Monotonic slit with angular oscillation in the interval $[\pi/4, 3\pi/8]$

Then $l(\alpha, \beta)$ is a Jordan slit in $\mathbb{H} \cup \{0\}$ oscillating in the interval $[\alpha\pi, \beta\pi]$. Since $l(\alpha, \beta)$ has corners, it is not smooth. Rounding corners with circular arcs of small radius, one can get an example of a smooth slit oscillating in the same interval $[\alpha\pi, \beta\pi]$.

Our main result for slits with angular oscillation is the following.

Theorem 4 *Let l be a monotonic slit oscillating in the interval $[\alpha\pi, \beta\pi]$ and let ζ_k, $k = 1, 2, \ldots$, be a sequence of points on l such that $\lim_{k\to\infty} \zeta_k = 0$ and $\lim_{k\to\infty} \arg \zeta_k = \sigma\pi$, $\alpha \le \sigma \le \beta$. Then for every $a \in \mathbb{H}$, the following sharp inequalities hold true:*

$$\frac{\frac{\partial}{\partial y} u_2^-(0, \sigma, \beta)}{\frac{\partial}{\partial y} u_2^+(0, \sigma, \beta)} \le \liminf_{k\to\infty} \frac{\omega^-(a, \zeta_k)}{\omega^+(a, \zeta_k)} \le \limsup_{k\to\infty} \frac{\omega^-(a, \zeta_k)}{\omega^+(a, \zeta_k)} \le \frac{\frac{\partial}{\partial y} u_1^-(0, \alpha, \sigma)}{\frac{\partial}{\partial y} u_1^+(0, \alpha, \sigma)}.$$

$$(7.9)$$

Proof   To prove the upper bound, we use comparison slits $j_{1,k} = \{z = te^{i\alpha\pi} : 0 \le t \le \varepsilon_k\} \cup \{s = \varepsilon_k e^{i\theta} : \alpha\pi \le \theta \le \arg \zeta_k\}$, $k = 1, 2, \ldots$ Carleman's extension principle for harmonic measure implies that

$$\frac{\omega^-(a, \zeta_k)}{\omega^+(a, \zeta_k)} \le \frac{\omega(a, j_{1,k}^-, \mathbb{H} \setminus j_{1,k})}{\omega(a, j_{1,k}^+, \mathbb{H} \setminus j_{1,k})} \qquad (7.10)$$

for all a such that $|a| \ge \varepsilon_k$.

Changing variables in the right-hand side of (7.10) via $z \to \varepsilon_k/\bar{z}$ and using the invariance property of the harmonic measure, we obtain

$$\frac{\omega^-(a, \zeta_k)}{\omega^+(a, \zeta_k)} \le \frac{\omega(\varepsilon_k/\bar{a}, J_1^-(\alpha, \tau_k), \mathbb{H} \setminus J_1(\alpha, \tau_k))}{\omega(\varepsilon_k/\bar{a}, J_1^+(\alpha, \tau_k), \mathbb{H} \setminus J_1(\alpha, \tau_k))} \qquad (7.11)$$

Now, passing to the lim sup in (7.11) as $k \to \infty$ and arguing as in the proofs of Theorems 1 and 3, we obtain the upper bound in (7.9). The proof of the lower bound is similar.

To establish sharpness of the upper bound in (7.9), we truncate the slit $l(\alpha, \beta)$ defined in Example 2 at the points $\zeta'_n = 2^{-n(2n-1)} e^{i\sigma\pi}$, $\alpha \leq \sigma \leq \beta$. Then we convert the ζ'_n-tails of $l(\alpha, \beta)$ to unbounded slits changing variables via $z \to 2^{(2n-1)(n-1)}/\overline{z}$. After that our standard procedure of passing to the limit and using the Carathéodory convergence theorem leads to the desired sharpness result for any given σ.

To prove sharpness of the lower bound in (7.9), we truncate $l(\alpha, \beta)$ at the points $\zeta''_n = 2^{-(2n-1)(n-1)} e^{i\sigma}$ and then repeat our previous argument.   □

(c) Slits with full trace.   Here we present two examples of oscillating slits whose traces coincides with the interval $[-1, 1]$.

Example 3 A radially oscillating slit with full trace. Let $0 < \alpha \leq 1/2$. For $n \geq 1$, let I'_n and I''_n be line segments defined by

$$I'_n = \left\{ z = e^{i\alpha\pi} t : \ 2^{-(2n-1)n} \leq t \leq 2^{-(2n-1)(n-1)} \right\},$$
$$I''_n = \left\{ z = e^{i\alpha\left(1 - \frac{(-1)^n}{2n}\right)\pi} t : \ 2^{-(2n+1)(n+1)} \leq t \leq 2^{-(2n-1)(n-1)} \right\}.$$

Let C'_n be a circular arc, which is tangent to I'_n and I''_n at its endpoints

$$z'_{n,1} = e^{i\alpha\pi} 2^{-(2n-1)(n-1)} \quad \text{and} \quad z''_{n,1} = e^{i\alpha\left(1 - \frac{(-1)^n}{2n}\right)\pi} 2^{-(2n-1)(n-1)}.$$

Similarly, let C''_n be a circular arc, which is tangent to I''_n and I'_{n+1} at its endpoints

$$z''_{n,2} = e^{i\alpha\left(1 - \frac{(-1)^n}{2n}\right)\pi} 2^{-(2n+1)(n+1)} \quad \text{and} \quad z'_{n,2} = e^{i\alpha\pi} 2^{-(2n+1)(n+1)}.$$

We assume here that the arcs C'_n and C''_n form angles of opening π with corresponding segments at their endpoints.

Finally, we define l_r as

$$l_r = \left(\cup_{n=1}^{\infty} \left(I'_n \cup C'_n \cup I''_n \cup C''_n \right) \right) \cup \{0\}.$$

Then l_r is a Jordan arc in $\mathbb{H} \cup \{0\}$ joining the points $z = 0$ and $z = 1/2$, which has a tangent vector $\overrightarrow{v}(\zeta)$ at every point $\zeta \in l_r$ including the point at $\zeta = 0$. Moreover, this tangent vector is continuous at every point $\zeta \neq 0$ but is not continuous at $\zeta = 0$.

Let $l_{0,n}$ denote the tail of l truncated at the point $z'_{n,2}$ and let $H_{0,n} = \mathbb{H} \setminus l_{0,n}$. If we convert $l_{0,n}$ into unbounded slits via the mapping $z \to 2^{-(2n-1)(n-1)}/\overline{z}$ and then pass to the limit as $n \to \infty$, the limit configuration will consist of the domain H^{α} and the slit I^{α} defined in Sect. 4. Furthermore, one can easily see that if $n = 2k - 1$ is odd, then the image set of boundary points of the right shore $l^+_{0,2k-1}$ of the slit $l_{0,2k-1}$ will converge to the set of all boundary points of the radial slit I^α. Similarly, if $n = 2k$ is even, the image set of boundary points of the left shore $l_{0,2k}^-$ of the slit $l_{0,2k-1}$ will converge to the set of all boundary points of the radial slit I^α. These observations imply that the trace of the slit l_r (as defined in Sect. 1) coincides with the interval $[-1, 1]$.

Example 4 A monotonic angularly oscillating slit with full trace. Let C_{2k-1} and C_{2k}, $k = 1, 2, \ldots$, denote circular arcs defined as follows:

$$C_{2k-1} = \left\{ z = \tfrac{1}{2k-1} e^{i\theta} : \tfrac{\pi}{2k+2} \leq \theta \leq \tfrac{\pi(2k+3)}{2k+4} \right\},$$
$$C_{2k} = \left\{ z = \tfrac{1}{2k} e^{i\theta} : \tfrac{\pi}{2k+4} \leq \theta \leq \tfrac{\pi(2k+3)}{2k+4} \right\}.$$

Also, let I_{2k-1} and I_{2k}, $k = 1, 2, \ldots$, denote the radial segments defined by

$$I_{2k-1} = \left\{ z = e^{i\frac{\pi(2k+3)}{2k+4}} t : \tfrac{1}{2k+1} \leq t \leq \tfrac{1}{2k} \right\},$$
$$I_{2k} = \left\{ z = e^{i\frac{\pi}{2k+4}} t : \tfrac{1}{2k} \leq t \leq \tfrac{1}{2k-1} \right\}.$$

Consider the slit l_a defined by

$$l_a = \left(\cup_{k=1}^\infty \left(C_{2k-1} \cup I_{2k-1} \cup C_{2k} \cup I_{2k} \right) \right) \cup \{0\}.$$

Then l_a is a piecewise smooth monotonic slit in $\mathbb{H} \cup \{0\}$ with endpoints at $z = 0$ and $z = e^{i\frac{\pi}{4}}$.

To show that l_a has full trace, one can truncate l_a at the points $z_k^+ = \tfrac{1}{2k-1} e^{i\frac{\pi}{2k+2}}$ to obtain a sequence $l_{a,k}'$ of z_k^+-tails of l_a and at the points $z_k^- = \tfrac{1}{2k} e^{i\frac{\pi(2k+3)}{2k+4}}$ to obtain a sequence $l_{a,k}''$, $k = 1, 2, \ldots$, of z_k^--tails of l_a. Using these two sequences of tails of l_a and arguing as in our previous example, one can see that the trace of l_a coincides with the interval $[-1, 1]$.

8 Further Questions

We conclude this paper with a few suggestions for future study. Some of the questions stated below may be easy to solve, or their solutions might already be known to experts in the theory of Löwner's differential equation.

(a) **Differentiation formulas.** If l is a slit in $\mathbb{H} \cup \{0\}$ whose arc-length parametrization $\zeta = \zeta(s)$ is sufficiently smooth, then it reasonable to expect that the harmonic measures $\omega^+(a, \zeta(s))$ and $\omega^-(a, \zeta(s))$ are differentiable functions of s. It would be interesting to find formulas for $\frac{d}{ds}\omega^+(a, \zeta(s))$ and $\frac{d}{ds}\omega^-(a, \zeta(s))$ similar to the differentiation formula for the module of a doubly-connected domain with a slit and to the differentiation formula for the reduced module of a simply connected domain with a slit, which are given in Theorems 5.2 and 5.3 in our paper [8].

(b) Monotonicity problem for the quotient of harmonic measures. Each of the quotients in the left-hand side and in the right-hand side of (7.9) depends on the parameter σ, $\alpha \le \sigma \le \beta$. To obtain estimates for the limit behavior of the quotient $\frac{\omega^-(a,\zeta)}{\omega^+(a,\zeta)}$, which *do not depend* on σ, we have to show that the left-hand side of (7.9) is a decreasing function of σ and the right-hand side of (7.9) is an increasing function of σ. We have not been able to prove these technical but useful monotonicity properties. Thus, we suggest the following problem.

For $0 < \alpha < \sigma < 1$, let $J_1(\alpha, \sigma)$ be the slit in $\mathbb{H}$ defined by the first formula in (7.8). Prove that for fixed α, $0 < \alpha \le 1/2$, and fixed a such that $|a| > 1$, the quotient

$$\frac{\omega(a, J_1^-(\alpha, \sigma), \mathbb{H} \setminus J_1(\alpha, \sigma))}{\omega(a, J_1^+(\alpha, \sigma), \mathbb{H} \setminus J_1(\alpha, \sigma))}$$

strictly increases as σ runs from α to 1.

(c) Almost circular slits. Let l be a smooth slit in $\mathbb{H} \cup \{0\}$ which is tangent to the positive real axis at $\zeta_0 = 0$. We say that l is almost circular at ζ_0 if l has continuous curvature $cur_l(\zeta)$ and $cur_l(0) = \kappa > 0$. For almost circular slits D. Prokhorov (see [4]) conjectured the following limit relation for the harmonic measures:

$$\lim_{l \ni \zeta \to 0} \frac{\omega^-(a, \zeta)}{(\omega^+(a, \zeta))^2} = \frac{\kappa}{2\pi}. \tag{8.1}$$

Under additional assumptions on the smoothness of the slit l, relation (8.1) was proved in [4], where the authors used the Löwner differential equation. For a general almost circular slit, Eq. (8.1) remains a conjecture. The method used in this paper does not allow us to prove (8.1) in its full generality.

To explain why our method fails, consider reflected slits $\bar{l}_\zeta = \{z : \bar{z} \in l_\zeta\}$ and extended domains $\Omega_\zeta = \mathbb{C} \setminus (l_\zeta \cup \bar{l}_\zeta)$. In an attempt to imitate the proof of Theorem 1, we have to change variables and transform domains Ω_ζ into unbounded domains $\widetilde{\Omega}_\zeta$ such that the family of domains $\widetilde{\Omega}_\zeta$ converges in the Carathéodory sense to a model domain Ω_0 as $\zeta \to 0$. For an almost circular slit l, the point $\zeta \in l$ can be represented as

$$\zeta = \frac{i}{\kappa}\left(1 - (1 - \delta(\varepsilon))e^{i\varepsilon}\right) = \frac{\varepsilon}{\kappa}(1 - i\delta_1(\varepsilon))$$

with some $\delta_1(\varepsilon) \to 0$ as $\varepsilon \to 0$. Therefore, the required change of variables in this case can be done via the function

$$f_\varepsilon(x) = \frac{2}{\kappa}\frac{1}{\bar{z}} - \cot\frac{\varepsilon}{2}. \tag{8.2}$$

Let $\widetilde{\Omega}_\zeta$ denote the image of Ω_ζ under the mapping (8.2). Then it is not difficult to see that the family of domains $\widetilde{\Omega}_\zeta$ converges in the Carathéodory sense to the domain $\Omega_0 = C \setminus \{z = \pm i + t : t \geq 0\}$ as $\varepsilon \to 0$, which is a good sign. Unfortunately, for any fixed point $a \in \mathbb{H} \setminus l$, the image point $f_\varepsilon(a)$ approaches the point at ∞ as $\varepsilon \to 0$. Since the limit point $z = \infty$ is a *boundary point* of the limit domain Ω_0 (and not an *interior point* as it was in the proof of Theorem 1!) our version of the Carathéodory convergence theorem stated in Corollary 2 cannot be applied to estimate harmonic measures in the problem under consideration. This is exactly the point where our method fails. There is a chance that a more elaborate version of the Carathéodory convergence theorem, which takes into account the continuity of the curvature of the slit l, may help to overcome these difficulties.

(d) Scaled driving term. The driving term $\lambda(t)$ in Eq. (1.2) together with the initial condition $f(z, 0) \equiv 0$ defines $f(z, t)$ uniquely. The latter suggests that the scaled driving term $\Lambda(\zeta)$ given by (1.5) defines the slit $l \subset \mathbb{H}$ uniquely up to scaling and translation in the horizontal direction. This may already be known to experts in Löwner theory.

Also, it would be interesting to study the correspondence between the geometric properties of l and the functional properties of its scaled driving term $\Lambda(\zeta)$.

(e) Escaping paths. Let $g(z, \zeta)$ be defined by (1.4). Then, for every fixed $z \in \mathbb{H} \setminus l$, the set $E_z(l) = \{g(z, \zeta(t)) : 0 \leq t \leq T\}$ represents a continuous path along which the point $g(z, T)$ *escapes* to ∞ as t varies from T to 0.

It is an easy exercise using Schwarz's lemma (applied to the class of analytic functions from the domain $\overline{\mathbb{C}} \setminus I_0$ to itself) to show that the path $E_z(l)$ is simple and has at most one point of intersection with each ellipse with foci at ± 1. Furthermore, since $g(z, \zeta(t))$ is normalized by the condition $g'(\infty, \zeta(t)) > 0$, it follows that each escaping path has a well-defined "escaping direction" at ∞. It would be interesting to know whether or not a single escaping path $E_z(l)$ determines the slit l uniquely and to establish relations between geometric properties of $E_z(l)$ and geometric properties of l.

To give an example of escaping paths, consider the vertical slit $I_i = [0, i]$ parameterized by arc length. Then the mapping functions are $g(z, \tau) = \sqrt{1 + (z/\tau)^2}$, $0 < \tau \leq 1$. Let $z_0 = r_0 e^{i\theta_0}$, $r_0 > 0$, $0 < \theta_0 < \pi/2$. One easily calculates that the escaping path of z_0 is given by the equation

$$u^2 - 2uv \cot(2\theta_0) - v^2 = 1, \quad \text{where } w = u + iv. \tag{8.3}$$

Thus, in the case under consideration, all points of the ray $\{z = re^{i\theta_0} : r > 0\}$ escape to ∞ along the same path (starting at different initial points of this path) given by Eq. (7.7). The latter equation defines a hyperbola with asymptotes $\{w : \Im(e^{-i\theta_0}w) = 0\}$ and $\{w : \Re(e^{-i\theta_0}w) = 0\}$. We also note that all these hyperbolic escaping paths start at $w = 1$ and they do not intersect each other; see Fig. 9. For $z_0 = r_0 e^{i\theta_0}$ with $\pi < \theta_0 < \pi/2$, we obtain similar hyperbolic escaping paths while in case $\theta_0 = \pi/2$, the escaping path is along the positive imaginary axis. It

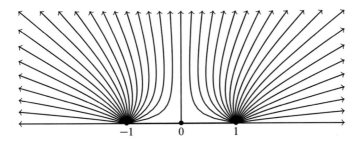

Fig. 9 Escaping paths for vertical slit $I_i = [0, i]$

is interesting that in the case of the slit $l = I_i$, the escaping paths foliate the upper half-plane $\mathbb{H}$. We do not know any other example of slits l with this foliation property. So, it would be interesting to prove (or disprove) the following:

If the escaping paths of a slit l having its initial point at $z = 0$ foliate some domain $D \subset \mathbb{H}$, then l is a vertical segment.

(f) Randomness. During the last two decades, two-dimensional random walks and their relation to the theory of Löwner's differential equation attracted the attention of many complex analysts. We suggest the following problem. Let l_ζ be a Brownian walk in $\mathbb{H}$ starting at some point $z_0 \in \partial\mathbb{H}$ and terminating at $\zeta \in \mathbb{H}$. Then the corresponding scaled driving term $\Lambda(\zeta)$ represents a random walk on the interval $I_0 = [-1, 1]$. It would be interesting to find some relations between the properties of these two random processes.

References

1. G.M. Goluzin, *Geometric Theory of Functions of a Complex Variable* (American Mathematical Society, Providence, 1969). MR0247039
2. G.F. Lawler, *Conformally Invariant Processes in the Plane* (American Mathematical Society, Providence, 2005). MR2129588
3. Ch. Pommerenke, *Boundary Behaviour of Conformal Maps* (Springer, Berlin, 1992). MR1217706
4. D. Prokhorov, D. Ukrainskii, Asymptotic ratio of harmonic measures of slit sides. Izv. Sarat. Univ. Write. Ser. Ser. Math. Mech. Comput. **15**(2), 160–167 (2015)
5. D. Prokhorov, A. Zakharov, Harmonic measures of sides of a slit perpendicular to the domain boundary. J. Math. Anal. Appl. **394**(2), 738–743 (2012). MR2927494
6. A. Sebbar, Th. Falliero, Equilibrium point of Greens function for the annulus and Eisenstein series. Proc. Am. Math. Soc. **135**, 313–328 (2007). MR 2255277 (2007h:30011)
7. A.Y. Solynin, Polarization and functional inequalities. Algebra i Analiz **8**, 148–185 (1996). English transl., St. Petersburg Math. J. **8**, 1015–1038 (1997). MR1458141 (98e:30001a)
8. A.Y. Solynin, Modules and extremal metric problems. Algebra i Analiz **11**, 3–86 (1999). English transl., St. Petersburg Math. J. **11**(1), 1–65 (2000). MR1691080 (2001b:30058) (98e:30001a)
9. A.Y. Solynin, A note on equilibrium points of Green's function. Proc. Am. Math. Soc. **136**, 1019–1021 (2008). MR2361876

Coupling of Gaussian Free Field with General Slit SLE

Alexey Tochin and Alexander Vasil'ev

Abstract We consider a coupling of the Gaussian free field with slit holomorphic stochastic flows, called (δ, σ)-SLE, which contains known SLE processes (chordal, radial, and dipolar) as particular cases. In physical terms, we study a free boundary conformal field theory with one scalar bosonic field, where Green's function is assumed to have some general regular harmonic part. We establish which of these models allow coupling with (δ, σ)-SLE, or equivalently, when the correlation functions induce local (δ, σ)-SLE martingales (martingale observables).

Keywords Löwner equation • Stochastic flows • General Löwner theory • SLE • Conformal field theory • Boundary conformal field theory • Gaussian free field • Coupling

2010 Mathematics Subject Classification 30C35, 34M99, 60D05, 60G57, 60J67

1 Introduction

1.1 A Simple Example of Coupling

The paper is focused on the coupling of the Gaussian free field (GFF) with the (δ, σ)-SLE, and studies in detail some special cases. Let us explain briefly the case of the chordal SLE_4 and its coupling with the GFF subject to the Dirichlet boundary conditions, see [19] and [27] for details. We review some known generalizations, and then, explain the main results of the paper.

Let $G_t : \mathbb{H} \setminus K_t \to \mathbb{H}$ be a conformal map from a subset $\mathbb{H} \setminus K_t$ of the upper half-plane $\mathbb{H} := \{z \in \mathbb{C} : \operatorname{Im} z > 0\}$ onto $\mathbb{H}$ defined by the initial value problem for the

A. Tochin (✉) • A. Vasil'ev
Department of Mathematics, University of Bergen, P.O. Box 7803, Bergen N-5020, Norway
e-mail: alexey.tochin@gmail.com; alexander.vasiliev@math.uib.no

© Springer International Publishing AG 2018

301

M. Agranovsky et al. (eds.), *Complex Analysis and Dynamical Systems*,
Trends in Mathematics, https://doi.org/10.1007/978-3-319-70154-7_15

stochastic differential equation

$$d^{\text{Itô}} G_t(z) = \frac{2}{G_t(z)} dt - \sqrt{\kappa}\, d^{\text{Itô}} B_t, \quad G_0(z) = z, \quad z \in \mathbb{H}, \quad t \geq 0, \quad \kappa > 0,$$

(1.1)

where $d^{\text{Itô}}$ denotes the Itô differential and B_t stands for the standard Brownian motion This initial value problem is known as the chordal SLE whose solution G_t is a random conformal map which defines a strictly growing bounded random set $K_t := \mathbb{H} \setminus G_t^{-1}(\mathbb{H})$, called the SLE hull, ($K_t \subset K_s$, $0 \geq t < s$). In particular, it is a random simple curve for $\kappa \leq 4$ that generically tends to infinity as $t \to \infty$ a.s. It turns out that the random law defined this way is related to various problems of mathematical physics.

It is straightforward to check that the following random processes are local martingales for $\kappa = 4$

$$M_{1t}(z) := \arg G_t(z),$$

$$M_{2t}(z, w) := -\log \frac{|G_t(z) - G_t(w)|}{|G_t(z) - \overline{G_t(w)}|} + \arg G_t(z) \arg G_t(w),$$

$$M_{3t}(z, w, u) := -\log \frac{|G_t(z) - G_t(w)|}{|G_t(z) - \overline{G_t(w)}|} \arg G_t(u) + \arg G_t(z) \arg G_t(w) \arg G_t(u) +$$

$$+\{z \leftrightarrow w \leftrightarrow u\},$$

$$\cdots$$

for $z, w, u \in \mathbb{H}$ until a stopping time t_z, which is geometrically defined by the fact that K_t touches z for the first time, i.e. $G_t(z)$ is no longer defined for $t > t_z$. The Wick pairing structure can be recognized in this collection of martingales. It appears in higher moments of multivariant Gaussian distributions as well as in the correlation functions in the free quantum field theory. We consider the GFF Φ which is a random real-valued functional over smooth functions in $\mathbb{H}$. Let us use the heuristic notation '$\Phi(z)$' ($z \in \mathbb{H}$) until a rigorous definition is given in Sect. 2.7.

For some special choice of GFF the moments are

$$S_1(z) := \mathbb{E}[\Phi(z)] = \arg z,$$

$$S_2(z, w) := \mathbb{E}[\Phi(z)\Phi(w)] = -\log \frac{|z - w|}{|z - \overline{w}|} + \arg z \arg w,$$

$$S_3(z, w, u) := \mathbb{E}[\Phi(z)\Phi(w)\Phi(u)] = -\log \frac{|z - w|}{|z - \overline{w}|} \arg u + \arg z \arg w \arg u$$

$$+ \{z \leftrightarrow w \leftrightarrow u\},$$

$$\cdots$$

Roughly speaking, $\Phi(z)$ is a Gaussian random variable for each z with the expectation $\arg z$, and with the covariance $-\log \left|\frac{z-w}{z-\bar{w}}\right|$. The connection to (1.1) can be obtained if we consider stochastic processes

$$M_{nt} := S_n(G_t(z_1), G_t(z_2), \ldots G_t(z_n)) \tag{1.2}$$

that are turned to be martingales for all $n = 1, 2, \ldots$ (see also [8]).

This last fact is closely related to the observation that the random variable $\Phi(G_t(z))$ agrees in law with $\Phi(z)$, where Φ and G_t are sampled independently. It was also obtained, see [23], that the properly defined zero-level line of the random distribution $\Phi(z)$ starting at the origin in the vertical direction agrees in law with SLE$_4$.

This connection, i.e., coupling, can be generalized for an arbitrary $\kappa > 0$, see [30]. In order to explain its geometric meaning let us associate another field $J(z)$ of unit vectors $|J(z)| = 1$ with $\Phi(z)$ by

$$J(z) := e^{i\Phi(z)/\chi}$$

for some $\chi > 0$. A unit vector field J transforms from a domain D to $D \backslash K$ according to the rule

$$J(z) \rightarrow \tilde{J}(\tilde{z}) = \sqrt{\frac{\overline{G'(\tilde{z})}}{G'(\tilde{z})}} J(G(\tilde{z})),$$

where G is the conformal map $G : D \backslash K \rightarrow D$ (the coordinate transformation). Thereby, $\Phi(z)$ transforms as

$$\Phi(z) \rightarrow \tilde{\Phi}(\tilde{z}) = \Phi(G(\tilde{z})) - \chi \arg G'(\tilde{z}). \tag{1.3}$$

The coupling for $\kappa > 0$ is the agreement in law of $\Phi(G_t(z)) - \chi \arg G'(z)$ with $\Phi(z)$, where

$$\chi = \frac{2}{\sqrt{\kappa}} - \frac{\sqrt{\kappa}}{2}. \tag{1.4}$$

Besides, the flow line of $J(z)$ starting at the origin agrees in law with the SLE$_\kappa$ curve.

The statement above can be extended from the chordal equation (1.1) to its radial version, see [2, 18], or to the dipolar one, see [5, 17, 20], see also [16] for the SLE-(κ, ρ) case.

In [14, 15], we considered slit holomorphic stochastic flows, called in this paper (δ, σ)-SLE, which contain all the above mentioned SLEs as special cases except SLE-(κ, ρ). In the present paper, we consider a general case of the coupling of (δ, σ)-SLE with GFF. In particular, we study the known cases of the couplings in a systematic way as well as consider some new ones.

There is also another type of coupling described in [30] for the chordal case. The usual forward Löwner equation is replaced by its reverse version with the opposite sign at the drift term proportional to dt in (1.1). The corresponding solution is a conformal map $G_t : \mathbb{H} \to \mathbb{H} \setminus K_t$, and $\Phi(z)$ agrees in law with $\Phi(G_t(z)) + \gamma \log |G_t'(z)|$ for some real κ-dependent constant γ.

In addition to the coupling, there are also interesting connections to other aspects of conformal field theory (CFT) such as the highest weight representations of the Virasoro algebra [3], and the vertex algebra [19]. On the other hand, the crossing probabilities, such as touching the real line by an SLE curve, are connected with the CFT stress tensor correlation functions, see [3, 11]. Both of these directions, as well as the reverse coupling, exceed the scope of this paper.

1.2 (δ, σ)-SLE Overview

The (δ, σ)-SLE is a unification of the well-known chordal, radial, and dipolar stochastic Löwner equations described first in [14]. We will give a rigorous definition in Sect. 2 and will use a simplified version in Introduction. Let $D \subset \mathbb{C}$ be a simply connected hyperbolic domain. A (δ, σ)-SLE or a *slit holomorphic stochastic flow* is the solution $\{G_t\}_{t \geq 0}$ to the stochastic differential equation

$$\mathrm{d}^S G_t(z) = \delta(G_t(z))dt + \sigma(G_t(z))\, \mathrm{d}^S B_t, \quad G_0(z) = z, \quad z \in D, \quad t \geq 0, \qquad (1.5)$$

where B_t is the standard Brownian motion, $\sigma: D \to \mathbb{C}$ and $\delta: D \to \mathbb{C}$ are some fixed holomorphic function (vector fields) defined in such a way that the solution G_t is always a conformal map $G_t: D \setminus K_t \to D$ for some random curve-generated subset K_t. The differential d^S is the Stratonovich differential, which is more convenient in our setup than the more frequently used Itô differential $\mathrm{d}^{\mathrm{It\hat{o}}}$, see for example [9]. The same equation in the Itô form is

$$\mathrm{d}^{\mathrm{It\hat{o}}} G_t(z) = \left(\delta(G_t(z)) + \frac{1}{2}\sigma'(G_t(z))\sigma(G_t(z)) \right) dt + \sigma(G_t(z))\, \mathrm{d}^{\mathrm{It\hat{o}}} B_t,$$

$$G_0(z) = z, \quad t \geq 0, \qquad (1.6)$$

see "Appendix 3: Some Formulas from Stochastic Calculus" for details. Equation (1.6) can be easily reformulated for any other domain $\tilde{D} \subset \mathbb{C}$ using the conformal transition map $\tau: \tilde{D} \to D$. Define $\tilde{G}_t := \tau^{-1} \circ G_t \circ \tau$. The equation for $\tilde{G}_t$ is of the same form as (1.5), but with new fields $\tilde{\delta}$ and $\tilde{\sigma}$ defined as

$$\tilde{\delta}(\tilde{z}) = \frac{1}{\tau'(\tilde{z})}\delta(\tau(\tilde{z})), \quad \tilde{\sigma}(\tilde{z}) = \frac{1}{\tau'(\tilde{z})}\sigma(\tau(\tilde{z})).$$

The general form of $\delta(z)$ and $\sigma(z)$ for $D = \mathbb{H}$ (the upper half-plane) is

$$\delta^{\mathbb{H}}(z) = \frac{\delta_{-2}}{z} + \delta_{-1} + \delta_0 z + \delta_1 z^2,$$

$$\sigma^{\mathbb{H}}(z) = \sigma_{-1} + \sigma_0 z + \sigma_1 z^2, \tag{1.7}$$

$$\delta_{-1}, \delta_0, \delta_1, \sigma_{-1}, \sigma_0, \sigma_1 \in \mathbb{R}, \quad \sigma_{-1} \neq 0, \quad \delta_{-2} > 0.$$

The values of δ and σ for the classical SLEs (in the upper half-plane) are summarized in the following table.

SLE type	$\delta^{\mathbb{H}}(z)$	$\sigma^{\mathbb{H}}(z)$
Chordal	$2/z$	$-\sqrt{\kappa}$
Radial	$1/(2z) + z/2$	$-\sqrt{\kappa}(1 + z^2)/2$
Dipolar	$1/(2z) - z/2$	$-\sqrt{\kappa}(1 - z^2)/2$
ABP SLE, see [13]	$1/(2z)$	$-\sqrt{\kappa}(1 + z^2)/2$

We use the half-plane formulation in this table just for simplicity. In order to obtain the commonly used form of the radial equation in the unit disk one applies the transition map

$$\tau_{\mathbb{H},\mathbb{D}}(z) := i\frac{1-z}{1+z}, \quad \tau_{\mathbb{H},\mathbb{D}} : \mathbb{D} \to \mathbb{H}, \tag{1.8}$$

for the unit disk $\mathbb{D} := \{z \in \mathbb{C} : |z| < 1\}$. For more details, see Example 2 and Sect. 4.3. The same procedure with the transition map

$$\tau_{\mathbb{H},\mathbb{S}}(z) = \text{th}\frac{z}{2}, \quad \tau_{\mathbb{H},\mathbb{S}} : \mathbb{S} \to \mathbb{H}, \tag{1.9}$$

for the strip $\mathbb{S} := \{z \in \mathbb{C} : 0 < \text{Im}\, z < \pi\}$, gives the common form for the dipolar SLE, see also Sect. 4.2.

It was shown in [14], that the choice of δ and σ given by (1.7) guarantees that the random set K_t is curve-generated similarly to the standard known types of SLE$_\kappa$. Moreover, K_t has the same local behaviour (the fractal dimension, dilute phases, and etc.).

The main difference between the classical SLEs and the case of the general (δ, σ)-SLE is the absence of fixed normalization points, in general, e.g., in the classical cases the solution G_t to the radial SLE equation always fixes an interior point (0 in the unit disk formulation), the dipolar SLE preserves two boundary points ($-\infty$ and $+\infty$ in the strip coordinates $\mathbb{S}$), the chordal equation is normalized at the boundary point at infinity in the half-plane formulation. The chordal case can be considered as the limiting case of the dipolar one as both fixed boundary points collide, or the limiting case of the radial equation when the fixed interior point approaches the boundary. In the general case of (δ, σ)-SLE, there is no such a simple normalization. Numerical experiments show that the curve (or the curve-generated hull for $\kappa \geq 4$) K_t tends to a random point inside the disk as $t \to \infty$. A version of the domain Markov property still holds due to an autonomous form of Eq. (1.6).

1.3 Overview and Purpose of This Paper

We consider the general form of two-dimensional massless GFF Φ with some expectation

$$\eta(z) := \mathbb{E}\left[\Phi(z)\right],$$

and the covariance

$$\Gamma(z, w) := \mathbb{E}\left[\Phi(z)\Phi(w)\right] - \mathbb{E}\left[\Phi(z)\right]\mathbb{E}\left[\Phi(w)\right].$$

We postulate that $\Gamma(z, w)$ is symmetric fundamental solution to the Laplace equation,

$$\Delta_z\Gamma(z, w) = 2\pi\delta(z, w), \quad \Gamma(z, w) = \Gamma(w, z),$$

but imposing no boundary conditions. In other words, $\Gamma(z, w)$ has the from

$$\Gamma(z, w) = -\frac{1}{2}\log|z - w| + \text{ symmetric harmonic function of } z \text{ and } w.$$

Generalizing [30], we address the following question: which of so defined GFF can be coupled with some (δ, σ)-SLE in the sense that $\Phi(z)$ and $\Phi(G_t(z))$ agree in law? In particular, we show (Theorem 1) that this is possible if and only if the following system of partial differential equations is satisfied

$$\begin{aligned}
\left(\mathcal{L}_\delta + \tfrac{1}{2}\mathcal{L}_\sigma^2\right)\eta(z) &= 0, \\
\mathcal{L}_\delta\,\Gamma(z, w) + \mathcal{L}_\sigma\,\eta(z)\,\mathcal{L}_\sigma\,\eta(w) &= 0, \\
\mathcal{L}_\sigma\,\Gamma(z, w) &= 0,
\end{aligned} \tag{1.10}$$

where, following [19], we understand under $\mathcal{L}$ a generalized version of the Lie derivatives defined by

$$\begin{aligned}
\mathcal{L}_v\,\eta(z) &:= v(z)\,\partial_z + \overline{v(z)}\,\partial_{\bar{z}} + \chi\,\mathrm{Im}\,v'(z), \\
\mathcal{L}_v\,\Gamma(z, w) &:= v(z)\,\partial_z + \overline{v(z)}\,\partial_{\bar{z}} + v(w)\,\partial_w + \overline{v(w)}\,\partial_{\bar{w}}.
\end{aligned}$$

Here, $v = \delta$ or $v = \sigma$ and χ refers to the SLE parameter κ by (1.4). These equations generalize the corresponding relation from [30], see the table on page 45, and they are versions of (M-cond') and (C-cond') from [16]. The second equation in (1.10) is known as a version of Hadamard's variation formula, and the third states that the covariance Γ must be invariant with respect to the Möbius automorphisms generated by the vector field σ.

The paper is a continuation and extension of [15] and is organized as follows. The rest of Introduction provides some remarks about the relations between Eq. (1.10) and the BPZ equation considered first in [6]. In Sect. 2, we give all necessary definitions for (δ, σ)-SLE as well as a very basic definitions and properties of GFF that we will need in what follows. Further, in Sect. 3, we prove the coupling Theorem 1. It states that for any (δ, σ)-SLE a proper pushforward of S_n denoted by $G_t^{-1}{}_* S_n$ (a generalisation of (1.2) with an additional χ-term similar to (1.3)) is a local martingale if and only if the system (1.10) is satisfied for given δ, σ, Γ, and η. The same theorem also states that both: the local martingale property and the system of equations are equivalent to a *local coupling* which is a weaker version of the coupling from [30] discussed above. We expect, but do not consider in this paper, that the local coupling leads to the same property of the flow lines of $e^{i\Phi(z)/\chi}$ to agree in law with the (δ, σ)-SLE curves.

The general solution to the system (1.10) gives all possible ways to couple (δ, σ)-SLEs with the GFF at least in the frameworks of our assumptions of the pre-pre-Schwarzian behaviour of η and of the scalar behaviour of Γ.

In Sect. 4, we assume the simplest choice of the Dirichlet boundary conditions for Γ and study all (δ, σ)-SLEs that can be coupled. It turns out that only the classical SLEs with drift are allowed plus some exceptionals cases, see Sects. 4.5 and 4.6, that define the same measure as for the chordal SLE up to time reparametrization Observe however, that the coupling with the classical SLEs with drift has not been considered in the literature so far.

Further, in Sects. 5 and 6, we assume less trivial choices of Γ and obtain that among all (δ, σ)-SLEs only the dipolar and radial SLEs with $\kappa = 4$ can be coupled. The first case was considered in [17]. The second one requires a construction of a specific ramified GFF Φ_{twisted}, that changes its sign to the opposite while being turned once around some interior point in D. This construction was considered before and it is called the 'twisted CFT' as we were informed by Nam-Gyu Kang, see [21].

2 Preliminaries

Each version of the Löwner equations and holomorphic flows are usually associated with a certain canonical domain $D \subset \mathbb{C}$ in the complex plane specifying fixed interior or boundary points, for example, in the case of the upper half-plane, the unit disk, etc. It is always possible to map these domains one to another if necessary. In this paper, we focus on conformally invariant properties, i.e., those which are not related to a specific choice of the canonical domain. This can be achieved by considering a generic hyperbolic simply connected domain and conformally invariant structures from the very beginning. It could be also natural to go further and work with a simply connected hyperbolic Riemann surface $\mathcal{D}$ (with the boundary $\partial \mathcal{D}$). In what follows, $\mathcal{D}$ is thought of as a generic domain with a well-defined boundary or a Riemann surface.

We use mostly global chart maps $\psi : \mathcal{D} \to D^\psi \subset \mathbb{C}$ from $\mathcal{D}$ to a domain of the complex plane, writing ψ for a chart (D, ψ) for simplicity. The charts $\psi_\mathbb{H} : \mathcal{D} \to \mathbb{H}$ corresponding to the upper half-plane or $\psi_\mathbb{D} : \mathcal{D} \to \mathbb{D}$ to the unit disk are related by (1.8). Another example is a multivalued map $\psi_\mathbb{L} : \mathcal{D} \to \mathbb{H}$ onto the upper half-plane related to $\psi_\mathbb{D}$ by (4.11). A single-valued branch of $\psi_\mathbb{L} : \mathcal{D} \to \mathbb{H}$ is not a global chart map.

2.1 Vector Fields and Coordinate Transform

Consider a holomorphic vector field v on $\mathcal{D}$, which can be defined as a holomorphic section of the tangent bundle. We also can define it as a map $\psi \mapsto v^\psi$ from the set of all possible global chart maps $\psi : \mathcal{D} \to D^\psi$ to the set of holomorphic functions $v^\psi : D^\psi \to \mathbb{C}$ defined in the image of these maps $D^\psi := \psi(\mathcal{D})$. For the vector fields, the following coordinate change holds. For any $\tilde{\psi} : \mathcal{D} \to D^{\tilde{\psi}}$,

$$v^{\tilde{\psi}}(\tilde{z}) = \frac{1}{\tau'(\tilde{z})} v^\psi(\tau(\tilde{z})), \quad \tau := \psi \circ \tilde{\psi}^{-1}, \quad \tilde{z} \in \tilde{D}. \tag{2.1}$$

We consider a conformal homeomorphism $G : \mathcal{D} \setminus \mathcal{K} \to \mathcal{D}$ (inverse endomorphism) that in a given chart ψ is given as

$$G^\psi := \psi \circ G \circ \psi^{-1} : D^\psi \setminus K^\psi \to D^\psi.$$

A vector field in other words is a $(-1, 0)$-differential and the *pushforward of a vector field* $G_*^{-1} : v^\psi \mapsto \tilde{v}^{\tilde{\psi}}$ is defined by the rule

$$G_*^{-1} v^\psi(z) := v^{\psi \circ G}(z) = v^{\psi \circ G \circ \psi^{-1} \circ \psi}(z) = v^{G^\psi \circ \psi}(z) =$$
$$= \frac{1}{(G^\psi)'(z)} v^\psi\left(G^\psi(z)\right), \quad z \in D^\psi, \tag{2.2}$$

thereby G^ψ plays the role of τ.

It is easy to see that for any two given maps G and $\tilde{G}$ as above

$$\tilde{G}_*^{-1} G_*^{-1} = \left(\tilde{G}^{-1} \circ G^{-1}\right)_*,$$

which motivates the upper index -1 in the definition of G_*, because in this case we are working with a left module.

The pushforward map G_*^{-1} also can be obtained as a map from the tangent space to $\mathcal{D}$ induced by G. We follow the way above because it can be generalized then to sections of tangent and cotangent spaces and their tensor products, see Sect. 2.3.

Let v_t and $\tilde{v}_t$ be two holomorphic vector fields depending on time continuously such as the following differential equations has continuously differentiable solutions F_t and $\tilde{F}_t$ in some time interval

$$\dot{F}_t = v_t \circ F_t,$$

$$\dot{\tilde{F}}_t = \tilde{v}_t \circ \tilde{F}_t.$$

Then, we can conclude that

$$\frac{\partial}{\partial t}(F_t \circ \tilde{F}_t) = \left(v_t + F_{t*}\tilde{F}_{t*}\tilde{v}_t\right) \circ F_t \circ \tilde{F}_t$$

$$\dot{F}_t^{-1} = -\left(F_t^{-1}{}_* v_t\right) \circ F_t^{-1}.$$

in the same t-interval and in the region of $\mathcal{D}$ where F_t and $\tilde{F}_t$ are defined. The latter relation can be reformulated in a fixed chart ψ as

$$\dot{F}_t^{-1}{}^\psi(z) = -\left(\left(F_t^{-1}\right)^\psi\right)'(z)\, v_t^\psi(z).$$

2.2 (δ, σ)-SLE Basics

Here we repeat briefly some necessary material about the slit holomorphic stochastic flows SLE(δ, σ) considered first in the paper [14] that we advice to follow for more details.

A holomorphic vector field σ on $\mathcal{D}$ is called *complete* if the solution $H_s[\sigma]^\psi(z)$ of the initial valued problem

$$\dot{H}_s[\sigma]^\psi(z) = \sigma^\psi(H_s[\sigma]^\psi(z)), \quad H_0[\sigma]^\psi(z) = z, \quad z \in D^\psi \tag{2.3}$$

is defined for $s \in (-\infty, \infty)$. The solution $H_s[\sigma]^\psi : D^\psi \to D^\psi$ is a conformal automorphism of $D^\psi = \psi(\mathcal{D})$. Here and further on, we denote the partial derivative with respect to s as $\dot{H}_s := \frac{\partial}{\partial s} H_s$. It is straightforward to see that the differential equation is of the same form in any chart ψ. So it is reasonable to drop the index ψ and variable z as follows

$$\dot{H}_s[\sigma] = \sigma \circ H_s[\sigma], \quad H_0[\sigma] = \mathrm{id}. \tag{2.4}$$

Let $\psi_{\mathbb{H}}$ be a chart map onto the upper half-plane $D^\psi = \mathbb{H}$. We denote $X^{\mathbb{H}} := X^{\psi_{\mathbb{H}}}$ if X is a vector field, a conformal map, or a pre-pre-Schwarzian form (defined below) on $\mathcal{D}$.

A complete vector field σ in the half-plane chart admits the form

$$\sigma^{\mathbb{H}}(z) = \sigma_{-1} + \sigma_0 z + \sigma_1 z^2, \quad z \in \mathbb{H}, \quad \sigma_{-1}, \sigma_0, \sigma_1 \in \mathbb{R}. \tag{2.5}$$

A vector field δ is called *antisemicomplicate* if the initial valued problem

$$\dot{H}_s[\delta] = \delta \circ H_s[\delta], \quad H_0[\delta] = \mathrm{id}.$$

has a solution $H_t[\sigma]$, which is a conformal map $H_t[\sigma] : \mathcal{D} \backslash \mathcal{K}_t \to \mathcal{D}$ for all $t \in [0, +\infty)$ for some family of subsets $\mathcal{K}_t \subset \mathcal{D}$.

The linear space of all antisemicomplete fields is essentially bigger and infinite-dimensional, but we restrict ourselves to the from

$$\delta^{\mathbb{H}}(z) = \frac{\delta_{-2}}{z} + \delta_{-1} + \delta_0 z + \delta_1 z^2, \quad z \in \mathbb{H}, \quad \delta_{-1}, \delta_0, \delta_1 \in \mathbb{R}, \quad \delta_{-2} > 0, \tag{2.6}$$

which guarantees that the set $\mathcal{K}_t$ is curve-generated, see [14]. The first term gives just a simple pole in the boundary at the origin with a positive residue. The sum of the last three terms is just a complete field.

Let us consider first a continuously differentiable *driving function* function $u_t : t \mapsto \mathbb{R}$ for motivation. Then the solution G_t of the initial value problem

$$\dot{G}_t = \delta \circ G_t + \dot{u}_t \sigma \circ G_t, \quad G_0 = \mathrm{id}, \quad t \geq 0, \tag{2.7}$$

is a family of conformal maps $G_t : \mathcal{D} \backslash \mathcal{K}_t \to \mathcal{D}$, where the family of subsets $\{\mathcal{K}_t\}_{t \geq 0}$ depends on the driving function u. To avoid the requirement of continuous differentiability we use the following method.

Define a conformal map

$$g_t := H_{u_t}[\sigma]^{-1} \circ G_t, \quad t \geq 0.$$

It satisfies the equation

$$\dot{g}_t = (H_{u_t}[\sigma]_*^{-1} \delta) \circ g_t, \quad g_0 = \mathrm{id}, \tag{2.8}$$

where $H_{u_t}[\sigma]_*^{-1} \delta$ is defined in (2.2) for $G = H_{u_t}[\sigma]$. Reciprocally, (2.7) can be obtained from (2.8), although (2.8) is defined for a continuous function u_t, not necessary continuously differentiable. This motivates the following definition.

Definition 1 Let σ and δ be a complete and a semicomplete vector fields as in (2.5) and (2.6), and let u_t be a continuous function $u : [0, \infty) \to \mathbb{R}$. Then the solution g_t to the initial value problem (2.8) is called the *(forward) general slit Löwner chain.*

Respectively, the map G_t is defined by

$$G_t := H_{u_t}[\sigma] \circ g_t, \quad t \geq 0.$$

The stochastic version of Eq. (2.8) can be set up by introducing the Brownian measure on the set of driving functions, or equivalently, as follows.

Definition 2 Let σ and δ be as in (2.5) and (2.6), and let B_t be the standard Brownian motion (the Wiener process). Then the solution to the stochastic differential equation

$$d^S G_t = \delta \circ G_t dt + \sigma \circ G_t \, d^S B_t, \quad G_0 = \mathrm{id}, \quad t \geq 0 \qquad (2.9)$$

(where d^S is the Stratonovich differential) is called a *slit holomorphic stochastic flow* or (δ, σ)-SLE.

In order to formulate (2.9) in the Itô form we have to chose some chart ψ:

$$d^{\text{Itô}} G_t^\psi(z) = \left(\delta^\psi + \frac{1}{2} \sigma^\psi \left(\sigma^\psi \right)' \right) \circ G_t^\psi(z) dt + \sigma^\psi \circ G_t^\psi(z) \, d^{\text{Itô}} B_t, \quad G_0^\psi(z) = z,$$

$$(2.10)$$

see, for example, [9, Section 4.3] for the definition of the Stratonovich and Itô's differentials and for the relations between them. A disadvantage of Itô's form is that the coefficient at dt transforms from chart to chart in a complicated way.

Example 1 (Chordal Löwner Equation)

Let us show how the construction (2.7)–(2.10) works in the case of the chordal Löwner equation. Define

$$\delta^{\mathbb{H}}(z) = \frac{2}{z}, \quad \sigma^{\mathbb{H}}(z) = -1,$$

in the half-plane chart. Then $H[\sigma]_s$ can be found from Eq. (2.4) as

$$\dot{H}_s[\sigma]^{\mathbb{H}}(z) = -1 , \quad H_0[\sigma]^{\mathbb{H}}(z) = z, \quad z \in \mathbb{H},$$

$$H_s[\sigma]^{\mathbb{H}}(z) = z - s \quad z \in \mathbb{H}.$$

The equation for g_t becomes

$$\dot{g}_t^{\mathbb{H}} = \frac{2}{g_t^{\mathbb{H}} - u_t},$$

and it is known as the chordal Löwner equation.

For a differentiable u_t we can write

$$\dot{G}_t^{\mathbb{H}}(z) = \frac{2}{G_t^{\mathbb{H}}(z)} - \dot{u}_t$$

The stochastic version in the Stratonovich form can be obtained by substituting $u_t = \sqrt{\kappa} B_t$:

$$d^{\mathrm{S}} G_t^{\mathbb{H}}(z) = \frac{2}{G_t^{\mathbb{H}}(z)} dt - \sqrt{\kappa}\, d^{\mathrm{S}} B_t, \qquad (2.11)$$

and it is of the same form in the Itô case in the half-plane chart because $\sigma^{\mathbb{H}\,\prime}(z) \equiv 0$,

$$d^{\mathrm{It\hat{o}}} G_t^{\mathbb{H}}(z) = \frac{2}{G_t^{\mathbb{H}}(z)} dt - \sqrt{\kappa}\, d^{\mathrm{It\hat{o}}} B_t.$$

In other charts $(\sigma^{\chi})'(z) \neq 0$, and the Stratonovich and Itô forms differ. The space $\mathcal{G}[\delta, \sigma]$ consists of endomorphisms that in the half-plane chart are normalized as

$$G_t^{\mathbb{H}}(z) = z + O\left(\frac{1}{z}\right).$$

Example 2 (Radial Leowner Equation)
    Let $\psi_{\mathbb{D}} : \mathcal{D} \to \mathbb{D}$ be the chart map defined by (1.8), and let $\tau_{\mathbb{H},\mathbb{D}} = \psi_{\mathbb{H}} \circ \psi_{\mathbb{D}}^{-1}$. The transition function $\tau_{\mathbb{H},\mathbb{D}}$ maps the point 1 in the unit disk chart to the point 0 in the half-plane chart, the point -1 to a point at infinity, and the point 0 to $+i$. Similarly to what we have done in the half-plane case, we define $X^{\mathbb{D}} := X^{\psi_{\mathbb{D}}}$ and call it *the unit disk chart*.
    Let

$$\delta^{\mathbb{D}}(z) = -z\frac{z+1}{z-1}, \qquad \sigma^{\mathbb{D}}(z) = -iz, \qquad (2.12)$$

that corresponds to

$$\delta^{\mathbb{H}}(z) = \frac{1}{2}\left(\frac{1}{z} + z\right), \qquad \sigma^{\mathbb{H}}(z) = -\frac{1}{2}(1 + z^2) \qquad (2.13)$$

in the half-plane chart.
    The equation for $H_s[\sigma]$ in the unit disk chart becomes

$$\dot{H}_s[\sigma]^{\mathbb{D}}(z) = -iH_s[\sigma]^{\mathbb{D}}(z), \qquad H_0[\sigma]^{\mathbb{D}}(z) = z, \qquad z \in \mathbb{D}.$$

The solution takes the form

$$H_s[\sigma]^{\mathbb{D}}(z) = e^{-is}z,$$

and

$$\dot{G}_t^{\mathbb{D}}(z) = -G_t^{\mathbb{D}}(z)\frac{G_t^{\mathbb{D}}(z)+1}{G_t^{\mathbb{D}}(z)-1} - iG_t^{\mathbb{D}}(z)\dot{u}_t.$$

Thus, the equation for g_t becomes

$$\dot{g}_t^{\mathbb{D}}(z) = \left(\frac{1}{H_{u_t}[\sigma]^{\mathbb{D}\prime}}\left(-H_{u_t}[\sigma]^{\mathbb{D}}\frac{H_{u_t}[\sigma]^{\mathbb{D}}+1}{H_{u_t}[\sigma]^{\mathbb{D}}-1}\right)\right) \circ g_t^{\mathbb{D}}(z) =$$

$$= \frac{1}{e^{-iu_t}}\left(-e^{-iu_t}g_t^{\mathbb{D}}(z)\frac{e^{-iu_t}g_t^{\mathbb{D}}(z)+1}{e^{-iu_t}g_t^{\mathbb{D}}(z)-1}\right)$$

or

$$\dot{g}_t^{\mathbb{D}}(z) = g_t^{\mathbb{D}}(z)\frac{e^{iu_t}+g_t^{\mathbb{D}}(z)}{e^{iu_t}-g_t^{\mathbb{D}}(z)}.$$

Its stochastic version in the Stratonovich form is

$$d^S G_t^{\mathbb{D}}(z) = G_t^{\mathbb{D}}(z)\frac{G_t^{\mathbb{D}}(z)+1}{G_t^{\mathbb{D}}(z)-1}dt - i\sqrt{\kappa}G_t^{\mathbb{D}}(z)\,d^S B_t. \qquad (2.14)$$

or in the half-plane chart

$$d^S G_t^{\mathbb{H}}(z) = \frac{1}{2}\left(\frac{1}{G_t^{\mathbb{H}}(z)}+G_t^{\mathbb{H}}(z)\right)dt - \frac{\sqrt{\kappa}}{2}\left(1+(G_t^{\mathbb{H}}(z))^2\right)d^S B_t.$$

The collection $\mathcal{G}[\delta,\sigma]$ consists of maps that in the unit disk chart are normalized by

$$G^{\mathbb{D}}(0) = 0.$$

It turns out that some different combinations of δ and σ induce measures that can be transformed one to another in a simple way. For example, let $m : \mathcal{D} \to \mathcal{D}$ be a Möbius automorphism fixing a point $a \in \partial\mathcal{D}$ where δ has a pole. Then

$$\delta \to m_*\delta, \qquad \sigma \to m_*\sigma \qquad (2.15)$$

are also vector fields of the form (2.5) and (2.6). For instance, let δ and σ be as in Example 2, and let $m^{\mathbb{D}} : \mathbb{D} \to \mathbb{D}$ map the center point 0 of the unit disk to another point inside it. Then we come to an equation defined by $m_*\delta$ and $m_*\sigma$, which is still in fact, the radial equation written in different coordinates.

Another example of such a transformation preserving the form (2.5) and (2.6) can be constructed as follows.

$$\delta^{\mathbb{H}}(z) \to c^2\delta^{\mathbb{H}}(z), \quad \sigma^{\mathbb{H}}(z) \to c\sigma^{\mathbb{H}}(z), \quad t \to c^{-2}t, \quad c > 0. \tag{2.16}$$

The solution is not changed as a random law, because cB_{t/c^2} and B_t agree in law.

It is important to know which of the equations defined by the parameters δ_{-2}, $\delta_{-1}, \delta_0, \delta_1, \sigma_{-1}, \sigma_0$, and σ_0 are 'essentially different'. A systematic analysis of this question was presented in [14]. Without lost of generality we can restrict ourselves to the from

$$\delta^{\mathbb{H}}(z) = \frac{2}{z} + \delta_{-1} + \delta_0 z + \delta_1 z^2, \quad \sigma^{\mathbb{H}}(z) = -\sqrt{\kappa}(1 + \sigma_0 z + \sigma_1 z^2), \quad \kappa > 0. \tag{2.17}$$

Transformations (2.16) and

$$\delta \to r_{c*}\delta, \quad \sigma \to r_{c*}\sigma, \quad r_c^{\mathbb{H}}(z) := \frac{z}{1 - cz}, \quad c \in \mathbb{R} \tag{2.18}$$

preserve (2.17) and keep κ unchanged. Thus, we can fix some 2 of 6 parameters in (2.17). Besides, the transform

$$\delta \to \delta + \frac{\nu}{\sqrt{\kappa}}\sigma, \quad \nu \in \mathbb{R},$$

can be interpreted as an insertion of a drift to the Brownian measure. Thus, all 6 parameters can be fixed by κ (responsible for the fractal dimension of the slit), ν (the drift), and some 2 parameters that set the equation type (for example, chordal, radial, and dipolar).

Due to the autonomous form of Eq. (2.7) the solution G_t possesses the following property.

Proposition 1 ([14]) *Let $\tilde{u}_{t-s} := u_t - u_s$ for some fixed $s : t \geq s \geq 0$, and let $\tilde{G}_{\tilde{t}}$ be defined by (2.7) with the driving function $\tilde{u}_{\tilde{t}}$. Then*

$$G_t = \tilde{G}_{t-s} \circ G_s.$$

In the stochastic case the process $\{G_t\}_{t \geq 0}$, taking values in the space of inverse endomorphisms of $\mathcal{D}$, is a continuous homogeneous Markov process. In particular,

Proposition 2 ([14]) *If G_t and $\tilde{G}_s$ are two independently sampled (δ, σ)-SLE maps, then $G_t \circ \tilde{G}_s$ has the same law as G_{t+s}.*

2.3 Pre-pre-Schwarzian

A collection of maps $\eta^{\psi} : \psi(\mathcal{D}) \to \mathbb{C}$, each of which is given in a global chart map $\psi : \mathcal{D} \to \psi(\mathcal{D}) \subset \mathbb{C}$, is called a *pre-pre-Schwarzian* form of order $\mu, \mu^* \in \mathbb{C}$ if for any chart map $\tilde{\psi}$

$$\eta^{\tilde{\psi}}(\tilde{z}) = \eta^{\psi}(\tau(\tilde{z})) + \mu \log \tau'(\tilde{z}) + \mu^* \log \overline{\tau}'(\tilde{z}), \quad \tau = \psi \circ \tilde{\psi}^{-1}, \quad \tilde{z} \in \tilde{D}, \quad \forall \psi, \tilde{\psi}, \tag{2.19}$$

for any chart map $\tilde{\psi}$. If η is defined for one chart map, then it is automatically defined for all chart maps. We borrowed the term 'pre-pre-Schwarzian' from [19]. In [30], an analogous object is called 'AC surface'.

Analogously to vector fields in Sect. 2.1 we define the *pushforward of a pre-pre-Schwarzian* by

$$G_*^{-1} \eta^{\psi}(z) := \eta^{\psi \circ G}(z) = \eta^{\psi}(G^{\psi}(z)) + \mu \log (G^{\psi})'(z) + \mu^* \log \overline{(G^{\psi})'(z)} \tag{2.20}$$

We are interested in two special cases. The first one corresponds to $\mu = \mu^* = \gamma/2 \in \mathbb{R}$, and

$$G_*^{-1} \eta^{\psi}(z) = \eta^{\psi}(G^{\psi}(z)) + \gamma \log \left| G^{\psi \prime}(z) \right|. \tag{2.21}$$

The second one is $\mu = i\chi/2$, $\mu^* = -i\chi/2$, $\chi \in \mathbb{R}$, and

$$G_*^{-1} \eta^{\psi}(z) = \eta^{\psi}(G^{\psi}(z)) - \chi \arg G^{\psi \prime}(z). \tag{2.22}$$

In both cases η can be chosen real in all charts. Moreover, if the pre-pre-Schwarzian is represented by a real-valued function it is one of two above forms in all charts.

A (μ, μ^*)-pre-pre-Schwarzian can be obtained from a vector field v by the relation

$$\eta^{\psi}(z) \equiv -\mu \log v^{\psi}(z) - \mu^* \log \overline{v^{\psi}(z)}. \tag{2.23}$$

For two special cases above we have

$$\eta = -\gamma \log |v|$$

and

$$\eta = \chi \arg v,$$

where we drop the upper index ψ.

In Sect. 2.5, we obtain the transformation rules (2.21) by taking logarithm of a $(1, 1)$-differential. The second type of the real pre-pre-Schwarzian is connected to a sort of an imaginary analog of the metric.

We can define the *Lie derivative* of X as

$$\mathcal{L}_v X^\psi(z) := \left. \frac{\partial}{\partial \alpha} H_t^{-1}[v]_* X^\psi(z) \right|_{t=0},$$

where X can be a pre-pre-Schwarzian, a vector field, or even an object with a more general transformation rule, see (2.3) for any two holomorphic vector fields v. If X is a holomorphic vector field w, then

$$\mathcal{L}_v w^\psi = [v, w]^\psi(z) := v^\psi(z) w^{\psi\prime}(z) - v^{\psi\prime}(z) w^\psi(z), \tag{2.24}$$

see (2.2).

If X is a pre-pre-Schwarzian, then

$$\mathcal{L}_v \eta^\psi(z) = v^\psi(z) \, \partial_z \, \eta^\psi(z) + \overline{v^\psi(z)} \, \partial_{\bar{z}} \, \eta^\psi(z) + \mu \, v^{\psi\prime}(z) + \mu^* \overline{v^{\psi\prime}(z)}. \tag{2.25}$$

Here and further on, we use notations

$$\partial_z := \frac{1}{2} \left(\frac{\partial}{\partial x} - i \frac{\partial}{\partial y} \right), \quad \partial_{\bar{z}} := \frac{1}{2} \left(\frac{\partial}{\partial x} + i \frac{\partial}{\partial y} \right),$$

and $f'(z) := \partial_z f(z)$ for a holomorphic function f.

If $\mu = \mu^* = 0$, then η is called a *scalar*. It is remarkable that if η is a pre-pre-Schwarzian, then $\mathcal{L}_v \eta$ is a scalar anyway, which is stated in the following lemma.

Lemma 1 *Let η be a pre-pre-Schwarzian of order μ, μ^*, let v be a holomorphic vector field, and let G be a conformal self-map. Then*

$$G_*^{-1}(\mathcal{L}_v \eta)^\psi(z) = (\mathcal{L}_v \eta)^\psi \circ G^\psi(z)$$

or in the infinitesimal form,

$$\mathcal{L}_w(\mathcal{L}_v \eta)^\psi(z) = w^\psi(z) \, \partial_z (\mathcal{L}_v \eta)^\psi(z) + \overline{w^\psi(z)} \, \partial_{\bar{z}} (\mathcal{L}_v \eta)^\psi(z). \tag{2.26}$$

Proof The straightforward calculations imply

$$G_*^{-1}(\mathcal{L}_v\,\eta)^\psi(z) = G_*^{-1}\left(v^\psi(z)\,\partial_z\,\eta^\psi(z) + \overline{v^\psi(z)}\,\partial_{\bar z}\,\eta^\psi(z) + \mu v^{\psi\prime}(z) + \mu^*\overline{v^{\psi\prime}(z)}\right) =$$

$$= \frac{v^\psi\circ G^\psi(z)}{G^{\psi\prime}(z)}\,\partial_z\left(\eta^\psi\circ G^\psi(z) + \mu\log G^{\psi\prime}(z) + \mu^*\log\overline{G^{\psi\prime}(z)}\right) +$$

$$+ \frac{\overline{v^\psi\circ G^\psi(z)}}{\overline{G^{\psi\prime}(z)}}\,\partial_{\bar z}\left(\eta^\psi\circ G^\psi(z) + \mu\log G^{\psi\prime}(z) + \mu^*\log\overline{G^{\psi\prime}(z)}\right) +$$

$$+ \mu\,\partial_z\,\frac{v^\psi\circ G^\psi(z)}{G^{\psi\prime}(z)} + \mu^*\,\partial_{\bar z}\,\frac{\overline{v^\psi\circ G^\psi(z)}}{\overline{G^{\psi\prime}(z)}} =$$

$$= v^\psi\circ G^\psi(z)(\partial\,\eta^\psi)\circ G^\psi(z) + \mu\frac{v^\psi\circ G^\psi(z)}{G^{\psi\prime}(z)}\,\partial_z\log G^{\psi\prime}(z) + 0 +$$

$$+ \overline{v^\psi\circ G^\psi(z)}(\bar\partial\eta^\psi)\circ G^\psi(z) + 0 + \mu^*\frac{\overline{v^\psi\circ G^\psi(z)}}{\overline{G^{\psi\prime}(z)}}\,\partial_{\bar z}\log\overline{G^{\psi\prime}(z)} +$$

$$+ \mu v^{\psi\prime}\circ G^\psi(z) - \mu\frac{v^\psi\circ G^\psi(z)G^{\psi\prime\prime}(z)}{G^{\psi\prime}(z)^2} + \mu^*\overline{v^{\psi\prime}\circ G^\psi(z)} - \mu^*\frac{\overline{v^\psi\circ G^\psi(z)G^{\psi\prime\prime}(z)}}{\overline{G^{\psi\prime}(z)^2}} =$$

$$= \left(v^\psi\,\partial\,\eta^\psi\right)\circ G^\psi(z) + \mu v^{\psi\prime}\circ G^\psi(z) +$$

$$+ \left(\overline{v^\psi}\bar\partial\eta^\psi\right)\circ G^\psi(z) + \mu^*\overline{v^{\psi\prime}}\circ G^\psi(z) =$$

$$= \left(v^\psi\,\partial\,\eta^\psi + \overline{v^\psi}\bar\partial\eta^\psi + \mu v^{\psi\prime} + \mu^*\overline{v^{\psi\prime}}\right)\circ G^\psi(z) =$$

$$= (\mathcal{L}_v\,\eta)^\psi\circ G^\psi(z)$$

$\square$

2.4 Test Functions

We will define the Schwinger functions S_n and the Gaussian free field Φ in terms of linear functionals over some space of smooth test functions defined in what follows.

Let $\mathcal{H}_s^\psi$ be a linear space of real-valued smooth functions $f\colon D^\psi\to\mathbb{R}$ in the domain $D^\psi := \psi(\mathcal{D})\subset\mathbb{C}$ with compact support equipped with the topology of homogeneous convergence of all derivatives on the corresponding compact, namely, the topology is generated by following collection of neighborhoods of the zero function

$$U_K^\psi := \bigcap_{n,m=0,1,2,\dots}\{f(z)\in C^\infty(D)\colon \operatorname{supp}f\subseteq K \wedge \left|\partial^n\,\bar\partial^m f^\psi(z)\right| < \varepsilon_{n,m},\quad z\in K\},$$

$$\varepsilon_{n,m} > 0,\quad n,m = 0,1,2\dots,$$

where $K\subset D^\psi$ is any compact subset of D^ψ.

We call $f^\psi \in \mathcal{H}_s^\psi$ the *test functions* and assume that they are $(1, 1)$-differentials

$$f^{\tilde\psi}(\tilde z) = \tau'(\tilde z)\overline{\tau'(\tilde z)}f^\psi(\tau(\tilde z)), \quad \tau := \psi \circ \tilde\psi^{-1}, \tag{2.27}$$

It is straightforward to check that any transition map τ induces a homeomorphism between $\mathcal{H}_s^\psi$ and $\mathcal{H}_s^{\tilde\psi}$. Thereby, we will drop the index ψ at $\mathcal{H}_s$, and consider the space $\mathcal{H}_s$ as a topological space of smooth $(1, 1)$-differentials with compact support.

The space $\mathcal{H}_s$ does not match all cases of coupling. For the couplings with radial SLE we use spaces $\mathcal{H}_{s,b}$ and $\mathcal{H}_{s,b}^\pm$ defined in corresponding Sects. 4.3 and 6. Henceforth, we denote by $\mathcal{H}$ any of those nuclear spaces $\mathcal{H}_s$, $\mathcal{H}_{s,b}$, or $\mathcal{H}_{s,b}^\pm$ for shortness. An important property of $\mathcal{H}$ is *nuclearity*, see [10, 12, 25] which is necessary and sufficient to admit the uniform Gaussian measure on the dual space $\mathcal{H}'$ (the GFF).

Constructing such a uniform Gaussian measure on a finite dimensional linear space is a trivial problem, however, it is not possible on an infinite-dimensional Hilbert space. On the other hand, if a space $\mathcal{H}$ is nuclear as $\mathcal{H}_s$, then the dual space $\mathcal{H}'$ admits a uniform Gaussian measure. A general recipy holds not only for Gaussian measures and is given by the following theorem.

Proposition 3 (Bochner-Minols [1, 10, 12]) *Let $\mathcal{H}$ be a nuclear space, and let $\hat\mu\colon \mathcal{H} \to \mathbb{C}$ be a functional (non-linear). Then the following 3 conditions*

1. $\hat\mu$ is positive definite

$$\forall\{z_1, z_2, \ldots z_n\} \in \mathbb{C}^n, \ \forall\{f_1, f_2, \ldots f_n\} \in \mathcal{H}^n \quad \Rightarrow \quad \sum_{1 \le k,l \le n} z_k \bar z_l \hat\mu[f_k - f_l] \ge 0;$$

2. $\hat\mu(0) = 1$;
3. $\hat\mu$ is continuous

are satisfied if and only if there exists a unique probability measure P_Φ on $(\Omega_\Phi, \mathcal{F}_\Phi, P_\Phi)$ for $\Omega_\Phi = \mathcal{H}'$, with $\hat\mu$ as a characteristic function

$$\hat\mu[f] := \int\limits_{\Phi \in \mathcal{H}'} e^{(i\Phi[f])} P_\Phi(d\Phi), \quad \forall f \in \mathcal{H}. \tag{2.28}$$

The corresponding σ-algebra $\mathcal{F}_\Phi$ is generated by the cylinder sets

$$\{F \in \mathcal{H}'\colon \quad F[f] \in B\}, \quad \forall f \in \mathcal{H}, \quad \forall \text{ Borel sets } B \text{ of } \mathbb{R}.$$

The random law on $\mathcal{H}'$ is called uniform with respect to a bilinear functional $B\colon \mathcal{H} \times \mathcal{H} \to \mathbb{R}$ if the characteristic function $\hat\mu$ is of the form

$$\hat\mu[f] = e^{-\frac{1}{2}B[f,f]}, \quad f \in \mathcal{H}.$$

We consider the class of bilinear functionals we work with in Sect. 2.6. First, we study the linear and bilinear functionals over $\mathcal{H}_s$ and their transformation properties.

2.5   Linear Functionals and Change of Coordinates

In this section, we consider linear functionals over $\mathcal{H}_s$ and $\mathcal{H}$ that transform as pre-pre-Schwarzians.

Let $\eta^\psi \in \mathcal{H}_s^{\psi}{}'$ be a linear functional over $\mathcal{H}_s^\psi$ for a given chart ψ. The functional is called regular if there exists a locally integrable function $\eta^\psi(z)$ such that

$$\eta^\psi[f] := \int_{D^\psi} \eta^\psi(z) f^\psi(z) l(dz),$$

where l is the Lebesgue measure on $\mathbb{C}$. We use the brackets $[\cdot]$ for functionals and the parentheses $(\cdot)$ for corresponding functions (kernels).

We assume that f transforms according to (2.27). If $\eta^\psi(z)$ is a scalar, then the number $\eta^\psi[f] \in \mathbb{R}$ does not depend on the choice of the chart ψ. Indeed, for any choice of another chart $\tilde{\psi}$, we have

$$\eta^{\tilde{\psi}}[f] := \int_{D^{\tilde{\psi}}} \eta^{\tilde{\psi}}(\tilde{z}) f^{\tilde{\psi}}(\tilde{z}) l(d\tilde{z}) = \int_{D^{\tilde{\psi}}} \eta^\psi(\tau(\tilde{z})) f^\psi(\tau(\tilde{z})) |\tau'(\tilde{z})|^2 l(d\tilde{z}) =$$

$$= \int_{D^\psi} \eta^\psi(z) f^\psi(z) l(dz) = \eta^\psi[f],$$

If $\eta^\psi(z)$ is a pre-pre-Schwarzian, then

$$\eta^{\tilde{\psi}}[f] = \int_{D^{\tilde{\psi}}} \eta^{\tilde{\psi}}(\tilde{z}) f^{\tilde{\psi}}(\tilde{z}) l(d\tilde{z}) =$$

$$= \int_{D^{\tilde{\psi}}} \left(\eta^\psi(\tau(\tilde{z})) + \mu \log \tau'(z) + \mu^* \overline{\log \tau'(z)} \right) f^\psi(\tau(\tilde{z})) |\tau'(\tilde{z})|^2 l(d\tilde{z}) =$$

$$= \int_{D^\psi} \left(\eta^\psi(z) - \mu \log \tau^{-1\prime}(z) - \mu^* \overline{\log \tau^{-1\prime}(z)} \right) f^\psi(z) l(dz) =$$

$$= \eta^\psi[f] - \int_{D^\psi} \left(\mu \log \tau^{-1\prime}(z) + \mu^* \overline{\log \tau^{-1\prime}(z)} \right) f^\psi(z) l(dz) \qquad (2.29)$$

according to (2.20).

If η^ψ is not a regular pre-pre-Schwarzian but just a functional from $\mathcal{H}'_s$ we can consider the last line of (2.29) as a definition of the transformation rule for $\eta[f]$ from a chart ψ to a chart $\tilde\psi$.

Let us denote by $\mathcal{H}'_s$ the linear space of pre-pre-Schwarzians as above. Consider now the pushforward operation G_*^{-1} on $(1,1)$-differentials f defined by

$$G_*^{-1}f^\psi(z) := \left|G^{\psi\prime}(z)\right|^2 f^\psi\left(G^\psi(z)\right).$$

The right-hand side is well-defined only for $\psi(\mathcal{D}\setminus\mathcal{K})$. Here we define F_* only on a subset of $\mathcal{H}_s$ of test functions that are supported in $\mathcal{D}\setminus\mathcal{K} = F^{-1}(\mathcal{D})$.

Define the pushforward operation by

$$G_*^{-1}\eta^\psi[f] = \eta^{\psi\circ G}[f] =$$

$$= \eta^\psi[G_*f] + \int\limits_{\mathrm{supp}f^\psi}\left(\mu\log G^{\psi\prime}(z) + \mu^*\overline{\log G^{\psi\prime}(z)}\right)f^\psi(z)l(dz), \tag{2.30}$$

$$f\in\mathcal{H}_s:\quad \mathrm{supp}f\subset G^{-1}(\mathcal{D}).$$

It can be understood as a pushforward $F_* : \mathcal{H}'_s \to \mathcal{H}'_s$ in the dual space.

Functionals over the space $\mathcal{H}_s$ are differentiable infinitely many times. According to (2.25) the Lie derivative is defined by

$$\mathcal{L}_v\,\eta[f] = \frac{\partial}{\partial s}H_s^{-1}[v]_*\eta^\psi[f]\bigg|_{s=0} =$$

$$= -\eta^\psi[\mathcal{L}_v f] + \int\limits_{\mathrm{supp}f^\psi}\left(\mu v^{\psi\prime}(z) + \mu^*\overline{v^{\psi\prime}(z)}\right)f^\psi(z)l(dz),$$

where

$$\mathcal{L}_v f^\psi(z) = \frac{\partial}{\partial s}H_s^{-1}[v]_*f^\psi\bigg|_{s=0} =$$

$$= v^\psi(z)\,\partial_z f^\psi(z) + \overline{v^\psi(z)}\,\partial_{\bar z}f^\psi(z) + v^{\psi\prime}(z)f^\psi(z) + \overline{v^{\psi\prime}(z)}f^\psi(z).$$

2.6 Fundamental Solution to the Laplace-Beltrami Equation

In this section, we consider linear continuous functionals with respect to each argument in $\mathcal{H}_s$. An important example is the Dirac functional

$$\delta_\lambda[f,g] := \int\limits_{\psi(\mathcal{D})}f^\psi(z)g^\psi(z)\frac{1}{\lambda^\psi(z)}l(dz), \quad f,g\in\mathcal{H}_s, \tag{2.31}$$

where $\lambda(z)l(dz)$ is some measure on D^ψ, which is absolutely continuous with respect to the Lebesgue measure $l(dz)$. The Radon-Nikodym derivative $\lambda^\psi(z)$ transforms as a $(1,1)$-differential:

$$\lambda^{\tilde\psi}(\tilde z) = \tau'(\tilde z)\overline{\tau'(\tilde z)}\lambda^\psi(\tau(\tilde z)), \quad \tau := \psi \circ \tilde\psi^{-1}.$$

It is easy to see that the right-hand side of (2.31) does not depend on the choice of ψ.

We call the functional regular if there exists a function $B^\psi(z,w)$ on $\psi^D \times \psi^D$ such that

$$B^\psi[f,g] := \int\limits_{\psi(D)} \int\limits_{\psi(D)} B^\psi(z,w)f^\psi(z)g^\psi(w)l(dz)l(dw), \quad f,g \in \mathcal{H}_s. \qquad (2.32)$$

Let us use the same convention about the brackets and parentheses as for the linear functionals. We consider only scalar regular bilinear functionals and require the transformation rules

$$B^{\tilde\psi}(\tilde z, \tilde w) = B^\psi(\tau(\tilde z), \tau(\tilde w)), \quad \tau = \psi \circ \tilde\psi^{-1}, \quad z, w \in \tilde\psi(D).$$

Thus, the right-hand side of (2.32) does not depend on the choice of the chart ψ and we can drop the index ψ in the left-hand side.

The pushforward is defined by

$$F_*B^\psi(z,w) = B^{\psi \circ F}(z,w) := B^\psi\left(\left(F^\psi\right)^{-1}(z), \left(F^\psi\right)^{-1}(w)\right), \quad z, w \in \operatorname{Im}\left(F^\psi\right),$$
$$(2.33)$$

which becomes

$$F_*B^\psi[f,g] = B^{\psi \circ F}[f,g] := B^\psi[F_*^{-1}f, F_*^{-1}g], \quad f,g \in \mathcal{H}_s : \operatorname{supp} f \subset \operatorname{Im}(F),$$

for an arbitrary functional F, The same remarks remain true in this case as in the previous section for η.

Define now the Lie derivative in the same way as before

$$\mathcal{L}_v B^\psi(z,w) := \left.\frac{\partial}{\partial s} H_s[v]_*^{-1} B^\psi(z,w)\right|_{s=0} =$$
$$= v^\psi(z)\partial_z B^\psi(z,w) + \overline{v^\psi(z)}\partial_{\bar z} B^\psi(z,w) + v^\psi(w)\partial_w B^\psi(z,w) + \overline{v^\psi(w)}\partial_{\bar w} B^\psi(z,w). \qquad (2.34)$$

We remark that $\mathcal{L}_v B$ is also scalar in two variables. Functionals δ_λ and B are both scalar and continuous with respect to each variable.

Define the Laplace-Beltrami operator Δ_λ as

$$\Delta_{\lambda 1} B^\psi(z, w) := -\frac{4}{\lambda^\psi(z)}\, \partial_z\, \partial_{\bar{z}}\, B^\psi(z, w),$$

where the lower index '1' means that the operator acts only with respect to the first argument.

Let a regular bilinear functional Γ_λ be a solution to the equation

$$\Delta_{\lambda 1}\Gamma_\lambda[f, g] = 2\pi\, \delta_\lambda[f, g], \quad \Gamma_\lambda[f, g] = \Gamma_\lambda[g, f], \quad f, g \in \mathcal{H}_s. \tag{2.35}$$

The boundary conditions will be fixed later. This equation is conformally invariant in the sense that if $\Gamma_\lambda^\psi(z, w)$ is a solution on a chart ψ, then

$$\Gamma_\lambda^\psi(\tau(\tilde{z}), \tau(\tilde{w})) = \Gamma_\lambda^{\tau^{-1}\circ\psi}(z, w)$$

is a solution in the chart $\tau^{-1}\circ\psi$.

The solution $\Gamma_\lambda^\psi(z, w)$ is a collection of smooth and harmonic functions on $\psi(\mathcal{D}) \times \psi(\mathcal{D}) \setminus \{z \times w : z = w\}$ of general form

$$\Gamma_\lambda^\psi(z, w) = -\frac{1}{2}\log(z - w)(\bar{z} - \bar{w}) + H^\psi(z, w), \tag{2.36}$$

where $H^\psi(z, w)$ is an arbitrary symmetric harmonic function with respect to each variable that is defined by the boundary conditions and will be specified in what follows.

It is straightforward to verify that the function $\Gamma_\lambda^\psi(z, w)$ does not depend on the choice of λ because the identity (2.35) in the integral form becomes

$$\int\limits_{\psi(\mathcal{D})} \int\limits_{\psi(\mathcal{D})} -\frac{4}{\lambda^\psi(z)}\, \partial_z\, \partial_{\bar{z}}\, \Gamma^\psi(z, w) f^\psi(z) g^\psi(w) l(dz) l(dw) =$$

$$= \int\limits_{\psi(\mathcal{D})} f^\psi(z) g^\psi(z) \frac{1}{\lambda^\psi(z)} l(dz).$$

The change $\lambda \to \tilde{\lambda}$ is equivalent to the change $f^\psi(z) \to \frac{\lambda^\psi(z)}{\tilde{\lambda}^\psi(z)} f^\psi(z)$. We will drop the lower index λ in Γ_λ in what follows. The fundamental solutions to the Laplace equation are also known as *Green's functions* (for the free field).

Example 3 (Dirichlet Boundary Conditions) Let us denote by Γ_D the solution Γ to (2.35) satisfying the zero boundary conditions, namely,

$$\Gamma_D^\mathbb{H}(z, w)\big|_{z\in\mathbb{R}} = 0, \quad \lim_{z\to\infty}\Gamma_D^\mathbb{H}(z, w) = 0, \quad w \in \mathbb{H}.$$

Then, Γ_D admits the form

$$\Gamma_D^{\mathbb{H}}(z, w) := -\frac{1}{2} \log \frac{(z - w)(\bar{z} - \bar{w})}{(z - \bar{w})(\bar{z} - w)}, \tag{2.37}$$

and possesses the property of symmetry with respect to all Möbious automorphisms $H : \mathcal{D} \to \mathcal{D}$,

$$H_* \Gamma_D = \Gamma_D$$

or

$$\mathcal{L}_\sigma \Gamma_D(z, w) = 0 , \quad \forall \text{ complete vector field } \sigma . \tag{2.38}$$

Example 4 (Combined Dirichlet-Neumann Boundary Conditions) Let Γ_{DN} denote the solution to (2.35) satisfying the following boundary conditions in the strip chart

$$\Gamma_{DN}^S(z, w)\big|_{z \in \mathbb{R}} = 0, \quad \partial_y \Gamma_{DN}^S(x + iy, w)\big|_{y = \pi} = 0, \quad x \in \mathbb{R},$$

$$\lim_{z \to \infty \wedge \mathrm{Re}\, z > 0} \Gamma_{DN}^S(z, w) = 0, \quad \lim_{z \to \infty \wedge \mathrm{Re}\, z < 0} \Gamma_{DN}^S(z, w) = 0, \quad w \in \mathbb{H} .$$

We consider this case in Sect. 5 and the exact form of Γ_{DN} is given by (5.1). It is not invariant with respect to all Möbious automorphisms but it is invariant if the automorphism preserves the points of change of the boundary conditions, which are $\pm\infty$ in the strip chart.

We will consider another example ($\Gamma_{tw,b}$) in Sect. 6.

2.7 Gaussian Free Field

Definition 3 For some nuclear space of smooth functions $\mathcal{H}$, let the linear functional η and some Green's functional Γ be given. Assume in Theorem 3

$$\hat{\mu}[f] := \exp\left(-\frac{1}{2}\Gamma[f,f] + i\eta[f]\right), \quad f \in \mathcal{H} . \tag{2.39}$$

Then the $\mathcal{H}'$-valued random variable Φ is called the *Gaussian free field (GFF)*. We will denote it by $\Phi(\mathcal{H}, \Gamma, \eta)$.

For convenience, we change the definition of the characteristic function from (2.28) to

$$\hat{\phi}[f] := \int_{\Phi \in \mathcal{H}'_s} e^{\Phi[f]} P^\Phi(d\Phi), \quad \forall f \in \mathcal{H},$$

and (2.39) changes to

$$\hat{\phi}[f] = e^{\left(\frac{1}{2}\Gamma[f,f]+\eta[f]\right)}, \tag{2.40}$$

which is possible for the Gaussian measures.

The *expectation* of a random variable $X[\Phi]$ $(X : \mathcal{H}' \to \mathbb{C})$ is defined as

$$\mathbb{E}[X] := \int_{\Phi \in \mathcal{H}'_s} X[\Phi] P^{\Phi}(d\Phi).$$

An alternative and equivalent (see, for example [12]) definition of GFF can be formulated as follows:

Definition 4 The *Gaussian free field* Φ is a $\mathcal{H}'$-valued random variable, that is a map $\Phi : \mathcal{H} \times \Omega \to \mathbb{R}$ (measurable on Ω and continuous linear on the nuclear space $\mathcal{H}$), or a measurable map $\Phi : \Omega \to \mathcal{H}'$, such that $\text{Law}[\Phi[f]] = N\left(\eta[f], \Gamma[f,f]^{\frac{1}{2}}\right)$, $f \in \mathcal{H}$, i.e., it possesses the properties

$$\mathbb{E}[\Phi[f]] = \eta[f], \quad \forall f \in \mathcal{H},$$

$$\mathbb{E}[\Phi[f]\Phi[f]] = \Gamma[f,f] + \eta[f]\eta[f], \quad \forall f \in \mathcal{H}$$

for Green's bilinear positively defined functional Γ, and for a linear functional η.

The random variable Φ introduced this way transforms from one chart to another according to the pre-pre-Schwarzian rule

$$\Phi^{\tilde{\psi}}[f] = \Phi^{\psi}[f] - \int_{\psi(\mathcal{D})} \left(\mu \log \tau^{-1\prime}(z) + \mu^* \overline{\log \tau^{-1\prime}(z)}\right) f^{\psi}(z) l(dz), \quad \tau := \psi \circ \tilde{\psi}^{-1}, \tag{2.41}$$

due to the corresponding property (2.29) of η.

The pushforward can also be defined by

$$G_*^{-1} \Phi^{\psi}[f] = \Phi^{\psi}[G_* f] + \int_{\text{supp} f^{\psi}} \left(\mu \log G^{\psi\prime}(z) + \mu^* \overline{\log G^{\psi\prime}(z)}\right) f^{\psi}(z) l(dz),$$

$$f \in \mathcal{H}_s: \quad \text{supp} f \subset G^{-1}(\mathcal{D})$$

as well as the Lie derivative becomes

$$\mathcal{L}_v \Phi[f] = \frac{\partial}{\partial s} H_s^{-1}[v]_* \Phi^{\psi}[f]\Big|_{s=0} =$$

$$= -\Phi^{\psi}[\mathcal{L}_v f] + \int_{\text{supp} f^{\psi}} \left(\mu v^{\psi\prime}(z) + \mu^* \overline{v^{\psi\prime}(z)}\right) f^{\psi}(z) l(dz),$$

Example 5 Let $\mathcal{H} := \mathcal{H}_s$, $\Gamma := \Gamma_D$ (as in Example 2.37), and let $\eta^\psi(z) := 0$ in all charts ψ ($\mu = \mu^* = 0$). Then we call $\Phi(\mathcal{H}_s, \Gamma_D, 0)$ the Gaussian free field with *zero boundary condition*.

Example 6 Relax the previous example. Let η^ψ be a harmonic function in D^ψ continuously extendable to the boundary ∂D^ψ if the chart map ψ can be extended to ∂D. Then we call Φ the Gaussian free field with the *Dirichlet boundary condition*.

We can define the Laplace-Beltrami operator Δ_λ over Φ as well as the Lie derivative by

$$(\Delta_\lambda \Phi)[g] := \Phi[\Delta_\lambda g], \quad g \in \mathcal{H},$$

where Δ_λ on a $(1, 1)$-differential is defined by

$$\Delta_\lambda g^\psi(z) := -4\, \partial_z\, \partial_{\bar{z}}\, \frac{g^\psi(z)}{\lambda^\psi(z)}$$

in any chart ψ. If η is harmonic the identity

$$\mathbb{E}\left[(\Delta_\lambda \Phi)[g]\Phi[f_1]\Phi[f_2]\ldots\Phi[f_n]\right] =$$

$$= \sum_{i=1,2,\ldots,n} \delta_\lambda[g, f_i]\, \mathbb{E}\left[\Phi[f_1]\Phi[f_2]\ldots\Phi[f_{i-1}]\Phi[f_{i+1}]\ldots\ldots\Phi[f_n]\right] \qquad (2.42)$$

is satisfied. Thereby, one can write heuristically

$$\Delta_\lambda \Phi(z) = 0, \quad z \notin \mathrm{supp} f_1 \cup \mathrm{supp} f_2 \cup \ldots \cup \mathrm{supp} f_n.$$

It turns out that the characteristic functional $\hat{\phi}$ is also a derivation functional for the correlation functions. Define the variational derivative over some functional v as a map $\frac{\delta}{\delta f} : v \mapsto \frac{\delta}{\delta f} v$ to the set of functionals by

$$\left(\frac{\delta}{\delta f} v\right)[g] := \frac{\partial}{\partial \alpha} v[g + \alpha f]\Big|_{\alpha=0}, \quad \forall f, g \in \mathcal{H}.$$

If v is such that $v[g + \alpha f]$ is an analytic function in a neighbourhood of $\alpha = 0$ for each f and g, like $\hat{\phi}$, it is straightforward to see that for each $g, f_1, f_2, \ldots \in \mathcal{H}$,

$$v[g] = v[0], \quad g \in \mathcal{H} \quad \Leftrightarrow \quad \left(\frac{\delta}{\delta f_1} \frac{\delta}{\delta f_2} \cdots \frac{\delta}{\delta f_n} v\right)[0] = 0, \quad n = 1, 2, \ldots .$$

Define the Schwinger functionals as

$$S_n[f_1, f_2, \ldots, f_n] := \mathbb{E}\left[\Phi[f_1]\Phi[f_2]\ldots\Phi[f_n]\right] = \left(\frac{\delta}{\delta f_1} \frac{\delta}{\delta f_2} \cdots \frac{\delta}{\delta f_n} \hat{\phi}\right)[0] \qquad (2.43)$$

where $\hat{\phi}: \mathcal{H} \to \mathbb{R}$ is defined in (2.40).

The identity (2.42) can be reformulated as

$$\mathbb{E}\left[(\Delta_\lambda \Phi)[g]e^{\Phi[f]}\right] = (\delta_\lambda[g,f] + \eta[\Delta_\lambda f])\,\hat{\phi}[f], \quad f,g \in \mathcal{H}.$$

2.8 The Schwinger Functionals

In this section, we consider the Schwinger functionals defined by (2.43) and their derivation functional $\hat{\phi}$ in detail.

For any finite collection $\{f_1, f_2, \dots, f_n\}$ of functions from $\mathcal{H}_s$ or H_Γ, the collection of random variables $\{\Phi[f_1], \Phi[f_2], \dots, \Phi[f_n]\}$ has the multivariate normal distribution. Thus, we have

$$\mathbb{E}\left[\Phi[f_1]\Phi[f_2]\dots\Phi[f_n]\right] = \sum_{\text{partitions}} \prod_k \Gamma[f_{i_k}, f_{j_k}],$$

for $\eta(z) \equiv 0$, where the sum is taken over all partitions of the set $\{1, 2 \dots, n\}$ into disjoint pairs $\{i_k, j_k\}$. In particular, the expectation of the product of an odd number of fields is identically zero. For the general case ($\eta \not\equiv 0$) the Schwinger functionals are

$$S[f_1, f_2, \dots, f_n] := \mathbb{E}\left[\Phi[f_1]\Phi[f_2]\dots\Phi[f_n]\right] = \sum_{\text{partitions}} \prod_k \Gamma[f_{i_k}, f_{j_k}] \prod_l \eta[f_{i_l}],$$

where the sum is taken over all partitions of the set $\{1, 2 \dots, n\}$ into disjoint non-ordered pairs $\{i_k, j_k\}$, and non-ordered single elements $\{i_l\}$. In particular,

$$S_1[f_1] = \eta[f_1],$$
$$S_2[f_1, f_2] = \Gamma[f_1, f_2] + \eta[f_1]\eta[f_2],$$
$$S_3[f_1, f_2, f_3] = \Gamma[f_1, f_2]\eta[f_3] + \Gamma[f_3, f_1]\eta[f_2] + \Gamma[f_2, f_3]\eta[f_1] + \eta[f_1]\eta[z_2]\eta[f_3],$$
$$S_4[f_1, f_2, f_3, f_4] = \Gamma[f_1, f_2]\Gamma[f_3, f_4] + \Gamma[f_1, f_3]\Gamma[f_2, f_4] + \Gamma[f_1, f_4]\Gamma[f_2, f_3] +$$
$$+ \Gamma[f_1, f_2]\eta[f_3]\eta[f_4] + \Gamma[f_1, f_3]\eta[f_2]\eta[f_4] + \Gamma[f_1, f_4]\eta[f_2]\eta[f_3] +$$
$$+ \eta[f_1]\eta[f_2]\eta[f_3]\eta[f_4].$$

Such correlation functionals are called the *Schwinger functionals*. Their kernels

$$S_n(z_1, z_2, \dots, s_n)$$

are known as Schwinger functions or *n*-point functions. For regular functionals Γ and η, the Schwinger functions are also regular but it is still reasonable to understand

S_n as a functional because the derivatives are not regular. For example,

$$\Delta_{\lambda 1} S_2^\psi(z, w) = 2\pi \delta_\lambda(z - w).$$

The transformation rules for S_n (the behaviour under the action of G_*) are quite complex. We present here only the infinitesimal ones

$$\mathcal{L}_v S_n^\psi[f_1, f_2, \dots] = -\sum_{1 \le k \le n} S_n^\psi[f_1, f_2, \dots \mathcal{L}_v f_k, \dots, f_n] -$$

$$- \sum_{1 \le k \le n} S_{n-1}^\psi[f_1, f_2, \dots f_{k-1}, f_{k+1}, \dots f_{n-1}] \int_{\psi(\mathcal{D})} \left(\mu v^{\psi\prime}(z) + \mu^* \overline{v^{\psi\prime}(z)} \right) f_k^\psi(z) l(dz)$$

We prefer to work with the characteristic functional $\hat{\phi}$, rather than with S_n. For instance, for any inverse endomorphism $G: \mathcal{D} \setminus \mathcal{K} \to \mathcal{D}$, we can define the pushforward $G_*^{-1}: \hat{\phi}(\Gamma, \eta) \mapsto \hat{\phi}(F_*\Gamma, F_*\eta)$ that maps the functionals on $\mathcal{D}$ to functionals on $\mathcal{D} \setminus \mathcal{K}$. Equivalently,

$$\left(G_*^{-1} \hat{\phi}(\Gamma, \eta) \right) [\tilde{f}] := \hat{\phi}(G_*^{-1}\Gamma, G_*^{-1}\eta)[\tilde{f}], \quad \tilde{f} \in \mathcal{H}_s[\tilde{\mathcal{D}}] \tag{2.44}$$

(we need to mark the dependence on the functionals Γ and on η here).

The Lie derivative $\mathcal{L}_v$ over an arbitrary nonlinear functional $\rho: \mathcal{H}_s \to \mathbb{C}$ can be also defined as

$$\mathcal{L}_v \rho[f] := (\mathcal{L}_v \rho)[f] = \frac{\partial}{\partial t} (H_\alpha[v]_*^{-1} \rho)[f] \big|_{t=0}$$

(if the partial derivative w.r.t. α is well-defined).

For example,

$$\mathcal{L}_v \exp(\rho[f]) = (\mathcal{L}_v \rho[f]) \exp(\rho[f]),$$

$$\mathcal{L}_v^2 \exp(\rho[f]) = \left(\mathcal{L}_v^2 \rho[f] + (\mathcal{L}_v \rho[f])^2 \right) \exp(\rho[f]).$$

In our case $\rho[f] = \hat{\phi}[f] = \exp(\frac{1}{2}\Gamma[f, f] + \eta[f])$. We remind that the Lie derivative of η and Γ are defined in (2.25) and (2.34) respectively.

The operations G_*^{-1} and $\frac{\delta}{\delta f}$ or $\mathcal{L}$ and $\frac{\delta}{\delta f}$ commute. Thus, for example, we have

$$\mathcal{L}_v S_n[f_1, f_2, \dots, f_n] = \left(\frac{\delta}{\delta f_1} \frac{\delta}{\delta f_2} \cdots \frac{\delta}{\delta f_n} \mathcal{L}_v \hat{\phi} \right) [0].$$

We use this to deduce the martingale properties of $G_t^{-1} {}_* S_n$ and of all their variational derivatives from the martingale property of $G_t^{-1} {}_* \hat{\phi}$, which will be discussed in the next section.

3 Coupling Between SLE and GFF

Let $(\Omega^{\Phi}, \mathcal{F}^{\Phi}, P^{\Phi})$ be the probability space for GFF Φ and let $(\Omega^{B}, \mathcal{F}^{B}, P^{B})$ be the independent probability space for the Brownian motion $\{B_t\}_{t \in [0, +\infty)}$, which governs some (δ, σ)-SLE $\{G_t\}_{t \in [0, +\infty)}$. In this section, we consider a coupling between these random laws.

The pushforward $G_t^{-1}{}_*\Phi[f]$ of the GFF $\Phi[f]$ is well-defined if $\text{supp} f \in \text{image}[G_t^{-1}]$. In order to handle this, we introduce a stopping time $T[f]$, for which the hull $\mathcal{K}_t$ of (δ, σ)-SLE touches some small neighborhood $U(\text{supp} f)$ of the support of f for the first time:

$$T[f] := \sup\{t > 0 : \mathcal{K}_t \cap U(\text{supp} f) = \emptyset\}, \quad f \in \mathcal{H}. \tag{3.1}$$

The neighborhood $U(\text{supp} f)$ can be defined, for example, as the set of points from $\text{supp} f$ with the Poincare distance less than some $\varepsilon > 0$. Thus, $T[f] > 0$ a.s. We consider a stopped process $\{G_{t \wedge T[f]}\}_{t \in [0, +\infty)}$. This approach was also used in [16]. The most important property of the process $\{G_{t \wedge T[f]}^{-1}{}_*\Phi[f]\}_{t \in [0, +\infty)}$ is that it is a local martingale. A stopped local martingale is also a local martingale. That is why a stopping of $\{G_t\}_{t \in [0, +\infty)}$ does not change our results. However, we lose some information, which makes the proposition of coupling less substantial than one possibly expects.

We present here two definitions of the coupling. The first one is similar to [16, 30]. The second one is a weaker statement that we shall use in this paper.

Definition 5 A GFF $\Phi(\mathcal{H}, \Gamma, \eta)$ is called *coupled* to the forward or reverse (δ, σ)-SLE, driven by $\{B_t\}_{t \in [0, +\infty)}$, if the random variable $G_{t \wedge T[f]}^{-1}{}_*\Phi^{\psi}[f]$ obtained by independent sampling of Φ and G_t has the same law as $\Phi^{\psi}[f]$ for any test function $f \in \mathcal{H}$, chart map ψ, and $t \in [0, +\infty)$.

If the coupling holds for a fixed chart map ψ and for any $f \in \mathcal{H}$, then it also holds for any chart map $\tilde{\psi}$, due to (2.41). We also give a weaker version of the coupling statement that we plan to use here. To this end, we have to consider a stopped versions of the stochastic process $\{G_{t \wedge T[f]}\}_{t \in [0, +\infty)}$.

A collection of stopping times $\{T_n\}_{n=1,2,\ldots}$ is called a *fundamental sequence* if $0 \le T_n \le T_{n+1} \le \infty$, $n = 1, 2, \ldots$ a.s., and $\lim_{n \to \infty} T_n = \infty$ a.s.

A stochastic process $\{x_t\}_{t \in [0, +\infty)}$ is called a *local martingale* if there exists a fundamental sequence of stopping times $\{T_n\}_{n=1,2,\ldots}$, such that the stopped process $\{x_{t \wedge T_n}\}_{t \in [0, +\infty)}$ is a martingale for each $n = 1, 2, \ldots$.

Let now the statement of coupling above be valid only for the process $\{G_{t \wedge T[f] \wedge T_n}\}_{t \in [0, +\infty)}$ stopped by T_n for each $n = 1, 2, \ldots$. Namely, $G_{t \wedge T[f] \wedge T_n}^{-1}{}_*\Phi^{\psi}[f]$ has the same law as $\Phi^{\psi}[f]$ for each $n = 1, 2, \ldots$.

We are ready now to define the local coupling.

Definition 6 A GFF $\Phi(\mathcal{H}, \Gamma, \eta)$ is called *locally coupled* to (δ, σ)-SLE, driven by $\{B_t\}_{t \in [0, +\infty)}$, if there exists a fundamental sequence $\{T_n[f, \psi]\}_{n=1,2,\ldots}$, such that the random variable $G_{t \wedge \mathbb{T}[f] *}^{-1} \Phi^{\psi}[f]$ obtained by independent sampling of Φ and G_t has the same law as $\Phi^{\psi}[f]$ until the stopping time $T_n[f, \psi]$ for each $n = 1, 2, \ldots$, for any test function $f \in \mathcal{H}$, and a chart map ψ.

Remark 1 If $\mathbb{T}[f, \psi] = +\infty$ a.s. for each $f \in \mathcal{H}$, then the coupling is not local.

The following theorem generalizes the result of [30].

Theorem 1 *The following three statements are equivalent:*

1. *GFF $\Phi(\mathcal{H}, \Gamma, \eta)$ is locally coupled to (δ, σ)-SLE;*
2. *$G_{t \wedge \mathbb{T}[f] *}^{-1} \hat{\phi}^{\psi}[f]$ is a local martingale for $f \in \mathcal{H}$ in any chart ψ;*
3. *The system of the equations*

$$\mathcal{L}_\delta \, \eta[f] + \frac{1}{2} \mathcal{L}_\sigma^2 \, \eta[f] = 0, \quad f \in \mathcal{H}, \tag{3.2}$$

$$\mathcal{L}_\delta \, \Gamma[f, g] + \mathcal{L}_\sigma \, \eta[f] \, \mathcal{L}_\sigma \, \eta[g] = 0, \quad f, g \in \mathcal{H}, \tag{3.3}$$

and

$$\mathcal{L}_\sigma \, \Gamma[f, g] = 0, \quad f, g \in \mathcal{H}. \tag{3.4}$$

is satisfied.

We start the proof after some remarks. Just for clarity (but not for applications) we reformulate the system (3.2)–(3.4) directly in terms of partial derivatives using (2.25), (2.26), (2.34), and (1) as

$$\delta(z) \, \partial_z \, \eta(z) + \overline{\delta(z)} \, \partial_{\bar{z}} \, \eta(z) + \mu \delta'(z) + \mu^* \overline{\delta'(z)} +$$

$$+ \frac{1}{2} \sigma^2(z) \, \partial_z^2 \, \eta(z) + \frac{1}{2} \overline{\sigma^2(z)} \, \partial_{\bar{z}}^2 \, \eta(z) + \sigma(z) \overline{\sigma(z)} \, \partial_z \, \partial_{\bar{z}} \, \eta +$$

$$+ \frac{1}{2} \sigma(z) \sigma'(z) \, \partial_z \, \eta(z) + \frac{1}{2} \overline{\sigma(z) \sigma'(z)} \, \partial_{\bar{z}} \, \eta(z) + \mu \sigma(z) \sigma''(z) + \mu^* \overline{\sigma(z) \sigma''(z)} = 0;$$

$$\delta(z) \, \partial_z \, \Gamma(z, w) + \delta(w) \, \partial_w \, \Gamma(z, w) + \overline{\delta(z)} \, \partial_{\bar{z}} \, \Gamma(z, w) + \overline{\delta(w)} \, \partial_{\bar{w}} \, \Gamma(z, w) +$$

$$+ \left(\sigma(z) \, \partial_z \, \eta(z) + \overline{\sigma(z)} \, \partial_{\bar{z}} \, \eta(z) + \mu \sigma'(z) + \mu^* \overline{\sigma'(z)} \right) \times$$

$$\times \left(\sigma(w) \, \partial_w \, \eta(w) + \overline{\sigma(w)} \, \partial_{\bar{w}} \, \eta(w) + \mu \sigma'(w) + \mu^* \overline{\sigma'(w)} \right) = 0;$$

$$\sigma(z) \, \partial_z \, \Gamma(z, w) + \sigma(w) \, \partial_w \, \Gamma(z, w) + \overline{\sigma(z)} \, \partial_{\bar{z}} \, \Gamma(z, w) + \overline{\sigma(w)} \, \partial_{\bar{w}} \, \Gamma(z, w) = 0,$$

where we drop the upper index ψ for shortness.

The first equation (3.2) is just a local martingale condition for η. The second one (3.3) is a special case of Hadamard's variation formula, where the variation is concentrated at one point at the boundary. The third equation means that Γ should be invariant under the one-parametric family of Möbius automorphisms generated by σ.

Proof of Theorem 1 Let us start with showing how the statement 1 about the coupling implies the statement 2 about the local martingality.

1.⇔2. Let $G_{t \wedge T_f \wedge T_n[f, \psi]}$ be a stopped process $G_{t \wedge T_f}$ by the stopping times $T_n[f, \psi]$ forming some fundamental sequence. The coupling statement can be reformulated as an equality of characteristic functions for the random variables $G^{-1}_{t \wedge \tilde{T}_n[f, \psi]\,*} \Phi^\psi[f]$ and $\Phi^\psi[f]$ for all test functions f. Namely, the following expectations must be equal

$$\mathbb{E}_B \left[\mathbb{E}_\Phi \left[e^{G^{-1}_{t \wedge T[f] \wedge T_n[f, \psi]\,*} \Phi^\psi[f]} \right] \right] = \mathbb{E}_\Phi \left[e^{\Phi[f]} \right], \quad f \in \mathcal{H}, \quad t \in [0, +\infty), \quad n = 1, 2, \dots,$$

which in particular, means the integrability of $e^{G^{-1}_{t \wedge \tilde{T}_n[f, \psi]\,*} \Phi[f]}$ with respect to Ω^B and Ω^Φ. We used $\mathbb{E}_B[\cdot]$ for the expectation with respect to the random law of $\{B_t\}_{t \in [0, +\infty)}$ (or $\{G_t\}_{t \in [0, +\infty)}$) and $\mathbb{E}_\Phi[\cdot]$ for the expectation with respect to Φ. Let us use (2.40) and (2.44) to simplify this identity to

$$\mathbb{E}_B \left[G^{-1}_{t \wedge T[f] \wedge T_n[f, \psi]\,*} \hat{\phi}^\psi[f] \right] = \hat{\phi}^\psi[f], \quad f \in \mathcal{H}, \quad t \in [0, +\infty), \quad n = 1, 2, \dots.$$

After substituting $f \to \tilde{G}_{s \wedge T[f]\,*} f$ for some independently sampled $\tilde{G}_s$ and $s \in [0, +\infty)$, we obtain

$$\mathbb{E}_B \left[G^{-1}_{t \wedge T[\tilde{G}_{s \wedge T[f]}*f] \wedge T_n[\tilde{G}_{s \wedge T[f]}*f, \psi]\,*} \hat{\phi}^\psi[\tilde{G}_{s \wedge T[f]}*f] \right] = \hat{\phi}^\psi[\tilde{G}_{s \wedge T[f]}*f],$$

$$f \in \mathcal{H}, \quad t \in [0, +\infty), \quad n = 1, 2, \dots.$$

Multiplying both sides by

$$\underset{\text{supp} \, e}{\int}^{\text{supp} f^\psi} \left(\mu \log \left(\tilde{G}^\psi_{s \wedge T[f]} \right)'(z) + \mu^* \overline{\log \left(\tilde{G}^\psi_{s \wedge T[f]} \right)'(z)} \right) f(z) l(dz)$$

and by making use of (2.30) and (2.40), we conclude that

$$\mathbb{E}_B \left[\tilde{G}^{-1}_{s \wedge T[f]} * G^{-1}_{t \wedge T[\tilde{G}_{s \wedge T[f]}*f] \wedge T_n[\tilde{G}^{-1}_{s \wedge T[f]}*f, \psi]\,*} \hat{\phi}^\psi[f] \right] = \tilde{G}_{s \wedge T[f]} * \hat{\phi}^\psi[f], \tag{3.5}$$

$$f \in \mathcal{H}, \quad t \in [0, +\infty), \quad n = 1, 2, \dots.$$

Defined now the process

$$\tilde{G}_{t+s} := G_t \circ \tilde{G}_s, \quad s, t \in [0, +\infty),$$

which has the law of (δ, σ)-SLE. Its stopped version possesses the identity

$$G_{t+s\wedge T[f]} = G_{t\wedge T[\tilde{G}_{s\wedge T[f]}*f]} \circ \tilde{G}_{s\wedge T[f]}, \quad s, t \in [0, +\infty), \quad f \in \mathcal{H}.$$

The left-hand side of (3.5) is equal to

$$\mathbb{E}_B\left[\left(G_{t\wedge T[\tilde{G}_{s\wedge T[f]}*f]\wedge T_n[\tilde{G}_{s\wedge T[f]}^{-1}*f,\psi]} \circ \tilde{G}_{s\wedge T[f]}\right)_*^{-1}\hat{\phi}^{\psi}[f]\right] =$$

$$=\mathbb{E}_B\left[\left(\tilde{G}_{t+s\wedge T[f]\wedge T_n[\tilde{G}_{s\wedge T[f]}^{-1}*f,\psi]+s}\right)_*^{-1}\hat{\phi}^{\psi}[f] \mid \mathcal{F}_{s\wedge T[f]}^B\right].$$

We use now the Markov property of (δ, σ)-SLE and conclude that $T_n'[f, \psi] := T_n[\tilde{G}_{s\wedge T[f]}^{-1}*f, \psi] + s$ is a fundamental sequence for the pair of f and ψ. Thus, (3.5) simplifies to

$$\mathbb{E}_B\left[G_{t+s\wedge T[f]\wedge T_n'[f,\psi]}^{-1}*\hat{\phi}^{\psi}[f] \mid \mathcal{F}_{s\wedge T[f]\wedge T_n'[f,\psi]}^B\right] = G_{t+s\wedge T[f]\wedge T_n'[f,\psi]}^{-1}*\hat{\phi}[f],$$

hence, $\{G_{t\wedge T[f]}^{-1}*\hat{\phi}[f]\}_{t\in[0,+\infty)}$ is a local martingale.

The inverse statement can be obtained by the same method in the reverse order.

2.⇔3. According to Proposition 8, Appendix 1: Nature of Coupling, the drift term, i.e., the coefficient at dt, vanishes identically when

$$\mathcal{A}\,W[f] + \frac{1}{2}(\mathcal{L}_\sigma\,W[f])^2 = 0, \quad f \in \mathcal{H}. \tag{3.6}$$

The left-hand side is a functional polynomial of degree 4. We use the fact that a regular symmetric functional $P[f] := \sum_{k=1,2,\ldots n} p_k[f, f, \ldots, f]$ of degree n over such spaces as $\mathcal{H}_s$, $\mathcal{H}_s^*$, $\mathcal{H}_{s,b}$, or $\mathcal{H}_{s,b}^{\pm}{}^*$ is identically zero if and only if

$$p_k[f_1, f_2, \ldots, f_n] = 0, \quad k = 1, 2, \ldots n, \quad f \in \mathcal{H}.$$

Thus, each of the following functions must be identically zero:

$$\mathcal{A}\,\eta[f] = 0, \quad \frac{1}{2}\mathcal{A}\,\Gamma[f, g] + \frac{1}{2}\mathcal{L}_\sigma\,\eta[f]\,\mathcal{L}_\sigma\,\eta[g] = 0,$$

$$\mathcal{L}_\sigma\,\eta[f]\,\mathcal{L}_\sigma\,\Gamma[g, h] + \text{symmetric terms} = 0, \tag{3.7}$$

$$\mathcal{L}_\sigma\,\Gamma[f, g]\,\mathcal{L}_\sigma\,\Gamma[h, l] + \text{symmetric terms} = 0,$$

$$f, g, h, l \in \mathcal{H}.$$

We can conclude that $\mathcal{L}_\sigma\,\Gamma[f, g] = 0$, $\mathcal{A}\,\Gamma[f, g] = \mathcal{L}_\delta\,\Gamma[f, g]$ for any $f, g \in \mathcal{H}$, and this system is equivalent to the system (3.2)–(3.4). For the case $\mathcal{H} = \mathcal{H}_s$ we can write (3.7) in terms of functions on $\psi(\mathcal{D})$:

$$\mathcal{A}\,\eta(z) = 0, \quad \frac{1}{2}\,\mathcal{A}\,\Gamma(z,w) + \frac{1}{2}\,\mathcal{L}_\sigma\,\eta(z)\,\mathcal{L}_\sigma\,\eta(w) = 0,$$

$$\mathcal{L}_\sigma\,\eta(z)\,\mathcal{L}_\sigma\,\Gamma(w,u) + \text{symmetric terms} = 0,$$

$$\mathcal{L}_\sigma\,\Gamma(z,w)\,\mathcal{L}_\sigma\,\Gamma(u,v) + \text{symmetric terms} = 0,$$

$$z,w,u,v \in \psi(\mathcal{D}), \quad z \neq w, \ u \neq v, \ldots .$$

Remark 2 Fix a chart ψ. The coupling and the martingales are not local if in addition to the Proposition 3 in Theorem 1 the relation

$$\mathbb{E}_B\left[\left|\int_0^t \exp\left(G^{-1}_{t\wedge T[f]*}W^\psi[f]\right)G^{-1}_{t\wedge T[f]*}\mathcal{L}_\sigma\,W^\psi[f]\,\mathrm{d}^{\mathrm{Itô}}\,B_\tau\right|\right] < \infty, \quad t \geq 0,$$

holds. This is the condition that the diffusion term at $\mathrm{d}^{\mathrm{Itô}}\,B_t$ in (12) is in $L_1(\Omega^B)$. However, this may not be true, in general, in another chart $\tilde{\psi}$. Meanwhile, if the local martingale property of $G^{-1}_{t\wedge T[f]*}\hat{\phi}^\psi[f]$ is satisfied in one chart ψ for any $f \in \mathcal{H}$, then it is also true in any chart due to the invariance of the condition (3.6) in the proof.

The study of the general solution to (3.2)–(3.4) is an interesting and complicated problem. Take the Lie derivative $\mathcal{L}_\sigma$ in the second equation, the Lie derivative $\mathcal{L}_\delta$ in the third equation, and consider the difference of the resulting equations. It is an algebraically independent equation

$$\mathcal{L}_{[\delta,\sigma]}\,\Gamma[f,g] = -\mathcal{L}_\sigma^2\,\eta[f]\,\mathcal{L}_\sigma\,\eta[g]\,\mathcal{L}_\sigma^2\,\eta[f]\,\mathcal{L}_\sigma^2\,\eta[g].$$

Continuing by induction we obtain an infinite system of a priori algebraically independent equations because the Lie algebra induced by the vector fields δ and σ is infinite-dimensional. Thereby, the existence of the solution to the system (3.2)–(3.4) is a special event that is strongly related to the properties of this algebra.

Before studying special solutions to the system (3.2)–(3.4), let us consider some of its general properties. We also reformulate it in terms of the analytic functions η^+, Γ^{++} and Γ^{+-}, which is technically more convenient.

Lemma 2 *Let δ, σ, η, and Γ be such that the system (3.2)–(3.4) is satisfied, let Γ be a fundamental solution to the Laplace equation (see (2.36)), which transforms as a scalar, see (2.33), and let η be a pre-pre-Schwarzian. Then,*

1. *η is a $(i\chi/2, -i\chi/2)$-pre-pre-Schwarzian (2.22) given by a harmonic function in any chart with χ given by (1.4).*
2. *The boundary value of η undergoes a jump $2\pi/\sqrt{\kappa}$ at the source point a, namely, its local behaviour in the half-plane chart is given by (3.21) up to a sign;*

3. *The system (3.2)–(3.4) is equivalent to the system (3.8), (3.13), (3.14), (3.17), and (3.16).*

Proof The system (3.2)–(3.4) defines η only up to an additive constant C that we keep writing in the formulas for η below. The condition for the pre-pre-Schwarzian η to be real leads to only two possibilities:

1. $\mu = -\mu^*$ and is pure imaginary as in (2.22);
2. $\mu = \mu^*$ and is real as in (2.21).

Equation (3.3) shows that the functional $\mathcal{L}_\sigma \eta$ has to be given by a harmonic function as well as $\mathcal{L}_\sigma^2 \eta$ in any chart. On the other hand, (3.2) implies that $\mathcal{L}_\delta \eta$ is also harmonic. The vector fields δ and σ are transversal almost everywhere. We conclude that η is harmonic. We used also the fact that the additional μ-terms in (2.25) are harmonic.

The harmonic function $\eta^\psi (z)$ can be represented as a sum of an analytic function $\eta^{+\,\psi} (z)$ and its complex conjugate in any chart ψ

$$\eta^\psi (z) = \eta^{+\,\psi} (z) + \overline{\eta^{+\,\psi} (z)}. \tag{3.8}$$

Below in this proof, we drop the chart index ψ, which can be chosen arbitrarily.

Let us define η^+ and $\overline{\eta^+}$ to be pre-pre-Schwarzians of orders $(\mu, 0)$ and $(0, \mu^*)$ respectively by (2.25). Thus, η^+ is defined up to a complex constant C^+. We denote

$$j^+ := \mathcal{L}_\sigma \eta^+. \tag{3.9}$$

and

$$j := \mathcal{L}_\sigma \eta = \mathcal{L}_\sigma \eta^+ + \overline{\mathcal{L}_\sigma \eta^+}. \tag{3.10}$$

The reciprocal formula is

$$\eta^+ (z) := \int \frac{j^+ (z) - \mu \sigma' (z)}{\sigma (z)} dz. \tag{3.11}$$

This integral can be a ramified function if $\sigma(z)$ has a zero inside of $\mathcal{D}$ (the elliptic case). We consider how to handle this technical difficulty in Sect. 4.3.

Let us reformulate now (3.2) in terms of j^+. Using the fact that

$$\mathcal{L}_v^2 (\eta^+ + \overline{\eta^+}) = \mathcal{L}_v^2 \eta^+ + \mathcal{L}_v^2 \overline{\eta^+},$$

we arrive at

$$\mathcal{L}_\delta \eta^+ + \frac{1}{2} \mathcal{L}_\sigma{}^2 \eta^+ = C^+. \tag{3.12}$$

Here $C^+ = i\beta$ for some $\beta \in \mathbb{R}$ for the forward case. For the reverse case, $C^+ = -\beta + i\beta'$ for some $\beta, \beta' \in \mathbb{R}$ because (3.12) is an identity in sense of functionals over $\mathcal{H}_s^*$.

The relation (3.12) is equivalent to

$$\frac{\delta}{\sigma} \mathcal{L}_\sigma \eta^+ + \frac{\sigma \mathcal{L}_\delta \eta^+ - \delta \mathcal{L}_\sigma \eta^+}{\sigma} + \frac{1}{2} \mathcal{L}_\sigma^2 \eta^+ = C^+ \quad \Leftrightarrow$$

$$\frac{\delta}{\sigma} j^+ + \frac{\sigma \delta \partial \eta^+ + \mu \sigma \delta' - \delta \sigma \partial \eta^+ - \mu \delta \sigma'}{\sigma} + \frac{1}{2} \mathcal{L}_\sigma j^+ = C^+ \quad \Leftrightarrow$$

$$\frac{\delta}{\sigma} j^+ + \mu \frac{[\sigma, \delta]}{\sigma} + \frac{1}{2} \mathcal{L}_\sigma j^+ = C^+ \tag{3.13}$$

by (2.25) and (2.24).

Consider now the function $\Gamma^{\mathbb{H}}(z, w)$. It is harmonic with respect to both variables with the only logarithmic singularity. Hence, it can be split as a sum of four terms

$$\Gamma^{\mathbb{H}}(z, w) := \Gamma^{++\mathbb{H}}(z, w) + \overline{\Gamma^{++\mathbb{H}}(z, w)} - \Gamma^{+-\mathbb{H}}(z, \bar{w}) - \overline{\Gamma^{+-\mathbb{H}}(z, \bar{w})}, \tag{3.14}$$

where $\Gamma^{++\mathbb{H}}(z, w)$ and $\Gamma^{+-\mathbb{H}}(z, w)$ are analytic with respect to both variables except the diagonal $\underline{z = w}$ for $\Gamma^{++\mathbb{H}}(z, w)$.

So, e.g., $\Gamma^{+-\mathbb{H}}(z, \bar{w})$ is anti-analytic with respect to z and analytic with respect to w. We can assume that both $\Gamma^{++}(z, w)$ and $\Gamma^{+-}(z, w)$ transform as scalars represented by analytic functions in all charts and symmetric with respect to $z \leftrightarrow w$. Observe that these functions are defined at least up to the transform

$$\Gamma^{++\mathbb{H}}(z, w) \to \Gamma^{++\mathbb{H}}(z, w) + \epsilon^{\mathbb{H}}(z) + \epsilon^{\mathbb{H}}(w),$$

$$\Gamma^{+-\mathbb{H}}(z, w) \to \Gamma^{+-\mathbb{H}}(z, w) + \epsilon^{\mathbb{H}}(z) + \epsilon^{\mathbb{H}}(w)$$

for any analytic function $\epsilon^{\mathbb{H}}(z)$ such that

$$\overline{\epsilon^{\mathbb{H}}(z)} = \epsilon^{\mathbb{H}}(\bar{z}).$$

These additional terms are canceled due to the choice of minus in the pairs '$\mp$' in (3.14). In the reverse case, the contribution of these functions is equivalent to zero bilinear functional over $\mathcal{H}_s^*$.

Consider Eq. (3.4). It leads to

$$\mathcal{L}_\sigma \Gamma^{++\mathbb{H}}(z, w) = \beta_2^{\mathbb{H}}(z) + \beta_2^{\mathbb{H}}(w), \quad \mathcal{L}_\sigma \Gamma^{+-\mathbb{H}}(z, w) = \beta_2^{\mathbb{H}}(z) + \beta_2^{\mathbb{H}}(w) \tag{3.15}$$

for any analytic function $\beta_2^{\mathbb{H}}(z)$ such that $\overline{\beta_2^{\mathbb{H}}(z)} = \beta_2^{\mathbb{H}}(\bar{z})$. One can fix this freedom, i.e., the function $\beta_2^{\mathbb{H}}$, by the conditions

$$\mathcal{L}_\sigma \Gamma^{++\mathbb{H}}(z, w) = 0, \quad \mathcal{L}_\sigma \Gamma^{+-\mathbb{H}}(z, w) = 0. \tag{3.16}$$

Thus, $\Gamma^{++\mathbb{H}}(z, w)$ and $\Gamma^{+-\mathbb{H}}(z, w)$ are fixed up to a non-essential constant.

The second equation (3.3) can be reformulated now as

$$\mathcal{L}_\delta \Gamma^{++\mathbb{H}}(z, w) + \mathcal{L}_\sigma \eta^{+\mathbb{H}}(z) \mathcal{L}_\sigma \eta^{+\mathbb{H}}(w) = \beta_1^{\mathbb{H}}(z) + \beta_1^{\mathbb{H}}(w),$$

$$\mathcal{L}_\delta \Gamma^{+-\mathbb{H}}(z, \bar{w}) + \mathcal{L}_\sigma \eta^{+\mathbb{H}}(z)\overline{\mathcal{L}_\sigma \eta^{+\mathbb{H}}(w)} = \beta_1^{\mathbb{H}}(z) + \overline{\beta_1^{\mathbb{H}}(\bar{w})} \tag{3.17}$$

for any analytic function $\beta_1^{\mathbb{H}}(z)$ such that $\overline{\beta_1^{\mathbb{H}}(z)} = \beta_1^{\mathbb{H}}(\bar{z})$ analogously to (3.15). We can conclude now that the system (3.2)–(3.4) is equivalent to the system (3.8), (3.13), (3.14), (3.17), and (3.16).

Use now the fact

$$\Gamma^{++\mathbb{H}}(z, w) = -\frac{1}{2} \log(z - w) + \text{analytic terms} \tag{3.18}$$

to obtain a singularity of $j^{+\mathbb{H}}$ about the origin in the half-plane chart. Relation (2.17) yields

$$\frac{2}{z} \partial_z \left(-\frac{1}{2} \log(z - w) \right) + \frac{2}{w} \partial_w \left(-\frac{1}{2} \log(z - w) \right) = \frac{1}{zw}, \tag{3.19}$$

hence,

$$j^{+\mathbb{H}}(z) = \frac{-i}{z} + \text{holomorphic part.} \tag{3.20}$$

The choice of the sign of $j^{+\mathbb{H}}(z)$ is irrelevant. We made the choice above to be consistent with [30]. The analytic terms in (3.18) can give a term with the sum of simple poles at z and w but in the form of the product $1/(zw)$.

From (3.11) we conclude that the singular part of $\eta^{+\mathbb{H}}(z)$ is proportional to the logarithm of z:

$$\eta^{+\mathbb{H}}(z) = \frac{i}{\sqrt{\kappa}} \log z + \text{holomorphic part.}$$

Thus, we have

$$\eta^{\mathbb{H}}(z) = \frac{-2}{\sqrt{\kappa}} \arg z + \text{non-singular harmonic part.} \tag{3.21}$$

We can chose the additive constant such that, in the half-plane chart, we have

$$\eta^{\mathbb{H}}(+0) = -\eta^{\mathbb{H}}(-0) = \frac{\pi}{\sqrt{\kappa}} \tag{3.22}$$

in the forward case. This provides the jump $2\pi/\sqrt{\kappa}$ of the value of η at the boundary near the origin, which is exactly the same behaviour of η needed for the flow line construction in [23] and [30]. However, the form (3.22) is not chart independent, and only the jump $2\pi/\sqrt{\kappa} = \eta^{\psi}(+0) - \eta^{\psi}(-0)$ does not change its value if the boundary of $\psi(\mathcal{D})$ is not singular in the neighbourhood of the source $\psi(a)$.

Substitute now (3.20) in (3.13) in the half-plane chart, use (2.17), and consider the corresponding Laurent series. We are interested in the coefficient near the first term $\frac{1}{z^2}$:

$$\frac{2}{z - \sqrt{\kappa}} \frac{1}{z} \frac{-i}{z} + \mu \frac{-2}{z^2} + \frac{1}{2}(-\sqrt{\kappa}) \frac{i}{z^2} + o\left(\frac{1}{z^2}\right) = C^+.$$

We can conclude that

$$\mu = i\frac{4 - \kappa}{4\sqrt{\kappa}}. \tag{3.23}$$

Thus, the pre-pre-Schwarzians (2.22) with χ given by (1.4) is only one that can be realized. $\qquad\square$

4   Coupling of GFF with the Dirichlet Boundary Conditions

In this section, we consider a special solution to the system (3.2)–(3.4) with the help of Lemma 2. We assume the Dirichlet boundary condition for Γ considered in Example 3, and find the general solution in this case. In other words, we systematically study which of (δ, σ)-SLE can be coupled with GFF if $\Gamma = \Gamma_D$.

Let us formulate the following general theorem, and then consider each of the allowed cases of (δ, σ)-SLE individually.

Theorem 2 *Let a (δ,σ)-SLE be coupled to the GFF with $\mathcal{H} = \mathcal{H}_s$, $\Gamma = \Gamma_D$, and let η be the pre-pre-Schwarzian (2.22) of order χ. Then the only special combinations of δ and σ for $\kappa \neq 6$ and $\nu \neq 0$ summarized in Table 1, and their arbitrary combinations when $\kappa = 6$ and $\nu = 0$ are possible.*

Table 1 consists of 6 cases, each of which is a one-parameter family of (δ,σ)-SLEs parametrized by the drift $\nu \in \mathbb{R}$, and by a parameter $\xi \in \mathbb{R}$. These cases may overlap for vanishing values of ν or ξ.

In other words, different combinations of δ and σ can correspond essentially to the same process in $\mathcal{D}$ but written in different coordinates. We give one example of such choices in each case of δ and σ.

Table 1 (δ,σ)-SLE types that can be coupled with CFT with the Dirichlet boundary conditions $(\Gamma = \Gamma_0)$

N	Name	$\delta^{\mathbb{H}}(z)$	$\sigma^{\mathbb{H}}(z)$	α	β
1	Chordal with drift	$\dfrac{2}{z} - \nu$	$-\sqrt{\kappa}$	$-\dfrac{\nu}{2}$	$\dfrac{-\nu^2}{2\sqrt{\kappa}}$
2	Chordal with fixed time change	$\dfrac{2}{z} + 2\xi z$	$-\sqrt{\kappa}$	0	$\dfrac{\xi(8-\kappa)}{2\sqrt{\kappa}}$
3	Dipolar with drift	$2\left(\dfrac{1}{z} - z\right) - \nu(1 - z^2)$	$-\sqrt{\kappa}(1 - z^2)$	$-\dfrac{\nu}{2}$	$\dfrac{4-\nu^2}{2\sqrt{\kappa}}$
4	One right fixed boundary point	$\dfrac{2}{z} + \kappa - 6 + {} + 2(3 - \kappa + \xi)z + {} + (\kappa - 2 - 2\xi)z^2$	$-\sqrt{\kappa}(1 - z^2)$	$\dfrac{1}{2}(\kappa - 6)$	$\dfrac{\xi(8-\kappa)}{2\sqrt{\kappa}}$
5	One left fixed boundary point	$\dfrac{2}{z} - (\kappa - 6) + {} + 2(3 - \kappa + \xi)z + {} - (\kappa - 2 - 2\xi)z^2$	$-\sqrt{\kappa}(1 - z^2)$	$-\dfrac{1}{2}(\kappa - 6)$	$\dfrac{\xi(8-\kappa)}{2\sqrt{\kappa}}$
6	Radial with drift	$2\left(\dfrac{1}{z} + z\right) - \nu(1 + z^2)$	$-\sqrt{\kappa}(1 + z^2)$	$-\dfrac{\nu}{2}$	$\dfrac{4-\nu^2}{2\sqrt{\kappa}}$

Each of the pairs of δ and σ is given in the half-plane chart for simplicity. For the same purpose we use the normalization (2.17). The details are in Sects. 4.1 and 4.3. See also the comments after Theorem 2

Some particular cases of CFTs studied here were considered earlier in the literature. The chordal SLE without drift (case 1 from the table with $\nu = 0$) was considered in [19], the radial SLE without drift (case 3 from the table with $\nu = 0$) in [18], and the dipolar SLE without drift (case 4 from the table with $\nu = 0$) appeared in [20]. The case 2 actually corresponds to the same measure defined by the chordal SLE but stopped at the time $t = 1/4\xi$ (see Sect. 4.5). The cases 5 and 6 are mirror images of each other. They are discussed in Sect. 4.6.

Remark 3 An alternative approach to the relation between CFT and SLE based on the highest weight representation of the Virasoro algebra was considered in [4] and [8]. We remark that such a representation can not be constructed for non-zero drift ($\nu \neq 0$).

Proof of Theorem 2 Let us use Theorem 1 and assume the Dirichlet boundary conditions for $\Gamma = \Gamma_D$.

$$\Gamma^{++\mathbb{H}}(z, w) = -\frac{1}{2}\log(z - w), \quad \Gamma^{+-\mathbb{H}}(z, \bar{w}) = -\frac{1}{2}\log(z - \bar{w})$$

in Theorem 2. The condition (3.15) is satisfied for any complete vector field σ and some σ-dependent β_2 which is irrelevant.

In order to obtain j^+ we remark first that due to the Möbious invariance (2.38) we can ignore the polynomial part of $\delta^{\mathbb{H}}(z)$

$$\mathcal{L}_\delta \Gamma^{\mathbb{H}}(z,w) = \left(\frac{2}{z} \partial_z + \frac{2}{\bar{z}} \partial_{\bar{z}} + \frac{2}{w} \partial_w + \frac{2}{\bar{w}} \partial_{\bar{w}} \right) \Gamma^{\mathbb{H}}(z,w).$$

Using (3.17), (2.37), and (3.19) we obtain that

$$j^{+\,\mathbb{H}}(z) = \frac{-i}{z} + i\alpha, \quad \alpha \in \mathbb{C}, \tag{4.1}$$

with

$$\beta_1(z) = \frac{\alpha}{z} - \frac{\alpha^2}{2}.$$

In order to satisfy all conditions formulated in Lemma 2 we need to check (3.13). Substituting (4.1) to (3.13) gives

$$\frac{\delta}{\sigma} j^+ + \mu \frac{[\sigma, \delta]}{\sigma} + \frac{1}{2} \mathcal{L}_\sigma^+ j^+ = i\beta \ \Leftrightarrow$$

$$\delta j^+ + \mu\,[\sigma, \delta] + \frac{1}{2}\sigma\,\mathcal{L}_\sigma^+ j^+ - i\beta\sigma = 0 \ \Leftrightarrow$$

$$\delta^{\mathbb{H}}(z) \left(\frac{-i}{z} + i\alpha \right) + \mu\,[\sigma, \delta]^{\mathbb{H}}(z) + \frac{1}{2}\left(\sigma^{\mathbb{H}}(z)\right)^2 \partial\left(\frac{-i}{z} + i\alpha \right) - i\beta\sigma^{\mathbb{H}}(z) = 0. \tag{4.2}$$

In what follows, we will use the half-plane chart in the proof. With the help of (2.16) and (2.18) we can assume without lost of generality that $\sigma^{\mathbb{H}}$ is one of three possible forms:

1. $\sigma^{\mathbb{H}}(z) = -\sqrt{\kappa}$,
2. $\sigma^{\mathbb{H}}(z) = -\sqrt{\kappa}(1 - z^2)$,
3. $\sigma^{\mathbb{H}}(z) = -\sqrt{\kappa}(1 + z^2)$.

Let us consider these cases turn by turn.

1. $\sigma(z) = -\sqrt{\kappa}$.

Inserting (2.17) to relation (4.2) reduces to

$$\frac{-2 + \frac{\kappa}{2} - 2i\sqrt{\kappa}\mu}{z^2} + \frac{2\alpha - \delta_{-1}}{z} + \left(\beta\sqrt{\kappa} + \alpha\delta_{-1} - \delta_0 + i\sqrt{\kappa}\mu\delta_0 \right) +$$

$$+ z\left(\alpha\delta_0 - \delta_1 + 2i\sqrt{\kappa}\mu\delta_1 \right) + z^2\alpha\delta_1 \equiv 0 \quad \Leftrightarrow$$

$$(3.23), \quad 2\alpha - \delta_{-1} = 0, \quad \beta\sqrt{\kappa} + \alpha\delta_{-1} - \delta_0 + i\sqrt{\kappa}\mu\delta_0 = 0,$$

$$\alpha\delta_0 - \delta_1 + 2i\sqrt{\kappa}\mu\delta_1 = 0, \quad \alpha\delta_1 = 0.$$

There are three possible cases:

a.

$$\delta_{-1} = 2\alpha, \quad \delta_0 = 0, \quad \delta_1 = 0, \quad \kappa > 0, \quad \beta = \frac{-2\alpha^2}{\sqrt{\kappa}}.$$

It is convenient to use the drift parameter v. Thus,

$$v = -2\alpha,$$

which is related to the drift in the chordal equation. This case is presented in the first line of Table 1.

b.

$$\delta_{-1} = 0, \quad \delta_0 = -\frac{4\beta\sqrt{\kappa}}{\kappa - 8}, \quad \delta_1 = 0, \quad \kappa > 0, \quad \alpha = 0.$$

This case is presented in the second line of Table ($\xi \in \mathbb{R}$) and discussed in details in Sect. 4.5.

c.

$$\delta_{-1} = 0, \quad \delta_0 = 2\sqrt{5}\beta, \quad \delta_1 \in \mathbb{R}, \quad \kappa = 6, \quad \alpha = 0.$$

This is a general case of δ with $\kappa = 6$ and $v = 0$.

2. $\sigma^{\mathbb{H}}(z) = -\sqrt{\kappa}(1 - z^2)$.

Relation (4.2) reduces to

$$\frac{-2 + \frac{\kappa}{2} + 2i\mu\sqrt{\kappa}}{z^2} + \frac{2\alpha - \delta_{-1}}{z}\left(\beta\sqrt{\kappa} - \kappa + 6i\sqrt{\kappa}\mu + \alpha\delta_{-1} - \delta_0 + i\sqrt{\kappa}\mu\delta_0\right) +$$

$$+z\left(2i\sqrt{\kappa}\mu\delta_{-1} + \alpha\delta_0 - \delta_1 + 2i\sqrt{\kappa}\mu\delta_1\right)z^2\left(-\beta\sqrt{\kappa} + \frac{\kappa}{2} + i\sqrt{\kappa}\mu\delta_0 + \alpha\delta_1\right) = 0 \quad \Leftrightarrow$$

$$(3.23), \quad 2\alpha - \delta_{-1} = 0, \quad \beta\sqrt{\kappa} - \kappa + 6i\sqrt{\kappa}\mu + \alpha\delta_{-1} - \delta_0 + i\sqrt{\kappa}\mu\delta_0 = 0,$$

$$2i\sqrt{\kappa}\mu\delta_{-1} + \alpha\delta_0 - \delta_1 + 2i\sqrt{\kappa}\mu\delta_1 = 0, \quad -\beta\sqrt{\kappa} + \frac{\kappa}{2} + i\sqrt{\kappa}\mu\delta_0 + \alpha\delta_1 = 0.$$

There are four solutions each of which is a two-parameter family. The first one corresponds to the dipolar SLE with the drift v, line 3 in Table 1. The second and the third equations are 'mirror images' of each other, as it can be seen from the lines 4 and 5 in the table. They are parametrized by $\xi := \frac{2\beta\sqrt{\kappa}}{8-\kappa}$ and discussed in details in Sect. 4.6. The fourth case is given by putting

$$\delta_{-1} = 0, \quad \delta_0 = 2(\sqrt{6}\beta - 3), \quad \delta_1 \in \mathbb{R}, \quad \kappa = 6, \quad \alpha = 0.$$

This is a general form of δ with $\kappa = 6$ and $v = 0$.

3. $\sigma^{\mathbb{H}}(z) = -\sqrt{\kappa}(1 + z^2)$.

Relation (4.2) reduces to

$$\frac{-2 + \frac{\kappa}{2} - 2i\sqrt{\kappa}\mu}{z^2} + \frac{2\alpha - \delta_{-1}}{z} +$$

$$+ \left(\beta\sqrt{\kappa} - \kappa + 6i\sqrt{\kappa}\mu + \alpha\delta_{-1} - \delta_0 + i\sqrt{\kappa}\mu\delta_0\right) +$$

$$+ z\left(2i\kappa\mu\delta_{-1} + \alpha\delta_0 - \delta_1 + 2i\sqrt{\kappa}\mu\delta_1\right) +$$

$$+ z^2\left(-\beta\sqrt{\kappa} + \frac{\kappa}{2} + i\sqrt{\kappa}\mu\delta_0 + \alpha\delta_1\right) = 0 \quad \Leftrightarrow$$

$$(3.23), \quad 2\alpha - \delta_{-1} = 0, \quad \beta\sqrt{\kappa} - \kappa + 6i\sqrt{\kappa}\mu + \alpha\delta_{-1} - \delta_0 + i\sqrt{\kappa}\mu\delta_0 = 0,$$

$$2i\kappa\mu\delta_{-1} + \alpha\delta_0 - \delta_1 + 2i\sqrt{\kappa}\mu\delta_1 = 0, \quad -\beta\sqrt{\kappa} + \frac{\kappa}{2} + i\sqrt{\kappa}\mu\delta_0 + \alpha\delta_1 = 0.$$

The first solution is presented in the line 6 of Table 1, where it is again convenient to introduce the parameter ν related to the drift in the radial equation. The second solution is

$$\delta_{-1} = 0, \quad \delta_0 = 2(\sqrt{6}\beta - 3), \quad \delta_1 \in \mathbb{R}, \quad \kappa = 6, \quad \alpha = 0.$$

This is a general form of δ with $\kappa = 6$ and $\nu = 0$. $\qquad\qquad\qquad\square$

4.1 Chrodal SLE with Drift

It is natural to study this case in the half-plane chart, where

$$\delta_c^{\mathbb{H}}(z) := \frac{2}{z} - \nu, \quad \sigma_c^{\mathbb{H}}(z) := -\sqrt{\kappa}, \quad \nu \in \mathbb{R}. \tag{4.3}$$

The form of η^+ can be found from (3.9) by substituting (4.1) as

$$-\kappa^{\frac{1}{2}}\partial_z \eta^{+\mathbb{H}}(z) + \mu \cdot 0 = \frac{-i}{z} + i\alpha.$$

Then

$$\eta^{+\mathbb{H}}(z) = \frac{i}{\sqrt{\kappa}}\log z - \frac{i\alpha z}{\sqrt{\kappa}} + C^+, \tag{4.4}$$

and taking into account that $\alpha = -\frac{\nu}{2}$ we obtain

$$\eta^{\mathbb{H}}(z) = \frac{-2}{\sqrt{\kappa}} \arg z - \frac{\nu}{\sqrt{\kappa}} \operatorname{Im} z + C. \tag{4.5}$$

Let us present here an explicit form of the evolution of the one-point function $S_1(z) = \eta(z)$

$$M_t^{\mathbb{H}}(z) = (G_t^{-1} {}_* \eta)^{\mathbb{H}}(z) = \frac{-2}{\sqrt{\kappa}} \arg G_t^{\mathbb{H}}(z) - \frac{\nu}{\sqrt{\kappa}} \operatorname{Im} G_t^{\mathbb{H}}(z) + \frac{\kappa - 4}{2\sqrt{\kappa}} \arg G_t^{\mathbb{H}'}(z) + C$$

This expression with $\nu = 0$ coincides (up to a constant) with the analogous one from [19, Section 8.5].

Now we need to work with a concrete form of the space $\mathcal{H}_s$ discussed in Sect. 2.4. It is convenient to define it in the half-plane chart. In other charts it can be obtained with the rule (2.27). We choose the subspace $C^{\infty \mathbb{H}}_0$ of C^∞-smooth functions with compact support in the half-plane chart. The function ϕ defining metric can be, for example, zero in the half-plane chart, $\phi^{\mathbb{H}}(z) \equiv 0$. This choice guarantees that the integrals in (9) and (11) are well-defined with η as above and $\Gamma = \Gamma_D$.

4.2 Dipolar SLE with Drift

The dipolar SLE equation is usually defined in the strip chart, see (1.9), as

$$d^{\text{Itô}} G_t^{\mathbb{S}}(z) = \operatorname{cth} \frac{G_t^{\mathbb{S}}(z)}{2} dt - \sqrt{\kappa} \, d^{\text{Itô}} B_t - \frac{\nu}{2} dt, \tag{4.6}$$

where we add the drift term $\frac{\nu}{2} dt$ in the Itô differentials with the same form in terms of Stratonovich

$$d^{\mathbb{S}} G_t^{\mathbb{S}}(z) = \operatorname{cth} \frac{G_t^{\mathbb{S}}(z)}{2} dt - \sqrt{\kappa} \, d^{\mathbb{S}} B_t - \frac{\nu}{2} dt \tag{4.7}$$

because $\sigma^{\mathbb{S}}(z)$ is constant. The vector fields δ and σ in the strip chart, see (1.9), are

$$\delta^{\mathbb{S}}(z) = 4 \operatorname{cth} \frac{z}{2} - \frac{\nu}{2}, \qquad \sigma^{\mathbb{S}}(z) = -\sqrt{\kappa}, \qquad \nu \in \mathbb{R},$$

The from of δ and σ in the half-plane can be obtained by (2.1) as

$$\delta^{\mathbb{H}}(z) = \frac{1}{\tau_{\mathbb{S},\mathbb{H}}(z)} \delta^{\mathbb{S}}(\tau_{\mathbb{S},\mathbb{H}}(z)) = \frac{1}{2}\left(\frac{1}{z} - z\right) - \frac{\nu}{2}(1 - z^2),$$

$$\sigma^{\mathbb{H}}(z) = \frac{1}{\tau_{\mathbb{S},\mathbb{H}}(z)} \sigma^{\mathbb{S}}(\tau_{\mathbb{S},\mathbb{H}}(z)) = -\frac{\sqrt{\kappa}}{2}(1 - z^2).$$

It is more convenient to use for our purposes the transform (2.16) with $c = 2$, and to define

$$\delta_d^{\mathbb{H}}(z) := 2\left(\frac{1}{z} - z\right) - \nu(1 - z^2), \quad \sigma_d^{\mathbb{H}}(z) := -\sqrt{\kappa}(1 - z^2), \quad \nu \in \mathbb{R} \qquad (4.8)$$

that possess normalization (2.17) used in Table 1.

Let us first find η^+, η and M_{1t} in the half-plane chart. The same way as in the previous subsection we calculate

$$-\sqrt{\kappa}(1 - z^2)\, \partial_z\, \eta^{+\mathbb{H}}(z) + \mu\left(-\sqrt{\kappa}(1 - z^2)\right)' = \frac{-i}{z} + i\alpha.$$

Taking into account (3.23) and $\alpha = -\nu/2$ we obtain

$$\eta^{+\mathbb{H}}(z) = \frac{i}{\sqrt{\kappa}}\log z + \frac{i(\kappa - 6)}{4\sqrt{\kappa}}\log(1 - z^2) + \frac{i\nu}{2\sqrt{\kappa}}\operatorname{arcth} z + C^+,$$

$$\eta^{\mathbb{H}}(z) = \frac{-2}{\sqrt{\kappa}}\arg z - \frac{(\kappa - 6)}{2\sqrt{\kappa}}\arg(1 - z^2) - \frac{\nu}{\sqrt{\kappa}}\operatorname{Im}\operatorname{arcth} z + C,$$

$$M_{ct}^{\mathbb{H}}(z) = (G_t^{-1}{}_*\eta)^{\mathbb{H}}(z) =$$

$$= \frac{-2}{\sqrt{\kappa}}\arg G_t^{\mathbb{H}}(z) - \frac{(\kappa - 6)}{2\sqrt{\kappa}}\arg(1 - G_t^{\mathbb{H}}(z)^2) -$$

$$- \frac{\nu}{\sqrt{\kappa}}\operatorname{Im}\operatorname{arcth} G_t^{\mathbb{H}}(z) + \frac{\kappa - 4}{2\sqrt{\kappa}}\arg\left(G_t^{\mathbb{H}}\right)'(z) + C$$

The corresponding relations in the strip chart are

$$\delta_d^{\mathbb{S}}(z) = 4\operatorname{cth}\frac{z}{2} - 2\nu, \quad \sigma_d^{\mathbb{S}}(z) = -2\sqrt{\kappa}, \quad \nu \in \mathbb{R},$$

obtained with the help of (2.19). Then

$$\eta^{\mathbb{S}}(z) = \eta^{\mathbb{H}}(\tau_{\mathbb{H},\mathbb{S}}(z)) + \frac{\kappa - 4}{2\sqrt{\kappa}}\operatorname{Im}(\tau_{\mathbb{H},\mathbb{S}})'(z),$$

where we used (1.4) and the expression for $\tau_{\mathbb{H},\mathbb{S}}(z) = \tau_{\mathbb{S},\mathbb{H}}^{-1}(z) = \psi_{\mathbb{H}} \circ \psi_{\mathbb{S}}^{-1}(z)$ that defines the strip chart (1.9). Alternatively $\eta^{\mathbb{S}}(z)$ can be found as the solution to (3.9) in the strip chart

$$\eta^{S}(z) = \frac{-2}{\sqrt{\kappa}} \arg \operatorname{sh} \frac{z}{2} - \frac{v}{2\sqrt{\kappa}} \operatorname{Im} z + C,$$

$$M_{1t}^{S}(z) = (G_t^{-1}{}_* \eta)^{S}(z) =$$

$$= \frac{-2}{\sqrt{\kappa}} \arg \operatorname{sh} \frac{G_t^{S}(z)}{2} - \frac{v}{2\sqrt{\kappa}} \operatorname{Im} G_t^{S}(z) + \frac{\kappa - 4}{2\sqrt{\kappa}} \arg \left(G_t^{S}\right)'(z) + C.$$

The expression for Γ_D in the strip chart becomes

$$\Gamma_D^{S}(z, w) = \Gamma_D^{H}\left(\tau_{H,S}(z), \tau_{H,S}(w)\right) = -\frac{1}{2} \log \frac{\operatorname{sh}(\frac{z-w}{2})\operatorname{sh}(\frac{\bar{z}-\bar{w}}{2})}{\operatorname{sh}(\frac{\bar{z}-w}{2})\operatorname{sh}(\frac{z-\bar{w}}{2})}.$$

We remark that η can be presented in a chart-independent form as a function of δ_d and σ_d using (2.23) as

$$\eta = -\frac{2}{\sqrt{\kappa}} \arg \frac{\sigma_d^{\frac{\kappa}{4}}}{\sqrt{\sigma_d^2 - \frac{\kappa}{4}\left(\delta_d - \frac{v}{\sqrt{\kappa}}\sigma_d\right)^2}} + \frac{v}{\sqrt{\kappa}} \operatorname{Im} \operatorname{arcth} \left(\frac{2}{\sqrt{\kappa}} \frac{\sigma_d}{\delta_d - \frac{v}{\sqrt{\kappa}}\sigma_d}\right) + C.$$

The expression under the square root vanishes only at the same points as σ_d. As before, the choice of the branch is irrelevant because σ can vanish at infinity only at the boundary, and η is defined up to a constant C.

Now we work with the concrete form of the space $\mathcal{H}_s$ discussed in Sect. 2.4. It is convenient to define it in the strip chart. We choose the subspace $C^{\infty S}_0$ of C^{∞}-smooth functions with compact support in the strip chart. The function ϕ defining the metric can be, for example, zero in the strip chart, $\phi^{S}(z) \equiv 0$, which guarantees that the integrals in (9) and (11) are well-defined with η as above and $\Gamma = \Gamma_D$.

4.3 Radial SLE with Drift

The radial SLE Eq. (2.14) is usually formulated in the unit disk chart. It can be defined with the vector fields (2.12), which admit the form (2.13) in the half-plane chart. By the same reasons as for the dipolar SLE, we can change normalization and define

$$\delta_r^{H}(z) := 2\left(\frac{1}{z} + z\right) - v(1 + z^2), \quad \sigma_r^{H}(z) := -\sqrt{\kappa}(1 + z^2), \quad v \in \mathbb{R} \qquad (4.9)$$

that coincides with the expressions in Table 1.

Let us give here the expressions for δ, σ, Γ_D, η and M_{1t} in three different charts: half-plane, logarithmic (see below for the details), and the unit disk using the same method as before. The calculations are similar to the dipolar case. In fact, it is enough to change some signs and replace the hyperbolic functions by trigonometric. In contrast to the dipolar case, η is multiply defined. We discuss this difficulty at the end of this subsection. As for now, we just remark that from the heuristic point of view this is not an essential problem. In any chart $\eta^\psi(z)$ just changes its value only up to an irrelevant constant after the harmonic continuation about the fixed point of the radial equation.

In the half-plane chart, we have

$$-\sqrt{\kappa}(1+z^2)\,\partial_z\,\eta^{+^\mathbb{H}}(z) + \mu\left(-\sqrt{\kappa}(1+z^2)\right)' = \frac{-i}{z} + i\alpha,$$

$$\eta^\mathbb{H}(z) = \frac{-2}{\sqrt{\kappa}}\arg z - \frac{(\kappa-6)}{2\sqrt{\kappa}}\arg(1+z^2) - \frac{\nu}{\sqrt{\kappa}}\operatorname{Im}\operatorname{arctg}z + C, \qquad (4.10)$$

$$M_{1t}^{\mathbb{H}}(z) = (G_t^{-1}{}_*\eta)^\mathbb{H}(z) =$$

$$= \frac{-2}{\sqrt{\kappa}}\arg G_t^\mathbb{H}(z) - \frac{(\kappa-6)}{2\sqrt{\kappa}}\arg(1+G_t^\mathbb{H}(z)^2) -$$

$$- \frac{\nu}{\sqrt{\kappa}}\operatorname{Im}\operatorname{arctg}z + \frac{\kappa-4}{2\sqrt{\kappa}}\arg\left(G_t^\mathbb{H}\right)'(z) + C$$

analogously to (4.4).

The unit disk chart is defined in (1.8), and

$$\eta^\mathbb{D}(z) = \frac{-2}{\sqrt{\kappa}}\arg(1-z) - \frac{\kappa-6}{2\sqrt{\kappa}}\arg z + \frac{\nu}{2\sqrt{\kappa}}\log|z| + C,$$

$$M_{1t}^\mathbb{D}(z) = (G_t^{-1}{}_*\eta)^\mathbb{D}(z) =$$

$$= \frac{-2}{\sqrt{\kappa}}\arg(1-G_t^\mathbb{D}(z)) - \frac{\kappa-6}{2\sqrt{\kappa}}\arg G_t^\mathbb{D}(z) + \frac{\nu}{2\sqrt{\kappa}}\log|G_t^\mathbb{D}(z)| + \frac{\kappa-4}{2\sqrt{\kappa}}\arg\left(G_t^\mathbb{D}\right)'(z) + C,$$

$$\Gamma_D^\mathbb{D}(z,w) = \Gamma_D^\mathbb{D}\left(\tau_{\mathrm{H,D}}(z), \tau_{\mathrm{H,D}}(w)\right) = -\frac{1}{2}\log\frac{(z-w)(\bar{z}-\bar{w})}{(\bar{z}-w)(z-\bar{w})}.$$

The third chart is called *logarithmic*, and it is defined by the transition map

$$\tau_{\mathrm{D,L}}(z) := e^{iz} : \mathbb{H} \to \mathbb{D}, \qquad \tau_{\mathrm{L,D}}(z) = \tau_{\mathrm{D,L}}^{-1}(z) = -i\log z. \qquad (4.11)$$

Therefore,

$$\tau_{\mathrm{H,L}}(z) = \tau_{\mathrm{H,D}} \circ \tau_{\mathrm{D,L}}(z) = \operatorname{tg}\frac{z}{2} : \mathbb{H} \to \mathbb{H}, \qquad \tau_{\mathrm{L,H}}(z) = \tau_{\mathrm{H,L}}^{-1}(z) = 2\operatorname{arctg}z,$$

and $\psi^{\mathbb{L}}$ is not a global chart map as we used before because there is a point (the origin in the unit-disc chart) which is mapped to infinity. Besides, the function log is multivalued and the upper half-plane contains infinite number of identical copies of the radial SLE slit ($\tau_{\mathbb{L},\mathbb{H}}(z+2\pi) = \tau_{\mathbb{L},\mathbb{H}}(z)$). The advantage of this chart is that the automorphisms $H_t[\sigma_r]^{\mathbb{L}}$ induced by σ_r (see 2.3) are horizontal translations because $\sigma_r^{\mathbb{L}}(z)$ is a real constant (see below). The corresponding relations for the radial SLE in the logarithmic chart can be easily obtained from the dipolar SLE in the strip chart just by replacing the hyperbolic functions by their trigonometric analogs as

$$\delta_r^{\mathbb{L}}(z) = 4 \, \mathrm{tg}\frac{z}{2} - 2v, \quad \sigma_r^{\mathbb{L}}(z) = -2\sqrt{\kappa}, \quad v \in \mathbb{R}.$$

$$\eta^{\mathbb{L}}(z) = \frac{-2}{\sqrt{\kappa}} \arg \sin \frac{z}{2} - \frac{v}{2\sqrt{\kappa}} \, \mathrm{Im} \, z + C,$$

$$M_{1t}^{\mathbb{L}}(z) = (G_t^{-1} {}_* \eta)^{\mathbb{L}}(z) =$$

$$= \frac{-2}{\sqrt{\kappa}} \arg \sin \frac{G_t^{\mathbb{L}}(z)}{2} - \frac{v}{2\sqrt{\kappa}} \, \mathrm{Im} \, G_t^{\mathbb{L}}(z) + \frac{\kappa - 4}{2\sqrt{\kappa}} \arg \left(G_t^{\mathbb{L}}\right)'(z) + C.$$

$$\Gamma_D^{\mathbb{L}}(z, w) = -\frac{1}{2} \log \frac{\sin(\frac{z-w}{2}) \sin(\frac{\bar{z}-\bar{w}}{2})}{\sin(\frac{\bar{z}-w}{2}) \sin(\frac{z-\bar{w}}{2})},$$

This relations above coincide up to a constant with the analogous ones established in [18, 21].

We remark again that η can be represented in a chart-independent form as a function of δ_r and σ_r with the help of (2.23) by the relation

$$\eta = -\frac{2}{\sqrt{\kappa}} \arg \frac{\sigma_r^{\frac{\kappa}{4}}}{\sqrt{\sigma_r^2 + \frac{\kappa}{4} \left(\delta_r - \frac{v}{\sqrt{\kappa}}\sigma_r\right)^2}} + \frac{v}{\sqrt{\kappa}} \, \mathrm{Im} \, \mathrm{arcth} \left(\frac{2}{\sqrt{\kappa}} \frac{\sigma_r}{\delta_r - \frac{v}{\sqrt{\kappa}}\sigma_r}\right) + C.$$

In order to define GFF for radial SLE carefully we need to generalize slightly the above approach. Let $b \in \mathcal{D}$ be a zero point of δ_r and σ_r simultaneously inside $\mathcal{D}$: $\delta_r^\psi(b) = \sigma_r^\psi(b) = 0$ (for any ψ). We have $\psi^{\mathcal{D}}(b) = 0$ in the unit disk chart, $\psi^{\mathbb{H}}(b) = i$ in the half-plane chart, and $\psi^{\mathbb{L}}(b) = \infty$ in the logarithmic chart.

Let $\hat{\mathcal{D}}_b$ be the universal cover of $\mathcal{D} \setminus \{b\}$. Then the logarithmic chart map $\psi^{\mathbb{L}} : \hat{\mathcal{D}} \to \mathbb{H}$ defines a global chart map of $\hat{\mathcal{D}}_b$. The space $\mathcal{H}_s[\hat{\mathcal{D}}_b]$ in the logarithmic chart is defined as in Sect. 2.4 with $\phi^{\mathbb{L}} \equiv 0$, and we require in addition, that the support is bounded. The last condition guaranties the finiteness of functionals such as $\eta[f]$, $\Gamma_D[f_1, f_2]$, and the compatibility condition of $\mathcal{H}_s$, Γ_D, η (as above), δ_r and σ_r on $\hat{\mathcal{D}}_b$. The map $G_t : \mathcal{D} \setminus K_t \to \mathcal{D}$ is lifted to $\hat{G}_t : \hat{\mathcal{D}}_b \setminus \hat{K}_t \to \hat{\mathcal{D}}_b$, where $\hat{K}_t \subset \hat{\mathcal{D}}_b$ is the corresponding union of the countable number of the copies of K_t.

Consideration of $\hat{\mathcal{D}}_b$ instead of $\mathcal{D}$ is possible thanks to a special property of radial Löwner equation to have a fixed point $b \in \mathcal{D}$. The branch point b is in fact eliminated from the domain of definition and the pre-pre-Schwarzian η is well-defined on $\hat{\mathcal{D}}_b$.

4.4   General Remarks

Here we are aimed at explaining why all three cases of η above has the same form for $\kappa = 6$ and $\nu = 0$. Besides, we explain the relations between the chordal case and other cases considered in the next two subsections.

Let G_t be a (δ, σ)-SLE driven by B_t, and let and $\tilde{G}_{\tilde{t}}$ be a $(\tilde{\delta}, \tilde{\sigma})$-SLE driven by $\tilde{B}_{\tilde{t}}$ with the same parameter κ. Then there exists a stopping time $\tilde{\tau} > 0$, a family of random Möbius automorphisms $M_{\tilde{t}} : \mathcal{D} \to \mathcal{D}$, $\tilde{t} \in [0, \tilde{\tau})$, and a random time reparametrization $\lambda : [0, \tilde{\tau}) \to [0, \tau)$ ($\tau := \lambda(\tilde{\tau})$), such that

$$\tilde{G}_{\tilde{t}} = M_{\tilde{t}} \circ G_{\lambda_{\tilde{t}}}, \quad \tilde{t} \in [0, \tilde{\tau})$$

and

$$d^{\text{Itô}} \tilde{B}_{\tilde{t}} = a_{\tilde{t}} d\tilde{t} + \left(\dot{\lambda}(\tilde{t})\right)^{-\frac{1}{2}} d^{\text{Itô}} B_{\lambda(\tilde{t})}, \quad \tilde{t} \in [0, \tilde{\tau})$$

for some continuous $a_{\tilde{t}}$. In particular, this means that the laws of $\mathcal{K}_t$ and $\tilde{\mathcal{K}}_t$ induced by the (δ, σ)-SLE and $(\tilde{\delta}, \tilde{\sigma})$-SLE correspondingly are absolutely continuous with respect to each other until some stopping time. We proved this fact in [14]. However, it is possible to show a bit more: if $\nu = 0$ for both (δ, σ)-SLE and $(\tilde{\delta}, \tilde{\sigma})$-SLE, then the coefficient $a_{\tilde{t}}$ is proportional to $\kappa - 6$. Here the drift parameter ν is defined by

$$\nu := \delta_{-1} + 3\sigma_0,$$

see (1.7). This definition agrees with (4.3), (4.8) and (4.9) and is invariant with respect to (2.15). Since $\dot{\lambda}_{\tilde{t}}^{\frac{1}{2}} B_{\lambda(\tilde{t})}$ agrees in law with $B_{\tilde{t}}$, the random laws of $\mathcal{K}_t$ and $\tilde{\mathcal{K}}_t$ are identical, not just absolutely continuous as above, at least until some stopping time.

It can be observed that η for the chordal (4.5), dipolar (4.5), and radial (4.10) cases are identical for $\kappa = 6$ and $\nu = 0$. This is a consequence of the above fact. Special cases of chordal and radial SLEs were considered in [28].

Besides, there are two special situations when $a_{\tilde{t}}$ is identically zero for all values of $\kappa > 0$, not only for $\kappa = 6$ as above. In order to study them, let us consider the chordal SLE G_t, see (2.11), and a differentiable time reparametrization λ, which possesses property (6). Set

$$\tilde{G}_{\tilde{t}} := s_{c_{\tilde{t}}} \circ G_{\lambda_{\tilde{t}}},$$

where $s_c : \mathcal{D} \to \mathcal{D}$ is the scaling flow ($s_c^{\mathbb{H}}(z) = e^{-c}z$, $c \in \mathbb{R}$). In the half-plane chart we have

$$\tilde{G}_{\tilde{t}}^{\mathbb{H}}(z) = e^{-c_{\tilde{t}}}G_{\lambda_{\tilde{t}}}^{\mathbb{H}}(z). \tag{4.12}$$

The Stratonovich differential of $\tilde{G}_{\tilde{t}}^{\mathbb{H}}(z)$ is

$$d^S \tilde{G}_{\tilde{t}}^{\mathbb{H}}(z) = (d^S e^{-c_{\tilde{t}}})G_{\lambda_{\tilde{t}}}^{\mathbb{H}}(z) + e^{-c_{\tilde{t}}} d^S G_{\lambda_{\tilde{t}}}^{\mathbb{H}}(z) =$$

$$= (d^S e^{-c_{\tilde{t}}})G_{\lambda_{\tilde{t}}}^{\mathbb{H}}(z) + e^{-c_{\tilde{t}}}\dot{\lambda}_{\tilde{t}}\left(\frac{2}{G_{\lambda_{\tilde{t}}}^{\mathbb{H}}(z)}d\tilde{t} - \sqrt{\kappa}\, d^S B_{\lambda_{\tilde{t}}}\right)$$

Due to (7), we have to assume that

$$e^{-c_{\tilde{t}}}\dot{\lambda}_{\tilde{t}} \equiv \dot{\lambda}_{\tilde{t}}^{\frac{1}{2}},$$

in order to have an autonomous equation. So

$$e^{-c_{\tilde{t}}} = \dot{\lambda}_{\tilde{t}}^{-\frac{1}{2}},$$

and, consequently,

$$d^S e^{-c_{\tilde{t}}} = -\frac{1}{2}e^{-3c_{\tilde{t}}}a_{\tilde{t}}d\tilde{t} - \frac{1}{2}e^{-3c_{\tilde{t}}}b_{\tilde{t}}\, d^S \tilde{B}_{\tilde{t}},$$

where we used (6). Eventually, we conclude that

$$d^S \tilde{G}_{\tilde{t}}^{\mathbb{H}}(z) = \left(-\frac{1}{2}e^{-3c_{\tilde{t}}}a_{\tilde{t}}d\tilde{t} - \frac{1}{2}e^{-3c_{\tilde{t}}}b_{\tilde{t}}\, d^S \tilde{B}_{\tilde{t}}\right)e^{c_{\tilde{t}}}\tilde{G}_{\tilde{t}}^{\mathbb{H}}(z)$$

$$+ \frac{2}{\tilde{G}_{\tilde{t}}^{\mathbb{H}}(z)}d\tilde{t} - \sqrt{\kappa}\, d^S \tilde{B}_{\tilde{t}} + \frac{1}{4}\sqrt{\kappa}e^{-2c_{\tilde{t}}}b_{\tilde{t}}d\tilde{t}. \tag{4.13}$$

In order to have time independent coefficients we assume that $a_{\tilde{t}}$ and $b_{\tilde{t}}$ are proportional to $e^{2c_{\tilde{t}}}$. Hence, define $\xi \in \mathbb{R}$ by

$$a_{\tilde{t}} = -4\xi e^{2c_{\tilde{t}}}.$$

Without lost of generality, we can assume that $b_{\tilde{t}}$ is of the following three possible forms

1. $b_{\tilde{t}} = 0$,
2. $b_{\tilde{t}} = 4\sqrt{\kappa}e^{2c_{\tilde{t}}}$,
3. $b_{\tilde{t}} = -4\sqrt{\kappa}e^{2c_{\tilde{t}}}$,

because all other choices can be reduced to these three by (2.16). The first case is considered in Sect. 4.5. Other two cases are discussed in Sect. 4.6.

4.5 Chordal SLE with Fixed Time Reparametrization

Let $\xi \in (-\infty, +\infty) \setminus \{0\}$, and let G_t be a chordal stochastic flow, i.e., the chordal SLE (4.3). Define

$$\tilde{G}_{\tilde{t}}^{\mathbb{H}}(z) = e^{2\xi\tilde{t}} G_{\lambda(\tilde{t})}^{\mathbb{H}}(z)$$

in the half-plane chart and assume that

$$\lambda(\tilde{t}) := \frac{1 - e^{-4\xi\tilde{t}}}{4\xi};$$

$$\lambda : [0, +\infty) \to [0, (4\xi)^{-1}), \quad \xi > 0;$$

$$\lambda : [0, +\infty) \to [0, +\infty), \quad \xi < 0.$$

This choice of λ corresponds to $c_{\tilde{t}} = -2\xi\tilde{t}$, in the previous subsection. We remark, that the time reparametrization here is not random.

The flow $\tilde{G}_{\tilde{t}}$ satisfies the autonomous equation (2.9) with

$$\delta^{\mathbb{H}}(z) = \frac{2}{z} + 2\xi z, \quad \sigma^{\mathbb{H}}(z) = -\sqrt{\kappa}, \tag{4.14}$$

which are the vector fields from the second string of Table 1 and a special case of (4.13) with $a_{\tilde{t}} = -4\xi e^{2c_{\tilde{t}}}$ and $b_{\tilde{t}} = 0$.

There is a common zero of δ and σ at infinity in the half-plane chart, so infinity is a stable point $\tilde{G}_{\tilde{t}}^{\mathbb{H}} : \infty \to \infty$. But in contrast to the chordal case the coefficient at z^{-1} in the Laurent series is not 1 but $e^{2\xi\tilde{t}}$. The vector field δ is of radial type if $\xi > 0$, and of dipolar type if $\xi < 0$. It is remarkable that if $\xi < 0$, then the equation induces exactly the same measure as the chordal stochastic flow but with a different time parametrization. If $\xi > 0$ the measures also coincide when the chordal stochastic flow is stopped at the time $t = (4\xi)^{-1}$.

By the reasons described above it is natural to expect that the GFF coupled with such kind of (δ, σ)-SLE is the same as in the chordal case, because it is supposed to induce the same random law of the flow lines. Indeed, σ from (4.14) coincides with that from the chordal case, hence, η defined by (3.10), with $\alpha = 0$ (see the table) also coincides with (4.5) with $\nu = 0$. Thus, the martingales are the same as in the chordal case.

4.6 SLE with One Fixed Boundary Point

Let the vector fields δ and σ be defined by the 5th and the 6th strings of Table 1. There are two 'mirror' cases. The (δ, σ)-SLE denoted here by G_t is characterized by the stable point at $z = 1$ (the 5th case) or $z = -1$ (the 6th case) in the half-plane chart. We will consider only the first (the 5th string) case, the second (the 6th string) is similar.

We will show below that this (δ, σ)-SLE coincides with the chordal SLE up to a random time reparametrization for all values of $\kappa > 0$. Let us apply a Möbius transform $r_c: \mathcal{D} \to \mathcal{D}$ defined in (2.18) with $c = -1$

$$r_{-1}^{\mathbb{H}}(z) = \frac{z}{1+z}.$$

In the half-plane chart, it maps the stable point $z = 1$ to infinity keeping the origin and the normalization (2.17) unchanged. It results in

$$\tilde{G}_t := r_{-1} \circ G_t \circ r_{-1}^{-1},$$

$$r_{-1*}\delta^{\mathbb{H}}(z) = \tilde{\delta}^{\mathbb{H}}(z) = \frac{2}{z} + \kappa + 2\xi z,$$

$$r_{-1*}\sigma^{\mathbb{H}}(z) = \tilde{\sigma}^{\mathbb{H}}(z) = -\sqrt{\kappa}(1 + 2z),$$

and the equation for $\tilde{G}_t$ becomes

$$d^S \tilde{G}_t^{\mathbb{H}}(z) = \left(\frac{2}{\tilde{G}_t^{\mathbb{H}}(z)} + \kappa + 2\xi \tilde{G}_t^{\mathbb{H}}(z) \right) dt - \sqrt{\kappa} \left(1 + 2\tilde{G}_t^{\mathbb{H}}(z) \right) d^S B_t, \qquad (4.15)$$

which is a special case of (4.13) with $a_{\tilde{t}} = -4\xi \tilde{t}$ and $b_{\tilde{t}} = 4\sqrt{\kappa}e^{c_{\tilde{t}}}$. In other words, the relation (4.15) can be obtained from (4.12) with $c_{\tilde{t}} = -2\xi \tilde{t} + 2\sqrt{\kappa}\tilde{B}_{\tilde{t}}$ under the random time reparametrization $\lambda_{\tilde{t}} = e^{4\xi \tilde{t} - 4\sqrt{\kappa}\tilde{B}_{\tilde{t}}}$.

It is remarkable that the subsurface $\tilde{\mathcal{I}} \subset \mathcal{D}$ defined in the half-plane chart as

$$\psi^{\mathbb{H}}(\tilde{\mathcal{I}}) = \{z \in \mathbb{H} : \mathrm{Re}(z) > -\frac{1}{2}\}$$

is invariant $(G_t^{-1}(\tilde{\mathcal{D}}) \subset \tilde{\mathcal{D}})$ if and only if $\xi \geq \kappa$. In order to see this, it is enough to calculate the real parts of

$$\tilde{\delta}^{\mathbb{H}}(z) = \frac{2}{z} + \kappa + 2\xi z,$$

$$\tilde{\sigma}^{\mathbb{H}}(z) = -\sqrt{\kappa}(1 + 2z),$$

which are actually the horizontal components of the vector fields at the boundary of $\psi^{\mathbb{H}}(\tilde{\mathcal{I}})$ in $\mathbb{H}$, $\{z \in \mathbb{H} : \mathrm{Re}(z) = -\frac{1}{2}\}$,

$$\mathrm{Re}\left(\tilde{\delta}^{\mathbb{H}}\left(-\frac{1}{2} + ih\right)\right) = \mathrm{Re}\left(\frac{2}{-\frac{1}{2} + ih} + \kappa + 2\xi\left(-\frac{1}{2} + ih\right)\right) =$$

$$= -\frac{1}{h^2 + \frac{1}{4}} + \kappa - \xi,$$

$$\mathrm{Re}\left(\tilde{\sigma}^{\mathbb{H}}\left(-\frac{1}{2} + ih\right)\right) = \mathrm{Re}\left(\sqrt{\kappa}\left(1 + 2\left(-\frac{1}{2} + ih\right)\right)\right) = 0, \quad h > 0.$$

The first number is negative for all values of h if and only if $\xi \geq \kappa$.

We remark that the r_{-1}-transform has the invariant subsurface $\mathcal{I} := r_{-1}(\tilde{\mathcal{I}}) \subset \mathcal{D}$ for the (δ, σ)-SLE above, which is an upper half of the unit disk

$$\psi^{\mathbb{H}}(\mathcal{I}) = \{z \in \mathbb{H}: |z| < 1\}$$

Similarly to the previous subsection it is reasonable to expect that the GFF coupled with this $(\tilde{\delta}, \tilde{\sigma})$-SLE is the same as in the chordal case, because it is supposed to induce the same random law of the flow lines. Indeed, the solution to (3.9), with σ and α as in the 5th string of the table, is

$$\eta^{+\mathbb{H}}(z) = \frac{i}{\sqrt{\kappa}} \log z + i\frac{\kappa - 6}{2\sqrt{\kappa}} \arg(1 - z) + C^+.$$

Thus,

$$\eta^{\mathbb{H}}(z) = \frac{-2}{\sqrt{\kappa}} \arg z - \frac{\kappa - 6}{\sqrt{\kappa}} \arg(1 - z) + C.$$

After the r_{-1}-transform for $\tilde{\delta}$ and $\tilde{\sigma}$, we have

$$\tilde{\eta}^{\mathbb{H}}(z) = \frac{-2}{\sqrt{\kappa}} \arg z + C.$$

The last relation coincides with (4.5) with $\nu = 0$. We remind that Γ_0 is invariant under Möbius transforms, in particular, under r_{-1}.

5 Coupling of GFF with Dirichlet-Neumann Boundary Conditions

We assume in this chapter that $\Gamma = \Gamma_{DN}$, see Example 4, which becomes

$$\Gamma_{DN}^{S}(z, w) = -\frac{1}{2} \log \frac{\text{th}\frac{z-w}{4}\,\text{th}\frac{\bar{z}-\bar{w}}{4}}{\text{th}\frac{\bar{z}-w}{4}\,\text{th}\frac{z-\bar{w}}{4}}, \quad z, w \in \mathbb{S} := \{z \,:\, 0 < \text{Im} z < \pi\}. \tag{5.1}$$

in the strip chart (1.9). It is exactly Green's function used in [17] (it is also a special case of [16]).

The function Γ_{DN}^{S} satisfies the boundary conditions

$$\Gamma_{DN}^{S}(x, w)\big|_{x\in\mathbb{R}} = 0, \quad \partial_y \, \Gamma_{DN}^{S}(x + iy, w)\big|_{x\in\mathbb{R},\, y=\pi} = 0,$$

the symmetry property, and

$$\mathcal{L}_\sigma \, \Gamma_{DN}(z, w) = 0.$$

The coupling of GFF with this Γ to the dipolar SLE is geometrically motivated. We also require that both zeros of δ and σ are at the same boundary points where Γ_{DN} changes the boundary conditions from Dirichlet to Neumann. In the strip chart these points are $\pm\infty$.

Proposition 4 *Let the vector fields δ and σ be as in (4.8), let $\Gamma = \Gamma_{DN}$, and let η be a pre-pre-Schwarzian. Then the coupling is possible only for $\kappa = 4$ and $\nu = 0$.*

Proof We use Lemma 2 in the strip chart. From (6.2) we obtain that

$$\Gamma_{DN}^{++S}(z, w) = -\frac{1}{2} \log \text{th}\frac{z-w}{4}, \quad \Gamma_{DN}^{+-S}(z, \bar{w}) = -\frac{1}{2} \log \text{th}\frac{z-\bar{w}}{4},$$

and relations (3.16) hold. From (3.17) we find that

$$j^{+S}(z) = \frac{-i}{\text{sh}\frac{z}{2}} + i\alpha, \quad \alpha \in \mathbb{R}. \tag{5.2}$$

Substituting (5.2) in (3.13) gives

$$-i\frac{\left(\beta\sqrt{\kappa} - \alpha\nu\right)\text{sh}^2\frac{z}{2} + \nu\,\text{sh}\frac{z}{2} - 2i\sqrt{\kappa}\mu\,\text{ch}\frac{z}{2}\left(4\,\text{sh}\frac{z}{2}\alpha + (\kappa - 4)\right)}{2\sqrt{\kappa}\,\text{sh}^2\frac{z}{2}} \equiv 0,$$

which is possible only for $\kappa = 4$, $\nu = 0$, $\beta = 0$ and $\mu = 0$, where the latter agrees with (3.23). $\qquad\square$

From (3.9) we obtain that

$$\eta^{+^{\mathbb{S}}}(z) = \frac{i}{2} \log \mathrm{th} \frac{z}{4} + C^+,$$

and

$$\eta^{\mathbb{S}}(z) = -2 \arg \mathrm{cth} \frac{z}{4} + C.$$

We also present here the relations in the half-plane chart

$$\eta^{\mathbb{H}}(z) = -2 \arg \frac{z}{1 + \sqrt{1 - z^2}} + C,$$

$$\Gamma_{DN}^{\mathbb{H}}(z) = -\frac{1}{2} \log \frac{(z - w)(\bar{z} - \bar{w})(1 - \bar{z}w + \sqrt{1 - \bar{z}^2}\sqrt{1 - w^2})(1 - z\bar{w} + \sqrt{1 - z^2}\sqrt{1 - \bar{w}^2})}{(\bar{z} - w)(z - \bar{w})(1 - zw + \sqrt{1 - z^2}\sqrt{1 - w^2})(1 - \bar{z}\bar{w} + \sqrt{1 - \bar{z}^2}\sqrt{1 - \bar{w}^2})}.$$

The pre-pre-Schwarzian η is scalar in this case, and in its chart-independent form is

$$\eta = -\arg \frac{\frac{\sqrt{\kappa}}{2}(\delta - \frac{\nu}{\kappa}\sigma) + \sqrt{\frac{\kappa}{4}(\delta - \frac{\nu}{\kappa}\sigma)^2 - \sigma^2}}{\sigma}.$$

6 Coupling of Twisted GFF

This model is similar to the previous one. As it will be shown below, it is enough at the algebraic level to replace formally all hyperbolic functions in the dipolar case in the strip chart $\mathbb{S}$ by the corresponding trigonometric functions in order to obtain the relations for the radial SLE in the logarithmic chart $\mathbb{L}$. But at the analytic level, we have to consider the correlation functions which are doubly defined on $\mathcal{D}$ and change their sign after the analytic continuation about the center point. This construction was considered before as we were informed by Num-Gyu Kang [21].

We have to generalize slightly the general approach similarly to Sect. 4.3 considering the double cover $\mathcal{D}_b^{\pm}$ instead of the infinitely ramified cover of $\mathcal{D}\backslash\{b\}$. Let us define the space $\mathcal{H}_s[\mathcal{D}_b^{\pm}]$ of test functions $f \colon \mathcal{D}_b^{\pm} \to \mathbb{R}$ as in Sect. 2.4 with $\phi^{\mathbb{L}}(z) \equiv 0$ and with an extra condition $f(z_1) = -f(z_2)$, where z_1 and z_2 are two points of $\mathcal{D}_b^{\pm}$ corresponding to the same point of $\mathcal{D}\backslash\{b\}$. Thus, in the logarithmic chart, we have

$$f^{\mathbb{L}}(z) = f^{\mathbb{L}}(z + 4\pi k) = -f^{\mathbb{L}}(z + 2\pi k), \quad k \in \mathbb{Z}, \quad z \in \mathbb{H}. \tag{6.1}$$

Such functions are 4π-periodic and 2π-antiperiodic. In particular, $f^{\mathbb{L}}$ is not of compact support, but we require in addition that

$$\sup \operatorname{Im}\left(\{z \in \mathbb{H} : f^{\mathbb{L}}(z) \neq 0\}\right) < \infty$$

in order to maintain compatibility. In some sense, the 'value' of Φ_{tv} changes its sign after horizontal translation by π.

The *twisted Gaussian free field* Φ_{tw} is defined similarly to the usual one but taking values in $\mathcal{D}_b^{\pm}{}'$.

In this section, we define Γ by

$$\Gamma_{\mathrm{tw}}^{\mathbb{L}}(z, w) = -\frac{1}{2} \log \frac{\operatorname{tg}\frac{z-w}{4} \operatorname{tg}\frac{\bar{z}-\bar{w}}{4}}{\operatorname{tg}\frac{\bar{z}-w}{4} \operatorname{tg}\frac{z-\bar{w}}{4}}, \quad z, w \in \mathbb{H}. \tag{6.2}$$

in the logarithmic chart. Observe that

$$\Gamma_{\mathrm{tw}}^{\mathbb{L}}(z, w) = \Gamma_{\mathrm{tw}}^{\mathbb{L}}(z + 4\pi k, w) = -\Gamma_{\mathrm{tw}}^{\mathbb{L}}(z + 2\pi k, w), \quad k \in \mathbb{Z}.$$

In the unit disk chart the covariance $\Gamma_{\mathrm{tw}}^{\mathbb{D}}$ admits the form

$$\Gamma_{\mathrm{tw}}^{\mathbb{D}}(z, w) = -\frac{1}{2} \log \frac{(\sqrt{z} - \sqrt{w})(\sqrt{\bar{z}} - \sqrt{\bar{w}})(\sqrt{z} + \sqrt{\bar{w}})(\sqrt{\bar{z}} - \sqrt{w})}{(\sqrt{z} + \sqrt{w})(\sqrt{\bar{z}} + \sqrt{\bar{w}})(\sqrt{z} - \sqrt{\bar{w}})(\sqrt{\bar{z}} + \sqrt{w})},$$

or in the half-plane chart,

$$\Gamma_{\mathrm{tw}}^{\mathbb{H}}(z) = -\frac{1}{2} \log \frac{(z - w)(\bar{z} - \bar{w})(1 + \bar{z}w + \sqrt{1 + \bar{z}^2}\sqrt{1 + w^2})(1 + z\bar{w} + \sqrt{1 + z^2}\sqrt{1 + \bar{w}^2})}{(\bar{z} - w)(z - \bar{w})(1 + zw + \sqrt{1 + z^2}\sqrt{1 + w^2})(1 + \bar{z}\bar{w} + \sqrt{1 + \bar{z}^2}\sqrt{1 + \bar{w}^2})}.$$

It is doubly defined because of the square root, and the analytic continuation about the center changes its sign.

The covariance $\Gamma_{\mathrm{tw}}^{\mathbb{L}}$ satisfies the Dirichlet boundary conditions and tends to zero as one of the variables tends to the center point b (or ∞ in the $\mathbb{L}$ chart)

$$\Gamma_{\mathrm{tw}}^{\mathbb{L}}(x, w)\big|_{x \in \mathbb{R}} = 0, \quad \lim_{y \to +\infty} \Gamma_{\mathrm{tw}}^{\mathbb{L}}(x + iy, w) = 0, \quad x \in \mathbb{R}, \quad w \in \mathbb{H};$$

$$\Gamma_{\mathrm{tw}}^{\mathbb{D}}(z, w)\big|_{|z|=1} = 0, \quad \lim_{z \to 0} \Gamma_{\mathrm{tw}}^{\mathbb{D}}(z, w) = 0, \quad w \in \mathbb{D}.$$

The σ-symmetry property

$$\mathcal{L}_{\sigma_r} \Gamma_{\mathrm{tw}}(z, w) = 0$$

holds.

As we will see below, η also possesses property (6.1). Thus, the construction of the level (flow) lines can be performed for both layers simultaneously and the lines will be identical. In particular, this means that the line can turn around the central point and appears in the second layer but can not intersect itself. This agrees with the property of the SLE slit which evoids self-intersections.

Similarly to the dipolar case in the previous section the following proposition can be proved.

Proposition 5 *Let the vector fields* δ *and* σ *be as in (4.9), let* $\Gamma = \Gamma_{tw}$*, and let* η *be a pre-pre-Schwarzian. Then the coupling is possible only for* $\kappa = 4$ *and* $v = 0$.

The *proof* in the logarithmic chart actually repeats the proof of Proposition 4.

We give here the expressions for η in the logarithmic, unit-disk and half-plane charts:

$$\eta^{\mathbb{L}}(z) = -2\arg \mathrm{tg}\frac{4}{z} + C.$$

$$\eta^{\mathbb{D}}(z) = -2\arg \frac{1 - \sqrt{z}}{1 + \sqrt{z}} + C = 4\,\mathrm{Im}\,\mathrm{arctgh}\sqrt{z} + C.$$

$$\eta^{\mathbb{H}}(z) = -2\arg \frac{z}{1 + \sqrt{1 + z^2}} + C.$$

From this relation it is clear that η is antiperiodic.

The pre-pre-Schwarzian η is scalar in this case and its chart-independent form becomes

$$\eta = -\frac{2}{\sqrt{\kappa}}\arg \frac{\frac{\sqrt{\kappa}}{2}(\delta - \frac{v}{\kappa}\sigma) + \sqrt{\frac{\kappa}{4}(\delta - \frac{v}{\kappa}\sigma)^2 + \sigma^2}}{\sigma}.$$

We defined the linear functional Φ_{tw} on the space of antiperiodic functions before, however, such functional can be also defined on the space of functions with bounded support in the logarithmic chart. Thus, we can use the same space $\mathcal{H}_s$ as in Sect. 4.3.

7 Perspectives

1. The coupling with the reverse (δ, σ)-SLE can also be established using the same classification as in Table 1.
2. We did not prove in this paper but our experience shows that we listed all possible ways of coupling of GFF with (δ, σ)-SLE if we assume that Γ transforms as a scalar, see (2.33), and that η is a pre-pre-Schwarzian. It would be useful to prove this.

3. The pre-pre-Schwarzian rule (2.20) is motivated by the local geometry of SLE curves [30]. In principle, one can consider alternative rules. Moreover, the scalar behaviour of Γ can also be relaxed because the harmonic part $H^\psi(z, w)$ can transform in many ways. Such more general coupling is intrinsic and can be thought of as a generalization of the coupling in Sects. 5 and 6 for arbitrary κ.
4. We considered only the simplest case of one Gaussian free field. It would be interesting to examine tuples of Φ_i, $i = 1, 2, \ldots n$ which transform into non-trivial combinations $\tilde{\Phi}_i = G_i[\Phi_1, \Phi_2, \ldots \Phi_n]$ under conformal transforms G.
5. The Bochner-Minols Theorem 3 suggests to consider not only free fields, but for example, some polynomial combinations in the exponential of (2.39). In particular, the quartic functional corresponds to conformal field theories related to 2-to-2 scattering of particles in dimension two.

Acknowledgements The authors gratefully acknowledge many useful and inspiring conversations with Nam-Gyu Kang and Georgy Ivanov.

The authors were supported by EU FP7 IRSES program STREVCOMS, grant no. PIRSES-GA-2013-612669, and by the grants of the Norwegian Research Council #239033/F20.

Appendix 1: Nature of Coupling

One of the approaches to CFT boils down to consideration of a probability measure in a space of function in D. The simplest choice is GFF. The chordal SLE/CFT correspondence is revealed in, for example, [8] and [19]. Here, we extend this treatments to a (δ, σ)-SLE/CFT correspondence. This section is less important for the general objective of the paper and is dedicated mostly to a reader with physical background.

We consider a boundary CFT (BCFT) defined in a domain $D \subset \mathbb{C}$ with one free scalar bosonic field Φ. One of the standard approach to define a quantum field theory is the heuristic functional integral formulation with the classical action $I[\Phi]$, see for instance, [7]. The following triple definition of the Schwinger functions $S(z_1, z_2, \ldots z_n)$ manifests the relation between the probabilistic notations, functional integral formulation, and the operator approach.

$$
S_n(z_1, z_2, \ldots z_n) := \mathbb{E}\left[\Phi(z_1)\Phi(z_2) \ldots \Phi(z_n)\right] =
$$

$$
= c^{-1} \int \Phi(z_1)\Phi(z_2) \ldots \Phi(z_n)e^{-I[\Phi]}\mathfrak{D}\Phi =
$$

$$
= \langle | \mathbb{T}\left[\hat{\Psi}(a)\hat{\Phi}(z_1)\hat{\Phi}(z_2) \ldots \hat{\Phi}(z_n)\right] | \rangle,
$$

$$
z_i \in D, \quad i = 1, 2, \ldots n, \quad n = 1, 2, \ldots
$$

(1)

Here in the second term, '$|\rangle$' is the vacuum state, $\hat{\Phi}(z)$ the primary operator field, $\hat{\Psi}$ a certain operator field taken at a boundary point $a \in D$, and $\mathbb{T}[\ldots]$ is the time-ordering, which is often dropped in the physical literature, we refer to [29] for details. The second string in (1) contains a heuristic integral over some space of functions $\Phi(z)$ on D, which corresponds to the operator $\hat{\Phi}(z)$.

The first string in (1) is a mathematically precise formulation of the second one. The expectation $\mathbb{E}[\ldots]$ can be understood as the Lebesgue integral over the space $\mathcal{D}'(D)$ of linear functionals over smooth functions in D with compact support. The expression $e^{-I[\Phi]}\mathfrak{D}\Phi$ can be in its turn understood as the differential w.r.t. the measure.

We emphasize here that the correlation functions are not completely defined by the action $I[\Phi]$ because one has to specify also the space of functions the integral is taken over and the measure. For instance, the Euclidean free field action

$$I[\Phi] = \frac{1}{2}\int_D \partial\,\Phi(z)\bar{\partial}\Phi(z)d^2z$$

defines only the singular part of the two-point correlation function, which is $-(2\pi)^{-1}\log|z|$. To illustrate the statement above let us consider the following example. Let the expectation of Φ be identically zero,

$$S_1(z) = \int \Phi(z)e^{-I[\Phi]}\mathfrak{D}\Phi = 0, \quad z \in D.$$

Depending on the choice of the space of functions $\mathcal{H}'_s \ni \Phi$ and of the measure on it, the two-point correlation function

$$\Gamma(z_1, z_2) = \int \Phi(z_1)\Phi(z_2)e^{-I[\Phi]}\mathfrak{D}\Phi$$

varies. For example in [18–20, 23], Γ was assumed to possess the Dirichlet boundary conditions, see (2.37). But in [17], Γ vanishes only on a part of the boundary and on the other part the boundary conditions are Neumann's. In [16], even more general boundary conditions were considered.

In fact, under CFT one can understand a probability measure on the space of functions $\mathcal{H}'_s$ in D, which is just a version of $\mathcal{D}'(D)$. In the case of the free field, the Schwinger functions (equivalently, correlation functions or moments) S_n are of the form

$$S_1[z_1] = \eta[z_1],$$
$$S_2[z_1, f_2] = \Gamma[f_1, f_2] + \eta[f_1]\eta[f_2],$$
$$S_3[z_1, f_2, f_3] = \Gamma[f_1, f_2]\eta[f_3] + \Gamma[z_3, z_1]\eta[f_2] + \Gamma[z_2, z_3]\eta[f_1] + \eta[f_1]\eta[z_2]\eta[f_3],$$
$$S_4[f_1, f_2, f_3, f_4] = \Gamma[z_1, z_2]\Gamma[z_3, z_4] + \Gamma[f_1, f_3]\Gamma[f_2, f_4] + \Gamma[f_1, f_4]\Gamma[z_2, z_3] +$$
$$+ \Gamma[z_1, z_2]\eta(z_3)\eta[z_4] + \Gamma[z_1, z_3]\eta[z_2]\eta[z_4] + \Gamma[z_1, z_4]\eta[f_2]\eta[z_3] +$$
$$+ \eta[z_1]\eta[z_2]\eta[z_3]\eta[z_4],$$

$$\cdots$$

In this case, the measure is characterized by the space $\mathcal{H}'_s$ equipped with certain topology, by the field expectation $\eta(z) = S_1(z)$, and by Green's function (covariance) $\Gamma(z, w) = S_2(z, w) - S_1(z)S_1(w)$.

Assume now that

$$\mathcal{L} S_n(z_1, z_2, \ldots, z_n) = 0, \quad n = 1, 2, \ldots \tag{2}$$

for some first order differential operator $\mathcal{L}$. Then, in the language of quantum field theory, the functions S_n are called 'correlation functions' and the relations (2) can be called the Ward identities. This identities usually correspond to invariance of S_n with respect to some Lie group. For example, if

$$\mathcal{L} := \sum_{i=1}^{\infty} \sigma(z_i)\,\partial_{z_i} + \overline{\sigma(z_i)}\,\partial_{\bar{z}_i} \tag{3}$$

for a vector field σ of the form (1.7), then (2) is equivalent to

$$S_n(z_1, z_2, \ldots, z_n) = S_n(H_s[\sigma](z_1), H_s[\sigma](z_2), \ldots, H_s[\sigma](z_n)), \quad s \in \mathbb{R},$$

where $\{H_t[\sigma]\}_{t \in \mathbb{R}}$ is a one parametric Lie group of Möbious automorphisms $H_t[\sigma] : D \to D$ induced by σ:

$$dH_t[\sigma](z) = \sigma(H_t[\sigma](z))dt, \quad z \in D, \quad t \in \mathbb{R}, \quad H_0[\sigma](z) = z, \tag{4}$$

because

$$dS_n(H_s[\sigma](z_1), H_s[\sigma](z_2), \ldots, H_s[\sigma](z_n))|_{s=0} = \mathcal{L} S_n(z_1, z_2, \ldots, z_n)ds. \tag{5}$$

The relations (2) with (3) are satisfied if and only if

$$\left(\sigma(z)\,\partial_z + \overline{\sigma(z)}\,\partial_{\bar{z}}\right)\eta(z) = 0,$$
$$\left(\sigma(z)\,\partial_z + \overline{\sigma(z)}\,\partial_{\bar{z}} + \sigma(w)\,\partial_w + \overline{\sigma(w)}\,\partial_{\bar{w}}\right)\Gamma(z, w) = 0. \tag{6}$$

Consider now the vector field δ in place of σ. It is still holomorphic but generates a semigroup $\{H_t[\sigma]\}_{t \in (-\infty, 0]}$ of endomorphisms $G_t : D \to D \setminus \gamma_t$ (γ_t is some growing curve in D) instead of the group of automorphisms of D. Then the second identity in (6) does not hold because

$$\left(\delta(z)\,\partial_z + \overline{\delta(z)}\,\partial_{\bar{z}} + \delta(w)\,\partial_w + \overline{\delta(w)}\,\partial_{\bar{w}}\right)\Gamma(z, w) \neq 0.$$

Geometrically this can be explained by the fact that $H_t[\delta]$ is not just a change of coordinates but also it necessarily shrinks the domain D.

Let now G_t satisfy (1.5), which is just a stochastic version of (4), where $v(z)dt$ is replaced by $\delta(z)dt + \sigma(z)\,d^S B_t$. The vector field δ induces endomorphisms, σ induces automorphisms, and the multiple of $d^S B_t$ can be understood as the white noise. Such a stochastic change of coordinates G_t in the infinitesimal form, according to the Itô formula, leads to the second order differential operator $\delta(z)\,\partial_z + \frac{1}{2}(\sigma(z)\,\partial_z)^2$ instead of the first order Lie derivative $v(z)\,\partial_z$.

The first two Ward identities become

$$\left(\delta(z)\,\partial_z + \frac{1}{2}(\sigma(z)\,\partial_z)^2 + \text{complex conjugate}\right)\eta(z) = 0,$$

$$\left(\delta(z)\,\partial_z + \delta(w)\,\partial_w + \frac{1}{2}\left(\sigma(z)\,\partial_z + \sigma(w)\,\partial_w\right)^2 + \text{complex conjugate}\right) \times \quad (7)$$

$$\times\left(\Gamma(z,w) + \eta(z)\eta(w)\right) = 0,$$

which is equivalent to (1.10) if $\chi = 0$. Due to a version of the Itô formula we have

$$d^{\text{Itô}} S_n(G_s(z_1), G_s(z_2), \dots, G_s(z_n))\big|_{s=0} =$$

$$= \left(\mathcal{L}_\sigma + \frac{1}{2}\mathcal{L}_\sigma^2\right) S_n(z_1, z_2, \dots, z_n)dt + \mathcal{L}_\sigma S_n(z_1, z_2, \dots, z_n)\,d^{\text{Itô}} B_t,$$

which is an analog to the relation (5). In other words, the system (1.10) represents the local martingale conditions.

For the case

$$\delta(z) = \frac{2}{z}, \quad \sigma(z) = -\sqrt{\kappa}, \quad \kappa > 0, \quad z \in D = \mathbb{H}, \quad (8)$$

the identities (7) is an analog to the BPZ equation (5.17) in [6]. The choice of (8) corresponds to the chordal SLE (see Example 1) and it was considered first in [8], and later in [19]. In this paper, we assume that the field δ is holomorphic, has a simple pole of positive residue at a boundary point $a \in \partial D$, and tangent at the rest of the boundary.

Appendix 2: Technical Remarks

In this appendix section, we prove some technical propositions needed in the proof of Theorem 1.

Consider an Itô process $\{X_t\}_{t\in[0,+\infty)}$ such that

$$d^{\text{Itô}} X_t = a_t dt + b_t\,d^{\text{Itô}} B_t, \quad t \in [0, +\infty),$$

for some continuous processes $\{a_t\}_{t\in[0,+\infty)}$ and $\{b_t\}_{t\in[0,+\infty)}$. We denote by $\{X_{t\wedge T}\}_{t\in[0,+\infty)}$ the stopped process by a stopping time T. It satisfies the following SDE:

$$d^{\text{Itô}} X_{t\wedge T} = \theta(T-t)a_t dt + \theta(T-t)b_t\, d^{\text{Itô}} B_t, \quad t \in [0, +\infty),$$

where

$$\theta(t) := \begin{cases} 0 & \text{if } t \leq 0 \\ 1 & \text{if } t > 0. \end{cases}$$

If $\{X_t\}_{t\in[0,+\infty)}$ is a local martingale ($a_t = 0$, $t \in [0, +\infty)$), then $\{X_{t\wedge T}\}_{t\in[0,+\infty)}$ is also a local martingale.

We consider below the stopped processes $Y(G_{t\wedge T})\}_{t\in[0,+\infty)}$ instead of $\{Y(G_t)\}_{t\in[0,+\infty)}$. for some functions $Y : \mathscr{G} \to \mathbb{R}$, and the corresponding Itô SDE. In order to make the relations less cluttered, we usually drop the terms '$\ldots \wedge T[f]$' and $\theta(T-t)$. However, in the places where it is essential to remember them, e.g., the proof of Theorem 1, we specify the stopping times explicitly.

Define the *diffusion operator*

$$\mathcal{A} := \mathcal{L}_\delta + \frac{1}{2}\mathcal{L}_\sigma^2. \tag{1}$$

and consider how a regular pre-pre-Schwarzian η changes under the random evolution G_t. We also define the stopping time $T(x)$, $x \in \bar{\mathcal{D}}$ analogously to (3.1) using a neighborhood $U(x)$ of single point $x \in \mathcal{D}$. The functions $G_t^{-1}{}_*\eta^\psi(z)$ and $G_t^{-1}{}_*\Gamma^\psi(z, w)$ are defined by (2.20) and (2.33) until the stopping times $T(z)$ and $\min(T(z), T(w))$ respectively.

Proposition 6 *Let* $\{G_t\}_{t\in[0,+\infty)}$ *be a* (δ, σ)-*SLE.*

1. *Let* η *be a regular pre-pre-Schwarzian such that the Lie derivatives* $\mathcal{L}_\sigma \eta$, $\mathcal{L}_\delta \eta$, *and* $\mathcal{L}_\sigma^2 \eta$ *are well-defined. Then*

$$d^{\text{Itô}} G_t^{-1}{}_*\eta^\psi(z) = G_t^{-1}{}_*\left(\mathcal{A}\eta^\psi(z)\, dt + \mathcal{L}_\sigma \eta^\psi(z)d^{\text{Itô}} B_t\right). \tag{2}$$

2. *Let* Γ *be a scalar bilinear functional* ((2.33) *holds*) *such that the Lie derivatives* $\mathcal{L}_\sigma \Gamma$, $\mathcal{L}_\delta \Gamma$, *and* $\mathcal{L}_\sigma^2 \Gamma$ *are well-defined. Then*

$$d^{\text{Itô}} G_t^{-1}{}_*\Gamma^\psi(z, w) = G_t^{-1}{}_*\left(\mathcal{A}\Gamma^\psi(z, w)\, dt + \mathcal{L}_\sigma \Gamma^\psi(z, w)d^{\text{Itô}} B_t\right). \tag{3}$$

This can be proved by the direct calculation but we show a more preferable way, which is valid not only for pre-pre-Schwarzians but, for instance, for vector fields, and even more generally, for assignments whose transformation rules contain an arbitrary finite number of derivatives at a finite number of points. To this end let us prove the following lemma.

Lemma 3 *Let $X^i(t)$ $(i = 1, 2, \ldots, n)$ be a finite collection of stochastic processes defined by the following system of equations in the Stratonovich form*

$$d^S X_t^i = \alpha^i(X_t)dt + \beta^i(X_t)d^{Itô}B_t, \tag{4}$$

for some fixed functions $\alpha, \beta \colon \mathbb{R}^n \to \mathbb{R}^n$. Let us define Y_s^i, Z_s^i as the solution to the initial value problem

$$\begin{aligned}
\dot{Y}_s^i &= \alpha^i(Y_s), \quad Y_0^i = 0, \\
\dot{Z}_s^i &= \beta^i(Z_s), \quad Z_0^i = 0,
\end{aligned} \tag{5}$$

in some neighbourhood of $s = 0$. Let also $F \colon \mathbb{R}^n \to \mathbb{C}$ be a twice-differentiable function. Then, Itô's differential of $F(X_t)$ can be written in the following form

$$d^{Itô}F(X_t) = \left. \frac{\partial}{\partial s}F(X_t + Y_s)dt + \frac{\partial}{\partial s}F(X_t + Z_s)d^{Itô}B_t + \frac{1}{2}\frac{\partial^2}{\partial s^2}F(X_t + Z_s)dt \right|_{s=0}. \tag{6}$$

Proof The direct calculation of the right-hand side of (6) gives

$$F_i'(X_t)\left(\alpha^i(X_t) + \frac{1}{2}\beta_j'^i(X_t)\beta^j(X_t)\right)dt + F_i'(X_t)\beta^i(X_t)\,d^{Itô}B_t + \frac{1}{2}F_{ij}''(X_t)\beta^i(X_t)\beta^j(X_t)dt,$$

which is indeed Itô's differential of $F(X_t)$. We employed summation over repeated indices and used the notation $F_i'(X) := \frac{\partial}{\partial X^i}F(X)$. □

Proof of Proposition 6 We use the lemma above. Let $n = 4$, and let us define a vector valued linear map $\{\cdot\}$ for an analytic function $x(z)$ as

$$\{x(z)\} := \{\operatorname{Re}x(z), \operatorname{Im}x(z), \operatorname{Re}x'(z), \operatorname{Im}x'(z)\}. \tag{7}$$

For example,

$$X_t := \{G_t^\psi(z)\} = \{\operatorname{Re}G_t^\psi(z), \operatorname{Im}G_t^\psi(z), \operatorname{Re}G_t^{\psi'}(z), \operatorname{Im}G_t^{\psi'}(z)\}.$$

From (2.9) we have

$$\alpha(X_t) = \{\delta^\psi(G_t^\psi(z))\}, \quad \beta(X_t) = \{\sigma^\psi(G_t^\psi(z))\}.$$

Let also

$$F(X_t) = F(\{G_t^\psi(z)\}) := G_t^{-1}{}_* \eta^\psi(z) =$$

$$= \eta^\psi(G_t^\psi(z)) + \mu \log G_t^{\psi'}(z) + \mu^* \overline{\log G_t^{\psi'}(z)}.$$

Then

$$Y_s = \{H_s[\delta]^\psi(z) - z\}, \quad Z_s = \{H_s[\sigma]^\psi(z) - z\}$$

due to (2.4), (2.9), (4), and (5).

Now we can use Lemma 3 in order to obtain (2) for $t = 0$:

$$d^{\text{Itô}} G_t^{-1} {}_* \eta^\psi(z)\big|_{t=0} = d^{\text{Itô}} F[X_t]\big|_{t=0} =$$
$$= (\text{right-hand side of (6) with } t = 0) .$$

(8)

But

$$\frac{\partial}{\partial s} F(X_t + Y_s)\bigg|_{s=0,t=0} = \frac{\partial}{\partial s} F(\{z + H_s[\delta]^\psi(z) - z\})\bigg|_{s=0} =$$
$$= \frac{\partial}{\partial s} F(\{H_s[\delta]^\psi(z)\})\bigg|_{s=0} = \frac{\partial}{\partial s}\{H_s[\delta]_*^{-1}\eta^\psi(z)\bigg|_{s=0} = \mathcal{L}_\delta \eta^\psi(z).$$

A similar observation for other terms in (8) implies that

$$d^{\text{Itô}} G_t^{-1} {}_* \eta^\psi(z)\big|_{t=0} = \mathcal{L}_\delta \eta^\psi(z)dt + \mathcal{L}_\sigma \eta^\psi(z) d^{\text{Itô}} B_t + \frac{1}{2} \mathcal{L}_\sigma^2 \eta^\psi(z)dt =$$
$$= A \eta^\psi(z)dt + \mathcal{L}_\sigma \eta^\psi(z) d^{\text{Itô}} B_t.$$

For $t > 0$ we conclude that

$$d^{\text{Itô}} G_t^{-1} {}_* \eta^\psi(z) = d^{\text{Itô}} \left(\tilde{G}_{t-t_0} \circ G_{t_0}\right)_*^{-1} \eta^\psi(z) = d^{\text{Itô}} G_{t_0}^{-1} {}_* \tilde{G}_{t-t_0}^{-1} {}_* \eta^\psi(z)\big|_{t_0=t} =$$
$$= G_t^{-1} {}_* d^{\text{Itô}} \tilde{G}_s^{-1} {}_* \eta^\psi(z)\big|_{s=0} = G_t^{-1} {}_* \left(A \eta^\psi(z)dt + \mathcal{L}_\sigma \eta^\psi(z) d^{\text{Itô}} B_t\right).$$

The proof of (3) is analogous. The only difference is that we do not have the pre-pre-Schwarzian terms with the derivatives but there are two points z and w. We can assume

$$\{x\} := \{ \operatorname{Re} x(z), \operatorname{Im} x(z), \operatorname{Re} x(w), \operatorname{Im} x(w)\}$$

instead of (7) and the remaining part of the proof is the same. $\qquad\square$

We will obtain below the Itô differential of $G_t^{-1} {}_* \eta[f]$ and $G_t^{-1} {}_* \Gamma[f, g]$ for (δ, σ)-SLE $\{G_t\}_{t \in [0,+\infty)}$ and $f, g \in \mathcal{H}$. To this end we need the Itô formula for nonlinear functionals over $\mathcal{H}$. For linear functionals on the Schwartz space this has been shown in [22]. However, the authors are not aware of similar results for nonlinear functionals. The following propositions are special cases required for this paper. They are consequences of the proposition above, the classical Itô formula, and the stochastic Fubini theorem.

Proposition 7 *Under the conditions of Proposition 6 the following holds:*

1. The Itô differential is interchangeable with the integration over $\mathcal{D}$. Namely,

$$d^{\text{Itô}} \int_{\psi(\text{supp}f)} G_t^{-1}{}_* \eta^\psi(z) f^\psi(z) l(dz) =$$

$$= \int_{\psi(\text{supp}f)} G_t^{-1}{}_* \mathcal{A}\, \eta^\psi(z) f^\psi(z) l(dz)\, dt+ \tag{9}$$

$$+ \int_{\psi(\text{supp}f)} G_t^{-1}{}_* \mathcal{L}_\sigma\, \eta^\psi(z) f^\psi(z) l(dz)\, d^{\text{Itô}} B_t.$$

An equivalent shorter formulation is

$$d^{\text{Itô}} G_t^{-1}{}_* \eta[f] = G_t^{-1}{}_* \mathcal{A}\, \eta[f] dt + G_t^{-1}{}_* \mathcal{L}_\sigma\, \eta[f] d^{\text{Itô}} B_t. \tag{10}$$

2. The Itô differential is interchangeable with the double integration over $\mathcal{D}$, namely,

$$d^{\text{Itô}} \int_{\psi(\text{supp}f)} \int_{\psi(\text{supp}f)} G_t^{-1}{}_* \Gamma(x, y) f^\psi(x) f^\psi(y) l(dx) l(dy) =$$

$$= \int_{\psi(\text{supp}f)} \int_{\psi(\text{supp}f)} G_t^{-1}{}_* \mathcal{A}\Gamma(x, y) f^\psi(x) f^\psi(y) l(dx) l(dy)\, dt+ \tag{11}$$

$$+ \int_{\psi(\text{supp}f)} \int_{\psi(\text{supp}f)} G_t^{-1}{}_* \mathcal{L}_\sigma\, \Gamma(x, y) f^\psi(x) f^\psi(y) l(dx) l(dy)\, d^{\text{Itô}} B_t.$$

An equivalent shorter formulation is

$$d^{\text{Itô}} G_t^{-1}{}_* \Gamma[f, g] = G_t^{-1}{}_* \mathcal{A}\Gamma[f, g]\, dt + G_t^{-1}{}_* \mathcal{L}_\sigma\, \Gamma[f, g]\, d^{\text{Itô}} B_t.$$

Proof The relation (9) in the integral form becomes

$$\int_{\psi(\text{supp}f)} G_t^{-1}{}_* \eta^\psi(z) f^\psi(z) l(dz) = \eta[f]+$$

$$+ \int_0^t \int_{\psi(\text{supp}f)} G_\tau^{-1}{}_* \mathcal{A}\, \eta^\psi(z) f^\psi(z) l(dz) d\tau + \int_0^t \int_{\psi(\text{supp}f)} G_\tau^{-1}{}_* \mathcal{L}_\sigma\, \eta^\psi(z) f^\psi(z) l(dz)\, d^{\text{Itô}} B_\tau$$

The order of the Itô and the Lebesgue integrals can be changed using the stochastic Fubini theorem, see, for example [26]. It is enough now to use (2) to obtain (9).

The proof of 10 is analogous.   □

Proposition 8 *Let*

$$\hat{\phi}[f] = \exp\left(W[f]\right), \quad W[f] := \frac{1}{2}\Gamma[f,f] + \eta[f].$$

Then $G_t^{-1} *\hat{\phi}[f]$ *is an Itô process defined by the integral*

$$G_t^{-1} *\hat{\phi}[f] =$$

$$= \int_0^t \exp\left(G_\tau^{-1} *W[f]\right)\left(G_\tau^{-1} * \mathcal{A} W[f]d\tau + G_\tau^{-1} * \mathcal{L}_\sigma W[f]d^{\text{Itô}}B_\tau + \frac{1}{2}\left(G_\tau^{-1} * \mathcal{L}_\sigma W[f]\right)^2 d\tau\right).$$

$$(12)$$

Proof The stochastic process $G_t^{-1} *W^\psi[f]$ has the integral form

$$G_t^{-1} *W^\psi[f] = \frac{1}{2}G_t^{-1} *\Gamma^\psi[f,f] + G_t^{-1} *\eta^\psi[f] =$$

$$= \int_{\psi(\text{supp}f)}\int_{\psi(\text{supp}f)} G_t^{-1} *\Gamma^\psi(z,w)f^\psi(z)f^\psi(w)l(dz)l(dw) + \int_{\psi(\text{supp}f)} G_t^{-1} *\eta^\psi(z)f^\psi(z)l(dz).$$

due to Proposition 7. In terms of the Itô differentials it becomes

$$d^{\text{Itô}} G_t^{-1} *W^\psi[f] = G_t^{-1} * \mathcal{A} W^\psi[f]dt + G_t^{-1} * \mathcal{L}_\sigma W^\psi[f] d^{\text{Itô}} B_t.$$

In order to obtain the exponential function we can just use Itô's lemma

$$d^{\text{Itô}} G_t^{-1} * \exp\left(W^\psi[f]\right) = d^{\text{Itô}} \exp\left(G_t^{-1} *W^\psi[f]\right) =$$

$$= \exp\left(G_t^{-1} *W^\psi[f]\right)\left(G_t^{-1} * \mathcal{A} W^\psi[f]dt + G_t^{-1} * \mathcal{L}_\sigma W^\psi[f] d^{\text{Itô}} B_t + \frac{1}{2}\left(G_t^{-1} * \mathcal{L}_\sigma W^\psi[f]\right)^2 dt\right).$$

□

Appendix 3: Some Formulas from Stochastic Calculus

We refer to [9, 24], and [26] for the definitions and properties of the Itô and Stratonovich calculus and use the following relation between the Itô and Stratonovich integrals

$$\int_0^T F(x_t,t)\, d^S B_t = \int_0^T F(x_t,t)\, d^{\text{Itô}} B_t + \frac{1}{2}\int_0^T b_t\, \partial_1 F(x_t,t)dt.$$

The latter item can also be expressed in terms of the covariance

$$\int_0^T b_t \, \partial_1 F(x_t, t) dt = \langle F(x_T), B_t \rangle. \tag{1}$$

In order to obtain (2.10) from (2.9), let us assume

$$x_t := G_t(z), \quad b_t := \sigma(G_t(z)), \quad F(x_t, t) := \sigma(x_t) = \sigma(G_t(z)). \tag{2}$$

Then

$$\partial_1 F(x_t, t) = \sigma'(G_t(z)), \tag{3}$$

and

$$\int_0^T \sigma(G_t(z)) \, \mathrm{d}^S B_t = \int_0^T \sigma(G_t(z)) \, \mathrm{d}^{\mathrm{Itô}} B_t + \frac{1}{2} \int_0^T \sigma(G_t(z)) \sigma'(G_t(z)) dt. \tag{4}$$

It is enough now to add $\int_0^T \delta(G_t(z)) dt$ to both parts to obtain the right-hand sides of the integral forms of (2.10) and (2.9).

We also use in this paper that

$$\tilde{B}_{\tilde{t}} := \int_0^{\tilde{t}} \dot{\lambda}_{\tilde{t}}^{\frac{1}{2}} \, \mathrm{d}^{\mathrm{Itô}} B_{\lambda_{\tilde{t}}} = \int_0^{\lambda_{\tilde{t}}} \dot{\lambda}_{\lambda^{-1}_t}^{\frac{1}{2}} \, \mathrm{d}^{\mathrm{Itô}} B_t$$

has the same law as $B_{\tilde{t}}$ for any monotone and continuously differentiable function $\lambda : [0, \tilde{T}] \to [0, T]$. In differential form this relation becomes

$$\mathrm{d}^{\mathrm{Itô}} \tilde{B}_{\tilde{t}} = \dot{\lambda}_{\tilde{t}}^{\frac{1}{2}} \, \mathrm{d}^{\mathrm{Itô}} B_{\lambda_{\tilde{t}}}. \tag{5}$$

We need to reformulate relation (5) in the Stratonovich form. Let now λ satisfy

$$\mathrm{d}^S \dot{\lambda}_{\tilde{t}} = a_{\tilde{t}} d\tilde{t} + b_{\tilde{t}} \, \mathrm{d}^S \tilde{B}_{\tilde{t}}. \tag{6}$$

$$\int_0^{\tilde{t}} \mathrm{d}^S \tilde{B}_{\tilde{t}} = \int_0^{\tilde{t}} \mathrm{d}^{\mathrm{Itô}} \tilde{B}_{\tilde{t}} = \int_0^{\lambda_{\tilde{t}}} \dot{\lambda}_{\lambda^{-1}_t}^{\frac{1}{2}} \, \mathrm{d}^{\mathrm{Itô}} B_t = \int_0^{\lambda_{\tilde{t}}} \dot{\lambda}_{\lambda^{-1}_t}^{\frac{1}{2}} \, \mathrm{d}^S B_t - \frac{1}{2} \langle \dot{\lambda}_{\tilde{t}}^{\frac{1}{2}}, B_{\lambda_{\tilde{t}}} \rangle =$$

$$= \int_0^{\tilde{T}} \dot{\lambda}_{\tilde{t}}^{\frac{1}{2}} \, \mathrm{d}^S B_{\lambda_{\tilde{t}}} - \frac{1}{2} \langle \dot{\lambda}_{\tilde{T}}^{\frac{1}{2}}, \int_0^{\tilde{T}} \lambda_{\tilde{t}}^{-\frac{1}{2}} d\tilde{B}_{\tilde{t}} \rangle = \int_0^{\tilde{T}} \dot{\lambda}_{\tilde{t}}^{\frac{1}{2}} \, \mathrm{d}^S B_{\lambda_{\tilde{t}}} - \frac{1}{2} \int_0^{\tilde{T}} \frac{1}{2} \dot{\lambda}_{\tilde{t}}^{-\frac{1}{2}} b_{\tilde{t}} \lambda_{\tilde{t}}^{-\frac{1}{2}} d\tilde{t} =$$

$$= \int_0^{\tilde{T}} \dot{\lambda}_{\tilde{t}}^{\frac{1}{2}} \, \mathrm{d}^S B_{\lambda_{\tilde{t}}} - \frac{1}{4} \int_0^{\tilde{T}} \frac{b_{\tilde{t}}}{\dot{\lambda}_{\tilde{t}}} d\tilde{t}.$$

We conclude that

$$\mathrm{d}^S \tilde{B}_{\tilde{t}} = \dot{\lambda}_{\tilde{t}}^{\frac{1}{2}} \, \mathrm{d}^S B_{\lambda_{\tilde{t}}} - \frac{1}{4} \frac{b_{\tilde{t}}}{\dot{\lambda}_{\tilde{t}}} d\tilde{t}. \tag{7}$$

References

1. S. Albeverio, J. Jost, S. Paycha, S. Scarlatti, *A Mathematical Introduction to String Theory: Variational Problems, Geometric and Probabilistic Methods* (Cambridge University Press, London, 1997), 144 pp.
2. M. Bauer, D. Bernard, CFTs of SLEs: the radial case. Phys. Lett. B **583**(3–4), 324–330 (2004)
3. M. Bauer, D. Bernard, Conformal transformations and the SLE partition function martingale. Ann. Henri Poincaré **5**(2), 289–326 (2004)
4. M. Bauer, D. Bernard, 2D growth processes: SLE and Loewner chains. Phys. Rep. **432**(3–4), 115–221 (2006)
5. M. Bauer, D. Bernard, J. Houdayer, Dipolar stochastic Loewner evolutions. J. Stat. Mech. Theory Exp. **2005**(3), P03001, 18 pp. (2005)
6. A.A. Belavin, A.M. Polyakov, A. Zamolodchikov, Infinite conformal symmetry in two-dimensional quantum field theory. Nucl. Phys. B **241**(2), 333–380 (1984)
7. L.D. Faddeev, A.A. Slavnov, *Gauge Fields. Introduction to Quantum Theory*. Frontiers in Physics, vol. 83 (Addison-Wesley Publishing Company/Advanced Book Program, Redwood City, CA, 1991), 217 pp.
8. R. Friedrich, W. Werner, Conformal restriction, highest-weight representations and SLE. Commun. Math. Phys. **243**(1), 105–122 (2003)
9. C.W. Gardiner, *Stochastic Methods: A Handbook for the Natural and Social Sciences*. Springer Series in Synergetics (Springer, Heidelberg/New York, 2009), 447 pp.
10. I.M. Gel'fand, N.Y. Vilenkin, *Generalized Functions, IV: Some Applications of Harmonic Analysis* (Fizmatgiz, Moscow, 1961); Engl. transl.: Academic Press, New York, 1964
11. I. Gruzberg, Stochastic geometry of critical curves, Schramm-Loewner evolutions and conformal field theory. J. Phys. A **39**(41), 12601–12655 (2006)
12. T. Hida, S. Si, *Lectures on white noise functionals* (World Scientific, Singapore, 2008), 266 pp.
13. G. Ivanov, A. Vasil'ev, Löwner evolution driven by a stochastic boundary point. Anal. Math. Phys. **1**(4), 387–412 (2011)
14. G. Ivanov, A. Tochin, A. Vasil'ev, General slit Löwner chains, arXiv: 1404.1253 [math.CV] (2014), 44 pp.
15. G. Ivanov, N.-G Kang, A. Vasil'ev, Slit Holomorphic stochastic flows and gaussian free field. Complex Anal. Oper. Theory **10**(7), 1591–1617 (2016)
16. K. Izyurov, K. Kytölä, Hadamard's formula and couplings of SLEs with free field. Prob. Theory Relat. Fields **155**(1–2), 35–69 (2013)

17. N.-G. Kang, Conformal field theory of dipolar SLE(4) with mixed boundary condition. J. Korean Math. Soc. **50**(4), 899–916 (2013)

18. N.-G. Kang, N. Makarov, Radial SLE martingale-observables, arXiv: 1208.2789 [math.PR] (2012, to be revised)

19. N.-G. Kang, N.G. Makarov, Gaussian free field and conformal field theory. Astérisque **353**, 136 pp. (2013)

20. N.-G. Kang, H.-J. Tak, Conformal field theory of dipolar SLE with the Dirichlet boundary condition. Anal. Math. Phys. **3**(4), 333–373 (2013)

21. N.-G. Kang, N. Makarov, D. Zhan, In preparation

22. N.V. Krylov, *Controlled Diffusion Processes*. Stochastic Modelling and Applied Probability, vol. 14 (Springer, Berlin, 2009), 308 pp.

23. J. Miller, S. Sheffield, Imaginary Geometry I: Interacting SLEs, arXiv: 1201.1496 [math.PR], 154pp.

24. B. Øksendal, *Stochastic Differential Equations: An Introduction with Applications* (Springer, Heidelberg/New York, 2003), 352 pp.

25. A. Pietsch, *Nuclear Locally Convex Spaces*, Ergebnisse der Mathematik und ihrer Grenzgebiete. 2. Folge, vol. 66 (Springer, Berlin/New York 1972), 196 pp.

26. P.E. Protter, *Stochastic Integration and Differential Equations*. Stochastic Modelling and Applied Probability, vol. 21, 2nd edn. (Springer, Berlin/Heidelberg, 2005), 415 pp.

27. O. Schramm, S. Sheffield, Harmonic explorer and its convergence to SLE$_4$. Ann. Probab. **33**(6), 2127–2148 (2005)

28. O. Schramm, D.B. Wilson, SLE coordinate changes. N. Y. J. Math. **11**, 659–669 (2005)

29. M. Schottenloher, *A Mathematical Introduction to Conformal Field Theory*. Lecture Notes in Physics, VOL. 759, 2nd edn. (Springer, Berlin, 2008), 249 pp.

30. S. Sheffield, Conformal weldings of random surfaces: SLE and the quantum gravity zipper. Ann. Probab. **44**(5), 3474–3545 (2016)

A Marx-Strohhacker Type Result for Close-to-Convex Functions

Nikola Tuneski, Mamoru Nunokawa, and Biljana Jolevska-Tuneska

Abstract Let $f(z)$ be a close-to-convex function in the open unit disk $\mathbb{D}$. In this paper we use a result of Nunokawa et al. to obtain sufficient conditions which ensures $\operatorname{Re}[f(z)/z] > 0$ for all $z \in \mathbb{D}$. This result addresses a problem from complex dynamical systems.

Keywords Analytic • Close-to-convex • Real part • Marx-Strohhacker • Open problem

2010 Mathematics Subject Classification 30C45

1 Introduction and Preliminaries

Let $\mathcal{A}$ denote the class of analytic functions f in the open unit disk $\mathbb{D} = \{z : |z| < 1\}$ that are normalized such that $f(0) = f'(0) - 1 = 0$.

The class of *strongly starlike functions of order* α, $0 < \alpha \le 1$, is defined by

$$
\widetilde{S}^*(\alpha) = \left\{ f \in \mathcal{A} : \left| \arg \frac{z f'(z)}{f(z)} \right| < \frac{\alpha \pi}{2}, z \in \mathbb{D} \right\} .
$$

N. Tuneski (✉)
Faculty of Mechanical Engineering, Ss. Cyril and Methodius University in Skopje, Karpoš II b.b., 1000 Skopje, Republic of Macedonia
e-mail: nikola.tuneski@mf.edu.mk

M. Nunokawa
University of Gunma, Hoshikuki-cho 798-8, Chuou-Ward, Chiba 260-0808, Japan
e-mail: mamoru-nuno@doctor.nifty.jp

B. Jolevska-Tuneska
Faculty of Electrical Engineering and Informational Technologies, Ss. Cyril and Methodius University in Skopje, Karpoš II b.b., 1000 Skopje, Republic of Macedonia
e-mail: biljanaj@feit.ukim.edu.mk

M. Agranovsky et al. (eds.), *Complex Analysis and Dynamical Systems*, Trends in Mathematics, https://doi.org/10.1007/978-3-319-70154-7_16

For $\alpha = 1$ we receive the class of *starlike functions* $\widetilde{S}^* \equiv S^*(1)$ consisting of univalent functions f that map the unit disk onto a starlike region, i.e., if $\omega \in f(\mathbb{D})$, then $t\omega \in f(\mathbb{D})$ for all $t \in [0, 1]$.

Next,

$$K = \left\{ f \in \mathcal{A} : \operatorname{Re}\left[1 + \frac{zf''(z)}{f'(z)} \right] > 0, z \in \mathbb{D} \right\}$$

is the class of *convex functions* such that $f \in K$ if and only if $f(\mathbb{D})$ is a convex region, i.e., if for any $\omega_1, \omega_2 \in f(\mathbb{D})$ follows $t\omega_1 + (1 - t)\omega_2 \in f(\mathbb{D})$ for all $t \in [0, 1]$.

A generalization of the class of starlike functions is the class of *close-to-convex* functions denoted by $\mathcal{C}$ and introduced by Ozaki [10] and Kaplan [2]. A function $f \in \mathcal{A}$ belongs to $\mathcal{C}$ if and only if there exists a starlike function g (not necessarily normalized) such that

$$\operatorname{Re}\left[\frac{zf'(z)}{g(z)} \right] > 0 \quad (z \in \mathbb{D}).$$

Kaplan in [2] also proved that close-to-convex functions are univalent. Later, Lewandowski [3, 4] gave the following geometric characterisation: $f \in \mathcal{C}$ if and only if the complement of $f(\mathbb{D})$ is the union of rays (half lines) that are disjoint (except that the origin of one ray may lie on another one of the rays). If in the definition of close-to-convexity we restrict to functions $g \in \widetilde{S}^*(\alpha)$, $\alpha \in (0, 1]$, we receive a new class that we refer to as *close-to-convex with respect to strongly starlike functions of order α* and we denote by $\mathcal{C}_\alpha$. Obviously $\mathcal{C}_\alpha \subseteq \mathcal{C} = \mathcal{C}_1$.

All the above classes are in the class of univalent (analytic and one-to-one) functions in $\mathbb{D}$ and even more, $K \subset S^*$, $\widetilde{S}^*(\alpha) \subset \mathcal{C}_\alpha \subseteq \mathcal{C}$ and $\mathcal{C}_\alpha \subset \mathcal{C}_\beta$ for $\alpha < \beta$. For details see [1].

In [5, 12] Marx and Strohhacker proved that if $f \in \mathcal{A}$, then the following implication is sharp

$$f \in K \qquad \Rightarrow \qquad \operatorname{Re} \frac{f(z)}{z} > \frac{1}{2} \quad (z \in \mathbb{D}).$$

Similar implication hasn't been studied for the class of close-to-convex functions. This motivates the study of the class $\mathcal{C}_\alpha$ (the class of functions that are close-to-convex with respect to strongly starlike functions of order α) and a value of α such that

$$f \in \mathcal{C}_\alpha \qquad \Rightarrow \qquad \operatorname{Re} \frac{f(z)}{z} > 0 \quad (z \in \mathbb{D}).$$

Remark 1 Note that in the terms of complex dynamical systems the condition

$$\operatorname{Re} \frac{f(z)}{z} > \delta \geq 0 \tag{1}$$

means that f determines the so-called semi-complete vector field on $\mathbb{D}$ or, which is the same, f is the infinitesimal generator of a one-parameter semigroup $\Phi = \{F_t\}_{t \geq 0}$ of holomorphic self-mappings of $\mathbb{D}$ (see, for example, [11]). Moreover, δ is the exponential squeezing coefficient for Φ, i.e., the uniform rate of convergence of the semigroup Φ to its Denjoy-Wolff point ($z = 0$) is given by

$$|F_t(z)| \leq e^{-\delta t} |z|, \ t \geq 0.$$

Therefore, in these terms the Marx-Strohhacker result can be interpreted as follows:

Each convex function $f \in A$ generates a semigroup $\Phi = \{F_t\}_{t \geq 0}$ with the squeezing coefficient $\delta = \frac{1}{2}$.

This leads us to the following question.

Problem 1 Given $\delta \in [0, 1/2]$ find $\alpha \in [0, 1]$ such that for each $f \in C_\alpha$ condition (1) holds.

For $\delta = 0$ this problem is solved in the main result (Theorem 1) of this paper.

For the proof of the main result we will use methods from the theory of first order differential subordinations which was introduced by Miller and Mocanu in [7] and [8]. Namely, if $\phi : \mathbb{C}^2 \to \mathbb{C}$ (where $\mathbb{C}$ is the complex plane) is analytic in a domain D, $h(z)$ is univalent in $\mathbb{D}$, and $p(z)$ is analytic in $\mathbb{D}$ with $(p(z), zp'(z)) \in D$ when $z \in \mathbb{D}$, then $p(z)$ is said to satisfy a first-order differential subordination if

$$\phi(p(z), zp'(z)) \prec h(z). \tag{2}$$

Here, "$\prec$" denotes the usual subordination (if $g(z)$ is univalent in $\mathbb{D}$ then $f(z) \prec g(z)$ if and only if $f(0) = g(0)$ and $f(\mathbb{D}) \subseteq g(\mathbb{D})$). Further, an univalent function $q(z)$ is said to be a *dominant* of the differential subordination (2) if $p(z) \prec q(z)$ for all $p(z)$ satisfying (2). If $\widetilde{q}(z)$ is a dominant of (2) and $\widetilde{q}(z) \prec q(z)$ for all dominants of (2), then we say that $\widetilde{q}(z)$ is the *best dominant* of the differential subordination (2).

From the theory of first order differential subordinations we will use the following two results.

Lemma 1 ([8]) *Let $q(z)$ be univalent in the unit disk $\mathbb{D}$, and let $\theta(w)$ and $\phi(w)$ be analytic in a domain D containing $q(\mathbb{D})$, with $\phi(w) \neq 0$ when $w \in q(\mathbb{D})$. Set $Q(z) = zq'(z)\phi(q(z))$, $h(z) = \theta(q(z)) + Q(z)$, and suppose that*

(i) *$Q(z)$ is starlike in the unit disk $\mathbb{D}$, and*

(ii) *$\mathrm{Re}\, \dfrac{zh'(z)}{Q(z)} = \mathrm{Re}\left\{ \dfrac{\theta'(q(z))}{\phi(q(z))} + \dfrac{zQ'(z)}{Q(z)} \right\} > 0, \ z \in \mathbb{D}.$*

If $p(z)$ is analytic in $\mathbb{D}$, with $p(0) = q(0)$, $p(\mathbb{D}) \subseteq D$ and

$$\theta(p(z)) + zp'(z)\phi(p(z)) \prec \theta(q(z)) + zq'(z)\phi(q(z)) = h(z), \tag{3}$$

then $p(z) \prec q(z)$, and $q(z)$ is the best dominant of (3).

Lemma 2 (Corollary 4.1a.1 From [9, p. 189]) *Let $B(z)$ and $C(z)$ be complex-valued functions defined in $\mathbb{D}$, with*

$$|\operatorname{Im} C(z)| \leq \operatorname{Re} B(z) \quad (z \in \mathbb{D}). \tag{4}$$

If $p(z)$ is analytic in $\mathbb{D}$ with $p(0) = 1$, and if

$$\operatorname{Re}\left[B(z)zp'(z) + C(z)p(z)\right] > 0 \quad (z \in \mathbb{D}),$$

then

$$\operatorname{Re} p(z) > 0 \quad (z \in \mathbb{D}).$$

2   Main Result

For proving our main result first we will use Lemma 1 to show the following.

Lemma 3 *If $g \in \widetilde{S}^*(\alpha)$, $0 < \alpha \leq 1$ and $\beta = \tan(\alpha\pi/2)$, then*

$$\left|\arg\left[\frac{g(z)}{z}\right]\right| < \frac{\beta\pi}{2} \quad (z \in \mathbb{D}). \tag{5}$$

The implication is sharp, i.e., for given α, the value $\beta = \tan(\alpha\pi/2)$ is the smallest constant so that the implication holds.

Proof Let choose $p(z) = g(z)/z$, $q(z) = \left(\frac{1+z}{1-z}\right)^\beta$ (the power is taken with its principal value), $\phi(\omega) = 1/\omega$ and $\theta(\omega) = 1$. Then, $q(z)$ is convex univalent in $\mathbb{D}$, $p(z)$ is analytic in $\mathbb{D}$ and $p(0) = q(0) = 1$. Also, $\theta(w)$ and $\phi(w)$ are analytic in the domain $D = \mathbb{C} \setminus \{0\}$ containing $q(\mathbb{D}) = \{\omega : |\arg\omega| < \alpha\pi/2\}$, with $\phi(w) \neq 0$ when $w \in q(\mathbb{D})$ and $p(\mathbb{D}) \subseteq D$. Now, conditions (i) and (ii) form Lemma 1 hold because $Q(z) = zq'(z)\phi(q(z)) = \frac{zq'(z)}{q(z)} = \frac{2\beta z}{1-z^2}$, $h(z) = 1 + Q(z)$ and

$$\frac{zQ'(z)}{Q(z)} = \frac{zh'(z)}{Q(z)} = \frac{1+z^2}{1-z^2}.$$

Finally, from $g \in \widetilde{S}^*(\alpha)$ and from the definition of subordination we receive

$$\theta(p(z)) + zp'(z)\phi(p(z)) = 1 + \frac{zp'(z)}{p(z)} = \frac{zg'(z)}{g(z)} \prec h_1(z) \equiv \left(\frac{1+z}{1-z}\right)^\alpha$$

$$\prec h(z) \equiv 1 + \frac{2\beta z}{1-z^2} = \theta(q(z)) + zq'(z)\phi(q(z)). \tag{6}$$

Here, $h_1(z) \prec h(z)$ since $h_1(0) = h(0) = 1$, h is univalent (by its definition it is close-to-convex, even more) and $h(e^{i\theta}) = 1 + \frac{\beta i}{\sin \theta}$, which implies

$$h_1(\mathbb{D}) = \{\omega : |\arg \omega| < \alpha\pi/2\} \subset \mathbb{C} \setminus \{1 + ix : x \in \mathbb{R}, |x| \geq \beta\} = h(\mathbb{D}).$$

So, $p(z) \prec q(z)$, i.e., inequality (5) holds and $q(z)$ is the best dominant of (6).

Now, we want to show that the implication is sharp. Let β_1 be such that (5) implies

$$\left| \arg \left[\frac{g(z)}{z} \right] \right| < \frac{\beta_1 \pi}{2} \quad (z \in \mathbb{D}),$$

i.e., $p(z) \prec q_1(z) \equiv \left(\frac{1+z}{1-z} \right)^{\beta_1}$. On the other hand, we have already shown that inequality (5) implies subordination (6) and that $q(z) = \left(\frac{1+z}{1-z} \right)^{\beta}$ is the best dominant of (6). This means that $q(z) \prec q_1(z)$, i.e. $\beta \leq \beta_1$. $\qquad \square$

The previous lemma, together with Lemma 2, will help us to receive a result over the real part of $f(z)/z$ for a function f, close-to-convex with respect to strongly starlike functions of certain order.

Theorem 1 *If $f \in \mathcal{C}_\alpha$, $0 < \alpha \leq \frac{2}{\pi} \arctan \frac{1}{2} = 0.295\ldots$, then $\operatorname{Re} \frac{f(z)}{z} > 0$ for all $z \in \mathbb{D}$.*

Proof Let $f \in \mathcal{C}_\alpha$, $0 < \alpha \leq \frac{2}{\pi} \arctan \frac{1}{2}$. Then there exists $g \in \widetilde{S}^*(\alpha)$ such that

$$\left| \arg \left[\frac{zf'(z)}{g(z)} \right] \right| < \frac{\alpha\pi}{2} \quad (z \in \mathbb{D}).$$

From Lemma 3 we receive that (5) holds for $\beta \leq \frac{1}{2}$, i.e.

$$\left| \arg \left[\frac{g(z)}{z} \right] \right| < \frac{\pi}{4} \quad (z \in \mathbb{D}). \tag{7}$$

Now, let us choose in Lemma 2 choose $B(z) = C(z) = z/g(z)$ and $p(z) = f(z)/z$. These functions are analytic in $\mathbb{D}$ because $f(z)$ and $g(z)$ are univalent and vanish only for $z = 0$. Also $p(0) = 1$. For such choice of functions (when $B(z) = C(z)$) the inequalities (4) and (7) are equivalent since for $\omega = x + iy$, $x > 0$, we have

$$|\operatorname{Im} \omega| \leq \operatorname{Re} \omega \quad \Leftrightarrow \quad \left| \frac{y}{x} \right| \leq 1 \quad \Leftrightarrow \quad |\operatorname{tg}(\arg \omega)| \leq 1 \quad \Leftrightarrow \quad |\arg \omega| \leq \frac{\pi}{4}.$$

So, all conditions of Lemma 2 are satisfied and therefore

$$\operatorname{Re} p(z) = \operatorname{Re} \frac{f(z)}{z} > 0 \quad (z \in \mathbb{D}).$$

$\qquad \square$

An Open Problem Finding the largest α (ultimately $\alpha = 1$) such that $f \in \mathcal{C}_\alpha$ implies Re $\frac{f(z)}{z} > 0$, $z \in \mathbb{D}$, is an **open problem**. From the sharpness of the Lemma 3 it follows that the result given in Theorem 1 is the best possible that can be obtained by the approach used in this paper.

References

1. P.L. Duren, *Univalent Functions* (Springer, New York, 1983)
2. W. Kaplan, Close to convex schlicht functions. Mich. Math. J. **1**, 169–185 (1952)
3. Z. Lewandowski, Sur l'identité de certaines classes de fonctions univalentes. I. Ann. Univ. Mariae Curie-Skl. Sect. A. **12**, 131–145 (1958)
4. Z. Lewandowski, Sur l'identité de certaines classes de fonctions univalentes. II. Ann. Univ. Mariae Curie-Skl. Sect. A. **14**, 19–46 (1960)
5. A. Marx, Untersuchungen über schlichte Abbildungen. Math. Ann. **107**, 40–65 (1932/33)
6. S.S. Miller, P.T. Mocanu, Differential subordinations and univalent functions. Mich. Math. J. **28**, 157–171 (1981)
7. S.S. Miller, P.T. Mocanu, On some classes of first-order differential subordinations. Mich. Math. J. **32**, 185–195 (1985)
8. S.S. Miller, P.T. Mocanu, Differential subordination and inequalities in the complex plane. J. Differ. Equ. **67**(2), 199–211 (1987)
9. S.S. Miller, P.T. Mocanu, *Differential Subordinations, Theory and Applications* (Marcel Dekker, New York, 2000)
10. S. Ozaki, On the theory of multivalent functions. Sci. Rep. Tokyo Bunrika Daigaku, Sect. A. **2**, 167–188 (1935)
11. D. Shoikhet, *Semigroups in Geometrical Function Theory* (Kluwer Academic, Dordrecht, 2001)
12. E. Strohhacker, Beitrage zür Theorie der schlichten Funktionen. Math. Z. **37**, 356–380 (1933)

Printed in the United States
By Bookmasters